Straw and other fibrous by-products as feed

OTHER TITLES IN THIS SERIES

1. Guide Dogs for the Blind: their Selection, Development and Training
   by Clarence J. Pfaffenberger et al.
   1976 xii + 225 pp.

2. Ethology of Free-Ranging Domestic Animals
   by G.W. Arnold and M.L. Dudziński
   1978 xii + 198 pp.

3. Bacterial Infection and Immunity in Domestic Animals
   by J.B. Woolcock
   1979 xii + 254 pp.

4. Social Structure in Farm Animals
   by G.J. Syme and L.A. Syme
   1979 xii + 196 pp.

5. Physiology and Control of Parturition in Domestic Animals
   by F. Ellendorff, M. Taverne and D. Schmidt (Editors)
   1979 xii + 348 pp.

6. Trends in Veterinary Pharmacology and Toxicology
   Proceedings of the First European Congress on Veterinary Pharmacology and
   Toxicology, held in Zeist, 25—28 September, 1979
   edited by A.S.J.P.A.M. van Miert
   1980 x + 363 pp.

7. Veterinary Toxicology
   edited by M. Bartik and A. Piskac
   1981 346 pp.

8. Livestock Production in Europe
   edited by R. Politiek
   1982 viii + 336 pp.

9. Advances in Veterinary Immunology 1981
   edited by F. Kristensen and D.F. Antczak
   1982 vi + 282 pp.

10. Genetics and Animal Breeding
    Part A: Biological and Genetical Foundations of Animal Breeding
    Part B Stock Improvement Methods
    by J. Maciejowski and J. Zieba (Co-edition with PNW — Polish Scientific Publishers, Warszawa)
    1982 viii + 284 pp.
    1982 viii + 206 pp.

11. Constitutional Disorders and Hereditary Diseases in Domestic Animals
    by D. Hámori (Co-edited with Akadémiai Kiadó, Budapest)
    1983 730 pp.

12. Advances in Veterinary Immunology 1982
    edited by F. Kristensen and D.F. Antczak
    1983 v + 312 pp.

13. Prostaglandins in Animal Reproduction
    edited by L.E. Edqvist and H. Kindahl
    1984 viii + 304 pp.

Developments in Animal and Veterinary Sciences, 14

Straw and other fibrous by-products as feed

edited by

F. Sundstøl

Department of Animal Nutrition, Agricultural University of Norway, 1432-Ås-NLH, Norway

and

E. Owen

Department of Agriculture and Horticulture, University of Reading, Early Gate, Reading, RG6 2AT, Great Britain

ELSEVIER

Amsterdam — Oxford — New York — Tokyo   1984

ELSEVIER SCIENCE PUBLISHERS B.V.
Molenwerf 1
P.O. Box 211, 1000 AE Amsterdam, The Netherlands

Distributors for the United States and Canada:

ELSEVIER SCIENCE PUBLISHING COMPANY INC.
52, Vanderbilt Avenue
New York, NY 10017

Library of Congress Cataloging in Publication Data
Main entry under title:

Straw and other fibrous by-products as feed.

(Developments in animal and veterinary sciences ; 14)

1. Straw as feed. 2. Crop residues as feed.
3. Feed processing. I. Sundstøl, F. II. Owen, E.
III. Series.
SF99.S8S77 1984   636.2'08554   84-1491
ISBN 0-444-42302-8

ISBN 0-444-42302-8 (Vol. 14)
ISBN 0-444-41703-6 (Series)

Printed in The Netherlands

PREFACE

In folk-lore and metaphor, straw is regarded as an inferior substance. According to the Oxford English Dictionary, "a man of straw" is a "person without substantial financial means" and "to catch at a straw" is to "resort to an utterly inadequate expedient". Of other by-products, chaff provides a synonym for "worthless outside part of anything". Yet when grain and straw were harvested together, a good use was found for all the straw. Indeed, man was surrounded by straw; it held together his brick or plaster walls, it covered his roof, and it filled his mattress. Straw lined his cradle and straw ropes lowered him into his grave. It bedded his livestock, and as feed, it ensured their survival through cold or dry seasons.

For some time now, the combine harvester has threatened to change man's utilisation of straw, at least in the developed countries. The straw can be left in the field, to be burnt or incorporated in the soil. To cereal breeders, straw longer than 30 cm is indeed "worthless stuff", so they have introduced dwarf cultivars. Yet in many less-developed countries, straw and related by-products have continued to be essential materials for both man and his livestock. Now we see in the developed countries a renewed appreciation of the value of straw, and although some of this new interest is occasioned by the use of straw as fuel or manufacturing feedstock, much of it is concerned with the purely agricultural use of straw as an animal feed.

Why the new interest in straw for feeding? New feeding strategies for ruminants have played an important part. Thus the greater use of high-energy cereals in ruminant diets has created "space" in diet formulation for low-energy roughages like straw. In addition, the more precise determination of nutrient requirements has identified situations in which feeding on low-energy diets may be alternated with periods of better nutrition (in suckler cows, for example). Another factor is improved knowledge of the supplements needed by ruminants to promote digestion in the rumen. But the most significant stimulus to the use of straw as a feed has probably been the introduction of new methods - mechanical, physical and chemical - for improving its nutritive value.

This book comes at an important stage in the development of straw feeding, when enough is known about nutritional improvement to justify nine chapters reviewing processing methods. Around this core of the book are included several chapters on the chemistry of straw and related feeds, and on the assessment of their nutritive value. Further chapters deal with storage, supplementation and use of straw in practical feeding systems.

The economics of straw treatment and the practical implications of using more straw for feeding are discussed in two final chapters. The economics of straw utilization is, of course, a particularly important subject. One of the problems of straw and similar feeds is that collection, transport and treatment can convert a cheap by-product into an expensive feed. The developed countries may be able to justify economically the employment of industrial processing to improve straw, but ironically the developing countries - with their greater need for roughage improvement - may lack the resources needed for even farm-scale processing.

The contributors to this book, who have been drawn from many countries, have much experience of straw utilisation and thus have a great deal of information to hand on to their readers. I believe the book will stimulate both action and thought.

Professor J.F.D. Greenhalgh
School of Agriculture
University of Aberdeen
Scotland

ACKNOWLEDGEMENTS

We wish to thank the Agricultural University of Norway and Reading University, United Kingdom, for providing facilities and allowing us to undertake the project. One of us (E. Owen) wishes to thank the Norwegian Agricultural Research Council and Hargreaves Fertilizers Limited, United Kingdom, for financial support. We wish to thank chapter authors for their contributions, Professor Greenhalgh for the Preface and Mr. K.C. Plaxton of Elsevier Scientific Publishing Company for his advice and help. We are grateful to Mrs. Grethe Tuven for preparing many of the line drawings. Mrs. Marit Ensby deserves special thanks for typing and preparing the camera-ready copy. We owe Turi and Jane much for their help and patience.

ABBREVIATIONS

A	=	angstrom unit(s)
AA	=	Amino acids
ADF	=	acid detergent fibre
ADIN	=	acid detergent insoluble nitrogen
ADL	=	acid detergent lignin
atm	=	atmosphere(s)
ATP	=	adenosine-tri-phosphate
B-P	=	by-pass protein
BW	=	body weight
^{o}C	=	degree Celcius
C_2	=	acetic acid
C_3	=	propionic acid
C_4	=	butyric acid
C_6	=	glucose
C_6-E	=	synthesis energy
ca	=	circa
CF	=	crude fibre
CHO	=	carbohydrates
cm	=	centimetre
CP	=	crude protein
CWC	=	cell wall content
d	=	day
DM	=	dry matter
DDM	=	digestible dry matter
DMD	=	dry matter digestibility
DP	=	digestible (crude) protein
DOM	=	digestible organic matter
EDC	=	enzymatic digestible carbohydrate
FMR	=	fasting metabolic rate
FU	=	feeding unit (net energy)
g	=	gram
GI	=	gastro intestinal
h	=	hour(s)
ha	=	hectare
iu	=	abbrnational unit(s)

IVDMD	=	in vitro DMD
IVOMD	=	in vitro OMD
J	=	joule (0.239 cal)
k	=	kilo (thousand)
l	=	litre
LU	=	livestock unit(s)
M	=	mega (million = 10^6)
m	=	metre or milli (10^{-3})
mM	=	millimole
μM	=	micromole
MADF	=	modified acid detergent fibre
M-ATP	=	mainteanance-ATP-requirements
ME	=	metabolisable energy
MF	=	modulus of fineness
MFO	=	metabolic faecal output
min	=	minutes
MOG	=	material other than grain
MU	=	modulus of uniformity
M_w	=	molecular weight
N	=	nitrogen
n	=	number (of replicates, samples)
NADPH	=	reduced nicotineamide adenine dinucleotide phosphate
NDF	=	neutral detergent fibre
NIR	=	near infra red reflectance
nm	=	nanometre (10^{-9})
NMR	=	nuclear magnetic resonance spectroscopy
NPN	=	non protein nitrogen
OM	=	organic matter
OMD	=	organic matter digestibility
ppm	=	parts per million
psi	=	pounds per square inch
RSD	=	residual standard deviation
s	=	seconds
Sc.f.U	=	Scandinavian feed Units
SCP	=	single cell protein
SD	=	standard deviation
SE	=	standard error
t	=	tonne (= 1000 kg)
TCA	=	tricarboxylic acid

TDN   = total digestible nutrients
TSAE   = total soluble after enzymes
UK   = United Kingdom (Britain)
USSR   = Union of Soviet Socialist Republics
UV   = ultra violet
VFA   = volatile fatty acids
vs   = versus
vit   = vitamin
W   = liveweight
w   = watt

CONTENTS

Chapter 1

INTRODUCTION

by

Frik Sundstøl and Emyr Owen

Straw and other fibrous by-products are inevitably produced during cereal production. It is also inevitable that this will be the case in the future since cereals and other plant products will be needed for human consumption. Some will also be needed for animal production. Weight for weight, almost as much straw is produced as grain; the quantity involved is therefore enormous and will continue to be so. Straw and other by-products have traditionally been used for many purposes including feeding animals. The potential of these by-products as a feed resource for ruminants and herbivores is being increasingly appreciated (FAO, 1977).

In densely populated areas of the world (eg. Egypt, Bangladesh) little cultivable land can be allocated to animal-forage production because of the need to grow human food. These areas are likely to expand in the future due to population increases and reduced crop yield caused by scarcity of support-energy to manufacture fertilizers etc. For the same reasons, the amount of cereals fed to farm animals (currently about one third of world production) will have to be reduced in the future. Thus livestock will have to rely more on by-products such as straw. Because of their ability to utilize fibrous feeds, ruminants and other herbivores will be particularly valuable as human-food producers.

In cereal production to date interest has centered mainly on grain quantity and quality; straw value has received little attention. Cereal breeders have improved yield and grain:straw ratio, but in doing so have produced shorter and stiffer (coarser) straw with lower feeding value. The question may arise whether cereal breeders need to consider quantity and quality of straw as well as that of grain.

The present book is concerned with straw etc. as a feed. Alternative uses of straw are not considered; interested readers are referred to reviews elsewhere (FAO, 1977; Staniforth, 1982a, 1982b; Grossbard, 1979).

There is currently world-wide interest in utilizing low quality roughages as feed and in upgrading their nutritive value. Extensive research has been carried out in this field during the last 10-15 years. The need for a gathering-together of this information has been expressed on several occasions. This book is therefore an attempt to do this and to assess the information available. Hopefully the book will help curtail wasteful repetition of research already done and identify areas needing further investigation. Although a global approach is taken, it should be emphasised that this is a subject area in which it is difficult to generalise. Conditions of biology, technology, sociology and economics vary widely between countries and even within countries. Year to year variations also occur. All these factors demand that research findings be developed and tested under local conditions.

The following five questions form the basis of the book:

- How much straw and other fibrous by-products are produced and how best can they be handled?
- What is the nature of these materials and what are their properties?
- What is the feeding value of these materials and how can this be improved?
- What role can they play in the nutrition of animals and subsequently man?
- What is the economic importance and potential use of straw and other fibrous by-products in the future?

REFERENCES

FAO, 1977. New Feed Resources. FAO Production and Health Paper 4.

Grossbard, E. (ed.), 1979. Straw Decay and Effect on Disposal and Utilization. John Wiley and Sons, Chichester, New York, Brisbane, Toronto.

Staniforth, A.R., 1982a (ed.). Report on 6th Straw Utilisation Conference, Oxford 24 and 25 Nov. Ministry of Agriculture, Fisheries and Food. HMSO Birmingham.

Staniforth, A.R., 1982b. Straw for Fuel, Feed and Fertilizer? Farming Press, Ipswich.

Chapter 2

LOCATION AND POTENTIAL
FEED USE

by

Vappu L. Kossila

Food and Agriculture Organization
Via delle Terme di Caracalla
00100 Rome, Italy

2.1. INTRODUCTION

Most plants grown on farms with the purpose of production of commodities yield considerable amounts of crop by-products, which are not consumed by man. Such by-products usually contain quite high amounts of fibrous substances. The by-product yield per unit area of land is related positively to the yield of the commodity. Thus with increasing commodity yields, there will be increasing amounts of crop by-products available also for livestock as feed.

In this study the amounts of crop by-products have been estimated from the on-farm perspective. The majority of the so-called fibrous crop by-products still remain on farm. Such crops as sugarcane, sugarbeet, cassava, potatoes, nuts, citrus fruit, cereal grains, oilplant seeds, etc. which are processed outside the farm, produce off-farm products. In this study i.e. such off-farm byproducts as molasses, brans, millings, maltspent grains, oilcakes, have not been taken into account because they are not considered as fibrous by-products, (although brans and malt spent grains may have a rather high fibre content). The parts of the plants which have been considered as fibrous by-products are described later on.

Obviously a great deal of farm by-products are already fed to livestock, especially in intergrated livestock systems. However, much is still wasted for various reasons. The most important limitations for the use of by-products as feed are poor digestibility and palatability or infavourable location in respect to location of livestock.

The results of this study have been observed as regards various regions and socioeconomic groups of contries. Although separate figures have also been estimated for each individual country, it is not possible, within the given space of this article, to go into each individual country separately. The problem of unfavourable location of by-products is however relevant in many contries and upgrading of poor quality by-products is a worldwide problem.

The estimates of the fibrous by-product quantities in this study are based on data collected yearly by FAO (FAO, 1981). The commodities which have been included, are listed in Appendix 2.1.

2.2. THE METHOD OF ESTIMATING THE QUANTITIES AND NUTRITIVE VALUE OF FIBROUS BY-PRODUCTS AND NUMBER OF LIVESTOCK UNITS

2.2.1. Multipliers were formulated for calculation of the amounts of fibrous by-products (Tables 2.1 and 2.2) on the basis of the yearly production figures of various commodities. Calculations were made on an individual country basis and the by-product figures obtained were combined according to region, market economy and world total. Fibrous by-products of cereals were calculated using multipliers given in Table 2.1. The multipliers vary according to region. In Africa, for instance, the proportion of straw from the whole plant is larger compared to other regions. Table 2.2 lists the multipliers used in estimating the quantities of fibrous by-products of sugarcane, roots and tubers, pulses etc. Multipliers have large numerical value in such cases when dry commodity forms only a relatively small portion of the mother crop and small numerical value particularly when commodities contain large amounts of water.

Table 2.1. The multipliers used in converting cereal grain yields in different regions into fibrous by-product quantities

Species	Africa	North and Central America	South America	Asia	Europe	Oceania	USSR
Multipliers							
Wheat	2.0	1.5	1.2	1.3	1.0	1.3	1.0
Rice, paddy				1.3			
Barley	1.5	1.2	1.3	1.3	1.2	1.3	1.3
Maize	3.0	2.0	3.0	3.0	2.0	3.0	3.0
Rye				2.0			
Oats	1.5	1.3	1.3	1.3	1.3	1.3	1.3
Millet	5.0	4.0	4.0	4.0	4.0	4.0	4.0
Sorghum	5.0	4.0	4.0	4.0	4.0	4.0	4.0
Buckwheat				3.0			
Mixed grains				3.0			

Table 2.2. The multipliers used in converting various commodities into dry matter yields of their fibrous by-products

Products[1]	Multipliers
Sugarcane (fresh)	0.25
Roots and tubers (fresh)	0.20
Pulses (dry)	4.00
Nuts (dry)	2.00
Oilseeds, oilplant products (dry)	4.00
Vegetables, melons etc. (fresh)	0.25
Fruits, berries (fresh)	0.40

[1]Breakdown of various groups is given in Appendix 2.1.

2.2.2. Multipliers used in converting the animal numbers into livestock units (LU) are given in Table 2.3. Since the size of the animals vary from region to region, different multipliers were used for different regions.

Table 2.3. The multipliers used in converting the livestock numbers into livestock units (LU)[1]

Species	Africa	North and Central America	South America	Asia	Europe	Oceania	USSR
Horses	0.9	1.0	1.0	0.8	1.0	1.0	0.9
Mules	0.9	1.0	1.0	0.8	1.0	1.0	0.9
Asses	0.6	0.7	0.7	0.6	0.7	0.7	0.7
Cattle	0.6	0.9	0.8	0.7	0.9	0.8	0.8
Buffaloes	1.0	1.0	1.0	1.0	1.0	1.0	1.0
Camels	1.1	1.1	1.1	1.1	1.1	1.1	1.1
Pigs	0.16	0.30	0.20	0.18	0.30	0.18	0.30
Sheep	0.08	0.11	0.10	0.09	0.11	0.11	0.09
Goats	0.07	0.08	0.07	0.08	0.10	0.10	0.08
Chickens	0.012	0.015	0.012	0.012	0.015	0.015	0.015
Ducks	0.012	0.015	0.012	0.012	0.015	0.015	0.015
Turkeys				0.035			

[1] 1 LU corresponds to 500 kg live weight

2.2.3. Nutritive value of fibrous by-products were estimated on the basis of the analytical data presented by Morrison (1961), NJF (1969), Breirem and Homb (1970), Kellner and Becker (1971) Schiemann et al. (1971), Riviere (1977), Ensminger and Olentine (1978), NRC (1978), and Göhl (1981). However, analytical data is still not available for many by-products. Therefore the estimates which have been made are subject to readjustment when more information becomes available on this subject matter.

2.2.4. Limitations of the methods used in this study. At present it has been possible to make only rough estimates on the quantities and the nutritive value of fibrous crop by-products because amongst others of the following reasons:

- Information as regards the quantities of various by-products in relation to the amount of each commodity has not been established in practice for a great many crops.
- The proportion of by-products in relation to the mother crop is subject to variation due to different varieties among species, environmental conditions, cultivation practices etc. in different countries, the source and extent of variation being largely unknown at present.
- Chemical composition of a great many by-products is still not known.
- There are many minor commodities which yield fibrous by-products but which are not included in FAO data.
- FAO data are incomplete in the case of some commodities.
- Such commodities like coffee, tea, cocoa, jute, sisal, tobacco, were not included in this study although they may yield fibrous by-products.

2.3.   THE BY-PRODUCT GROUPS

Individual commodities have been gathered into eight main groups (Appendix 2.1). Cereals have been examined more thoroughly than other groups because the fibrous by-products of cereals are the most important as regards the quantities. Sugar cane is a special case which has been observed separately. Roots and tubers have been observed together and carrots and sugarbeets have been taken under this topic together with potatoes etc. Soybeans have been observed under pulses but it might also have been observed separately since it makes up a major part of the fibrous by-products of this group. Nuts are a fairly uniform group although FAO data is somewhat incomplete in respect of this group. Oilplants also make up a fairly uniform group with the exception of olives and palm kernels. Vegetables, melons etc. form a large group of commodities, which contain much water in general. Fruit and berries are also numerous and they also contain much water.

2.3.1. <u>Cereals</u>. Cereal crops yield large quantities of stems, stover, straw, husks, cobs, awns, chaffs etc. as their on-farm fiborus by-products. They are quite low in crude protein, about 3.0-6.5% in dry matter, but fairly high in metabolizable energy if fed to ruminants or equines (grass eaters). Palatability of cereal by-products is often rather low.

2.3.2. <u>Sugarcane</u>. Bagasse (off-farm by-product) and leaves and tops (on-farm by-products) are the main fibrous by-products of sugarcane. Molasses is a very important by-product of sugarcane but since it contains no fibre it is not considered in this study. Sugarcane by-products are very low in CP, 2.2% in DM, but fairly high in ME. Palatability of leaves and tops is very good and that of bagasse satisfactory.

2.3.3. <u>Roots and tubers</u>. The fibrous by-products of roots and tubers are mainly leaves, stems, vines (on-farm), peels pulp (off-farm). These by-products are fairly high in CP (13-15% in DM) and their energy value is also quite high. Palatability is usually excellent and these by-products make up very valuable diet components for livestock.

2.3.4. <u>Pulses</u>. The fibrous by-products of pulses consist generally the entire aerial part of the plant (on-farm by-products) except the seeds. CP content of such by-products is rather high, 10-15% in DM and energy content is higher compared to respective by-products of cereals and sugarcane. Palatability is also satisfactory.

2.3.5. <u>Nuts</u>. Fibrous by-products of nuts consist mainly of shells, hulls, husks, and skin of the kernel (off-farm by-products), but also in some cases the leaves and parts of the branches (tree) and aerial part of the plant (groundnut) are available on farm for animal feed. Hulls and shells are low in CP and energy whereas for instance groundnut hay is a fairly good energy and CP source for ruminants and equines. Palatability of shells, hulls and husks is not very good while stems and vines may be quite palatable.

2.3.6. <u>Oilplants</u>. Large quantities of on-farm fibrous by-products are produced by oilplants (stems, hulls, leaves, pods, husks etc.) in relation to oil seed yield. The CP content of fibrous byproducts is rather low, about 5% in DM, and the energy content corresponds closely to that of cereal straw.

2.3.7. <u>Vegetables, melons etc</u>. A large portion of the fibrous by-products of vegetables (stems, peels, leaves, skins) are usually high in CP and quite high in energy, 15% CP, and 9.2 MJ ME/kg DM. These by-products are very palatable and valuable feeds for livestock. Part of the by-products remain on farm, part is marketed with the product.

2.3.8. <u>Fruits, berries</u>. Fibrous by-products of fruit and berries are stems, peels, skins, seed capsules etc. Part of the by-products remain on farm and part is marketed with the commodities. The CP content of such by-products is quite low, usually only 4-8% in DM but the energy content is quite high - 8.4 to 10.9 MJ ME/kg DM. Such products are very palatable and besides grass eaters these can be fed also to pigs. Fruit trees and berry bushes also produce leaves which can, for instance, be fed to sheep.

2.4. RESULTS

2.4.1. <u>Fibrous by-products of cereals</u>. The total DM, CP, ME, and TDN yields of the fibrous by-products of cereals produced in different regions, market economies and whole world in 1970 (A) and 1981 (B) are summarized in Table 2.4. Asia was the leading producer while Oceania produced the smallest quantities. In 1981, more by-product DM was produced than in 1970. The rise was 61.1% in developed countries, 34.0 in developing countries and 22.7 in centrally planned market economies and 38.7 in the whole world. The highest absolute rise happened in North and Central America while there was actually a decrease in this respect in the USSR. CP, ME and TDN yields follow the trends seen in the development of DM yields (Table 2.4).

The changes in the estimated quantities of fibrous by-products of individual cereals over the period from 1970 to 1981 are depicted in Figure 2.1. It shows that maize is the most important one in this respect and wheat and rice + paddy rank more or less equally as the second. Most of the rise in the cereal by-product quantities is due to these three cereal crops.

2.4.2. <u>Other fibrous by-products</u>. The changes in the yield of other by-products i.e. those of pulses, sugarcane, oil plants, etc. from 1970 to 1981 are depicted in Figure 2.2. The figure shows that pulses produce nearly as much by-products as wheat or

Table 2.4.   Total amounts of dry matter (DM), crude protein (CP), metabolizable energy (ME) and total digestible nutrients (TDN) of the fibrous by-products of cereals according to region, market economy and world total in 1970 (A) and 1981 (B)

Area		DM $t.10^3$	% change (A to B)	CP $t.10^3$	ME $MJ.10^9$	TDN $t.10^3$
Africa	A	185 071		8 847	1456.6	96 251
	B	236 397	(27.7)	11 722	1892.8	124 948
North and Central America	A	466 252		24 472	3810.8	252 699
	B	819 073	(75.7)	44 198	6745.7	447 105
South America	A	124 900		6 960	1042.3	68 777
	B	183 052	(46.6)	10 020	1527.8	100 782
Asia	A	832 489		39 118	6044.6	401 725
	B	1 113 432	(33.7)	53 366	8144.6	541 962
Europe	A	256 767		12 194	1937.4	130 079
	B	339 827	(32.3)	16 336	2583.9	173 429
Oceania	A	19 584		818	138.0	9 377
	B	35 982	(83.7)	1 494	252.5	17 160
USSR	A	235 956		10 358	1685.9	114 016
	B	213 720	(-9.5)	9 332	1526.9	103 364
Developed	A	654 501		33 096	5177.1	344 861
	B	1 057 536	(61.6)	55 468	8501.3	565 589
Developing	A	745 458		35 781	5639.6	374 059
	B	999 429	(34.0)	48 006	7570.9	502 492
Centrally planned	A	721 059		33 895	5299.1	354 004
	B	884 518	(22.7)	42 999	6602.1	440 670
World	A	2 121 018		102 766	16115.7	1 072 924
	B	2 941 482	(38.7)	146 467	22674.3	1 508 751

rice + paddy (Figure 2.1). Sugarcane alone produces as much byproducts as roots and tubers together.

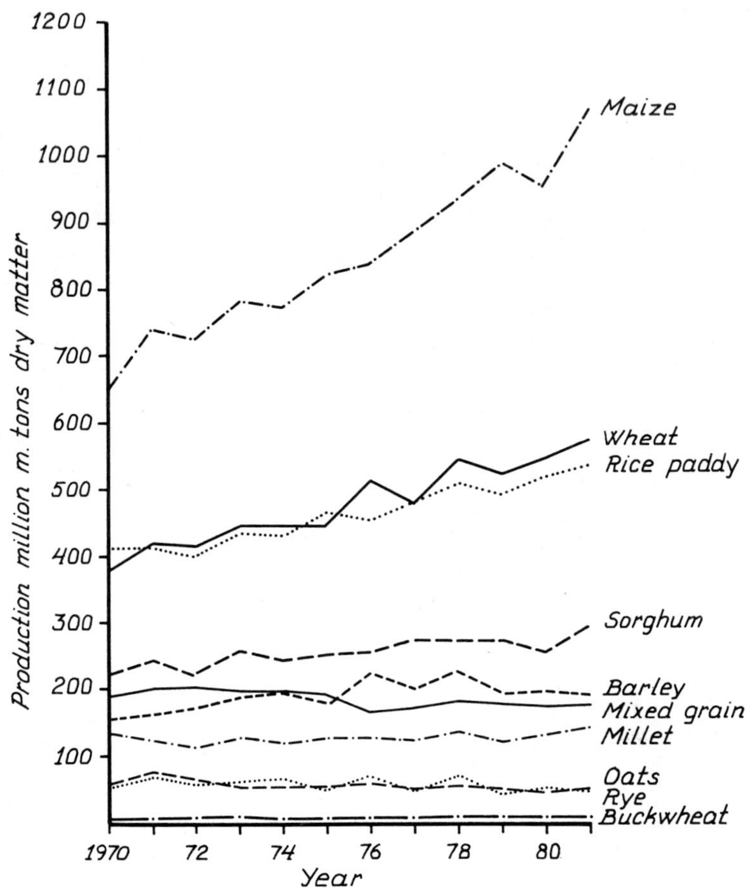

Figure 2.1. Changes in the estimated dry matter quantities of fibrous by-products of cereal crops form 1970 to 1981
(world total).

2.4.3. <u>Total by-products</u>. Total amounts of fibrous by-products
in different regions and market economies and in the whole world
in 1970 and 1981 are shown in Table 2.5. DM yields have increased
from 1970 to 1981 by 55.7% in developed, 34.4% in developing
countries, and 19% in the centrally planned countries. Absolute

Table 2.5. Total amounts of dry matter (DM), crude protein (CP), metabolizable energy (ME) and total digestible nutrients (TDN) contained in the fibrous by-products of cereals and other crops according to region, market economy and world total in 1970 (A) and 1981 (B)

Area		DM t.10^3	% change (A to B)	CP t.10^3	% change (A to B)	ME MJ.10^9	% change (A to B)	TDN t.10^3
Africa	A	278 318		17 256		2155.9		144 746
	B	343 564	(23.6)	21 147	(22.6)	2710.6	(25.7)	181 240
North and Central America	A	717 424		42 137		5583.8		376 024
	B	1 193 355	(66.3)	71 876	(70.6)	9375.1	(67.9)	629 834
South America	A	221 906		13 604		1759.6		118 739
	B	380 253	(71.4)	23 283	(71.1)	2958.8	(68.2)	200 669
Asia	A	1 244 955		76 711		9151.6		612 310
	B	1 628 882	(30.8)	98 116	(27.9)	11989.9	(31.0)	803 317
Europe	A	402 963		26 704		3075.8		209 433
	B	504 276	(25.1)	32 120	(20.3)	3830.3	(24.5)	260 086
Oceania	A	32 127		1 551		233.1		15 859
	B	53 199	(65.6)	2 497	(61.0)	380.5	(63.2)	25 909
USSR	A	358 065		22 932		2573.4		175 017
	B	320 390	(-10.5)	19 740	(-13.9)	2296.0	(-10.8)	156 385
Developed	A	972 901		59 577		7509.7		507 128
	B	1 515 161	(55.7)	92 541	(55.3)	11777	(56.8)	156 385
Developing	A	1 249 874		74 598		9407.7		631 040
	B	1 679 830	(34.4)	99 583	(33.5)	12638.3	(34.3)	849 700
Centrally planned	A	1 032 982		66 720		7624.2		513 959
	B	1 228 929	(19.0)	76 655	(14.9)	9126.1	(19.7)	614 499
World	A	3 255 757		200 895		24541.6		1 652 128
	B	4 423 919	(35.8)	268 779	(33.8)	33541.4	(36.7)	2 257 440

14

increase was the highest in North and Central America. CP, ME and
TDN yields followed the trends seen in the DM. The cereals account
for about 2/3 and other crops 1/3 of the total by-products under
observation in this study.

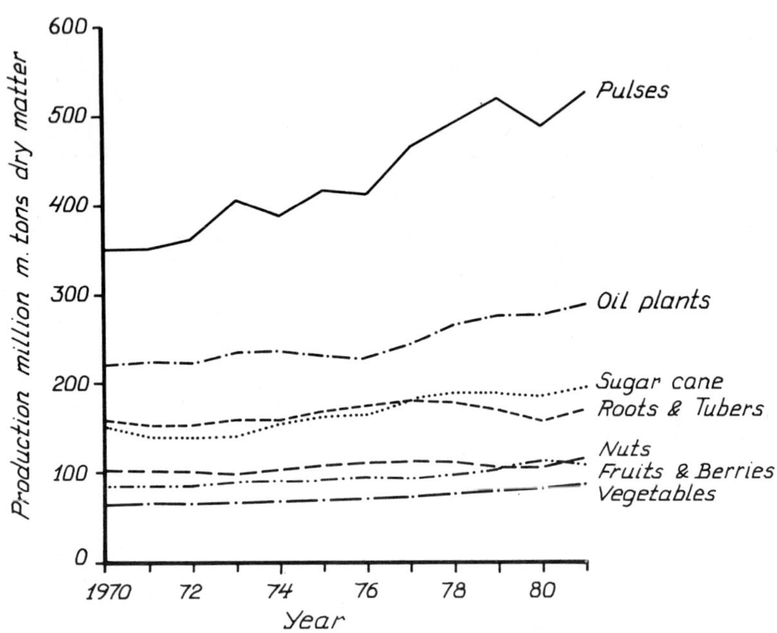

Figure 2.2. Changes in the estimated dry matter quantities of fib-
          rous by-products of different crop groups from 1970 to
          1981 (world total)

   2.4.4. <u>Livestock quantities in livestock units (LU)</u>. Livestock
was divided into grass eaters and grain eaters (Table 2.6). The
former group has much greater potential for utilizing fibrous by-
products as feed than the latter. Table 2.6 shows livestock quantities in different regions, market economies and the whole world
in 1970 and 1981. During this period, the grain eater population
has risen faster than than the grass eater population. The rise
was 10.0 and 38.2% in grass and grain eaters in the whole world
respectively. The LU-number of grass and grain eaters rose by 2.1

Table 2.6. Grass eaters[1] and grain eaters[2] in livestock units according to region, market economy and world total in 1970 (A) and 1981 (B) and the %-change in the development of the two populations

Area		Livestock units LU.10^3					Grain eaters % of total
		Grass eaters	% change (A to B)	Grain eaters	% change (A to B)	Total	
Africa	A	139 256		6 465		145 721	4.4
	B	156 576	(12.4)	9 257	(43.2)	165 834	5.5
North and Central America	A	168 290		34 294		202 584	16.9
	B	179 837	(6.8)	37 958	(10.7)	217 795	17.9
South America	A	172 598		13 191		185 789	7.1
	B	203 540	(17.9)	19 115	(44.9)	222 655	8.6
Asia	A	416 836		59 944		476 780	12.6
	B	457 198	(9.7)	94 520	(57.7)	551 717	17.1
Europe	A	136 261		56 814		193 076	29.4
	B	142 696	(4.7)	72 455	(27.5)	215 151	33.7
Oceania	A	52 173		1 334		53 507	2.4
	B	50 477	(-3.3)	1 738	(30.3)	52 215	3.3
USSR	A	94 902		26 039		120 942	21.5
	B	109 735	(15.6)	38 730	(48.7)	148 465	26.1
Developed	A	289 042		72 547		361 588	20.1
	B	295 221	(2.1)	85 259	(17.5)	380 479	22.4
Developing	A	667 965		39 649		707 614	5.6
	B	757 098	(13.3)	54 928	(38.5)	812 026	6.8
Centrally planned	A	223 310		85 885		309 195	27.8
	B	247 739	(10.9)	133 587	(55.5)	381 326	35.0
World	A	1 180 317		198 081		1 378 397	14.4
	B	1 300 058	(10.1)	273 773	(38.2)	1 573 831	17.4

[1] Grass eaters = horses + mules + asses + cattle + buffaloes + camels + sheep + goats

[2] Grain eaters = pigs + poultry

16

and 17.5% in developed countries, 13.3 and 38.5 in developing
countries and 10.9 and 55.5% in centrally planned economies respectively from 1970 to 1981. The proportion of grain eaters from
the total livestock population was still very low in Africa
(5.5%), South America (8.6%) and Oceania (3.3%) in 1981, while the
highest figure was obtained for Europe (33.7%). In the whole
world, grain eaters accounted for 14.4 and 17.4% of the total
livestock units in 1970 and 1981 respectively.

   2.4.5. <u>Total quantities of fibrous by-products per LU.</u> Table
2.7 summarizes the total quantities of DM, ME and TDN available on
the average per LU/year from all by-products under observation
according to region, market economies and the whole world in 1970
and 1981. In Oceania, where livestock is mainly kept on grass and
where crop production is smaller than in other areas, availability
of by-products per LU is low, 600 kg in 1970 and 1019 kg DM in
1981. On the other hand in North and Central America, respective
figures are as high as 3540 and 5480 kg DM/LU/year. Availability
of DM, CP, ME and TDN from total by-products/LU has also increased
in other regions during the same period except in USSR, where
grain and other crops gave poor yields in 1981.

   The whole world by-product figures per LU/year in 1970 and 1981
(Table 2.7) were: 145 and 171 kg CP, 17803 and 21313 MJ ME, 1199
and 1434 kg TDN. This shows that the by-product figures have risen
faster than numbers of LU (see also Tables 2.5, 2.6 and 2.7).

2.5. DISCUSSION

   The concept "fibrous by-products of plants" is stretchable because stage of growth of the plants is a very strong determinant
in this respect. Furthermore fibre content of plants (parts of the
plants) is determined by several methods each giving different results (Sundstøl et al., 1978b). There are many plant parts on
which analytical data are lacking or inadequate. Also, researchers
look at the by-product situation from different angles.

   Availability of number of by-products for livestock in different regions and countries has been discussed recently in several
papers (Horszczaruk, 1976; Adgebola, 1977; Balch, 1977; Chenost
and Mayer, 1977; Chicco and Schultz,1977; Skouri, 1977; Abou-Raya,
1978; Khajarern and Khajarern, 1980; Ranjhan, 1980; Anon, 1981;
Devendra, 1981; Sansoucy and Mahadevan, 1981; Osuji, 1982). A sop-

Table 2.7. Total quantities of dry matter (DM), crude protein (CP), metabolizable energy (ME) and total digestible nutrients (TDN) of all crop by-products under observation in relation to total livestock units and overall nutritive value of the crop-byproducts according to region, market economy and world total in 1970 (A) and 1981 (B)

Area		Quantities of nutrients from all by-products/LU.year				Average nutrient contents of by-products		
		kg DM	kg CP	MJ ME	kg TDN	CP g/kg	ME MJ/kg	TDN %
Africa	A	1 910	118	14 795	993	62	7.74	52
	B	2 070	128	16 347	1 093	62	7.91	53
North and Central America	A	3 540	208	27 564	1 856	59	7.78	52
	B	5 480	330	43 045	2 892	60	7.87	53
South America	A	1 190	73	9 473	639	61	7.95	54
	B	1 710	104	13 288	901	61	7.78	53
Asia	A	2 610	160	19 192	1 284	61	7.36	49
	B	2 950	178	21 732	1 458	60	7.36	49
Europe	A	2 090	138	15 928	1 085	66	7.61	52
	B	2 344	149	17 803	1 209	64	7.57	51
Oceania	A	600	28	4 356	296	47	7.28	49
	B	1 019	48	7 289	496	47	7.15	48
USSR	A	2 961	190	21 280	1 447	64	7.20	49
	B	2 158	133	15 464	1 053	62	7.15	49
Developed	A	2 961	165	20 769	1 403	61	7.70	52
	B	3 982	243	30 953	2 085	61	7.78	52
Developing	A	1 766	105	13 297	892	59	7.53	50
	B	2 069	123	15 564	1 046	59	7.53	51
Centrally planned	A	3 341	215	24 656	1 662	64	7.28	50
	B	3 222	201	23 932	1 612	62	7.41	50
World	A	2 362	145	17 803	1 199	61	7.53	51
	B	2 811	171	21 313	1 434	61	7.57	51

18

histicated statistical system has been developed by the Winrock
International Livestock Research and Training Centre (ref. Fitzhugh et al., 1978; Wheeler et al., 1981) for estimation of nutrient requirements of various types of livestock and availablility
of various feedstuffs. According to Wheeler et al. (1981) perma-
nent pastures and meadows had a potential of producing 20.1 trillion, arable lands 13.0 trillion, and forest and woodland 4.2
trillion of MJ feed ME in 1970. Crop residues from all sources
were estimated at 12.1 trillion ME; in the present study cereals
produced 16.1 trillion ME of fibrous by-products in 1970 (Table
2.4). According to Wheeler et al. (1981) pasture, forage and crop
residues (straw etc.) accounted for 49.4 MJ ME, in total, which
was double the total needed in the ruminant system in 1970. Furthermore, according to Wheeler et al. the total nutrient requirements of all world livestock was as shown in Table 2.8. Results of
the present study are given for comparison also.

Table 2.8. Estimated total requirements and available nutrients in
fibrous byproducts 1977-78

Area	Requirement (Wheeler et al., 1981)		Available (present study)	
	ME $MJ.10^9$	DCP $t.10^6$	ME $MJ.10^9$	DCP[1] $t.10^6$
Developed	10 121	73.8	A 9 991	59.3
			B 10 545	62.6
Developing	17 203	113.5	A 11 196	66.6
			B 11 513	68.1
Centrally planned	9 108	64.0	A 8 623	55.2
			B 9 469	59.9
Total world	36 432	251.4	A 29 810	181.0
			B 31 527	190.5

[1]DCP was calculated assuming that 75% of CP is digested.
A = 1977, B = 1978.

The above compilation indicates rather clearly that, in comparison with the nutrient requirements of animals there are more
fibrous by-products available in developed and centrally planned
than in developing countries. Using the figures of the above compilation, one can demonstrate that the total by-product quanti-

ties, obtained for the whole world in this study would have supplied in theory about 84 per cent of energy and 74 per cent of DCP requirements of all livestock. In practice, however, availability of nutrients from these by-products is much smaller. High-yielding milk animals and fast-growing young animals (cattle, buffaloes, lambs), and pigs and poultry particularly, need more concentrated feeds, especially in intensive production systems than most fibrous by-products are ever able to provide. There are also large losses of by-products, mainly on-farm, for various reasons:

- Losses occur in harvesting, preserving and storing of by-products;

- Losses of by-products occur when processing and marketing the respective commodities;

- A proportion of by-products is left in the fields to decay, to be destroyed by burning (purposely or accidentally), or to be ploughed into the ground. Reasons for this vary, either there are not enough livestock in the neighbourhood to consume them, their quality is not good enough, there is not enough labour to harvest the residues, or there is a lack of knowledge on the utililization of crop byproducts as animal feed and on improvement of their nutritive value;

- Part of the by-products are used as fuel, as paper raw material, as bedding, as building materials etc.

Roughly speaking then, losses of by-product may easily exceed 50 per cent of total yield. Limited digestibility and high bulk, particularly of straw, reduces the DM intake of by-products by ruminant animals (i.e. Rissanen et al., 1977; Klopfenstein and Owen, 1981; Rissanen et al., 1981). Mature cows may consume 6-7 kg straw DM and growing cattle 1.5 to 2.0 kg DM/day, accounting for about one third of total DM intake in intensive production systems. Even in extensive production systems, untreated straw alone cannot support maintenance; it has to be supplemented with more valuable by-products (legumes, soy, sunflower seeds, etc.) so that growth and milk secretion can be maintained even at a low level (Devendra, 1981).

There are many fibrous by-products such as those derived from forests (leaves, branches, bark, treetops), from the wood industry (sawdust, waste fibres, spent solids, waste paper etc.) (Söderhjelm, 1976; Söderhjelm and Lampila, 1976; Pigden, 1977; Scholler, 1977; Kommeri, 1980; Staples et al., 1981; Schingoethe et al., 1981), which have potential as animal feed, but which have not been accounted for because they are beyond the scope of this study. There are also great quantities of dried-up fibrous vegetation wasted each year on pasture lands. In addition bushes, woodlands, organic material of aquatic plants in open waters, peat and organic matter on swamps, etc., have potential as animal feed, but remain under-used from year to year. Large amounts of vegetation are also wasted in the wet tropics yearly. It thus seems that there is still a large feed resource potential, particularly for ruminants. The problem of how to make the best use of this remains to be solved.

Utililization of fibrous by-products as animal feeds can be increased, in general, by physical, chemical and biochemical treatments, by recycling with animal excreta, by ensiling various crop by-products together (Müller, 1977; 1980; Raininko et al., 1981) and by increasing the integration of plant and animal production systems (Preston, 1977; Müller, 1977) so that animals can have acess to crop residues.

That the quantity of fibrous crop by-products has risen faster than animal numbers (in LU) during 1970-81 is encouraging in the sense that more energy from crop by-products per LU is now available than before. The intensification of crop production yields has also made available increasing amounts of high-energy by-products (brans, cakes, millings, molasses, vinasse, etc.), particularly for intensive animal production systems.

2.6. SUMMARY

Using FAO data, the quantities of fibrous by-products of cereals, sugarcane, roots and tubers, pulses, nuts, oilseeds, vegetables and melons, and fruits and berries were estimated for different regions, market economies, and the whole world for the period

1970-81. DM, CP, ME, and TDN yields were estimated for all of the by-product groups. It was found that the total quantity of fibrous by-products under observation had increased faster than livestock numbers from 1970 to 1981. In the whole world there were 2362 and 2811 kg DM, 145 and 171 kg CP, 17 800 and 21 300 MJ ME and 1199 and 1434 kg TDN per livestock unit in 1970 and 1981, respectively. The average nutrient contents of all fibrous crop by-products were 61 kg CP/g DM, 7.53 MJ ME/kg DM and 51 per cent TDN.

2.7. REFERENCES

Abou-Raya, A.K., 1978. Preliminary survey of the feed resources of the Gulf and Arabian Peninsula countries along with possible means of developing them. FAO RNEA, Cairo, Egypt, 38 pp.

Adegbola, A.A., 1977. Utilization of agroindustrial by-products in Africa. FAO Anim. Prod. & Health Paper 4: 147-162.

Anonym. 1981. The role of crop residues in African livestock production. ARNAB Newsletter 1:1: 3-4.

Balch, C.C., 1977. The potential of poor-quality agricultural roughages for animal feeding. FAO Anim. Prod. & Health Paper 4: 1-6.

Breirem, K. and Homb, T., 1970. Fórmidler og fórkonservering. Gjøvik, Norway. pp. 450.

Chenost, M. and Mayer, L., 1977. Potential contribution and use of agro-industrial by-products in animal feeding. FAO Anim. Prod. & Health Paper 4: 87-110.

Chicco, C.F. and Schultz, T.A., 1977. Utilization of agro-industrial by-products in Latin America. FAO Anim. Prod. & Health Paper 4: 125-146.

Devendra, C., 1981. Non conventional feed resources in the S.E. Asian Region. World. Rev. Anim. Prod. 17: 3: 65-80.

Devendra, C., 1982. Perspectives in the utilization of untreated rice straw by ruminants in Asia. 2nd Ann. Meet. of Austr. Asian Fibrous Residues Res. Network, 3-7 May, Malaysia, 34 pp.

Ensminger, M.E. and Olentine, C.G., 1978. Feeds and nutrition. 1st ed. California, 1417 pp.

FAO, 1981. FAO Production Yearbook. Vol. 35. FAO Statistics Series no 40, Roma 1982, v+306 pp.

Fitzhugh, H.A., Hodgson, H.J., Scoville, O.J., Nguyen, T.D. and Byerly, T.C., 1978. The role of ruminants in support of man. Morrilton, Arkansas. Winrock Intern. Livest. Res. & Train. Center, 135 pp.

Göhl, B., 1981. Tropical Feeds. FAO Anim. Prod. & Health Ser. 12, Rome, xviii + 529 pp.

Horszczaruk, F., 1976. Ruminant nutrition in the Near East. Handbook no 12. UN development progr., Bagdad, 72 pp.

Kellner, O. and Becker, M., 1971. Grundzuge der Futterungslehre. 15 Aufl. 374 pp.

Khajarern, S. and Khajarern, J., 1980. Feed resources in Southeast Asia. IFI-APHCA seminar on Feed Composition. Data Doc. & Anim. Feeding Systems. 22-24 Jan.

Klopfenstein, T. and Owen, F.G., 1981. Value and potential use of crop residues and by-products in dairy ration. J. Dairy Sci. 64: 1250-1268.

Kommeri, M., 1980. Use of waste fibres from industrial wood processing as feed for ruminants. Finnish Agr. Res. Cent., Inst. Anim. Husb., Report to Finnish Academy, 57 pp. (in Finnish).

Morrison, F.B., 1961. Feeds and Feeding, Care and management of farm animals, including poultry. The Morrison Publ., Co., Iowa. 9th ed. 696 pp. (Table I, pp. 538-605).

Müller, Z.O., 1977. Economic aspects of recycled wastes. FAO Anim. Prod. & Health Paper 4: 265-294.

Müller, Z.O., 1980. Feed from animal waste. State of knowledge. FAO Anim. Prod. & Health Paper 18, XI + 190 pp.

NJF, 1969. Fodermiddeltabel. Nord. Jordbr. Forsk., Gjøvik, Norway, 40 pp.

NRC, 1978. Nutrient requirements of domestic animals. No. 3: Nutrient requirements of dairy cattle. 5th rev. ed., Washington, D.C., 76 pp.

Osuji, P.O., 1982. Agroindustrial by-products and animal feeding in the Carribean region. World. Rev. Anim. Prod. 18: 1: 43-56.

Pigden, W.J., 1977. Nutritional and economic aspects of utilizing wood processing by-products. FAO Anim. Prod. & Health Paper 4: 211-226.

Preston, T.R., 1977. Utilization of agro-industrial by-products in integrated systems of plant and animal production. FAO Anim. Prod. & Health Paper 4: 187-198.

Raininko, K., Heikkila, T., Lampila, M. and Kossila, V., 1981. Effect of chemical and physical tretment on the composition and digestibility of barley straw. Agric. Environm. 6: 261-266.

Ranjhan, S.K., 1980. Utilization of crop residues and agro-industrial by-products in animal feeding in Sri Lanka. FAO/UNDP project SRL/78/028. Tropical Pasture Development Report, 58 pp.

Riquelme, E., Dyer, I.A., Baribo, L.E. and Couch, B.Y., 1975. Wood cellulose as an energy source in lamb fattening rations. J. Anim. Sci. 40: 5: 977-981.

Rissanen, H., Kossila, V., Kommeri, M. and Lampila, M., 1981. Ammonia treated straw in the feeding of dairy cows and growing cattle. Agric. Environm. 6: 267-271.

Rissanen, H., Kossila V. and Tuomikoski, V., 1977. Untreated and ammonized straw with or without silage for dairy cows. NJF Grovfodersymposium, 20-22 April, Uppsala, Sweden.

Riviere, R., 1977. Manuel d'Alimentation des Ruminants Domestiques en Milieu Tropical. no. 9 Ministere de la Cooperation.

Sansoucy, R. and Mahadevan, P., 1981. Potential lignocellulose resources and their utilization by ruminants - tropical regions. Meeting on "Isotope-aided studies on non-protein nitrogen and agro-industrial by-products utilization by ruminants with particular reference to developing countries", 30 Nov. to 4 Dec., V.I.C., Vienna, Austria, 10 pp.

Schiemann, R., Nehring, K., Hoffmann, L., Jentsch, W. and Chudy, A., 1971. Energetische Futterbewertung und Energienormen. Berlin, 344 pp.

Schingoethe, D.J., Kipp, D.S. and Kamstra, L.D., 1981. Aspen pellets as partial roughage replacement for lactating dairy cows. J. Dairy Sci. 64: 698-702.

Scholler, H., 1977. Industrial wood processing by-products as potential sources of animal feed. FAO Anim. Prod. & Health Paper 4: 207-210.

Skouri, M., 1977. Utilization des sous-produits des industries agricoles et alimentaires das les pays mediterraneens et Proche-Orient. FAO Anim. Prod. & Health Paper 4: 163-172.

Staples, C.R., Fahey, G.C., Rindsig, R.B. and Berger, L.L., 1981. Evaluation of dairy waste fibre as a roughage source for ruminants. J. Dairy Sci. 64: 662-671.

Sundstøl, F., Kossila, V., Theander, O. and Vestergaard Thomsen, K., 1978. Evaluation of the feeding value of straw. A comparison of laboratory methods in the Nordic countries. Acta Agric. Scand. 28: 10-16.

Søderhjelm, L., 1976. Possible uses for fibrous sludges from the pulp and paper industry. Paperi ja Puu 58: 620-622, 625-626, 629.

Søderhjelm, L. and Lampila, M., 1976. The use of waste fibres as absorption material in grass silage. Paperi ja Puu 58: 2: 41-46.

Wheeler, R.O., Cramer, G.L., Young, K.B. and Ospina, E., 1981. The world livestock product, feedstuff and food grain system. Winrock Intern. Livestock Res. & Train. Cent., USA, Arkansas, vii + 85 pp.

APPENDIX 2.1

THE INDIVIDUAL COMMODITIES INCLUDED INTO THIS STUDY

<u>Cereals</u>:

Wheat	Rye	Sorghum
Rice, paddy	Oats	Buckwheat
Barley	Millet	Mixed grains
Maize		

<u>Sugar cane; Roots & tubers</u>:

Potatoes	Taro	Sugarbeet
Sweet potatoes	Yams	Carrot
Cassava	Nonspecific roots	

<u>Pulses etc.</u>:

Beans	Cow peas	Lupins
Broad beans	Pigeon peas	Pulses nonspecific
Peas	Lentils	Soy beans
Chick peas	Vegetables	Castor beans

<u>Nuts</u>:

Brazil nuts	Walnuts	Coconuts
Cashew nuts	Pistachios	Tung nuts
Chestnuts	Hazelnuts	Nuts, other
Almonds	Groundnuts in shell	

<u>Oilplants</u>:

Palmkernels	Saflower	Cottonseed
Olives	Sesame	Linseed
Sunflower seeds	Mustard	Hempseed
Rape seeds	Poppyseed	

<u>Vegetables, melons etc.</u>:

Cabbages	Pumpkins, squash,	Greem gram
Artichokes	gourds etc.	Other vegetables
Aspargus	Cucumbers	Melons, cantaloupes
Lettuce	Eggplants	Watermelons
Spinach	Chillies, peppers	Beans, green
Tomatoes	Onions	Pears, green
Cauliflower	Garlic	String beans
		Carobs

<u>Fruits & berries</u>:

Grape fruit	Bananas	Dates
Oranges	Plantains	Grapes
Mandarins etc.	Papayas	Pears
Lemons, limes	Strawberries	Peaches
Other citrus fruit	Raspberries	Plums
Apricots	Gooseberries	Apples
Avocados	Blueberries	Cherries
Mangoes	Cranberries	Figs
Pineapples		

Chapter 3

HANDLING AND STORING

by

Arne Hilmersen[1], Frands Dolberg[2] and Ottar Kjus[3]

[1]Department of Agricultural Economics
Agricultural University of Norway
P.O. Box 33, 1432 Aas-NLH, Norway

[2]Novembervej 17
8210 Aarhus, Denmark

[3]Norwegian Institute of Agricultural Engineering
P.O. Box 65, 1432 Aas-NLH, Norway

3.1. INTRODUCTION

The growing of grain, maize and rice is primarily done to produce high quality food. The by-products, i.e. the rest of the
plants, consist of leaves, stems and stalks. These parts contain
great portions of cellulose, hemicellulose and lignin. They are
voluminous, long loose straw from grain has a specific weight of
$30-40kg/m^3$. Straw from grain and rice and stalks from maize contain much energy and are in many countries burned to produce heat.
It is, however, more difficult to utilize the energy in these
roughages in animal feeding. Untreated, the digestibility is low.
For burning purposes and partly for feeding, the straw has to be
dry when stored. In many areas there are great difficulties in
obtaining straw with a water content of less than 10-12%. In many

parts of the world straw is therefore left in the field to be ploughed into the soil or burned in the open.

This chapter deals with the methods and equipment for the handling and storing of straw and stalks for feeding purposes. In most of the chapter problems and solutions for handling and storing of straw from grain are discussed. Some references, however, are also made for straw from rice and stalks from maize.

3.2. STRAW HANDLING AND STORING IN DEVELOPING COUNTRIES

Straw handling and storing in developing countries covers a very wide range of problems. For practical reasons the discussion deals mainly with the situation in India and Bangladesh. Some references, however, are made to harvesting systems used in some other countries.

3.2.1. <u>Handling</u>

Bangladesh has three harvest seasons. The "aus" harvest is in July/August, the "aman" in November/December and the "boro" in April/May.

The "aus" harvest is normally the most difficult as it coincides with the monsoon season. For this reason many paddy fields are under water, and it is not uncommon to see harvesting being done from a boat. In order to carry as much grain as possible, and because the submerged straw is wet and heavy, the farmers cut the straw with their sickle just on the surface of the water - often they purposely cut the upper part of the straw, which carries the grain. The rest of the straw is left in the fields.

In some cases, the straw left in the fields will rot and serve as a compost for the next crop. But in other cases, once the water recedes, the remaining part will be cut and carried home. Such straw is typically used as fuel, while the upper part brought home with the grain, may be used as a fodder. However, it must be noted, it is a difficult point to generalize on, as much depends on the needs of the individual household.

The other reason why straw of the "aus" season may not be useful as a fodder is that the rains, so typical of the season, may make drying of the straw impossible. Often, in spite of many efforts to dry the straw by placing it on elevated places like

village paths and canal embankments, the farmer will have to conclude his efforts by throwing the straw into the compost pit.

The "aus" harvest therefore does not provide much straw for
feeding, because most of the fields are under water and it is
often raining. Even if the weather is bright, it will never be
enough to make the fields dry, and the limited space available -
and the time available to the farmer - all mean that preference
will be given to drying and storing of the grain.

The "aman" harvest is the one where the stacks of straw for
feeding are seen everywhere in Bangladesh. It does not follow,
however, that all the straw is going for feeding. There are large
areas under water even in November/December and the situation can
briefly be categorized as follows:

Deep water aman rice. The straw of this type of rice has the
quality that it can get a length of up to two to three metres.
This is useful as it prevents the grain from being submerged under
water. But the straw is extremely coarse. Work reported by Maurya
(1978) for India has shown that straw resistant to water logging
had highly lignified cells, and work in Bangladesh has shown that
this type of straw, even after ammonia treatment according to the
method described by Dolberg et al. (1981), is not eaten well by
animals (C.H. Davis, personal communication).

The straw will be under water at the time of harvest and only
the head, with the grain, is cut and carried home. The rest will
be left to rot or carried home after the water has receded and
used as a fuel. Sometimes leaves that separate from the stem are
used as a fodder.

Rice in water of medium depth. In this situation, the upper
part of the plant of the straw is cut first and carried home with
the grain. The threshing - as with the crop of any season - normally takes place immediately in the courtyard or on a field close
to the homestead. The threshing may either be done by hand, bea-
ting the paddy against a wooden beam, or by trampling of the paddy
by oxen.

With hand-threshing, the straw is thrown into a heap after the
threshing of each bundle. Although water has not receded from every field, the facilities for drying "aman" straw are much better
than for "aus" straw. Many fields are dry by now, so that they can
be used in addition to village paths and embankments for straw

drying. As the straw is often green at the time of harvest and having a moisture content of around 50%, it may develop heat as well as mould, when lying in the heap during threshing. Spoilage may take place in this situation, and application of a preservative such as ammonia might be a measure to consider. When threshing by cattle-trampling, the threshing normally takes place in a place - often a field - spacious enough to allow the straw to be spread out for drying afterwards. In both methods the straw is turned by hand or stick until dry enough for stacking.

The lower (uncut) part of the straw left in the field remains there until the water has receded and the straw has dried. This is a very rational measure on the part of the farmers. A considerable quantity of the transport of straw from the fields to the homestead is by human labour. Dry straw is easier to carry than wet straw and, if labour has to be hired, cheaper. In distant fields, the owner may find it too expensive to carry the straw home, and he leaves it on the field or allows his workers to take it as a part of their salary. It also happens that such straw is cut at night by the landless people to be used mainly as a fuel, but also as a feed (Briscoe, 1979).

<u>Rice on Land Which is Dry at Harvest</u>. The depth at which the straw is cut at harvest is often decided by the distance from home. The further from home, the more may be left on the fields and some of the considerations vis-a-vis dry straw, labour cost and landless people discussed above will apply.

Sometimes a legume is sown in the standing paddy crop just before harvest to make for a winter crop for either people or animals. In these cases the stubble may be left relatively tall to allow the legume to climb. Sometimes it is also seen that all the straw is left in a corner of the field, so that after the field has been cultivated and sown to a winter vegetable such as water melon, the straw is returned to the field as a mulch.

The "boro" harvest relates to more than one type of straw. This is contrary to the "aus" and "aman" harvests, which relate to rice straw alone. In the "boro" season, in addition to rice, wheat, maize and also barley (on an experimental basis) are cultivated. However, the discussion will concentrate on rice and wheat straw as they are the only two straws of any significance to the farmers.

Much of the rice in the "boro" season is of the high yielding variety. This is also contrary to the "aus" and "aman" seasons, where mainly rice of the traditional varieties are cultivated. The high yielding varieties of rice are grown on land which can be irrigated and fertilized relatively heavily. The straw is shorter and stiffer than that of local varieties. Farmers do not consider it to be a good fodder, although it is used to some extent.

As a general rule, the farmers of Bangladesh do not consider wheat straw useful as a fodder. It is not uncommon to see wheat straw being burnt on the fields after harvest. One problem referred to by the farmers is that wheat straw, due to a stiff stem and no leaves, is very difficult to stack. This is not the case with rice straw. These problems are overcome in India by application of a different stacking technique and chopping of the straw before stacking. Stacking in India is done by blowing the chopped straw into a silo made of branches or long straw. Upon completion of threshing the stack is closed by a thatching of long wheat straw. However, probably because of much smaller farms in Bangladesh, machine threshing is uncommon and chaff cutters are practically absent. No farmer has considered it worth his while to do the cutting by sickle before stacking - probably for good reasons, as the harvest season is a busy season and other jobs are pressing. But even if they did chop the straw by sickle, this would probably not provide the same feed as wheat straw threshed by machine in India. The reason is that the straw receives a number of beatings during the passage through the machine, which breaks its surface and makes it softer for animals to eat. The effect of a broken straw surface is likely to be that it ensures a greater surface for the digestive enzymes and bacteria of the rumen to attack, once the straw is eaten. This would provide for a more rapid rate of digestion, which in turn would increase intake.

3.2.2. Storage of straw

Some indications of the ways in which straw is stored in India and Bangladesh have already been given. In Bangladesh, straw is stacked as long straw, close to the farm. There are farmers who provide a raised wooden platform for the stack to avoid straw loss due to running water or wet soil. Three leaves are placed on top of the stack to prevent rain water from entering the stack.

Often the stack is made around a tree or wooden pole. This is convenient from the point of view of keeping the stack standing in stormy weather, but moulding along the pole is seen. In the state of Kerala in India farmers place a clay pot turned upside down on top of the pole to avoid water entering.

Removal of straw for feeding. Whether placed under handling or storing, removal of the straw for feeding must be included to complete the picture. Removal for feeding does not present any particular problem when the straw is placed in a building, but care has to be taken with stacks standing in the open. The most convenient way to remove the straw would be to start from the top and remove it layer by layer. However this would leave the stack open to spoilage by rain, as the compact layers of straw eventually formed would be broken daily. For this reason, the straw is often pulled from the lower part of the stack by hand. This is very labourious. During daytime, farmers often overcome this prob-

Plate 3.1. Self-feeding of straw from a stack in Bangladesh (Courtesy F. Dolberg)

lem by tethering the cattle next to the stack, so that they eat from it directly. But this is not possible for straw to be fed in the shed during night-time. Likewise, this procedure is not too good in rainy weather, as straw dropped by the animals is soiled and wasted.

3.2.3. Some harvesting methods in other countries

Before the age of the combine harvester a wide range of harvesting systems had been developed in the more industrialized countries. In general the harvesting process was as described on the next page.

Plate 3.2. Grain harvesting at 67°N in Norway 1936 (Courtesy Eli Rønneberg Wik)

Cutting with hand tools or mowing machine. Depending on the drying method, the crop was left loose or bound. The self-binder combined the cutting and the binding in one operation.

Drying. After cutting the crop was dried in the field either lying on the ground or put up on racks or poles.

Transport of the dried crop to the threshing station. In Northern Europe the crop was very often stored inside the barn.

Threshing. Over many decades we had a development from manual threshing to very good threshing machines. To power the machine we had a similar development, from the "horse-circle" to the electric motor and tractor.

Storage. The threshing operation separated the straw and chaff from the grain. The straw was mostly stored inside the barn or stacked outside.

These harvesting systems are still very attractive in many countries. The crop is well taken care of and the straw is a natural part of most of the systems. In many cases it requires more effort to get rid of the straw than to store it for later use.

3.3. HANDLING AND STORING OF RESIDUES LEFT BY THE COMBINE HARVESTER

The greatest part of this section deals with the handling and storing of straw from grain. The handling of maize stover and maize cobs is discussed briefly only.

3.3.1. Swath characteristics of straw left by the combine harvester

The characteristics of the swath have a significant influence on the handling efficiency of the straw. Type and size of crop, straw/grain ratio, cutting width of combine and cutting height are all factors that determine the density of the swath. Under Norwegian conditions we have observed from 0.25 to 1.6 kg DM per m swath. In many countries it is necessary to combine even under humid conditions. On one farm combining was observed to be done with only 38% DM in the straw. To be stored, baled or loose-stacked without preservative, the straw has to have at least 83% DM. This means that on many occasions straw has to be field-dried.

<u>3.3.2. Raking</u>

Raking may be necessary to speed up the drying of the straw or
to produce swaths with higher density than left by the combine. If
the straw is to be baled it is important, however, to make swaths
with uniform size and density. This should be kept in mind when
operating the rake, the raking method being more important than
the type of rake. In one of our investigations the capacity was
2.9 ha/h using a fingerwheel rake with four wheels to double the
density of the swath. Jonsson (1974) has estimated a capacity of
2.4 ha/h for a 6-wheel fingerwheel rake.

<u>3.3.3. Handling and storing loose straw</u>

Long, loose straw is rather bulky for transporting and storing
$(30-40$ kg/m^3). On the other hand the harvesting of loose straw
does not require expensive machinery. Handforks and buckrake may
be the only equipment needed.

If a self-loading wagon is available on the farm, it may also
be used for handling straw. In a research project at the Norwegian
Institute of Agricultural Engineering 670 kg of straw were picked
up in 15 minutes with a medium-size wagon. If straw is picked up
with a chopper the density will increase 75-100% compared with
long, loose material. The storing of cut straw must, however, be
done in bins or silos, while long straw can be stacked more inde-
pendently of buildings.

<u>3.3.4. Baling (conventional bales)</u>

In Western Europe most straw is baled with the conventional
piston-type medium-pressure balers. The height and width of bales
is determined by the size of the bale-chamber (mostly from 30 x 40
to 36 x 46 cm). Length is variable and determined by the operator's setting of the baler.
The baling operation may be divided into four parts:
1. Actual baling (picking up the swath at full capacity)
2. Turning
3. Rearranging bales (when a bale collector is attached and requires extra time)
4. Breakdowns and other idle time

34

The actual baling capacity lies within wide limits. In our investigations (1976-1980) we have had an average of 104 kg straw/ min for the bigger balers (bale size 36 x 46 x 90 cm) and 59 kg straw/min for the smallest. In Swedish tests 50% faster rates have been observed (Jonsson, 1974). Density of swath has a marked in- fluence on baling capacity. Based on our results we have found the relationship shown in Figure 3.1.

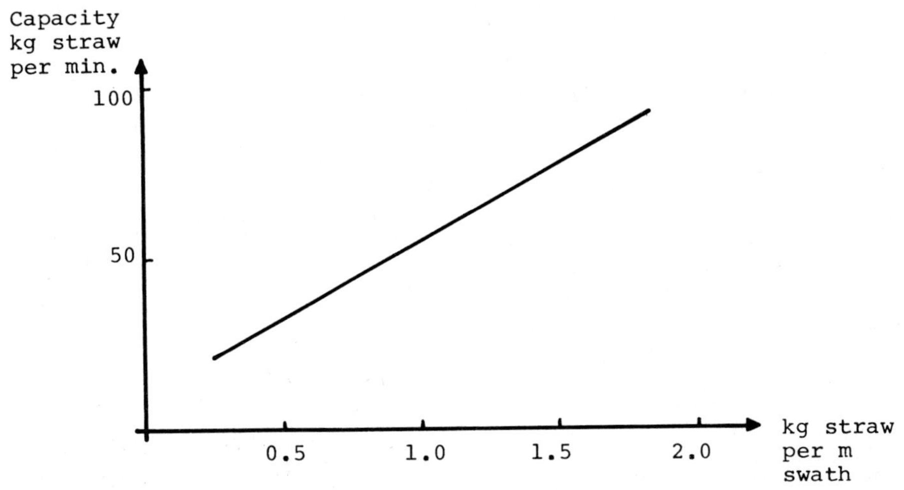

Figure 3.1. The influence of the density of the swath on actual baling performance

Time for turning varies much from field to field. On average our results have given 14-22% reduction in actual baling capacity because of turning.

Many bale-handling systems require some bale collection before the bales are left on the ground. This may hamper the operation of the baler itself and will therefore reduce baling capacity furt- her. The various bale collectors are described in the next sec- tion.

3.3.5. Handling (conventional bales)

The straw handling process consists of four different operations:

1. Bale-collection (may be omitted)
2. Loading for transport
3. Transport to storage
4. Unloading and stacking

In most bale-handling systems these operations are integrated. A general view of different bale-handling systems is given in Figure 3.2.

One of the most common bale-handling systems in Norway uses the buckrake designed for hay (Høysvans). During baling the bales are

Plate 3.3. Tractor with two buckrakes (Courtesy O. Kjus)

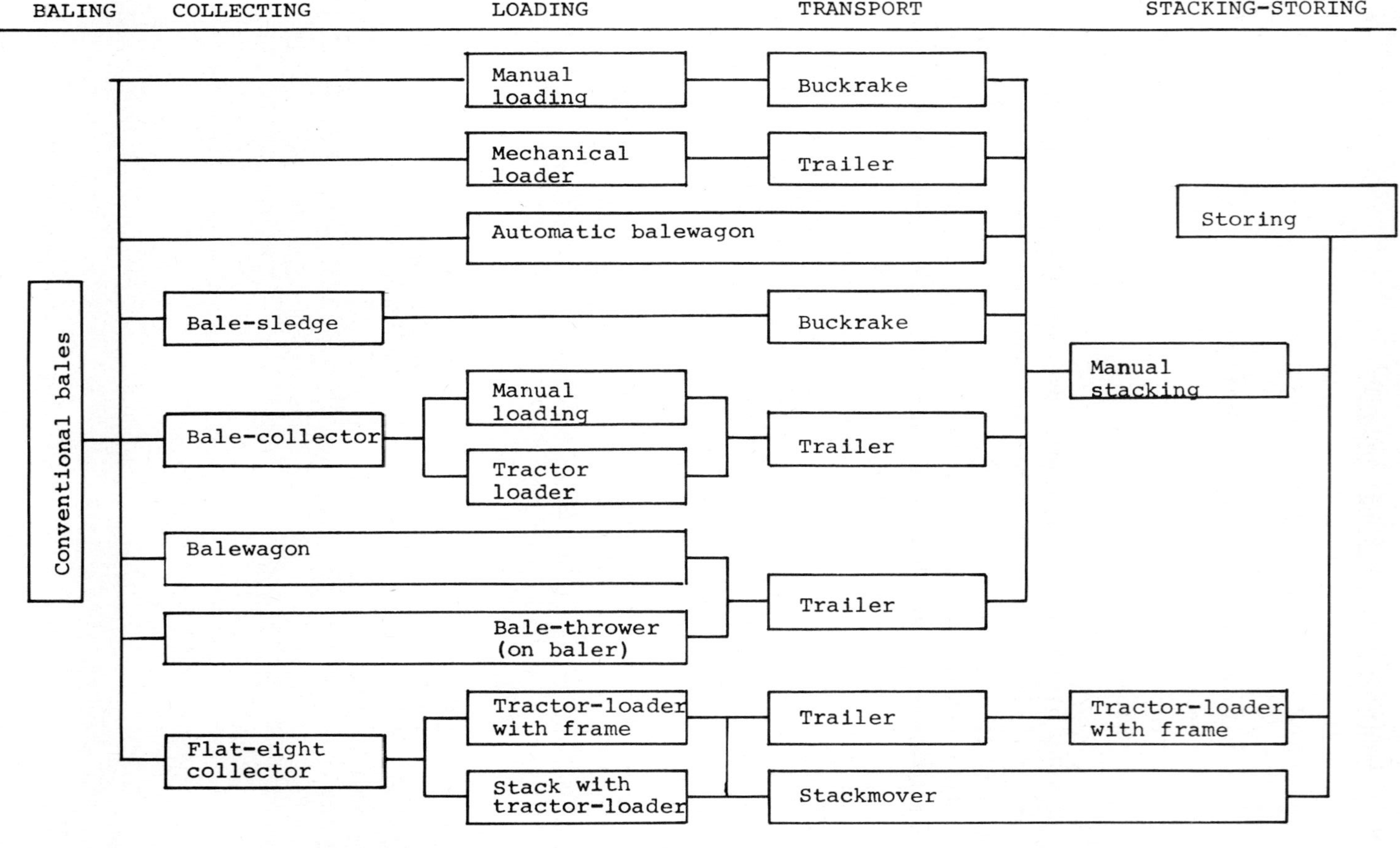

Figure 3.2. Handling systems for conventional medium density bales (KJUS, 1982).

mostly collected in piles with the use of a small balewagon trailed by the baler. The bales are loaded manually on the buckrake and transported to the storage place. If the straw is to be treated with ammonia, the bales are stacked directly on plastic sheets. If not, they are unloaded and stacked in the barn.

Another system uses trailers for transport. The bales are dropped individually and loaded manually as the trailer is pulled between rows of bales. With one person driving and two persons loading the bales, the capacity is rather good. For transport longer than 1 km it is necessary to stack properly on the trailer and for that, one extra person is needed. In this system it is also possible to load bales mechanically with the use of bale conveyors or separate bale throwers mounted to the tractor.

In the most mechanized handling system bales are loaded individually from the rows, transported and unloaded with the use of automatic bale wagons. One type keeps the bales on an endless conveyor belt inside the wagon (e.g. Kemper). The other type stacks the bales on the wagon-platform and can unload the bales individually or as a uniform stack (e.g. New Holland).

We also have group handling systems. Most are built on units of eight bales forming a flat, quadratic "cake" of bales. The length of the bales has to be double the width to make this system work. These "flat eights" may be formed in special collecting equipment attached to and trailed by the baler. The bales may also be collected to flat eights with the use of special frames on the front-end loader (separate operation). The further handling (loading, unloading, stacking) is then done with the tractor loader equipped with a loading frame with retractible claws. The transport of "flat eights" may be done on ordinary trailers or special stack-movers.

Examples of the different handling systems are given in Table 3.3 page 41.

3.3.6. <u>Baling big bales</u>

In agricultural production where manual labour is rather expensive compared to mechanical equipment there is a tendency to mechanize the processes. Big bales are a result of such a development. They represent a package suitable for tractor-handling, but are too heavy to be handled manually.

Plate 3.4. Roundbaler (Courtesy O. Kjus)

There are many different types of big balers. Some produce rectangular bales (Howard), but most make round bales. A typical round baler feeds the straw from the pickup into a set of parallel running belts which rolls the straw into a bigger and bigger straw cylinder. When the required diameter is achieved, twine is fed into the baler to fasten the outer layers of straw to the bale. The bale is then discharged from the baler by opening the rear of the bale chamber. Results from the investigations at the Norwegian

Institute of Agricultural Engineering 1976-80 are given in Table 3.1.

Table 3.1. Baling of straw with big balers. Results from Norwegian investigations 1976-1980 (Kjus, 1982)

	Round baler	Round baler	Rectangular baler
Bale size (diameter x length), cm	147x115	170x166	150x150x250
Bale weight, kg	185	408	441
Density of swath, kg/m	1.6	2.2	1.2
Density of bale, kg/m^3	95	111	78
Time-study results	------- min/t straw -------		
Baling (in swath)	6.4	4.2	7.3
Feeding (and tying) twine	5.3	2.7	0.3
Turn	1.2	0.4	1.0
Sum including 10% add. time	14.2	8.0	9.5
Calculated capacity (kg straw/hour)	4230	7500	6320

3.3.7. Handling big bales

Big bales have to be handled by tractor, and various kinds of equipment have been developed for this purpose. The most simple and cheapest tool is a big fork attached to the three-point linkage and/or to the front-end loader. The tool may even be a buckrake for grass with only 2-3 tines left. It is possible to haul 4 round bales of 1.2 m diameter at a time with a tractor equipped with one buckrake in front and one at the rear. Other equipment consists of hydraulically operated arms or frames that can grip and hold a round or rectangular big bale. Some results from Norwegian research projects are given in Table 3.2.

Table 3.2.   Handling of big bales. Results from Norwegian investi-
             gations 1976-1980 (Kjus, 1982)

	Round bales	Rectangular bales	Rectangular bales
Transport method	Tractor with buckrake, front and rear	Tractor with gripper front, buckrake rear	Tractor-trailer, loading with tractor gripper
Number of bales per load	4	2	5
Weight of bale, kg	225	416	453
Loading (min/load)	3.9	1.5	8.4
Unloading and stacking (min/load)	1.9	2.6	10.3

3.3.8. Capacity of handling systems

In Table 3.3 are shown examples of different handling systems
calculated to demonstrate capacity on a comparable basis. The calculations are based on:
1. All systems operated by one person (if possible)
2. Turning time equal to 20% of baling time (in swath)
3. Transport, all systems: 8 minutes per load
4. All systems end with bales stacked ready for ammonia-treatment i.e. bales stacked on plastic sheet
5. 3 t straw per hectare

3.3.9. Handling of maize stover, maize cobs

Three principal methods are used for the harvesting of maize.
The methods differ in the amount of the plant harvested.

3.3.9.1. Whole-plant silage
The entire crop is harvested with a precision-chop harvester
equipped with a row head. The crop is put into a clamp silo or
blown into a tower silo.

Table 3.3. Examples of bale handling systems (included baling). Calculations are based on a research project conducted at the Norwegian Institute of Agricultural Engineering 1976-1980

	Conventional bales				Big bales		
	Sep. bale-thrower on tractor Trailer	Automatic balewagon	Bale-wagon 2 Buckrakes Mechanical stacking	"Flat-eight" system. Trailer transport	Two buck-rakes (for grass)	Gripper front Buckrake rear	Gripper Trailer-trans-port
Bales per load (no)	71	109	48	96	4	2	6
Weight of bale (kg)	18.3	15.5	19.4	29.0	225	416	453
Weight of load (kg)	1218	1622	887	1978	781	731	1910
Time per load (min)							
Loading	9.5	8.8	7.4	15.6	3.9	1.5	8.4
Turning etc.	6.4	1.8		2.0			1.0
Transport	8.0	8.0	8.0	8.0	8.0	8.0	8.0
Unloading, stacking	18.7	10.9	5.5	15.7	1.9	2.6	10.3
Round trip	42.6	29.5	20.9	41.3	13.8	12.1	27.7
Time per ton of straw							
Baling etc.	12.1	12.1	12.1	12.1	12.9	11.2	11.2
Handling further	35.0	18.2	23.6	22.0	17.7	16.6	14.5
Sum incl. 10% idle time)	51.8	33.3	39.2	37.5	33.6	30.5	28.3
Capacity/ ha per 8 h day, 1 person	3.1	4.8	4.1	4.2	4.7	5.2	5.6

3.3.9.2. <u>Ear separated from the stalk</u>

Stone and Gulvine (1977) describe different types of machines for picking maize ears. The Corn Snapper removes or snaps the ear from the standing stalk but does not remove husks from the ear. Modifications of this harvests sweet corn. The Corn Picker snaps the ear from the standing stalk and removes the husks from the ear. Corn Snappers and Corn Pickers are equipped with an elevator for loading the ears into a trailing wagon. The machines can be pull-type, mounted to the tractor or self-propelled type. They can harvest one or two rows and the rate of work can be up to 13 hectares per day.

3.3.9.3. <u>The kernels shelled from the cobs</u>

The Corn Picker-Sheller removes the kernels from the cobs. The grain may be delivered to a trailing wagon or to a tank on the machine. However, most of shelled corn is now harvested with Combine Harvesters. The grain head is then replaced with a corn head (up to eight rows). The corn head snaps ears from the standing stalks and feeds snapped ears into the combine. There the combine cylinder shells the kernels from the cobs. The separating and cleaning systems function as they do in harvesting small grain.

The residues after grain removal are deposited on the ground by the corn-harvesting machines. After a drying period they can be picked up with a round baler or stack wagon. The latter one is a machine which makes weather resistant stacks.

Bargiel et al. (1982) describe a cob saver attachment designed to attach to the rear of a conventional combine. The separated cobs were elevated into a trailing forage wagon. The prototype of this equipment saved about 78 per cent of the cobs. About 11 per cent (by weight) of husks and stalks remained with the cobs under the best conditions. Dry cobs can be stored with minimum spoilage. However, the cob moisture content is directly related to moisture content of the grain. When grain moisture content is above 25 per cent, cob moisture is about 50 per cent. At approximately 12.5 per cent moisture, cob and grain have about the same moisture content.

3.4. SUMMARY

This chapter deals with the handling and storing of straw from grain and rice, and stalks from maize. The topic is very wide. The problems and solutions described are therefore limited mainly to situations in India and Bangladesh, and North Western Europe.

Harvesting of straw from rice in Bangladesh depends on season. Fields are flooded during the "aus" harvest in July/August and little straw is harvested. This season is also very rainy creating difficulties of drying the straw. During the "aman" harvest the situation is better and much of the straw is harvested and stored in stacks. In the "boro" harvest in April/May rice and other crop residues are harvested. Farmers in Bangladesh do not consider wheat straw a useful animal feed. Manually threshed wheat straw is also difficult to stack and is therefore mostly burned. In India wheat is mostly machine threshed with straw being chopped and blown into stacks. In many countries the grain crop is still harvested without using combine harvesters. These systems are briefly mentioned.

The problems and solutions connected with straw left by the combine harvester are discussed. Because of the low density of loose straw, most straw harvested in North Western Europe is baled. The influence of swath density on baler performance is considered. Different handling systems are discussed. Calculations show that mechanical stacking and transport with tractor mounted buckrakes can almost compete with the more expensive automatic balewagon system. Where manual labour is expensive the big-bale concept may be attractive. Handling capacity is generally higher for big bales than conventional bales.

The handling of maize stover and cobs is briefly discussed.

3.5. REFERENCES

Bargiel, D.A., Liljedahl, J.B. and Richey, C.B., 1982. A Combine Cob Saver. Transactions of the ASAE 25: 544-548.

Briscoe, J., 1979. Energy Use and Social Structure in a Bangladeshi Village. Development Review 5 (4). Washington.

Buchele, W.F., 1976. Research in Developing more Efficient Harvesting Machinery on Utilization of Crop Residues. Transactions of the ASAE 19: 809-811.

Davis, C.H., 1982. Personal communication. NIRDP/Danida, Noakhali, Bangladesh.

Dolberg, F., Saadullah, M., Haque, M. and Ahmed, R., 1981. Storage of Urea-Treated Straw using Indigenous Material. Wld. Anim. Rev. (FAO) 38: 37-41.

Hilmersen, A., Ohren, O., Bjartnes, P. and Kjus, O., 1977. Handteringssystemer i halmberginga (Straw Handling Systems). Norwegian Institute of Agric. Engineering, copy-print 549.

Jonsson, B., 1974. Olika metoder for halmbargning (Straw handling methods). SLA Arbetsekonomi 3.

Kepner, R.A., Bainer, R. and Barger, E.L., 1978. Principes of Farm Machinery. J. Wiley & Sons, New York.

Kjus, O., 1981. Handteringssystemer for halm (Straw Handling Systems). Norwegian Agric. Research Council, Project Report 376, Oslo.

Maurya, D.M., 1978. Anatomical Basis of Flood Resistance in Rice. Indian Journal Genetics Plant Breeding 38: 184-185.

Stone, A.A: and Gulvine, H.E., 1977. Machines for Power Farming. J. Wiley & Sons, New York.

Chapter 4

ANATOMICAL AND CHEMICAL CHARACTERISTICS

by

Olof Theander[1] and Per Åman[2]

[1]Department of Chemistry and Molecular Biology
Swedish University of Agricultural Sciences
S-750 07 Uppsala, Sweden

[2]Department of Animal Husbandry
Swedish University of Agricultural Sciences
S-750 07 Uppsala, Sweden

4.1. INTRODUCTION

This review will deal with the straw of wheat, barley, oats, rye, rice and to some extent corn (maize), and other fibrous by-products, which represent the main agricultural crop residues of the world and botanically belonging to the *Graminae* family. These residues are an enormous underutilized energy resource of great feed potential for ruminants as well as for fuel, and as raw material for chemicals and other technical products. A characteristic of straw is that it mainly consists of highly lignified cell wall material, which often constitutes up to 80% of the dry matter. These cell walls are mainly built up of structural polysaccharides and lignin. Readily available storage carbohydrates and proteins are present in much lower amounts than in most other forages and agricultural products. The mineral content of straws can sometimes be quite high.

The cell walls and the chemistry of individual components in cereal straw have not been so thoroughly studied as in wood but the present knowledge, particularly of the carbohydrate components is quite extensive.

This review will include the anatomy and morphology of straws, the contents and structures of organic and inorganic constituents in straw, the chemical composition of straw fractions, chemical changes during maturation, interrelations between cell wall constituents and, finally, the chemical modifications of straws with special reference to conditions used in technical feed-upgrading processes.

4.2. STRAW FRACTIONS

Straw consists of the above-ground fractions of cereal plants after removal of the grain. In many reports, however, the residual parts of flower heads are not included since they are commonly lost during the harvest and constitute a relatively small fraction of the straw.

The straw length varies greatly both within and between species. In many species a short and stiff straw is of great agricultural importance and is a quality selected for by plant breeders. Barley and wheat straw are generally 50-100 cm long, oat straw somewhat longer and rye straw even longer (up to 200 cm). Rice straw varies widely in length (30-500 cm) depending on the cultivation method. Irrigated rice usually has short straws while floating rice has long straws (Cobley and Steel, 1977). The average diameter of stalks of some oats, rye and wheat straws is about 3 mm (Muller, 1960), but large variations occur even within species and due to agricultural practice.

Straws of barley, oats, rice, rye and wheat belong to the slender stemmed type and have a similar botanical composition (Figure 4.1). The straw is composed of a variable number of internodes, separated by nodes - the points at which the leaves arise. The leaf consists of two parts, the leaf sheath and the leaf blade. The leaf sheath envelopes the lower part of the internode. The leaves are organs of limited growth and once they have reached their final size, they remain on the plant for a certain period of time and then die. The number of living leaves present on a given tiller varies within a relatively narrow limit. Senescence begins

at the tip of the leaf and spreads downwards. The cell constituents are   redistributed so that the leaf looses weight. Further weight losses are later caused by disintegration. The nodes
are small areas with meristematic activity in the living plant but
generally include the base of the leaf sheath and small parts of
the internodes when separated manually (Figure 4.2, Harper and
Lynch, 1981). Barley, oats, rye and wheat straws are commonly made up of 5-6 stem nodes while rice straws are reported to have 10-20 stem nodes (Grist, 1965). Straws of barley, rye and wheat include the rachis, the residual part of the flower head. The rachis is composed of a number of nodes with very short internodes. On the other hand,straws of oats and rice include remnants of the panicle.

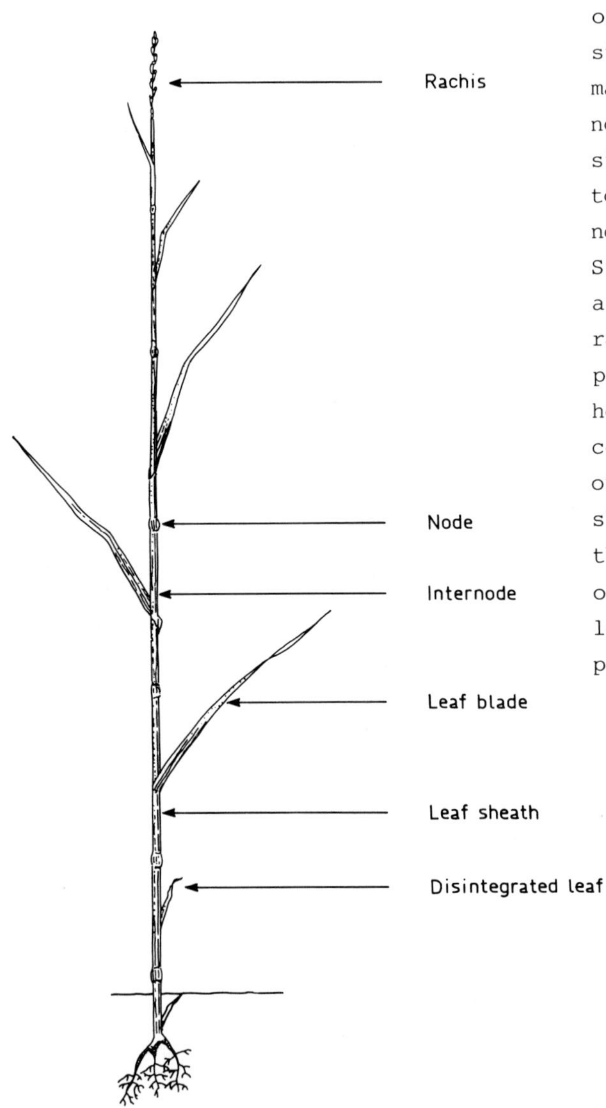

Figure 4.1. Anatomical fractions of wheat straw

48

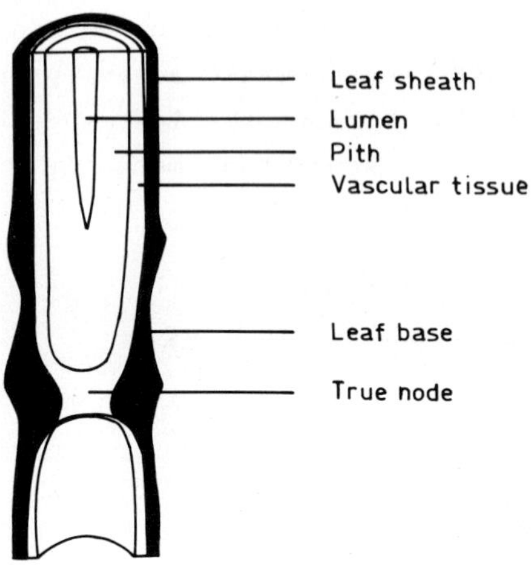

Figure 4.2. Schematic presentation of a longitudinal cross-section
of wheat straw, showing the node-area (after Harper
and Lynch, 1981)

Yields of botanical fractions have been reported in several
investigations and are summarized in Table 4.1. The stems, including internodes and nodes, make up 40-80% of the straw. The high
percentage of leaves in rice straw reflects the greater number of
stem nodes in that species. Barley and oat straws generally contain more leaves than rye and wheat straws. The percentage of leaves in cereal straws is of great agricultural importance since the
digestibility of leaves is generally higher than that of stems
(e.g. Åman and Nordkvist, 1983). The percentages of residual parts
of flower heads are not included in Table 4.1 but figures ranging
from 2% to higher than 10% are commonly reported. The botanical
composition of cereal straws depends on several factors, for
example, length of stubble, stage of ripeness and the way the
cereal is grown.

Table 4.1. Botanical composition of cereal straw (g/kg dry matter)

Crop	Internode	Node	Leaf	Reference
Barley	585	66	349	Åman and Nordkvist (1983)
	618[1]		382	Ernst et al. (1960)
Oats	528	36	436	Müller (1960)
	553	39	408	Thiago and Kellaway (1982)
Rice	396[1]		604	Ernst et al (1960)
Rye	722	50	228	Müller (1960)
	750[1]		250	Ernst et al. (1960)
Wheat				
spring	576	44	380	Müller (1960)
winter	540	48	410	Müller (1960)
winter	722	82	196	Thiago and Kellaway (1982)
winter	609	72	319	Harper and Lynch (1981)
winter	727	68	205	Åman and Nordkvist (1983)
winter	726[1]		274	Ernst et al. (1960)

[1] Internodes and nodes not separated

4.3. THE MORPHOLOGY OF STRAW

A schematic drawing of a typical slender-stemmed straw internode is presented in Figure 4.3.

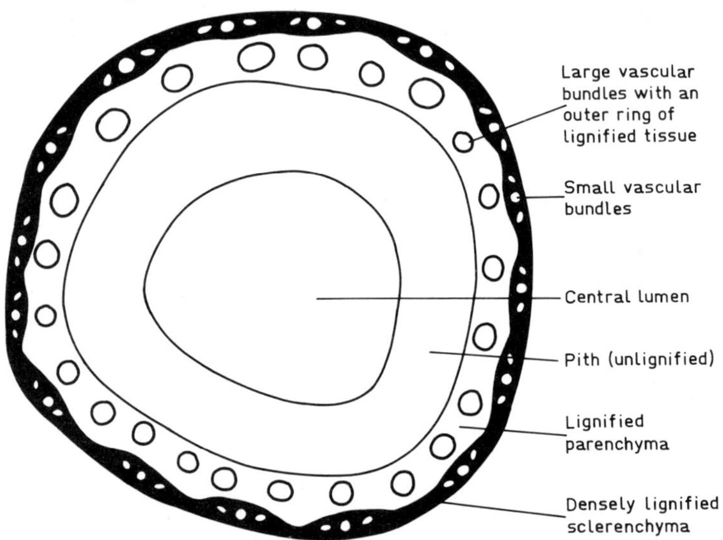

Figure 4.3. Schematic presentation of a transverse section of the internode of wheat straw.

The tubular part is made up of an outer layer of epidermis, a layer with numerous small vascular bundles set in mainly scleren-chymatous tissue (thick-walled often lignified cells) and an inner layer of large vascular bundles set in mainly parenchymatous tissue (thin-walled living cells). The central lumens are more or less filled with pith. The sclerenchymatous layer and the tissues of the vascular bundles are the most lignified tissues but lignin may also occur elsewhere and the degree of lignification increases when the plant grows older. Müller (1960) investigated various anatomical characeristics of internodes on rye, oats, spring wheat and winter wheat. As an average of the top, middle and bottom internodes, the walls of oats were about 0.3 mm thick and the walls of wheat and rye about 0.5 mm thick. A microscopic investi-gation revealed that the internode walls of the four species were composed of 5-7% epidermis, 25-27% sclerenchymatous tissue and 65-69% of parenchymatous tissue.

A tranverse section of typical leaf from a slender-stemmed straw is presented schematically in Figure 4.4.

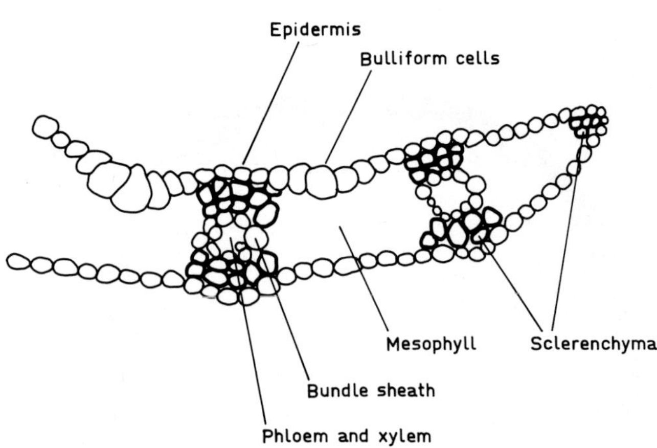

Figure 4.4. Schematic presentation of part of an oat leaf in tran-section

The leaf is also surrounded by a layer of epidermal cells including the bulliform cells, often described as the cells participating in involution and folding movements of the leaves. Vascular bundles, containing phloem and xylem, and surrounded by a bundle sheath, occur rather regularly. The vascular bundles are mostly embedded in mesophyll but sclerenchymatous tissue is often present above and below the bundles. Wheat straw (Harbers et al., 1982) and Bermuda grass (untreated and alkali-treated) (R.R. Spencer, L.L. Rigsby and D.E. Akin, personal communication, 1983) have been examined by scanning electron microscopy and the cellular organization in a transverse section of an internode (and leaf sheath) is shown in Plates 4.1, 4.2, 4.3 and 4.4. The swelling effect of sodium hydroxide on the cell walls of straw was first demonstrated by Watson (1941). For further reading about the anatomy of grasses, reference can be made to Esau (1977).

The lignified tissues in straw internodes are mainly located in an outer ring (Figure 4.3). This makes the straw internode very resistant to microbial degradation which predmoniantly occurs from the lumen side of the internode or from fracture areas. In the leaves, on the other hand, the lignified tissues are separated from each other, making the mesophyll cells more easily available for microbial degradation. Also the nodes are more easily degraded by micro-organisms than the internodes (Harper and Lynch, 1981).

4.4. POLYSACCHARIDES

Plant cell walls contain three types of structural polysaccharides, namely cellulose, hemicelluloses and pectic polysaccharides. This classification is based on the chemical composition and the solubility properties in various reagents. In addition, forages contain storage polysaccharides such as starch, fructans, some mannans, and important amounts of gums and mucilages.

In straws the polysaccharide composition is rather simple, with cellulose and xylans as the predominant components. Smaller amounts of polysaccharides containing mannose and galactose and probably pectic components are also present.

Cellulose is the most abundant molecule in nature. Although it is apparently a simple molecule, being a linear polymer composed of up to 10 000 β-1,4-linked glucopyranosyl units (Figure 4.5.), it is complicated by its three-dimensional structure. It occurs in

52

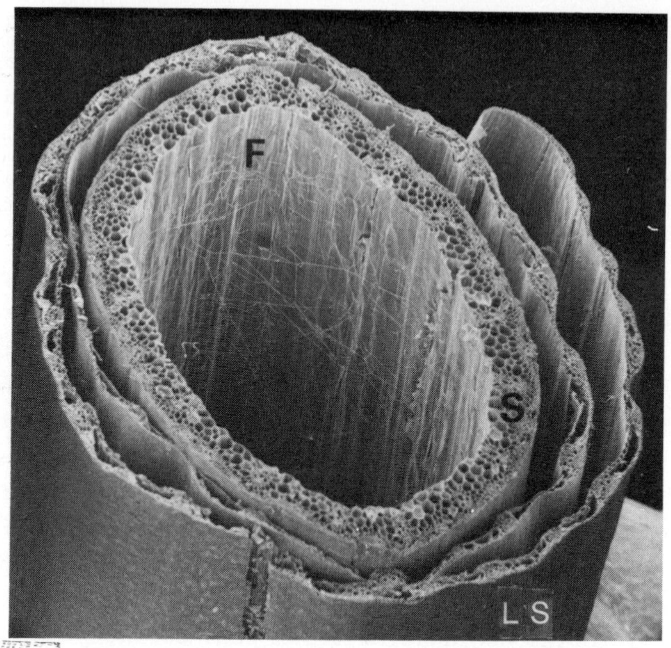

Plate 4.1. Cross section of wheat straw stem (S) and leaf sheath
(LS) with fungi (F) on inner cuticular surface
(Courtesy L.H. Harbers)

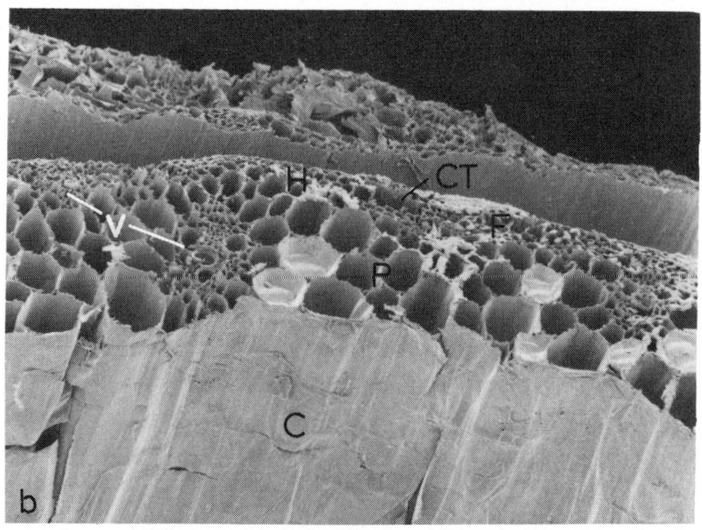

Plate 4.2. Stem showing arrangement of large vascular bundles near
the center and smaller vascular bundles near the outer
edge (V). An inner cuticular layer (C) encloses the
ground parenchymal cells (P). Thick walled fiber cells
(F) and chlorophyllous tissues (CT) are localized in
the hypoderm (H). (Courtesy L.H. Harbers)

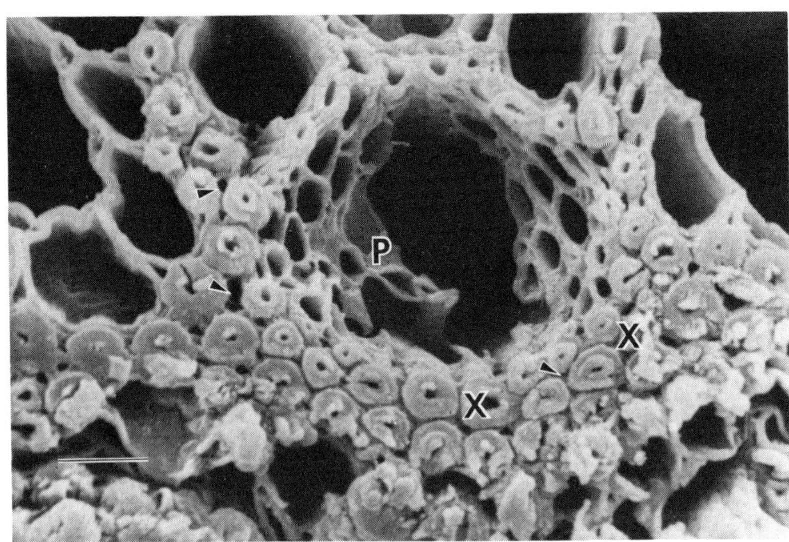

Plate 4.3. Untreated stem of *Cynodon dactylon* var. "Coastal". Scanning electron micrograph of vascular tissue showing intact, undistorted phloem (P) and lignified xylem (X) tissues. Bar = 10 µm

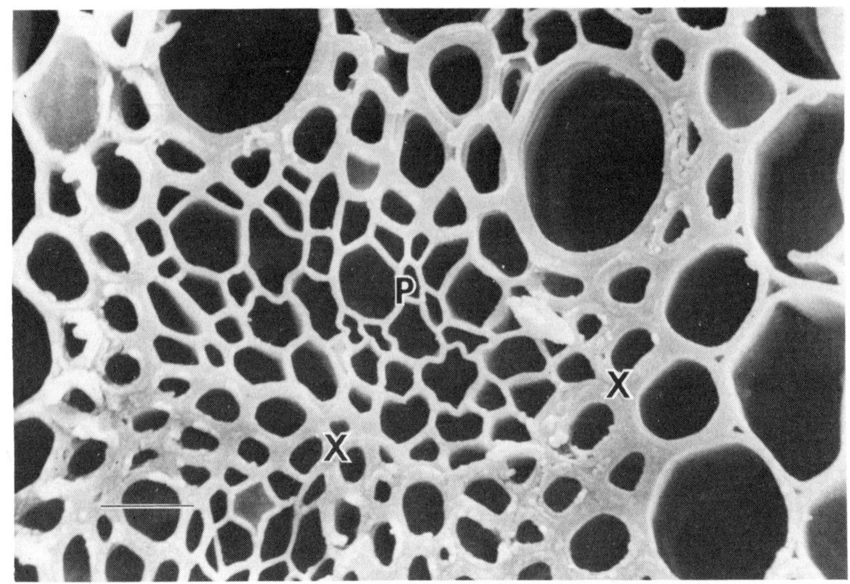

Plate 4.4. Stem of *Cynodon dactylon* var. "Coastal" treated with 10% KOH (w/v) at 65°C for 7 minutes. Scanning electron micrograph of vascular tissue showing disrupted and partially removed phloem tissue (P) and distortion and separation of lignified xylem tissues (X) into individual cells due to removal of intercellular substances (arrows. Bar = µm

Scanning electron micrographs supplied by Roland R. Spencer, Luanne L. Rigsby, and Danny E. Akin, Plant Structure and Composition Research Unit, Richard B. Russell, Agricultural Research Center, ARS-USDA, Athens, Georgia, 30613, USA.

54

nature in a largely crystalline form, organized as fibrils, where
the cellulose chain is tightly packed together in compact aggrega-
gates surrounded by a matrix of other cell wall constituents .
The glucan chains are held together by hydrogen bonds both between
sugar units in the chain and between adjacent chains. The conformation of cellulose (see the lower formula in Figure 4.5) favours
the formation of such bonds and explains the mechanical strength
of cellulose as well as its resistance both to biological degradation and acid hydrolysis. The accessibility of cellulose to hydrolysis can be increased by treatments such as milling to increase
the surface area, steaming, or treatment with swelling chemicals
to make the cellulose less crystalline and less hindered by associated components such as lignin or silica.

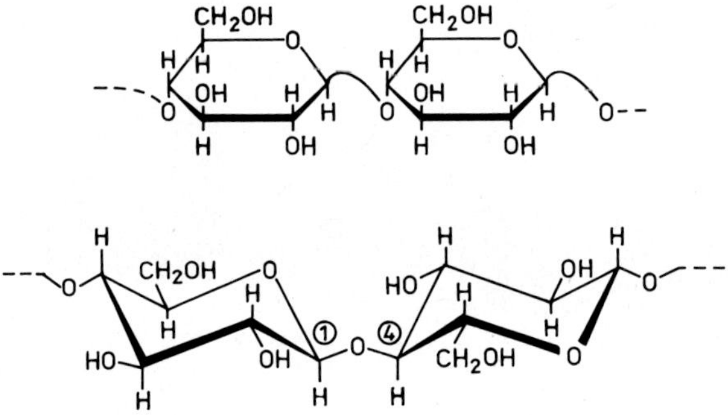

Figure 4.5. Cellulose, upper in the Haworth formula and lower in
the conformation formula.

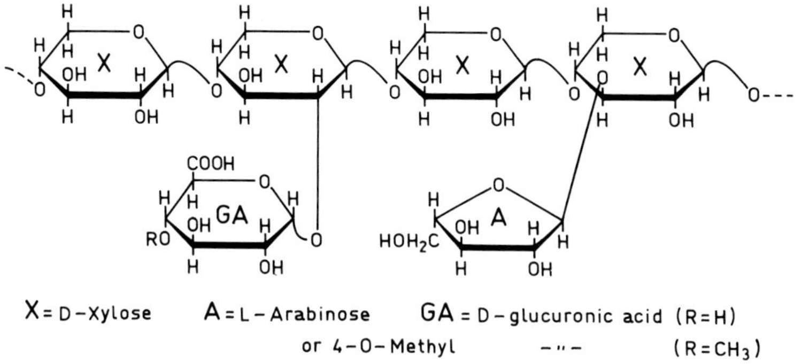

X = D-Xylose   A = L-Arabinose   GA = D-glucuronic acid (R=H)
or 4-O-Methyl   – " –   (R=CH₃)

Figure 4.6. Schematic Haworth formula of a xylan

Xylans from straw, cereals and grasses generally have a backbone of -1,4-linked xylopyranosyl units. Although an arabinose-
free xylan has been isolated from esparto grass (Chanda et al.,
1950), the xylans isolated from straws of wheat and oats are
heteroglycans with single arabinofuranosyl units attached to some
C-3 atoms of the main xylan backbone and glucuronic acid and/or
its 4-O-methyl ether linked to some xylose units (Figure 4.6). The
uronic acids are probably mainly linked to C-2 atoms, although
some results also indicate the presence of 1,3-linkages. Structural studies on isolated straw xylans have been reported in several
investigations (Aspinall, 1959; Aspinall and Mahomed,1954, Aspinall and Meek, 1956; Aspinall and Wilkie, 1956; Bishop, 1953; Wilkie, 1979).

The fact that the arabinose units generally seem to be linked
in the furanosidic form makes them very sensitive towards acid
hydrolysis. In corn hulls an unusual finding for land plants has
been reported, i.e., the presence of both D-galactose and L-galac-
tose as constituent sugars in a xylan (Whistler and Corbett, 1955;
Whistler and BeMiller, 1956).

It has long been known that xylans of angiosperms contain 0-acetyl groups (see in Timell, 1964) and Bouveng (1961) showed that they were distributed between the C-2 and C-3 positions of the xylose units. Bacon et al. (1975) demonstrated that acetyl groups account for 1-2% of the cell walls of *Graminaeous* plants and similar amounts have been found in barley straw (Lindberg et al., 1983). Cell walls of grasses also contain 1-2% of phenolic acids (El-Basyouni et al., 1964,Higuchi et al., 1967, Hartley, 1972 and 1973).

The degree of polymerization of xylans is much lower (50-200 residues) than that of cellulose. Having similar conformation to cellulose it can be strongly associated with that polysaccharide. The presence of side sugar units and substituents is a factor which may prevent tight association between xylan chains. For this reason and because of their lower molecular weight, xylans and other hemicelluloses can be extracted by alkali. The hemicelluloses of grasses and cereals have recently been treated in an extensive rewiew by Wilkie (1979).

Table 4.2. The chemical composition of some roughages as determined by the detergent fibre analysis scheme (Jackson, 1977)

Roughage	Cell walls	Hemicellulose	Cellulose	Lignin
		g/kg DM		
Barley straw	810	270	440	70
Oat straw	730	160	410	110
Paddy straw	790	260	330	70
Wheat straw	800	360	390	100
Sorghum stover	740	300	310	110
Chickpea straw	620	200	300	100
Lucerne straw	690	190	380	110
Sugarcane bagasse	820	290	400	130
Sugarcane trash	800	260	360	100
Paddy hulls	860	140	390	110
Cottonseed hulls	910	150	590	130

Table 4.2 gives some values from Jackson (1977) for cellulose, hemicellulose, as well as cell walls and lignin from straws and some other agricultural residues. These figures are based on the detergent fibre analysis scheme of Goering and Van Soest (1970).

This method is excellent for getting a rapid estimate of the fibre content and its three main components in fibre-rich animal feeds, although it has limitations, in particular regarding the accuracy of the cellulose and hemicellulose contents (cf. Theander and Åman, 1980).

The content of neutral sugar constituents from cellulose and hemicelluloses in straw can be determined by gas-liquid chromatography after acid hydrolysis and the uronic acid constituents can be analysed by a decarboxylation method. These techniques give much more information about the cell wall composition of various plant materials than gravimetric methods (Theander and Åman, 1979). The lignin content is often determined gravimetrically as the residue (Klason lignin) left after removal from the plant material of hydrophilic and lipophilic extractives and of the polysaccharides by strong sulphuric acid (Browning, 1967). This residue often over-estimates the true lignin content as, for instance, polyphenols, cutin and some nitrogeneous matter contribute to this. Table 4.3 gives values which have been obtained using this technique on some Swedish straws (Theander and Åman, 1978). In type of material like straw, calculations of the cellulose content can be made from the glucose value, and the hemicellulose content from the sum of the other sugar values plus the uronic acid value.

Table 4.3. Chemical composition of straw (Theander and Åman, 1978)

Cultivar	Hemicellulose	Cellulose	Klason lignin
Barley	------------------- g/kg DM ----------------------		
Cilla	270	290	200
Ingrid	240	300	220
Senat	230	280	240
Särla	270	280	220
Wing	260	330	230
Oats			
Titus	220	300	230
Winter wheat			
Holme	210	270	210
Spring wheat			
Drabant	210	270	190
Rye			
Petkus II	230	370	200

The table shows that there are significant differences between species and cultivars of the three main components. Of the hemicellulose fractions the uronic acid part varied between 2 and 3.7%. In their neutral parts the relative composition of the constituents varied as: xylose 82-78%, arabinose 19-13%, galactose 4-3% and mannose 1-0.5%. Similar carbohydrate compositions of straws have been reported more recently (Sinner et al., 1979; Chesson, 1981).

4.5. LIGNIN

Apart from the structural polysaccharides, lignin is a main component of straw. Lignin is a family of related polymers of a three-dimensional structure, made up of phenylpropane units. It is generally agreed that p-coumaryl alcohol (I), coniferyl alcohol (II) and sinapyl alcohol (III) are important precursors in the biosynthesis of lignin via a complex enzymatic dehydrogenation process.

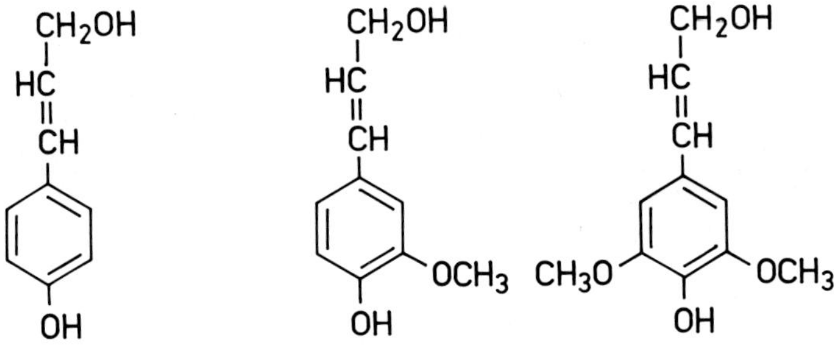

In the polymer these precursors are present as the corresponding structural elements called p-hydroxyphenyl (H), guaiacyl (G) and syringyl (S) respectively. Lignins are often classified according to the distribution of these three elements. Most structural studies so far have been on wood lignins, and Figure 4.7 shows how Adler (1977) presented the main units and linkage types in a softwood lignin as represented by their approximate proportions. Softwood (gymnosperm) lignin, is called a guaiacyl lignin because it is mainly a condensation product of coniferyl alcohol. There are, however, always small amounts of p-hydroxyphenyl- and syringyl-residues present in softwood lignin (see units 2 and 13 respectively in Figure 4.7). The lignins of angiosperms (hardwoods), belong to the guaiacyl-syringyl class, and are thus copolymers of coniferyl and sinapyl alcohols. The ratio between the two monomeric units varied from 4:1 to 1:2. Substantial amounts of all three aromatic residues are found in *Graminae* and for this reason grass lignin may be distinguished from other angiosperms and has been placed in a guaiacyl-syringyl-p-hydroxyphenyl class.

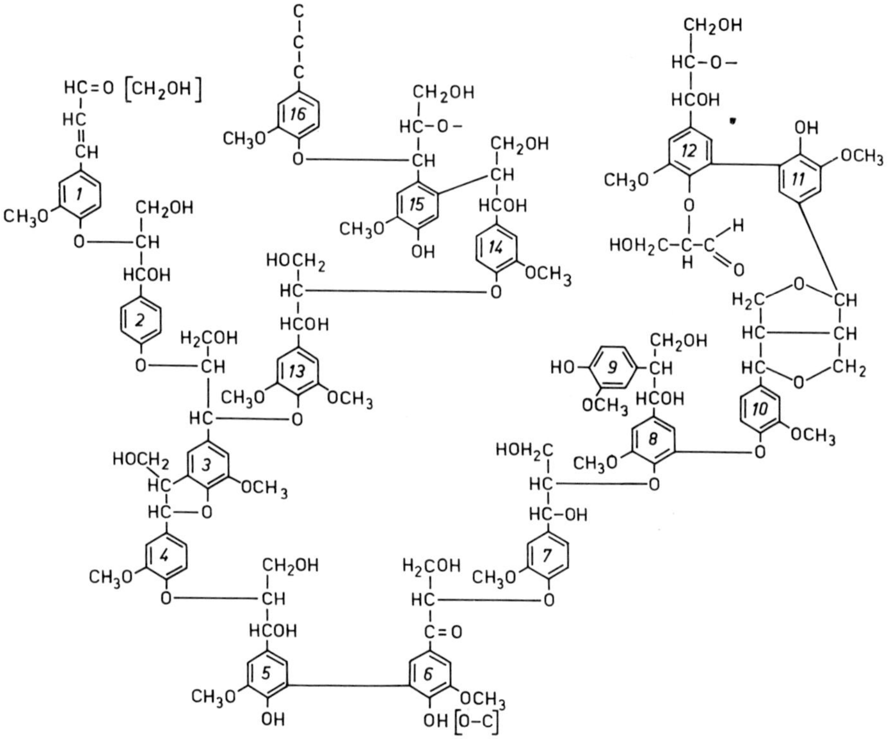

Figure 4.7. Schematic structure of main units in gymnosperm lignin (after Adler, 1977)

Stone et al. (1951) have determined the yield of aldehydes from young wheat plants, obtained after nitrobenzene oxidation. This is a classical method to get an estimate of the pattern of the aromatic residues in different lignins, in which H, G and S are oxidized to p-hydroxy-benzaldehyde, vanillin and syringaldehyde respectively. Also p-coumaric acid, which has been found to be esterified to grass lignin (Higuchi et al., 1967), and tyrosine from proteins should give p-hydroxy-benzaldehyde in this method.

Although the significance of the latter aldehyde is thus not quite clear, this study clearly indicated the presence of H, G and S units and rapid increase of the latter two units during the lignification phase. A more informative two-step oxidation method has been used for structural studies of the lignin in several monocotylidons, including wheat straw confirming the presence of the three residues and other structural features (Erickson et al., 1973).

In more recent years nuclear magnetic resonance spectroscopy (NMR) has become a very valuable tool for structural lignin studies and Nimz et al. (1981) have reported very interesting results from carbon-13 NMR on grass lignins (bamboo, wheat straw and maize stover). This study confirms that all three (H, G and S) residues are important constituents of these lignins. Their study also reveals that p-coumaric acid (IV) and ferulic acid (V) residues also are present. The results of Nimz et al. (1981) indicate that a part of p-coumaric and ferulic acid or both is linked by their phenolic groups via ether bonds to lignin. Immature grass lignins have also been studied by carbon-13 NMR (Himmelsbach and Barton II, 1980).

Lignin has a number of functions which are essential for the plants. Together with other components in cell walls it generates composite structures of outstanding resistance to microbial attack. By its presence as a bonding agent in the layers between cells it contributes to the mechanical strength properties of plants. The presence of this hydrophobic polymer in the cell walls of the conducting xylem tissues retards the permeation to water across the cell walls.

Tables 4.2 and 4.3 list some lignin determinations on straws, and Table 4.4 gives lignin determinations on a series of stem-fibre materials by sulfuric acid methods (Bagby et al., 1971). For an excellent review of lignin see Sarkanen and Ludwig (1971).

Table 4.4. Lignin determination with 72% and 80% sulphuric acid methods on stem fibres (Bagby et al., 1971)

Material	Source (variety)	Lignin 72%, 16h at 10°C	80%, 2h at 5°C
		--- g/kg DM ----	
Bagasse	Florida	196	194
Bagasse	Hawaii	213	218
Bagasse	Mexico	183	187
Bagasse	Louisiana	202	203
Bagasse	Florida	188	197
Bagasse	Phillippines	201	198
Bamboo	Missisippi (Arundinaria gigantea)	222	195
Maize stover	Iowa	147	147
Maize stover	Israel (hybrid)	153	137
Maize stover	Israel (yellow dent)	130	163
Broom maize stover	Illinois	173	185
Crotalaria	South Carolina	224	230
Kenaf fibre	Florida	109	101
Ramie fibre	Florida	206	214
Barley straw	Nebraska (Exond)	154	171
Oat straw	Illinois (Clinton)	177	174
Rice straw)	Louisiana (Zenith)	127	142
Rye straw	Minnesota (commercial)	180	191
Wheat straw	Illinois (Kawvale)	159	157
Wheat straw	Nebraska (Pawnee)	182	185
Wheat straw	Illinois (Pawnee)	200	201
Wheat straw	North Dakota (Premier)	163	174
Wheat straw	Washington (Rex)	164	178
Wheat straw	North Dakota (Stewart Durum)	162	165
Wheat straw	Kansas (Tenmarq)	170	163

4.6. PROTEIN

The content of crude protein in cereal straws is low (Table 4.5) but may vary within a large range. Barley and oat straws generally contain more crude protein than winter rye or winter wheat. In a Swedish investigation (Mattsson and Eriksson, unpublished results) the crude protein contents in 49 barley straws ranged from 3.9 to 9.0%, in 54 oat straws from 1.9 to 7.2%, in 8 rye straws from 2.3 to 4.3%, in 8 spring wheat straws from 4.9 to 5.9% and in 15 winter wheat straws from 2.4 to 5.8%. The wide variation, especially for the spring sown crops, is undoubtedly due to soil and fertilization conditions, time of cutting and other factors.

Table 4.5. Crude protein contents in cereal straw

Straw	Swedish[1]	American[2]	UK[3]
	-------------g/kg DM-----------		
Barley	54	41	38
Oats	45	44	34
Rice	–	42	40
Rye	32	32	36
Wheat, winter	39	36	24
" , summer	52	–	34

[1]Mattsson and Eriksson (unpublished)
[2]Joint United States - Canadian Tables of Feed Composition (1964)
[3]Staniforth (1979)

The major part of the protein is most likely associated with the cell-walls. Cell-wall proteins are known to have low digestibility. In some in vitro experiments (Åman and Nordkvist, 1983), crude protein of barley and wheat internodes was found to have an in vitro degradability of 41% and 27% respectively. After 20 hours of incubation, however, an accumulation of nitrogen on the residues of wheat internodes started, which may lead to erroneous digestibility results for crude protein in vitro.

4.7. OTHER ORGANIC CONSTITUENTS

In plants, hydrophobic layers are attached to the cell walls and act as diffusion barriers. Cuticle covers aerial organs and consists of a polymer, cutin, embedded in waxes. This polymer is attached to epidermal cell walls via pectinaecous materials. Underground organs and peridermis in all organs are protected from diffusion by another polymer, suberin, and associated waxes (Kolattukudy et al., 1981).

Cutin is an amorphous polymer primarily built up of a C_{16} and C_{18}-family of acids, including hydroxy- and epoxy-substituted acids. These acids are largely held together by ester linkages to a predominantly linear polymer. p-Coumaric acid has been found ester-linked to the cutin polymer (Kolattukudy et al., 1981). Aliphatic components of cutin from leaves of barley and maize have been investigated in detail (Espelie et al., 1979).

Suberin is a polymer tightly attached to the cell wall and therefore difficult to isolate. Suberin consists of 10-14% aliphatic alcohols and acids and 20-60% phenolic materials. Small amounts of esterified ferulic acid have been found in the suberinenriched polymers. The secondary structure of suberin is largely unknown (Kolattukudy et al., 1981). Aliphatic components of suberin from the bundle sheaths of maize have been investigated in detail (Espelie and Kolattukudy, 1979).

The waxes associated with cutin and suberin are usually a complex mixture of several classes of aliphatic components, as hydrocarbons, esters, free alcohols and acids (Kolattukudy, 1980). Cereal straw contains only small amounts of waxes; leaves of wheat (Tulloch and Weenik, 1969; Tulloch and Hoffman, 1971) and rye straw (Streibl et al., 1974) contain 0.2-0.5%. In these three studies it has been found that the usual wax components are also accompanied by substantial amounts of alpihatic β-diketones.

Cereal straws contain only traces of free phenolic acids extractable with refluxing 80% ethanol (Salomonsson et al., 1978). Combined phenolic acids are mainly ester-linked to cell wall polymers and can be released by alkaline extraction. Cereal straw contains 0.1-0.7% of combined phenolic acids (Table 4.6) of which trans-p-coumaric acid (IV) and trans-ferulic acid (V) dominate (Kuwatsuka and Shindo, 1973; Salomonsson et al., 1978).

Table 4.6. Content of combined phenolic acids in some cultivars of cereal straw
(Salomonsson et al., 1978)

| Phenolic acid | Barley | | | | | Oat, | Wheat | | | Rye, | Rice |
	Wing	Särla	Cilla	Ingrid	Senat	Titus	Starke II	Holme	Drabant	Petkus	
						mg/kg DM					
p-Hydroxybenzoic	92	26	37	28	170	290	26	42	91	62	Tr
Vanillic	110	93	13	20	120	79	13	24	180	33	17
cis-p-Coumaric	390	130	79	180	390	400	240	240	210	530	120
trans-p-Coumaric	2800	510	880	1700	2800	3100	920	1500	1700	2900	2600
cis-Ferulic	450	89	170	210	420	540	240	300	270	560	140
trans-Ferulic	2100	440	1000	790	3100	2400	620	1400	1400	2700	1200
identified phenolic acids	5942	1288	2179	2928	7000	6809	2059	3506	3851	6785	4077

In a study on nine different cereal straws (Theander and Åman, 1978) it was found that the amount of components extractable with refluxing 80% ethanol represented 5-13% of the dry weight. Low-molecular carbohydrates in these extracts amounted to 0.3-1.3% of the dry weight; fructose, glucose, sucrose and the sugar alcohols arabinitol and mannitol being the main constituents (Table 4.7).

Table 4.7. Low-molecular weight carbohydrates in straw (Theander and Åman, 1978)

Cultivar	Fruc-tose	Glu-cose	Suc-rose	Arabi-nitol	Manni-tol	Low-molecular weight carbohydr.
	------------------g/kg DM ---------------------					
Barley						
Cilla	3.4	4.8	1.5	1.9	1.8	13.4
Ingrid	0.4	0.2	0.8	0.5	1.9	3.8
Senat	1.3	1.5	0.8	1.4	0.7	5.9
Særla	4.1	3.6	1.0	1.5	1.2	11.4
Wing	0.3	0.3	0.2	0.9	1.2	2.9
Oats						
Titus	2.8	2.9	3.3	0.6	0.4	11.0
Winter wheat						
Holme	2.6	1.2	0.4	1.5	1.3	7.0
Spring wheat						
Drabant	2.5	1.8	4.4	2.1	1.8	12.6
Rye						
Petkus II	0.2	0.3	0.3	2.2	1.5	4.5

4.8. MINERALS

Average contents of minerals in cereal straw are presented in Table 4.8. It is important to remember that these figures sometimes only represent single samples. It is known that the mineral contents vary widely depending on agronomical factors and also with the amount of contaminating soil.

Table 4.8. Contents of minerals in cereal straws[1]

Mineral	Unit	Barley	Oats	Rice	Rye	Spring wheat	Winter wheat
Ash	g/kg	60	59	189	39	61	50
Silica	"	15	11	130	34	31	32
Ca	"	2.9	3.9	2.4	2.8	3.2	2.1
P	"	0.8	0.9	0.9	1.0	0.8	0.8
Mg	"	1.0	1.5	1.2	0.9	0.9	1.1
K	"	14.0	21.9	13.2	9.8	11.8	10.0
Na	"	–	–	–	0.5	0.5	0.5
Cl	"	7.7	8.1	–	2.5	6.1	3.5
S	"	1.4	2.5	1.3	1.2	1.4	1.6
Fe	mg/kg	305	214	347	300	420	230
Mn	"	27	89	–	44	–	36
Zn	"	60	138	–	25	–	54
Cu	"	3.9	6.5	–	3.0	–	3.1
Mo	"	0.4	0.6	–	0.1	–	0.4
J	"	0.4	0.3	–	0.5	0.8	0.6
Co	"	0.3	0.09	–	0.05	–	0.08
F	"	–	–	–	–	–	4.5

[1]From Joint United States - Canadian Tables of Feed Composition
(1964); Van Soest and Jones (1968); Goering and Van Soest (1970);
Nehring et al. (1970); Van Soest (1970); DLG-Futterwerttabellen
(1973); Theander and Åman (1978)

In rice straw the average content of ash is about three times
higher than in the other straws. As seen in Table 4.8, this is
mainly due to a higher silica content. Silica is taken up as mono-
silicic acid by the roots but is deposited as silica in the cell
walls. Silica in the walls of epidermal cells is mainly present in
the opaline form. By growing cereal plants in nutrient solutions,
either with or without added silica, it has been possible to obtain shoots with a silica content of approximately 12% of the dry
matter or less than 0.1% silica (Jones and Hartley, unpublished).
Silica has been negatively correlated with degradability of the
polysaccharides in the rumen (Van Soest and Jones, 1968; Van
Soest, 1970).

4.9. CHEMICAL COMPOSITION OF STRAW FRACTIONS AND CHANGES DURING MATURATION

Table 4.9 presents the chemical compositions of some fractions of barley and wheat straw (Åman and Nordkvist, 1983). Extractives soluble in 80% ethanol and chloroform constituted 7-11% of the fractions, with a somewhat lower content in the internode fractions compared to the other fractions. The crude protein contents were also lowest in the internode fractions (compare also Harper and Lynch, 1981; Thiago and Kellaway, 1982). Compared to the other fractions, the internodes contain more cellulose but similar amounts of hemicellulose and lignin. The internodes, on the other hand, have been reported by Thiago and Kellaway (1982) to be the most lignified fraction. It is, however, important to remember that the content of lignin varies with the analytical method used. The ash content is often high in leaf fractions, which is usually due to the high content of silica in leaf fractions of cereal straw (Goering and Van Soest, 1970; Thiago and Kellaway, 1982; Åman and Nordkvist, 1983).

Table 4.9. Chemical composition of straw fractions (Åman and Nordkvist, 1983)

Constituent	Wheat			Barley		
	Internode	Node	Leaf	Internode	Node	Leaf
			g/kg DM			
Extractives soluble in 80% ethanol and chloroform	68	81	81	98	95	112
Crude protein	29	45	48	17	40	37
Cellulose	411	327	323	433	332	364
Hemicellulose	245	286	256	242	331	283
Klason lignin	216	217	268	176	167	143
Ash	38	51	96	16	31	44
Silica	14	15	39	3	4	11

In Table 4.10 the relative composition of polysaccharide constituents in cereal straw fractions is presented. As expected, glucose
and xylose are the dominating polysaccharide constituents in all
fractions, but uronic acids, arabinose and small amounts of galactose and mannose are also present. The higher relative content of
glucose in the internode fractions reflects the higher cellulose
contents in these fractions. In the node fractions the relative
contents of uronic acids, arabinose and galactose are higher than
in the other botanical fractions, indicating an increased content
or the presence of a different pectic polymer.

Table 4.10. Sugar composition (relative percentage) and the ratio
of glucose to xylose in hydrolysates of 80% ethanol
and chloroform insoluble residues of botanical fractions from cereal straw (Åman and Nordkvist, 1983)

Constituent	Wheat			Barley		
	Internode	Node	Leaf	Internode	Node	Leaf
Arabinose	2.6	6.7	4.1	2.7	8.1	5.0
Xylose	31.5	30.0	35.8	29.6	31.4	32.6
Mannose	traces	1.7	traces	traces	1.4	traces
Galctose	traces	2.3	traces	traces	2.1	1.9
Glucose	62.5	52.6	55.7	64.1	50.0	55.5
Uronic acids	3.4	6.6	4.4	3.6	7.0	5.0
Glc/Xyl (ratio)	2.0	1.7	1.6	2.2	1.6	1.7

During maturation of grasses there is an increase in the stem-
to-leaf ratio and a secondary thickening and lignification of the
tion of the cell walls. This results in higher contents of structural polysaccharides (mainly cellulose and xylans) and lignin and
lower contents of extractives, soluble carbohydrates and crude
protein (Jones, 1970). Similar results (Table 4.11) have been
obtained by Garmo (1983) for barley straw.

Table 4.11. Chemical composition of barley straw at different cut-
          ting dates (Garmo, 1983)

Cutting date	Crude protein	NDF	ADF	Lignin	Hemi-cellulose	Cellulose
		----------------		g/kg DM	----------------	
27 June	160	563	304	55	259	250
4 July	149	655	370	57	285	314
11 July	118	617	356	57	261	299
18 July	102	669	397	80	273	317
25 July	86	682	420	73	262	348
1 August	75	729	430	68	290	363
8 August	57	819	519	98	300	421
11 August	51	806	486	88	320	398
29 August	52	868	540	97	328	443

Table 4.12 gives the chemical composition of wheat straw from
India harvested at three different times, at one week intervals
(Patel et al., 1971).

Table 4.12. Chemical composition of wheat straw at different cut-
           ting dates (Patel et al., 1971)

Constituent	Time of cutting		
	One week before	Normal	One week after
	-----------	g/kg DM	-----------
Crude protein	56	44	36
Ether-extract	19	15	15
Crude fibre	399	409	435
N-free extractives	414	420	410
Ash	111	112	104
Silica	54	53	46
Phosphorous	1.5	2.4	1.5
Calcium	3.5	3.6	3.7
Carotene, g/g	58.21	3.99	-
Yield, kg DM/ha	7110	7131	5174

The growing interest in using cereal straw as animal feed has
made the feeding value of the straw an important factor, especially since the grain yield suffers little from early cutting whereas the straw yield (Table 4.11) and its feeding value will fall
as a result of late cutting (Manley and Wood, 1978).

4.10. THE CELL WALL STRUCTURE AND INTERRELATIONS BETWEEN CELL WALL CONSTITUENTS

Traditionally, plant cell walls have been divided into three structural components, the middle lamella, the primary wall, and the secondary wall (Figure 4.8). The middle lamella, found in the space between the walls of two cells, contains large amounts of pectin but no cellulose. The primary cell wall is thin and formed first in the growing plant. It contains loosely bound cellulose fibrils together with pectic polymers, hemicelluloses and structural protein, sometimes containing high levels of the amino acid hyroxyproline (Albersheim, 1975; Darvill et al., 1980). During the thickening of the cell wall highly oriented cellulose and hemicelluloses are deposited inside the primary wall to form the secondary cell wall. At the end of the tickening phase, formation of lignin becomes noticeable, beginning around the primary walls at the cell corners and extending from there into the secondary wall. When lignification is completed, the plant cell dies. In cereal straw, the secondary cell walls constitute the major part of the plant material.

There is a close structural relationship between the xylans and the cellulose since D-xylose differs from D-glucose only in the absence of the C-6 CH_2OH units and the main chain has the same glycosidic linkages. Xylans therefore are able to form hydrogen bondings with cellulose in the plant cell walls.

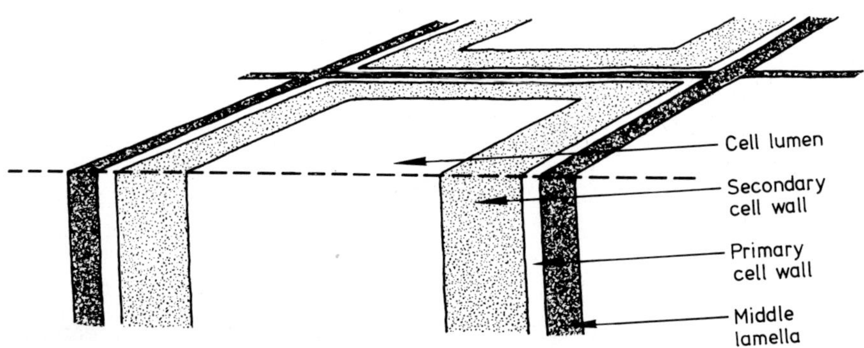

Figure 4.8. Schematic presentation of component layers in a straw cell wall

Lignin is encrusted into the cell wall and establishes covalent bondings to hemicelluloses, forming so-called ligno-carbohydrate complexes. Lignin may also be linked to other wall constituents by covalent bonds. Investigations of ligno-carbohydrate complexes in wood have indicated that lignin is bound to different types of sugar units and to uronic acid residues in the hemicelluloses. The linkages to the sugar residues are probably mainly benzyl ether bonds, while the linkages to the uronic acid residues are ester bonds, probably benzyl ester bonds (Eriksson et al., 1980). The ligno-carbohydrate complex between the hemicelluloses and lignin in rye grass (Morrison, 1974; Brice and Morrison, 1982), and in jute (Das et al., 1981) has also been studied.

Acetyl groups are found as substituents of plant cell walls (Bacon et al., 1975; Bacon et al., 1981). Phenolic acid monomers, mainly trans-ferulic acid and trans-p-coumaric acid have been found ester-linked to *Graminaeous* cell wall preparations (Higuchi et al., 1967; Hartley, 1973; Hartley, 1981; Theander et al., 1981). Probably all combined ferulic acid found in bran is ester-linked to terminal arabinofuranosyl residues in xylans (Smith and Hartley, unpublished). Diferulic acid in grain cell walls could be formed by dimerisation of two ferulic acid units forming a cross-link between polysaccharide chains (Markwalder and Neukom, 1976). Ferulic acid has also been reported to be ester-linked to noncarbohydrate cell wall polymers such as lignin (Higuchi et al., 1967) and suberin (Riley and Kolattukudy, 1975). p-Coumaric acid has been reported to be ester-linked to lignin (Nakamura and Higuchi, 1978; Nimz et al., 1981), cutin (Kolattukudi et al., 1981) and carbohydrates (Hartley et al., 1976).

Cutin and suberin have been reported to be closely associated to carbohydrates in the cell wall of plants (Kolattukudy et al., 1981). Some metal cat-anions have been shown to be bound or complexed with cell wall polymers (Jones, 1978).

There are strong indications that hemicelluloses, lignin, phenolic and acetyl substituents, and other bound compounds comprise one vast macromolecular matrix in the secondary cell walls. The interrelationships between the cell wall constituents must have a great influence on the biodegradability of plant cell walls.

4.11. SOME CHEMICAL MODIFICATIONS IN STRAW DURING DIFFERENT TECHNICAL TREATMENTS

Treatment of straws with alkaline reagents, in particular sodium hydroxide and ammonia, to increase the utilization of the energy by ruminants, has been extensively studied and used during recent years. The chemical composition and physiological properties of the treated straw is of course very much dependent on the type of alkali and treatment conditions used, and also if washing and/or neutralization is used afterwards. There are few studies which have reported on the effcts of technical alkali treatment of straw on its general chemical composition. Much information is available, however, on the more severe alkaline treatments, which form the basis of the alkaline pulping processes (soda and kraft) of wood and other lignocellulosic materials. Lignin and the cell wall polysaccharides would be expected to undergo similar reactions under the less severe conditions used in the alkaline upgrading conditions of straw but to a much lesser extent.

Saponification of ester linkages between acetic acid and phenolic acids, and polysaccharides and/or lignin as well as such linkages between uronic acid residues in the hemicelluloses and lignin would be expected during alkaline treatment of straw material. At elevated temperatures lignin undergoes a series of reactions under the influence of alkaline reagents, the most significant one being cleavage of ether linkages between phenylpropane units with simultaneous formation of free phenolic groups. As a result of the accompanying decrease in the molecular weight and cleavage of linkages to the hemicelluloses an increased solubility of lignin in the alkaline solution will occur. This is important in pulping for paper properties but in upgrading pretreatments of straw it is not necessary to employ energy-demanding, severe conditons for the removal of lignin. It seems to be enough to cleave linkages between lignin and hemicelluloses to make the latter and also the cellulose (which is embedded in the lignin-hemicellulose complex) more accessible for hydrolyzing enzymes.

During alkaline treatments of ligno-cellulosic materials the hemicelluloses are expected to be partly solubilized as in pulping conditions and the cellulose to be more accessible by an alkaline swelling effect. The alkaline liquor used during the treatment will be partly neutralized by acetic and phenolic acids released by cleavage of ester linkages and new phenolic groups formed when lignin-ether linkages are broken. A slower but continuing neutralization of the liquor will also occur because of the well-known peeling reaction of polysaccharides (for a review see: Whistler and BeMiller, 1958), starting from the reducing sugar end-unit. The released and transformed sugar residues will hereby form lactic, glycollic, so-called saccharinic and other aliphatic acids in the liquor.

Åman and Theander (1977) studied the chemical composition of both the resulting straw and the black-liquor in a Swedish closed sodium hydroxide, farm-scale process (Trolmen-process) where the liquors circulate through the straw. The values for the main components (% of dry matter) cellulose, hemicellulose (neutral part) and Klason lignin changed from 34, 21 and 27% respectively in the untreated straw to 39, 18 and 20% in the treated straw. These results indicate an enrichment of cellulose in the treated product and some removal of hemicellulose and lignin into the liquor. It is notable that in agreement with this, the four hemicellulose sugar components, which have survived the slow transformation process into acids discussed above, predominate over glucose from cellulose in the hydrolysate of the black-liquor (Table 4.13). In the black liquor a high content of silicate and substantial amounts of salt-bound acetic and phenolic acids were also formed.

In a recent study another notable result from technical alkaline treatments of straw was found (Lindberg et al., 1983), namely indication that xylans are partly translocated during aqueous sodium hydroxide treatments to a position in the straw material, where they are more available to rumen digestion. Thus when untreated or dry-sodium hydroxide treated straw was digested in nylon bags in the rumen the cellulose/pentosan ratio was constantly 1.5 and 1.65 respectively in the non-digested residue during the 72 h digestion. However, when the same straw, treated by two different aqueous sodium hydroxide systems, was digested, that ratio passed a marked maximum of 2.2 and 2.7 respectively after about 24 h in rumen. This could be caused by an effect similar to that which is known in the pulping technology, namely that hemicellulose polysaccharides removed in an early stage of the process can be partly absorbed again on the fibres if the pH of the liquor drops. Compare for instance results from pulping of bamboo Ahlgren et al., 1969).

Table 4.13. Analysis of "black-liquor" from wet treatment of straw (Åman and Theander, 1977)[1]

Constituent		Content
DM	g/l	123
Ash	g/kg DM	580
Arabinose	"	24
Xylose	"	73
Mannose	"	2
Galactose	"	8
Glucose	"	12

[1]Sugars analysed after acid hydrolysis

Chesson (1981) has studied the effect of sodium hydroxide treatments, under laboratory conditions, on straw composition in relation to enhanced degradation of the polysaccharides by rumen microorganisms.

Steaming at 160°C or higher temperature is another technical procedure used to improve the digestibility of lignocellulosic materials for ruminant food, which can be defined as a combination of heat treatment and autohydrolysis by acetic acid being generated from the acetyl groups. Depending on the conditions lignin can be melted by the thermal treatment and linkages in lignin - in particular the sensitive linkages in the -position to the aromatic rings - as well as glycosidic linkages in the polysaccharides, are cleaved by acid hydrolysis. Condensation of lignin residues, furfural or hydroxymethyl furfural, formed by acid degradation of polysaccharides, is also possible.

4.12. SUMMARY

A review is given of the anatomical and morphological composition of straw, its chemical composition regarding polymeric and low-molecular organic and inorganic components, interrelations between some of these components, differences in chemical composition between anatomical fractions and changes during the maturation of straw or by different technical treatments. Straw is characterised by the predominance of lignocellulosic cell wall material (fibre), with cellulose, hemicelluloses and lignin as main components, a high content of ash and a low content of protein and storage carbohydrates.

4.13. REFERENCES

Adler, E., 1977. Lignin Chemistry - past, present and future. Wood Sci. Technol. 11 (3): 169-218.

Ahlgren, P., Jaspal, N.S. and Hansson, J.Å., 1969. Sorption of xylan during Kraft pulping of bamboo. Ippta Souvenir, Vol. VI No. S: 134-138.

Albersheim, P., 1975. The walls of growing plant cells. Scientific American 232: 80-95.

Åman, P. and Nordkvist, E., 1983. Chemical composition and in vitro degradability of botanical fractions of cereal straw. Swedish J. Agric. Res. 13: 61-67.

Åman, P. and Theander, O., 1977. Chemical modification of straw by alkaline treatment. In: P.-G. Knutsson (ed.): Quality of forage, Proc. Seminar NJF, Uppsala 20-22 April, pp. 151-166.

Aspinall, G.O., 1959. Structural chemistry of the hemicelluloses. Adv. Carbohydr. Chem. 14: 437-442.

Aspinall, G.O. and Mahomed, R.S., 1954. The constitution of wheat straw xylan. J. Chem. Soc. 1731-1734.

Aspinall, G.O. and Meek, E.G., 1956. The constitution of wheat straw hemicellulose. J. Chem. Soc. 3830-3834.

Aspinall, G.O. and Wilkie, K.C.B., 1956. The constitution of an oat-straw xylan. J. Chem. Soc. 1072-1076.

Bacon, J.S.D., Chesson, A. and Gordon, A.H., 1981. Deacetylation and enhancement of digestibility. Agric. Environm. 6: 115-126.

Bacon, J.S.D., Gordon, A.H. and Morris, E.J., 1975. Acetyl groups in cell-wall preparations from higher plants. Biochem. J. 149: 485-487.

Bagby, M.O., Nelson, G.H., Helman, E.G. and Clark, T.F., 1971. Determination of lignin in non-wood plant fiber sources. Tech. Ass. Paper Pulp Ind. 54: 1876-1878.

Bishop, C.T., 1953. Crystalline xylans from straw. Can. J. Chem. 31: 793-800.

Bouveng, H.O., 1961. Phenylisocyanate derivatives of carohydrates. II. Location of the O-acetyl groups in birch xylan. Acta Chem. Scand. 15: 96-100.

Brice, R.E. and Morrison, I.M., 1982. The degradation of isolated hemicelluloses and lignin-hemicellulose complexes by cell-free, rumen hemicellulases. Carbohydr. Res. 101: 93-100.

Browning, B.L., 1967. Methods of Wood Chemistry, Vol. II. Interscience publishers, New York, 882 pp.

Chanda, S.K., Hirst, E.L., Jones, J.K.N. and Percival, E.G.U., 1950. Constitution of xylan from Esparto grass (*Stipa tenacissima L.*). J. Chem. Soc. 1289-1297.

Chesson, A., 1981. Effects of sodium hydroxide on cereal straw in relation to enhanced degradation of structural polysaccharides by rumen micro-organisms. J. Sci. Food Agric. 32: 745-758.

Cobley, L.S. and Steel, W.M., 1977. An introduction to the botany of tropical crops. Arnold, London, 371 pp.

Darvill, A., McNeil, M., Albersheim, P. and Delmer, D.P., 1980. The primary cell walls of flowering plants. Biochem. Plants 1: 91-162.

Das, N.N., Das, S.C., Dutt, A.S. and Roy, A., 1981. Lignin-xylan ester linkage in jute fiber (*Corchorus capsularis*). Carbohydr. Res. 94: 73-82.

DLG-Futterwerttabellen, 1973. Mineralstoffgehalte in Futtermitteln. 2., erweiterte und neugestaltete Auflage. Dokumentationsstelle der Universitat Hohenheim, DLG-Verlag, Frankfurt (Main), 199 pp.

El-Basyouni, S.Z., Neish, A.C. and Tower, G.H.N., 1964. The phenolic acids in wheat. III. Insoluble derivatives of phenolic cinnamic acids as natural intermediates in lignin biosynthesis. Phytochemistry 3: 627-639.

Erickson, M., Miksche, G.E. and Somfai, I., 1973. Charakterisierung der Lignine von Angiospermen durch oxydativen Abbau. II. Monocotyledonen. Holzforschung 27: 147-150.

Eriksson, Ö., Goring, D.A.I. and Lindgren, B.O., 1980. Structural studies on the chemical bonds between lignin and carbohydrates in spruce wood. Wood Sci. Technol. 14: 267-279.

Ernst, A.J., Fouad, Y. and Clark, T.F., 1960. Rice straw for bleached papers. Tech. Ass. Paper Pulp Ind. 43: 49-53.

Esau, K., 1977. Anatomy of seed plants. John Wiley and Sons Inc. New York, 550 pp.

Espelie, K.E., Dean, B.B. and Kolattukudy, P.E., 1979. Composition of lipid-derived polymers from different anatomical regions of several plant species. Plant Physiol. 64: 1089-1093.

Espelie, K.E. and Kolattukudy, P.E., 1979. Composition of the aliphatic components of suberin from the bundle sheaths of *Zea mays* leaves. Plant Sci. Lett. 15: 225-230.

Garmo, T.H., 1983. Avling og kvalitet av byggheilgrøde. Sci. Rep. Agric. Univ. Norway (in press).

Goering, H.K. and Van Soest, P.J., 1970. Forage fibre analysis (Apparatus, Reagents, Procedures and Some Applications). Agricultural Handbook No. 379, Washington, D.C., pp. 1-12.

Grist, D.H., 1965. Rice. Longmans, London, 548 pp.

Harbers, L.H., Kreitner, G.L., Davis, G.V., Rasmussen, M.A. and Corah, L.R., 1982. Ruminal digestion of ammonium hydroxide-treated wheat straw observed by scanning electron microscopy. J. Anim. Sci. 6: 1309-1319.

Harper, S.H.T. and Lynch, J.M., 1981. The chemical components and decomposition of wheat straw leaves, internodes and nodes. J. Sci. Food Agric. 32: 1057-1062.

Hartley, R.D., 1972. p-Coumaric and ferulic acid components of cell walls of ryegrass and their relationships with lignin and digestibility. J. Sci. Food Agric. 23: 1347-1354.

Hartley, R.D., 1973. Carbohydrate esters of ferulic acid as components of cell walls of *Lolium multiflorum*. Phytochemistry 12: 661-665.

Hartley, R.D., 1981. Chemical constitution, properties and processing of lignocellulosic wastes in relation to nutritional quality for animals. Agric. Environm. 6: 91-113.

Hartley, R.D., Jones, E.C. and Wood, T.M., 1976. Carbohydrates and carbohydrate esters of ferulic acid released from cell walls of *Lolium multiflorum* by treatment with cellulolytic enzymes. Phytochemistry 15: 305-307.

Higuchi, T., Ito, Y. and Kawamura, I., 1967. p-Hydroxyphenylpropane component of grass lignin and role of tyrosine-ammonia lyase in its formation. Phytochemistry 6: 875-881.

Higuchi, T., Ito, Y., Shimada, M. and Kawamura, I., 1967. Chemical properties of milled wood lignin of grasses. Phytochemistry 6: 1551-1556.

Himmelsbach, D.S. and Barton II, F.E., 1980. [13]C Nuclear magnetic resonance of grass lignins. J. Agric. Food Chem. 28: 1203-1208.

Jackson, M.G., 1977. Review article: the alkali treatment of straws. Anim. Feed Sci. Technol. 2: 105-130.

Joint United States-Canadian Tables of Feed Composition, 1964. Committee on animal nutrition, Agricultural Board and National Academy of Sciences - Committee on Animal Nutrition, Publication 1232, National Research Council, Washington, D.C., 167 pp.

Jones, D.I.H., 1970. Cell-wall constituents of some grass species and varieties. J. Sci. Food Agric. 21: 559-562.

Jones. L.H.P., 1978. Mineral components of plant cell walls. Am. J. Clin. Nutr. 31: 594-598.

Kolattukudy, P.E., 1980. Cutin, suberin and waxes. In: P.K. Stumpf and E.E. Conn (eds.): The biochemistry of plants Vol. 4. Lipids: structure and function. Academic Press, New York, pp. 571-645.

Kolattukudy, P.E., Espelie, K.E. and Soliday, C.L., 1981. Hydrophobic layers attached to cell walls and associated waxes. Encylopedia of Plant Physiology New Series, 13B: 225-254.

Kuwatsuka, S. and Shindo, H., 1973. Behaviour of phenolic substances in the decaying process of plants. Soil Sci. Plant Nutr. 19: 219-227.

Lindberg, E., Ternrud, I. and Theander, O., 1983. Degradation rate and chemical composition of different types of alkali treated straw during rumen digestion. J. Sci. Food Agric. (in press).

Manley, A.C. and Wood, R.S., 1978. The analysis of fractions of cereal crops on three dates prior to harvest. Agric. Prog. 53: 71-76.

Markwalder, H.U. and Neukom, H., 1976. Diferulic acid as a possible crosslink in hemicelluloses from wheat germ. Phytochemistry 15: 836-837.

Morrison, I.M., 1974. Structural investigation on the lignin-carbohydrate complexes of *Lolium perenne*. Biochem. J. 139: 197-204.

Muller, F.M., 1960. On the relation between properties of straw pulp and properties of straw. Tech. Ass. Paper Pulp Ind. 43, No. 2: 209A-218A.

Nakamura, Y. and Higuchi, T., 1978. Ester linkage of p-coumaric acid in bamboo lignin. Cellulose Chem. Technol. 12: 209-221.

Nehring, K., Beyer, M. and Hoffman, B., 1970. Futtermitteltabellenwerk. WEB Deutscher Landwirtschaftsverlag, Berlin, DDR, 460 pp.

Nimz, H.H., Robert, D., Faix, O. and Nemi, M., 1981. Carbon-13 NMR spectra of lignins, 8. Structural differences between lignins of hardwoods, grasses and compression wood. Holzforschung 35: 16-26.

Patel, B.M., Patel, R.B. and Shukla, P.C., 1971. Effect of stage of harvesting on the yield and composition of wheat straw and grain. Indian J. Nutr. Diet. 8: 264-267.

Riley, R.G. and Kolattukudy, P.E., 1975. Evidence of covalently attached p-coumaric acid and ferulic acid in cutins and suberins. Plant Physiol. 56: 650-654.

Salomonsson, A.-C., Theander O. and Åman, P., 1978. Quantitative determination by GLC of phenolic acids as ethyl derivatives in cereal straw. J. Agric. Food Chem. 26: 830-835.

Sarkanen, K.V. and Ludwig, C.H., 1971. Lignins, occurrence, formation, structure and reactions. Wiley-Interscience, New York, 916 pp.

Sinner, M., Puls, J. and Dietrichs, H., 1979. Carbohydrate composition of nut shells and some other agricultural residues. Starch 31: 267-269.

Staniforth, A.R., 1979. Cereal straw, Clarendon press, Oxford, pp. 16-37.

Stone, J.E., Blundell, M.J. and Tanner, K.G., 1951. Formation of lignin in wheat plants. Can. J. Chem. 29: 734-745.

Streibl, M., Konecny, K., Trka, A., Ubik, K. and Pazlar, M., 1974. Composition of the wax of rye straw. Collect. Czech. Chem. Commun. 39: 475-479.

Theander, O. and Åman, P., 1978. Chemical composition of some Swedish cereal straws. Swedish J. Agric. Res. 8: 189-194.
Theander, O. and Åman, P., 1979. Studies on dietary fibres. Part I. Analysis and chemical characterization of water-soluble and water-insoluble dietary fibres. Swedish J. Agric. Res. 9: 97-106.
Theander, O. and Åman, P., 1980. Chemical composition of some forages and various residues from feeding value determinations. J. Sci. Food Agric. 31: 31-37.
Theander, O., Udén, P. and Åman, P., 1981. Acetyl and phenolic acid substituents in timothy of different maturity and after digestion with rumen microorganisms or a commercial cellulase. Agric. Environm. 6: 127-133.
Thiago, L.R.L. and Kellaway, R.C., 1982. Botanical composition and extent of lignification affecting digestibility of wheat and oat straw and paspalum hay. Anim. Feed Sci. Technol. 7: 71-81.
Timell, T.E., 1964. Wood Hemicelluloses: Part I. Adv. Carbohydr. Chem. 19: 247-302.
Tulloch, A.P. and Hoffman, L.L., 1971. Leaf wax of durum wheat. Phytochemistry 10: 871-876.
Tulloch, A.P. and Weenik, R.O., 1969. Composition of the leaf wax on the Little Club wheat. Can J. Chem. 47: 3119-3126.
Van Soest, P.J., 1970. The role of silicon in the nutrition of plants and animals. Proc. Cornell Nutr. Conf. Feed Manufact. pp. 103-109.
Van Soest, P.J. and Jones, L.P.H., 1968. Effect of Silica in forages upon digestibility. J. Dairy Sci. 51: 1644-1648.
Watson, S.J., 1941. Increasing the feeding value of cereal straws. J. Roy. Agric. Soc. Engl. 101 Part II: 37-43.
Whistler, R.L. and BeMiller, J.N., 1956. Hydrolysis components from methylated corn fiber gum. J. Am. Chem. Soc. 78: 1163-1165.
Whistler, R.L. and BeMiller, J.N., 1958. Alkaline degradation of polysaccharides. Adv. Carbohydr. Chem. 13: 289-329.
Whistler, R.L. and Corbett, W.M., 1955. Oligosaccharides from partial acid hydrolysis of corn fiber hemicellulose. J. Am. Chem. Soc. 77: 6328-6330.
Wilkie, K.C.B., 1979. The hemicelluloses of grasses and cereals. Adv. Carbohydr. Chem. Biochem. 36: 216-264.

Chapter 5

PHYSICAL TREATMENT

by

Howard G. Walker

U.S. Department of Agriculture
Agriculture Research Service
Western Regional Research Center
Berkley, California 94710, USA

5.1. INTRODUCTION

A wide variety of agricultural residues have been fed to ruminant animals with varying degrees of success. Experience has shown
that although some of these residues may fulfill the maintenance
needs of mature animals, they have often given poor results when
used for production of meat, milk, or fiber. This is partly due to
their inadequate protein content; but, in addition, the abundant
energy stored in the cellulose and hemicellulose content of their
cell walls is not readily available for digestion by rumen microorganisms. Cowling and Kirk (1976) suggest at least eight factors related to fiber size, conformation, and composition which
can act as impediments to digestion of cellulose fibers and fibrous materials. The present review will deal with some of the
efforts made to overcome these digestive barriers by means of physical processing.

Even though a variety of lignocellulosic residues of specialty crops of local origin have been fed to ruminants, this survey will concentrate mainly on residues from the major food, feed, or fiber crops grown world-wide. Primarily because of space considerations, it will be impossible to consider two major sources of lignocellulosic wastes; namely wood residues, and animal fecal residues. These residues tend to differ compositionally, in degree if not in kind, from other major crop residues, and, as a result, have special problems of their own.

It is impossible, however, not to include some references to wood products, because much of our understanding of how to increase digestibility of lignified materials has been developed by innovative wood technologists and chemists (Hajny and Reese, 1969; Brown and Jurasek, 1979).

Roughages have almost always played an important part in ruminant feeding programs. It should be noted, however, that animal feeders face at least two different situations, and the feeding program selected may depend on the supply of available roughage and other feeds. If, as in many parts of the world, roughage is the major or perhaps only feedstuff avilable, then the goal of a feeding program has to be to maximize microbial digestion and utilization of all ingested material. Extensive processing may be quite practical. If, on the other hand, quality feedstuffs as well as low grade roughages are both available, it may be economical to feed roughage which is not completely utilized in the rumen. In this case the need for elaborate roughage processing may be harder to justify. As our knowledge of the ruminant digestion process expands, it becomes increasingly clear that we are dealing with a very dynamic system involving interaction of competing factors that include animal intake, particle size and retention time, balance of available nutrients, and inherent digestibility of individual feedstuff components (Demeyer, 1981). The complexities of the system compound the difficulties of evaluating what is really meant by the term "increasing the digestibility" of a low quality feedstuff, and this constraint must be borne in mind as the effects of physical processing methods are reviewed.

5.2. GRINDING, ETC.

One of the major ways to improve the nutritive value of lignocellulosic residues has been to reduce particle size by milling,
grinding, chopping or compaction. The techniques have been applied
widely to crop residues and wood products alike. Results of the
two types of studies are not interchangeable, however, in regard
to direct applicability to ruminant feeding systems. Much of the
work reported for wood has been directed toward increasing the
amount of glucose or other fermentable sugars that can be derived
from wood products by enzymatic or acidic hydrolysis in mechanical
digestors, and particle size considerations present no real problems. In the case of ruminant feeding, however, particle size
plays a significant role (Meyer et al., 1965). Nevertheless, some
of the concepts and data from studies on wood products will be
used as needed to fill out the picture of what happens when lignocellulosic materials are ground or milled.

A general finding has been that mechanical comminution of coarse agricultural residues increases their daily intake by animals.
This is due partly because the density of the feed may be increased and partly because chewing time required to reduce ingested
material to a particle size suitable for digestion by rumen microorganisms is decreased considerably. Partially offsetting this
beneficial effect, however, is the fact that the comminution process produces more or less fine material that may pass through the
digestive tract too rapidly for maximum nutrient utilization. For
a successful feeding program, a balance must be struck between
increased feed intake and decreased rate of passage time.

It seems unlikely that many of the widely used milling techniques actually increase the digestibility of cellulose or hemicellulose by rumen microorganisms per se. The situation in regard to
wood has been considered by Stone et al. (1969) and the reasoning
seems applicable to most other lignocellulosic residues as well.
These authors note that enzymatic digestion of cellulose depends
on the amount of cellulose surface accessible to the enzyme and
involves the relative pore size of substrate and the attacking
enzyme molecule (See also Cowling and Kirk, 1976). Grinding increases available surface but only moderately because of the
length-width relationships of fibers. Cell wall thickness of
fibers is very small compared to their length, and exposing cut

ends by grinding increases total surface only by the relatively small cross sectional area of the fiber. Stone et al. (1969) cite the example of an experiment with spruce wood which had an initial surface area (measured by nitrogen absorption) of about 1 m^2/g. When ground through an 80 mesh screen in a Wiley mill, little increase in surface area of the product was obtained, but wet grinding in a disk refiner increased the surface area of the product to over 40 m^2/g as measured by nitrogen absorption. This highly pulped material was still not digestible by rumen nicrobial enzymes. Even though the new surface generated by grinding was accessible to small nitrogen molecules, the openings in the cell walls were still too small to allow surface acess to the large protein molecules of carbohydrate-splitting enzymes.

It is believed that only extreme milling treatments which disrupt fiber structure of polymers at the molecular level are capable of increasing the digestibility of carbohydrate to any considerable extent (Stone et al., 1969). The most effective milling procedures of this type are ball milling, and vibratory ball milling, which produce dramatic increases in wood digestibility. Pew and Weyna (1962) suggest this is accomplished by disruption of lignin structure and Grohn (1958) showed cellulose chain length was actually degraded by a mechanical bond rupture hydrolytic mechanism during vibratory ball milling. Pew and Weyna (1962) found that 10 minutes of grinding in a vibratory ball mill permitted in vitro enzymatic digestion of about 67% of the carbohydrate material in spruce wood, and that up to 8 hours of milling was required to achieve almost total digestion (96%). Dehority and Johnson demonstrated rather convincingly that wet ball milling (porcelain jarstone balls) increased the in vitro dry matter digestibility (IVDMD) of mature forages, especially timothy hay (ripe seed stage). Both rate and extent of digestion were increased by the milling procedure. Grass hays responded more to this treatment than legume hay. Ball milling did not have a significant effect on chemical composition of the hay samples. Pigden and Heaney (1969) confirmed that ballmilling low quality roughages like brome straw or *Panicum maximum* increased their IVDMD. They also pointed out that this had little practical significance for animal production because of high processing costs and rapid rate of passage of the extremely fine particles. Dehority (1961) had voiced similar reservations at an earlier date.

Fan et al. (1974) subjected wheat straw to several milling pre-treatments as a possible way to increase the amount of sugars available for fermentation. Straw was ball milled for varying time periods (4 to 24 hours), ground in a stationary blade hammer mill through various screen sizes, and put through a roller mill for different milling times (15 and 30 min.). Judged by the relative extent of sugar production after 8 hours of enzymatic hydrolysis, all milling treatments gave some improvement over untreated control straw. Improvement ranged from about 4 fold for ball milling (8 hours) to 1.3 fold for roller milling. Several chemical pre-treatments were included in the study also, and the more successful ones were at least twice as effective as any of the milling treatments. Approximate energy costs for milling a ton of wheat straw were estimated to be as follows: ballmill (8 hours), $ 1300; roller mill (.25 hours), $ 2036, and stationary blade hammer mill (fine grind), $ 9.00.

Figure 5.1. (adapted from Millet et al., 1976) indicates that horse power requirements increase exponentially as particle size is decreased. Although the data are based on commercial production of wood flour from pine shavings by hammer milling and attrition milling procedures, they show that mechanical comminution of lignocellulosic materials to small particle size can be extrememly expensive at 1982 energy prices.

Owen (1978) has pointed out many of the problems associated with attempts to use mechanical processing of roughages as a means to increase their nutritive value. He notes that particle size reduction is a poorly defined term even when roughage is ground through a given size screen (see also Heaney et al., 1963. Most authors fail to indicate particle size objectively by measurement of modulus of fineness (MF) or modulus of uniformity (MU), or both (Pfost and Headly, 1976). Owen (1978) indicates that MF usually ranges from 1-2 for ground and pelleted roughages, up to 5-6 for chopped roughages. The same author cites results obtained by Swan and Clark (1974) in a beef fattening trial using a diet containing 30% ground barley straw. Straw ground through a 6.35 mm screen to give a product of MF 2.93 and MU 0:7:3 gave optimal animal performance in terms of diet digestibility, intake and growth rate compared to straw ground through a 3.18 mm screen (MF, 2.38; MU, 1:7:2) or a 12.7 mm screen (MU 3.68; MU 2:7:1).

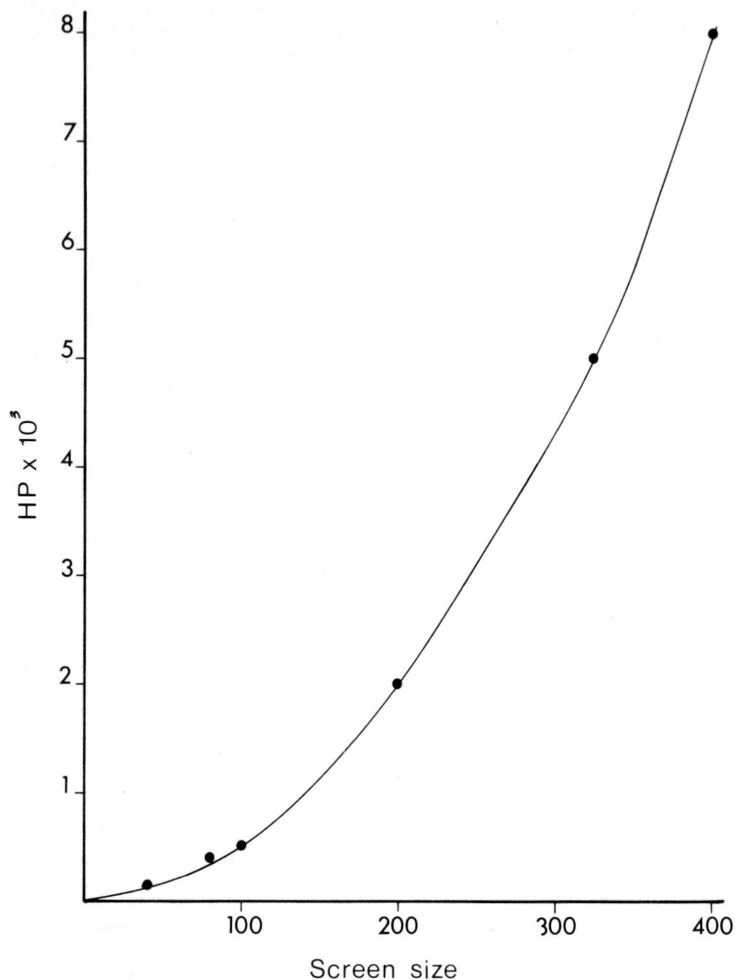

Figure 5.1. Approximate energy requirements (horsepower/ton) for production of wood flour from dry white pine shavings (adapted from Millet et al., 1976)

In another experiment Pickard et al. (1969) compared animal performance of diets containing ground barley straw (15 and 30%) fed
to calves from 12 weeks old to slaughter. The straw was ground
through either a 1.6 mm, 4.8 mm or 7.9 mm screen and was then mixed with concentrate and cubed to prepare a complete ration. Particle size distributions were determined on the ground straws and
complete rations but no MF or MU values were calculated. Results
of the study showed no differences in animal performance due to

particle size or level of straw in the diet. It is possible, however, that the original particle size distributions of the ground straws were not really reflected in the differences in the diets as eaten, since the cubing process tended to break down the size of the larger particles as they passed through the die.

White et al. (1971) fed rice straws to steers at rather low levels (5 and 20%) as a part of a fattening diet. Straw was ground through a 1.27 or a 3.81 cm screen (no MF or MU given). The size of the grinding screen had no significant influence on gain, carcass weight, or rumen volatile fatty acid content at either of the two levels fed.

Mathison (1981) showed that grinding barley straw through a hammer mill (4.75 mm, MF or MU unspecified) had no measurable influence on straw consumption or average daily gain compared to chopped straw when fed to mature pregnant beef cows.

Fernandez Gonzalez (1980) reported that no significant in vivo digestibility or voluntary intake differences by lambs could be detected for barley straw chopped to 20 mm, 33 mm, or 40 mm lengths, even though chemical analytical differences were noted for NDF, ADF, cellulose, lignin, and protein.

It has been noted that digestibility of the fibrous components of fresh, wet, turgid forage tissue is increased by mechanical comminution (Greenhalgh and Reid, 1975; Kohler et al., 1979). However, wet grinding of wheat or rice straw in a disk mill produced no significant increase in enzymatic digestibility presumably because rehydration conditions did not duplicate those of a living system (Walker and Kohler, 1976).

Ground roughages are frequently subjected to a compaction procedure such as cubing or pelleting before being fed (Beardsley, 1963; Moore, 1963) (see also Rexen and Knudsen, Chapter 6.2). Owen (1978) has noted that a clear advantage of grinding straw is that it can be readily incorporated into complete feeds for either meat or dairy diets. Stevens (1981) summarized a number of benefits of pelleting low grade roughage type ingredients. They include (1) more uniform and desirable appearance in most cases, (2) increased density, (3) less dust in most cases, (4) ease of handling, (5) reduced segregation, and (6) reduction of waste. However, these benefits are not achieved without cost; initial capital requirements of a complete pelleting system are high and energy costs alone are in the range of $ 4-5 per ton (1980 prices), depending

on amount of yearly production. Whether pelleting is justified or
not depends on the particular situation and involves careful evaluation of local conditions by management.

Many of the best studies on pelleting have been carried out on
forages, but the results are believed to be applicable to lignocellulosic residues. For example, Heany et al. (1963) studied the
effect of pelleting on digestion coefficients and digestible energy of forages cut at different stages of maturity and fed to
sheep. Table 5.1 shows the results of the trials with very mature
material (past bloom stage), hammer milled through a 3.2 mm screen
and pelleted. Digestibility of all components was decreased by the
pelleting process, more for the grasses than for alfalfa, but the
total daily intake of digestible energy increased notably in all
cases.Increases in digestible energy intake of the magnitude found
in this study (60-115%) are of major economic importance from an
animal production or maintenance standpoint. Just how applicable
these findings are to all low grade lignocellulosic residues is
open to question, however, since Weir (1962) showed that sheep
will not continue to consume unlimited quantities of pelleted low
grade roughages like barley straw or *Ceonothus thyrsiflorus*.

Table 5.1. Effect of pelleting on mature forages: % increase (+)
or decrease (-) of digestion coefficient or energy
intake due to pelleting chopped material (adapted from
Heany et al., 1963)

	Alfalfa	Orchard grass	Timothy
Digestibility:			
Dry matter	0	-16	-16
Organic matter	+2	-18	-17
Protein	+2	-14	-2
Fiber	+2	-25	-33
Nitrogen free extract	-2	-15	-15
Energy	+2	-19	-16
Daily digestible energy intake	+14	+64	+94

Greenhalgh and Wainman (1972) support the concept that the main
value of processing roughages is to increase energy intake. They
reported that the MF of ground, pelleted feeds is 1-2, 3 to 4 for
chopped and cubed feeds, and 5-6 for chopped but not cubed feeds,
and that daily intake increases as the MF falls. It is not clear

whether there is a critical limit for MF below which intake no longer increases. The authors note that other factors such as hardness or softness can also influence consumption of pellets.

Results of Haenlein and Holdren (1965) also indicate that MF and MU are important measures related to digestibility of ground cubed forages. They prepared a series of cubed samples from a single batch of alfalfa and fed them to sheep in digestibility trials. Samples having MF and MU of 5.7, 8:2:0; 4.2, 4:4:2; 3.0, 1:5:4; and 2.2, 0:4:6 respectively showed significant positive correlations with digestibility of dry matter, organic matter, crude protein and crude fiber. Even though coarsely ground material (chopped) had higher digestibility, the total nutritive value of cubes made from finer material (low MF) was 22.7% better because of increased intake.

Hackett et al. (1975) studied the effect of pelleting as a dust control measure in ground wheat straw (25%) - alfalfa hay (75%) diets fed to feeder lambs. When compared to unpelleted diets with and without other added dust suppressants, the pelleted material showed no significant difference in mean daily intake, suggesting that pelleting high roughage diets does not necessarily guarantee improved intake.

Owen (1978) summarized the use of cubed and uncubed barley straw in dairy diets. Cubing the diet increased dry matter intake compared to a diet with chopped straw when straw comprised 35% or 50% of the diet, but not when it was included at the 20% level. At all levels, chopped straw diets produced higher butter fat contents.

5.3. IONIZING RADIATION

Use of ionizing radiation to increase the digestibility of lignocellulosic materials has been investigated since work by Lawton et al. (1951) and Saeman et al. (1952) demonstrated that when wood was irradiated, cellulose chain length was decreased and insoluble carbohydrate components became more available to rumen bacteria. The nature of the ionizing radiation (x- or γ-ray, electron beam) is not particularly important since all radiation of this type has considerably more energy than is required to bring about the observed effects (Arthur, 1971). In fact, all the readily measurable degradations are caused by various reactions of secondary electrons released after initial capture of the high energy ionizing radiation. Radiative degradation of pure cellulose and cotton has been studied extensively to provide a fundamental understanding of the chemical degradations involved, while radiative degradation of wood has been studied widely as a practical means to increase fermentable carbohydrate obtainable by subsequent enzymatic or acid hydrolysis. The major reactions induced in cellulose by ionizing radiation have been shown to be (1) molecular depolymerization, (2) acid group formation, (3) reducing group formation, (4) gas production and (5) radical production (Arthur, 1971). Most of the available evidence indicates that cellulose crystallinity per se is not affected by high energy radiation. Although degradation of cellulose seems to be well understood, the situation with wood and other lignocellulosics like straws and bagasse is more complex due to the presence of relatively large amounts of hemicellulose and lignin.

Hemicellulose appears to degrade according to the usual carbohydrate mechanism, but yields different end products than does cellulose. Lignin, on the other hand, because of its high aromatic ring content, is more resistant to irradiation damage and has actually been shown to exert a mild protective effect on radiolysis of redwood holocellulose (Smith and Mixer, 1959).

Because of the expense involved in large scale irradation equipment, most of the irradiation work with lignocellulosic materials has been performed on a laboratory scale using chemical and in vitro tests to evaluate product changes. Only a limited number of irradation studies have been reported to upgrade crop residues for ruminant feed use. Most of the emphasis has been on increasing the level of glucose or glucose precursors from biomass for production of alcohol, single cell protein, and other fermentation products.

When wheat straw was irradiated with γ-rays from a cobalt-60 source at dosage levels of 0 to 1000 Mrad by Prichard et al. (1962) no significant changes in in vitro dry matter digestibility (IVDMD) were noted at dosage levels below 10 Mrad; IVDMD then more than doubled as dosage was increased to 100 and 1000 Mrad. Most of this apparent increase was due to solubilization of straw components, however, since almost as large DMD values were obtained when IVDMD was run without rumen inoculum as with it.

Organic acid production in the IVDMD test was greatest at 250 Mrad , and fell off at higher levels, suggesting that increased radiation was producing degradation products toxic to or unusable by rumen microorganisms.

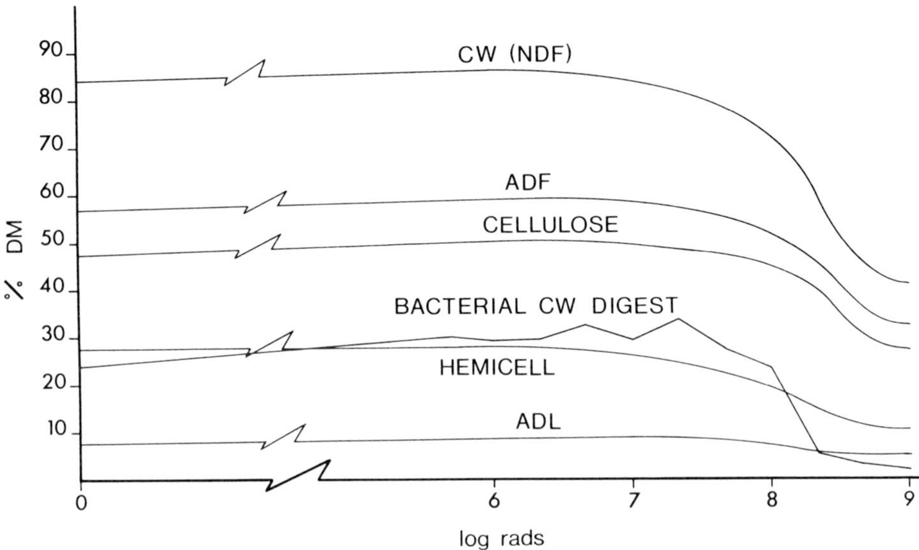

Figure 5.2. Fibrous composition and cell wall digestibility of irradiated wheat straw (from Yu et al., 1975)

Yu et al. (1975) reported analyses of cell wall constituents of a series of samples of electron beam irradiated wheat straw. These are summarized in Figure 5.2. The data show that actual percentages of all fibrous fractions decrease at radiation dosages above 21 Mrad due to solubilization. About 50% of the cell wall constituents (NDF) were solubilized at the highest irradiation level (1000 Mrad). In these experiments lignin and cellulose seemed to be degraded at about the same rate since the ADL/ADF ratio remained essentially constant for all irradiation levels. More hemicellulose (NDF-ADF) was solubilized than cellulose at all levels involving significant radiation damage. Cell wall digestibility by rumen microorganisms decreased sharply at dosages greater than 100 Mrad, even though overall apparent digestibility increased due to to solubilization. In was not established whether the observed decrease in digestibility by microorganisms at high radiation levels was due to the presence of inhibitory compounds generated by the irradiation process or to the general intractability of the non-solubilized cell wall residues. No treatment costs or estimates were included in this study.

In 1972, Australian researchers (McManus et al., 1972a) reported a fairly extensive irradiation study that included rice straw along with other materials. Both wet and dry samples of rice straw were irradiated at 0, 5, 10, 25, 100, and 200 Mrad from a γ-ray source. Treatment effects were measured by DMD of irradiated samples in terylene bags immersed for 72 hour periods in rumens of sheep fed diets based on either rice straw or alfalfa. DMD of irradiated samples was greatest for samples irradiated in the dry state, and, in general, increased with the radiation dosage although the 10 Mrad and some of the 25 Mrad materials had depressed DMD. When irradiated straw samples were shaken vigorously in water for 14 hours, apparent DMD values of up to 35% were noted for material irradiated at 200 Mrad. This was about 1/2 the DMD value obtained by the terylene bag method. When samples of straw irradiated at 100 and 200 Mrad were immersed in gently flowing water for 12 hours, DM losses of about 1% were obtained. It was thus concluded that the enhanced DMD values found for irradiated samples were largely due to fermentation in the rumen and not to increased solubility. To test whether the depressive effect of small doses of radiation (10 and some 25Mrad samples) was due to an altered nitrogen to energy ratio, base diets fed the test animals were supplemented with urea at the 1 and 2% levels. Results showed that urea slightly depressed the digestion rate of rice straw samples irradiated at 25 Mrad or less and that lack of available nitrogen did not account for the decreased digestibilities observed at the 5-25 Mrad levels.

The same research group also reported results of in vivo digestibility trials using sheep fed γ-irradiated wheat and rice straw (McManus et al., 1972b). In the first trial wheat straw was irradiated at 50 Mrad and fed with and without dietary supplementation to adult sheep. Voluntary feed intake of sheep given unsupplemented irradiated wheat straw was not significantly different from that of control animals fed non-irradiated material. However, irradiation decreased apparent digestibility by about 5 percentage points (39.4 vs. 44.5). Supplementation of the irradiated wheat straw with casein, vitamins and minerals restored its digestibility to that of untreated control straw. Mean time of passage of irradiated straw through the digestive tract was 1/3 less than that of non-irradiated straw, presumably because irradiation had decreased fiber strength appreciably. In a second experiment rice straw samples were irradiated at 25, 50 and 75 Mrad respectively. Table 5.2 shows that dry matter intake and digestibility were generally depressed by irradiation treatment. Decreased times of passage of irradiated materials were also noted. Slaughter evidence confirmed the enhanced rate of passage of irradiated straw because a strong tendency was noted for a greater proportion of ingesta to be located in the hind gut of animals fed irradiated straw compared to those fed non irradiated straw diets. There was a significant difference found in the acetic:propionic acid ratio attributed to irradiation treatment, with a trend toward increased propionic acid and decreased acetic acid in rumen liquor as the irradiation dosage increased. It was concluded that while irradiation at 25-75 Mrad levels may aid microbial attack on straw, any beneficial effects are more than offset by losses due to increased rate of passage and its associated nutrient digestion decreases. Whether or not irradiation of straws induces the formation of toxic substances was not clearly identified in these experiments.

Table 5.2. Mean values of intake and digestibility for sheep fed unsupplemented irradiated rice straw (from McManus et al., 1972b)

Irrad level (Mrad)	Dry matter	
	Intake, g/d	Digestibility (%)
0	560	47.0
25	577	26.8
50	474	38.1
75	508	19.9

Barley straw, pea straw, sugar cane bagasse, sunflower hulls and pine saw dust were irradiated (γ-rays) at dosages of 10, 100, 200, and 250 Mrad by Ibrahim and Pearce (1980). Results of in vitro organic matter digestibility (IVOMD) determinations showed the general patterns other researchers have reported. Significant increases in apparent digestibility were found at irradiation dosages of 200 Mrad or above for barley straw, pea straw and sugar cane bagasse, and at 100 Mrad or higher for sunflower hulls and pine saw dust. The results of Van Soest fiber analyses were somewhat different from those obtained by Yu et al. (1975). Solubilization of all fibrous components including lignin was noted when radiation dosage was increased. In all cases at the 100 Mrad level or higher, ADF values were higher than NDF, giving apparent negative hemicellulose values. From sequential NDF analyses followed by ADF analysis the authors conclude that irradiation has degraded cellulose and lignin into simpler molecules. An alternative explanation may be that irradiation, especially at intermediate levels, has created a considerable quantity of carboxylic acid material (Arthur, 1971) that is insoluble in the acid detergent solution, giving abnormally high ADF values. We have found this to be the case with kelp, where ADF values are higher than NDF values because of a high insoluble mannuronic and guluronic acid content (Hart and Kohler, 1980). Satisfactory Van Soest analyses can be made if the neutral detergent analysis is run first and acid detergent is then run on the NDF residue. Although no economic data are presented, the authors conclude that if irradiation has any role at all in treatment of lignocellulosic residues, it will only be with extremely refractory materials like sunflower hulls or sawdust because staw-like products respond appreciably to less expensive physical and chemical treatments.

Fan et al. (1981) investigated irradiation of wheat straw at a dosage level of 10, 20 and 50 Mrad to increase sugar production by enzymatic hydrolysis. They estimate the cost of energy to irradiate a ton of wheat straw to be about $ 90, which is somewhat less than an earlier estimate of $ 150 per ton for sawdust (Millet et al., 1976). Han and Ciegler (1982) have proposed a scheme for utilization of lignocellulosic wastes through treatment with irradiation and swelling agents like sodium hydroxide. Unfortunately no economic data are presented to indicate the cost effectiveness of the proposed process.

5.4. STEAM TREATMENT

The idea of treating lignocellulosic residues with high pressure steam to increase their digestibility has intrigued researchers for many years. When energy costs were low this process had considerable appeal because of its basic simplicity, even though initial capital requirements were high for suitable equipment. As energy costs have escalated, the economic attractiveness of the process has decreased and the present trend seems to be toward development of less expensive alternative treatment processes.

The technology is based on the hydrolytic action of high temperature steam that breaks chemical bonds, and causes various degradations that increase digestibility of residual product. Common characteristics of the various processes that have been reported are (1) the production of acetic and other acids, (2) production of furfurals and phenolic derivatives, (3) varying degrees of destruction of the hemicellulose fraction, (4) dry matter losses of 1 to 20% of starting material, depending on conditions. Most of the proposed treatment conditions for agricultural residues fall short of the very vigorous ones used for production of pulp from wood or other lignocellulosic sources. Early efforts to upgrade wood and other residues for feed use through steam treatment have been reported (Bender et al., 1970; Heaney and Bender, 1970) but this chapter will concentrate on selected examples of treated cereal straws and other non-woody lignocellulosic residues. Emphasis will be on treatment with steam alone, even though chemical treatment as a part of a steam treatment has been shown to be even more effective in some cases (Klopfenstein and Koers, 1973; Phoenix et al., 1974; Hart et al., 1981).

IVDMD data in Figure 5.3 reported by Klopfenstein and Bolsen (1971) for steam pressure treated corn cobs, milo stubble and wheat straw are typical of the results achieved with this method. Steam pressures varied from 7 to 28 kg/cm^2 with a constant heating time of 50 seconds. Enough water was added before treatment to assure a moisture of at least 60% to prevent charring at the end of the experiment. Figure 5.3 shows that IVDMD of corn cobs responded to this treatment more than that of wheat straw or milo stubble. Although not particularly pronounced in these data, there is a suggestion that DMD achieves a maximum value and then declines as further treatment proceeds. Corn cobs and wheat straw showed some continued improvement in DMD as pressure was increased from 10 to 21 kg/cm^2, but milo stubble DMD showed essentially no further response after an initial increase at 7 kg/cm^2.

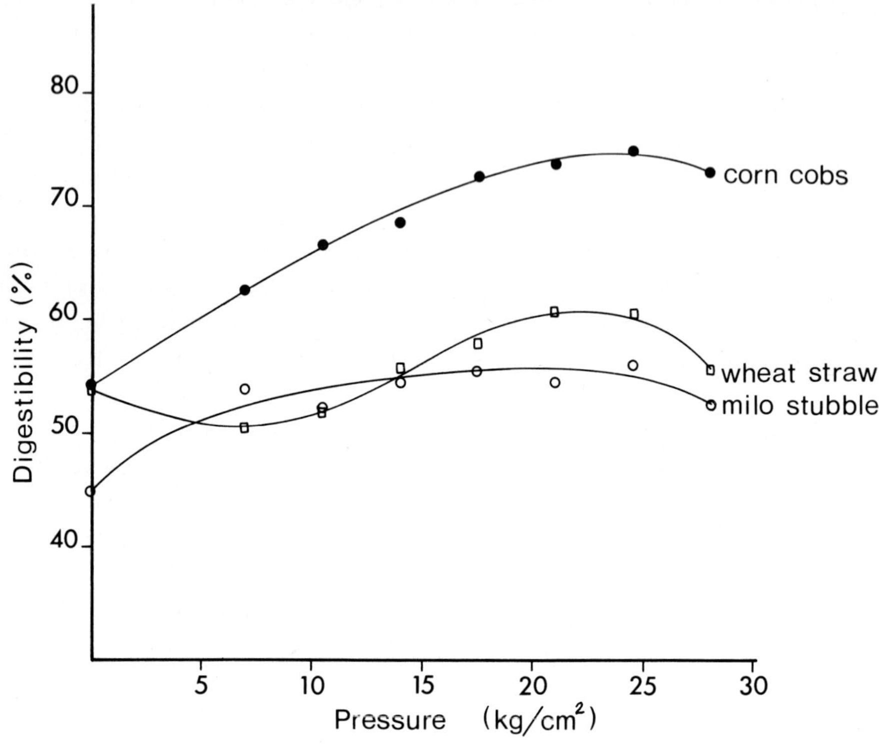

Figure 5.3. IVDMD of steam pressure treated agricultural residues
(from Klopfenstein and Bolsen, 1971)

Some of Klopfenstein's corn cob work (Klopfenstein and Koers,
1973) suggests that digestion inhibitors might be present in very
high pressure steam treated materials. In vivo and in vitro digestibilities were the same for material treated at 17.5 kg/cm^2,
but animal in vivo digestibility of material treated at 28
kg/cm^2 was more than 3 percentage points lower than the in vitro
value. Addition of sodium metabisulfite prior to steam treatment
tended to increase in vivo digestibility. A lamb growth trial comparing pressure treated (17.5 kg/cm^2, 50 s) corn cob diets with
untreated material (50% cobs in each case) showed that treatment
of cobs increased daily gain by 75% and feed efficiency by 35%
(Umunna et al., 1972). When these same treated and untreated products were fed at the 70% level in the diet, the effect of treatment was even more pronounced with daily gain on treated pro-

duct improved almost 200% and feed efficiency improved more than 50% over untreated control. When the content of treated and untreated corn cobs was raised to the 79% level in the diet, results were quite comparable to those obtained at the 50% level (Klopfenstein et al., 1974).

The tendency for IVDMD to decrease as a result of overtreatment was clearly demonstrated by Phoenix et al. (1974). Barley straw (68% moisture) was steam treated for 30 minutes at pressures ranging from 0 to 25.0 kg/cm^2. IVDMD increased from 47.8% at 0 kg/cm^2 (100° C) to a maximum of 60.3% at 8.1 kg/cm^2 (175°C) and then decreased sharply to 39.6% at 14.1 kg/cm^2 (200°C) and 36.6% at 25.0 kg/cm^2 (225 C).

Hart et al. (1981) reported a fairly extensive study of steam reatment of rice straw and sugar cane products. Treatment pressures ranged from 7 to 42.2 kg/cm^2 with treatment times varying from 10 to 600 seconds depending on the pressure used. Effects of processing were measured by an enzymatic assay. Enzymatic digestibility results of rice straw products treated with steam but no added chemicals are shown in Table 5.3. The treatment times and pressures chosen were ones considered to be economically feasible. Some of the samples had 36% of water added to make their moisture levels comparable to those of samples containing NaOH that were included in the study. Values reported as total solubles after enzyme (TSAE) represent total amount of material solubilized by the buffer medium and carbohydrase action (comparable to IVDMD by artificial rumen technique) and the enzymatically digestible carbohydrate values (EDC) represent materials solubilized by carbohydrase activity only. The data show the typical trade-off between process time and pressure used. The time needed to achieve a particular TSAE value decreased rapidly as pressure was increased. The main effect of these steam treatments seemed to be to increase the water solubility of some straw components and not to increase carbohydrate digestibility appreciably since the EDC values remained essentially constant. At the higher pressure levels, added water tended to help TSAE values from decreasing compared to drier samples. Data not included in Table 5.3 indicated that pressure treated samples containing NaOH had significantly higher TSAE values than samples without added caustic. Samples of rice straw which had an apparent phenolic content of 0.43% prior to steam treatment had a phenolic content ranging from .73% (7 kg/cm^2, 10 min) to 2.28% (42.2 kg/cm^2, 30 s) after treatment.

When diets containing 65% of steam treated (28.1 kg/cm^2, 20 or 90 s) rice straw were fed to lambs, animal performance was not improved over that of animals fed untreated material (Garrett et al., 1981). The diet containing material steamed for 20 s had lower organic matter digestibility (53.3%±1.0 vs. 56.9%±8), cellulose digestibility (41.0%±1.4 vs. 45.4%±1.7), nitrogen digestibility (47.0%±.9 vs. 63.9%±1.8) and digestible energy (1.92±.04 vs. 2.09±.04 Kcal/g) than untreated control, but daily weight gain, energy gain, feed intake and feed efficiency were not significantly different for the two materials. Digestibility of organic matter, cellulose and energy were even more depressed for the diet containing straw steamed for 90 s than for the material steamed 20 s. Animal performance was significantly poorer for the 90 s product than for the 20 s product or untreated control. Animals fed the product treated for 90 s actually suffered significant daily weight losses during the trial.

Experiments in the same pressure treating equipment showed that the TSAE of sugar cane bagasse was maximized by steam treatment at 28.1 kg/cm^2 for 30 s (Hart et al., 1981). TSAE was increased from 16.8% for untreated control to 43.5% for treated products. Treatment at lower or higher pressures gave lower TSAE values. A similar result was obtained when sugar cane field trash was steam treated in the same manner.

Enough sugar cane bagasse was steam pressure treated (28.1 kg/ cm^2 for 45 s) for an in vivo digestibility trial using lambs (Campbell et al., 1973). After drying, one half of the treated material was pelleted and the other half used as-is. Even though bagasse content of the trial diets was only about 40%, some palatability problems were noted. Digestibilities of all components of the two diets containing treated bagasse were superior to those of the diet containing raw bagasse, but the digestibilities of all three bagasse diets were inferior to those of a positive alfalfa control diet. IVDMD of either treated bagasse sample was about 35% compared to 23% for raw bagasse. Treated bagasse samples contained more than 12 times as much apparent phenolics as raw bagasse (5.28% vs 0.43%). Almost all the hemicellulose in the holocellulose fraction of treated product was destroyed by steam treatment but the ADL/ADF ratio remained unchanged, indicating no preferential destruction of lignin or cellulose.

Table 5.3. Enzymatic digestibility of rice straw, with and without added water, treated with steam for various times at different pressures (from Hart et al., 1981)

Pressure kg/cm^2	Time sec	Steam only		H_2O added (36%)	
		TSAE[1]	EDC[2]	TSAE[1]	EDC[2]
0	–	26.3	15.0	–	–
7	600	37.5	21.0	38.0	25.5
14.1	90	37.5	–	38.0	20.4
	120	45.1	–	43.1	20.2
	300	46.6	–	45.2	22.6
21.1	30	38.6	20.3	36.9	20.5
	60	41.8	20.4	42.0	22.2
	90	47.4	22.7	45.5	23.0
	120	45.0	20.3	48.3	22.2
	300	39.9	19.1	46.7	20.9
28.1	10	–	–	46.9	–
	20	42.9	19.7	40.0	–
	30	44.1	19.5	43.3	–
	60	47.1	20.0	46.5	–
	90	47.4	21.1	47.0	–
35.2	10	–	–	44.6	–
	20	46.4	–	44.8	–
	30	42.9	–	46.8	–
	60	41.6	–	58.1	–
42.2	20	40.5	–	48.1	–
	30	40.5	–	47.2	–
	60	35.5	–	39.1	–

[1]TSAE = total solubles after enzyme treatment
[2]EDC = enzyme digestible carbohydrate

Rangnekar et al. (1982) studied the effect of steam pressure treatment on sugar cane bagasse, paddy straw (rice) and sorghum straw at low pressures (5, 7 and 9 kg/cm^2) for relatively long treatment times (30 and 60 minutes). Results were generally comparable to those reported for higher pressures. For all three materials the most notable chemical changes found were large decreases in hemicellulose content as severity of treatment increased, and noticeable increases in volatile fatty acids under all treatment conditions. Average volatile fatty acid incease depended on material treated, ranging from about 50% for paddy straw to over 100% for bagasse and sorghum straw. IVDMD of all materials was increased by steam treatment but digestibility of cell wall contents was depressed at the severest treatment levels. Some dry matter loss occurred with all materials, under all treatment conditions with greatest losses at 9 kg/cm^2 for 60 minutes (20.8%, 39.5% and 22.5% for bagasse, paddy straw and sorghum straw respectively). There was a net gain in DDM as a result of treatment for bagasse and sorghum straw but a net loss in the case of paddy straw.

The steam treated products reported thus far were made in laboratory or pilot scale batch-type vessels that could be scaled up if economically justifiable. Equipment to process lignocelllosic residues continously for feed has been developed (Bender, 1979) and is now commercially available. This high pressure digester-extruder has been used primarily to make cattle feed from aspen wood, but use of other lignocelluloses has been investigated. According to the patent, material is subjected to steam at a pressure of at least 14 kg/cm^2 (198°C) for at least 15 s and then ejected and cooled adiabatically. With a final moisture content of 45-55% and a pH of 3.5-3.8, it is claimed that the processed material can be stored like silage without further preservation treatment. Unfortunately, no process cost data have been reported.

Sugar cane treated in this manner had an IVDMD of 55% compared to 36.1% for raw bagasse (Stake Technology, personal communication, 1979). Chemical analysis showed a drop in crude fiber due to treatment (33.9% vs 43.0% for untreated), and a negative value for hemicellulose (NDF-ADF) indicating considerable disruption of the holocellulose fraction. An in vivo trial (steers) gave a DMD of 53% for steam treated product compared to 42% for unprocessed bagasse. In a 91 day steer performance trial, steam treated bagasse was substituted for shelled corn at levels of 14, 30 and 46% in a balanced diet. Performance data indicated that treated bagasse could replace up to 30% of the corn in the diet but at the highest replacement level (46%) poorer animal performance was obtained than with corn alone. Oji and Mowat (1978) investigated the nutritive value of corn stover steam treated in this digester-extruder at 16.6 kg/cm^2 (205°C) for 15 minutes. Chemical analysis of steamed material showed the usual decreases in pH, NDF and hemicellulose, and increases in ADF, acid detergent insoluble nitrogen (ADIN), permanganate lignin and acetic acid. The considerable increase in apparent lignin (70%) was attributed to production of artifact lignin from nonenzymatic browning reaction. This is reflected by an increased ADIN value in the steamed product. Despite the low pH and acetic acid content, some moulds developed in the treated stover during storage.

A digestion trial using lambs was run with diets containing 87% steam treated or untreated corn stover. Steam treatment significantly increased dry matter intake, and apparent digestibility of organic matter, gross energy, cell contents, and cellulose but decreased digestibility of NDF and ADF (Oji and Mowat, 1979). The increased intake was partly attributed to decrease in particle size due to treatment, and partly to increased organic matter digestibility. The observed decrease in ADF digestibility was probably due to the previously mentioned artifact lignin production. Further experiments showed that steam pressure treatment increased the acetate:propionate ratio in rumen fluid compared to untreated control and decreased the rate of passage time of particulate matter by more than 20% (Oji et al., 1979). Overall results suggest that although the nutritive value of this steamed stover was generally superior to that of unsteamed material, processing conditions should be modified to reduce the amount of heat damage.

In connection with heat damage of steam treated lignocellulosic residues, Britton (1978) studied the growth inhibitor(s) produced from sawdust when it was treated with dilute sulfuric acid and subjected to high steam pressure hydrolysis. The exact nature of the inhibitor(s) was not established but rat feeding trials indicated that furfural was not the causative agent. Extraction of treated sawdust with either 80% or 95% ethanol removed the material(s) which inhibited in vitro cellulose digestion by rumen microorganisms, and extraction with 95% ethanol also removed the compound(s) depressing rat growth. Ultra violet absorption spectra of urine from rats fed the steam treated sawdust or a 95% ethanol extract of that product indicated that relatively high levels of phenolic type materials were being excreted compared to urine from rats on a control or extracted residue diet, but it was not determined that these UV absorbers were responsible for the growth depressing effects observed.

5.5. OTHER PROCESS TREATMENTS

Novel treatments like high temperature solvent cooking; repeated freezing and thawing cycles; very high pressure roll milling; etc have been investigated as ways to increase saccharification of wood products (Millet et al., 1976). No reports are available on use of similar procedures to upgrade feed quality of crop residues.

One technique which seems to offer some promise involves mechanical separation and collection of different parts of whole forage (see Chapter 10). The principle has been well developed for leaf-stem separation of high quality forage like alfalfa (Kohler et al., 1972) and appears applicable to straws. Rexen (1978) reports development of equipment to separate cereal straws into a stem fraction (43-63%) and a meal fraction composed of nodes, ears, and leaves (36-58%). In vitro estimates of the feed value of the meal fraction compared to whole straw show an increase of 117% for wheat straw and 45% for barley straw. Canadian workers (Coxworth et al., 1981) confirm that wheat straw stems have a lower IVDOM and crude protein content than other straw components. When stems and chaff were collected separately and fed to sheep, digestible energy intake was increased by about 33% for the chaff-fed animals compared to animals fed straw residue left after chaff removal. In this experiment, it was necessary to chop the straw fraction to increase animal intake, but this was not required for the chaff fraction.

Similar separation milling procedures have been developed for separation of pith and rind fractions for sugar cane bagasse (Lathrop, 1956) to provide a low fiber pith fraction that should be a useful feedstuff component.

5.6. SUMMARY

Various physical treatments of lignocellulosic residues can be used to upgrade their feed value. Some type of grinding or milling procedure followed by a compaction process (pelleting or cubing) is most commonly used for this purpose. Particle size must be carefully regulated to achieve a proper balance between increased feed intake and efficient rate of passage through the digestive tract for optimum feedstuff utilization. Exposure of lignocellulo-

sic residues to treatment with high steam pressure or high intensity ionizing radiation can increase digestibility, but considerable expenditure of expensive energy is required to achieve the
desired results. It seems most likely that a combination of some
grinding treatment, combined with a chemical treatment will prove
to be the most cost effective way to upgrade the nutritive value
of most major crop residues.

5.7. REFERENCES

Arthur, J.C. Jr., 1971. Reactions induced by high-energy radiation. In: N.M. Bickales and L. Segal (eds.): Cellulose and Cellulose Derivatives. Wiley Interscience. New York, pp. 937-975.

Beardsley, D.W., 1963. Symposium on forage utilization. Nutritive value of forage as affected by physical form. Part II. Beef cattle and sheep studies. J. Anim. Sci. 23: 239-245.

Bender, R., 1979. Method of treating lignocellulosic materials to produce ruminant feed. U.S. Pat. 4, 136, 207.

Bender, F., Heaney, D.P. and Bowden, A., 1970. Potential of steamed wood as a feed for ruminants. Forest Prod. J. 20: 36-44.

Britton, R., 1978. Removal of the growth inhibitor(s) from acid and pressure hydrolyzed sawdust. J. Agric. Food Chem. 26: 759-763.

Brown, R.D. Jr. and Jurasek, L., 1979 (eds). Hydrolysis of cellulose: mechanisms of enzymatic and acid catalysis. Advances in Chemistry series 181. Am. Chem. Soc. Washington. 399 pp.

Campbell, C.M., Wayman, O., Stanley, R.W., Kamstra, L.D., Olbrich, S.E., Ho-a, E.B., Nakayama, T., Kohler, G.O., Walker, H.G. and Graham, R.P., 1973. Effects of pressure treatment of sugar cane bagasse upon nutrient utilization. Proceedings, Western Section, Am. Soc. Anim. Sci. 24: 178-184.

Cowling, E.B. and Kirk, T.K., 1976. Properties of Cellulose and lignocellulosic materials as substrates for enzymatic conversion processes. In: E.L. Gadden (ed.): Enzymatic conversion of cellulosic materials: technology and applications. Biotechnol. and Bioeng. Symp. 6. Wiley-Interscience, New York, p. 95-123.

Coxworth, E., Kernan, J., Knipfel, O., Thorlacius, O. and Crowle, L., 1981. Review: Crop residues and forages in western Canada, potential for feed use either with or without chemical or physical processing. Agric. Environm. 6: 245-256.

Dehority, B.A., 1961. Effect of particle size on the digestion rate of purified cellulose by rumen cellulolytic bacteria in vitro. J. Dairy Sci. 44: 687-692.

Dehority, B.A. and Johnson, R.R., 1961. Effect of particle size upon the in vitro cellulose digestibility of forages by rumen bacteria. J. Dairy Sci. 44: 2242-2249.

Demeyer, D.I., 1981. Rumen microbes and digestion of plant cell walls. Agric. Environm. 6: 295-337.

Fan, L.T., Gharpuray, M.M. and Lee, Y.-H., 1981. Evaluation of pretreatments for enzymatic conversion of agricultural residues. In: C.D. Scott (ed.): Third symposium on biotechnology in energy production and conservation. Biotechnol. and Bioeng. Symp. 11, J. Wiley and Sons, New York, 29-45.

Fernandez Gonzales, E., 1980. Effects del trtamiento, con NaOH sobre la digestibiledad e ingestion de la paja de cebada en corderos sometida a diferentes grados de troceado y tiempo de actuacion 11. Advanves en Alementacion y Mejora Animal 21: 161-1973. In: Nutr. Abstr. Rev. Series B 1981. 51: 3478.

Garrett, W.N., Walker, H.G. Jr., Kohler, G.O., Hart, M.R. and Graham, R.P., 1981. Steam treatment of crop residues for increased ruminant digestibility. II. Lamb feeding response. J. Anim. Sci. 51: 409-413.

Greenhalgh, J.F.D. and Wainman, F.W. 1972. The nutritive value of processed roughages for fattening sheep and cattle. Proc. Br. Soc. Anim. Prod. 61-72.

Greenhalgh, J.F.D. and Reid, G.W., 1975. Mechanical processing of wet roughage. Proc. Nutr. Soc. 34: 74A.

Grohn, H., 1958. Mechanochemical degradation of wood by vibratory grinding. J. Polymer Sci. 30: 551-559.

Hackett, M.R., Hillers, J.K., Kromann, R.P. and Martin, E.L., 1975. Evaluation of wheat straw in feeder lamb rations. Proc. Western Section, Am. Soc. Anim. Sci. 26: 143-145.

Hajny, G.J. and Reese, E.T., 1969. (eds): Cellulases and their applications. Advances in chemistry series 95. Am. Chem. Soc. Washington, 479 pp.

Han, Y.W. and Ciegler, A., 1982. Use of nuclear wastes in utilization of lignocellulosic biomass. Process Research 17: 32-38, 42.

Hanlein, G.F.W. and Holdren, R.D., 1965. Response of sheep to wafered hay having different physical characteristics. J. Anim. Sci. 24: 810-818.

Hart, M.R. and Kohler, G.O., 1980. Unpublished communication.

Hart, M.R., Walker, H.G. Jr., Graham, R.P., Hanni, P.G., Brown, A.H. and Kohler, G.O., 1981. Steam treatment of crop residues for increased ruminant digestibility. 1. Effects of process parameters. J. Anim. Sci. 51: 402-408.

Heaney, D.P. and Bender, F., 1970. The feeding value of steamed aspen for sheep. Forest Prod. J. 20: 98-102.

Heaney, D.P., Pigden, W.J., Minson, D.J. and Pritchard, G.I., 1963. Effect of pelleting on energy intake of sheep fed from forages cut at three stages of maturity. J. Anim. Sci. 22: 752-757.

Ibrahim, M.N.M. and Pearce, G.R., 1980. Effects of gamma irradiation on the compositition and in vitro digestibility of crop by-products. Agric. Wastes 2: 253-259.

Klopfenstein, T.J. and Bolsen, K.K., 1971. High temperature pressure treated crop residues. J. Anim. Sci. 33: 290 (Abstr.).

Klopfenstein, T.J. and Koers, W., 1973. Agricultural cellulosic wastes for feed. In: G.E. Inglett (ed.): Symposium: Processing Agricultural and Municipal Wastes, Avi. Publishing Co., Westport, Conn. p. 38-54.

Klopfenstein, T.J., Graham, R.P., Walker, H.G. Jr. and Kohler, G.O., 1974. Chemicals with pressure treated cobs. J. Anim. Sci. 39: 243 (Abstr.).

Kohler, G.O., Walker, H.G. Jr. and Kuzmicky, D.D., 1979. Processing and use of crop residues including alfalfa press cake. Fed. Proc. 38: 1934-1938.

Kohler, G.O., Bickoff, E.M. and Beeson, W.M., 1972. Processed products for the feed and food industries. In: C.H. Hanson (ed.): Alfalfa: Science and Technology. Am. Soc. Agron. Madison, Wisc. p. 659-676.

Lathrop, E.C., 1956. Pith separation and sugar recovery from bagasse. U.S. Pat. 2: 744, 037.

Lawton, E.J., Bellamy, W.D., Hungate, R.E., Bryant, M.P. and Hall, E., 1951. Some effects of high velocity electrons on wood. Science 113: 380-382.

Mathison, G.W., 1981. Effect of grinding straw, and molasses additions to straw on voluntary consumption and weight changes in beef cows. 60th Annual Feeders Day Report. Agriculture and Forestry Bulletin (special issue) University of Alberta, Edmonton, p. 42-44.

McManus, W.R., Manta, L., McFarlane, J.D. and Gray, A.C., 1972a. The effects of diet supplements and gamma irradiation of dissimilation of low-quality roughages by ruminants. II. Effects of gamma irradiation and urea supplementation on dissimilation in the rumen. J. agric. Sci. (Camb.) 79: 41-53.

McManus, W.R., Manta, L., McFarlane, J.D. and Gray, A.C., 1972b. The effects of diet supplements and gamma irradiation on dissimilation of low-quality roughages by ruminants. III. Effects of feeding gamma irradiated base diets of wheaten straw and rice straw to sheep. J. agric. Sci (Camb.) 79: 55-66.

Meyer, J.H., Kromann, R. and Garret, W.N., 1965. Int. Symp. Physiol. Dig. Ruminant, 2nd, Ames, Iowa, 1964.

Millet, M.A., Baker, A.J. and Satter, L.D., 1976. Physical and chemical pretreatments for enhancing cellulose saccarification. In: E.L. Gadden (ed.): Enzymatic conversion of cellulosic materials: Technology and applications. Biotechnol. and Bioeng. Symp. 6. Wiley-Interscience, New York p. 125-153.

Moore, L.A., 1963. Symposium on forage utilization. Nutritive value of forage as affected by physical form. Part 1. General principles involved with ruminants and effect of feeding pelleted or wafered forage to dairy cattle. J. Anim. Sci. 23: 230-238.

Oji, V.I. and Mowat, D.N., 1978. Nutritive value of steam treated corn stover. Can. J. Anim. Sci. 58: 177-181.

Oji, V.I. and Mowat, D.N., 1979. Nutritive value of thermoammoniated and steam treated maize stover. 1. Intake, digestibility, and nitrogen retention. Anim. Feed Sci. Technol. 4: 179-186.

Oji, V.I., Mowat, D.N. and Buchanan-Smith, J.G., 1979. Nutritive value of thermoammoniated and steam treated maize stover. II. Rumen metabolites and rate of passage. Anim. Feed Sci. Tehcnol. 4: 187-197.

Owen, E., 1978. Processing of roughages. In: W. Haresign and D. Lewis (eds.): Recent advances in animal nutrition. Butterworths London, pp. 127-148.

Pew, J.C. and Weyna, P., 1962. Fine grinding, enzyme digestion, and lignin-cellulose bonds in wood. Tappi 45: 247-256.

Pfost, H.B. and Headly, V., 1976. Methods for determining and expressing particle size. In: H.B. Pfost (ed.): Feed Manufacturing Technology, American Feed Manufacturers Assn. Inc., Arlington, Va, 512.

Phoenix, S.L., Bilanski, W.K. and Mowat, D.N., 1974. In vitro digestibility of barley straw treated with sodium hydroxide at elevated temperatures. Transactions of the ASAE 17: 780-782.

Pickard, D.W., Swan, H. and Lamming, G.E., 1969. Studies on the nutrition of ruminants. 4. The use of ground straw of different particle sizes for cattle from twelve weeks of age. Anim. Prod. 11: 543-550.

Pigden, W.J. and Heaney, D.P., 1969. Lignocellulose in ruminant nutrition. In: G.J. Hayny and E.T. Reese (eds.): Cellulases and their applications. Advances in Chemistry series 95. American Chemical Society. Washington, 245-261.

Pritchard, G.I., Pigden, W.J. and Minton, D.J., 1962. Effect of gamma radiation on the utilization of wheat straw by rumen micro-organisms. Can J. Anim. Sci. 42: 215-217.

Ragnekar, D.V., Badve, V.C., Kharat, S.T., Sobale, D.N. and Joshi, A.L., 1982. Effect of high-pressure steam treatment on chemical composition and digestibility in vitro of roughages. Anim. Feed Sci. Technol. 7: 61-70.

Rexen, F., 1978. Recent Danish experience in straw utilization. In: Proceedings of the Fourth Straw Utilization Conference, Oxford, England. 30 Nov.-1 Dec. ADAS Ministry of Agric. Fisheries and Food. Oxford, p. 55-58.

Saeman, J.F., Millet, M.A. and Lawton, E.J., 1952. Effect of high energy cathode rays on cellulose. Ind. Eng. Chem. 44: 2848-2852.

Smith, D.M. and Mixer, R.Y., 1959. The effects of lignin on the degradation of wood by gamma irradiation. Radiation Research 11: 776-780.
Stevens, C., 1981. Pelleting: Emphasis on byproduct and roughage ingredients. Feedstuffs 53 (31), E1-E3.
Stone, J.E., Scallen, A.M., Donefer, E. and Ahlgren, E., 1969. Digestibility as a simple function of a molecule of similar size to a cellulase enzyme. In: G.J. Hajny and E.T. Reese (eds.): Cellulases and their applications. Advances in Chemistry Series 95. Am. Chem. Soc. Washington, p. 219-241.
Swan, H. and Clarke, V.J., 1974. In: H. Swan and D. Lewis (eds): University of Nottingham Nutrition Conference for feed manufacturers-8, Butterworths, London.
Ummuna, N.N., Klopfenstein, T.J. and Bolsen, K.K., 1972. Response of lambs fed pressure treated corn cobs. J. Anim. Sci. 35: 277-278 (Abstr.).
Walker, H.G. and Kohler, G.O., 1976. Unpublished observation.
Weir, W.C., 1962. Intake by sheep of pelleted feeds of varying composition. J. Anim. Sci. 21: 659 (Abstr.).
White, T.W., Reynolds, W.L. and Hembry, F.G., 1971. Level and form of rice straw in steer rations. J. Anim. Sci. 33: 1365-1370.
Yu, Y., Thomas, J.W. and Emery, R.S., 1975. Estimated nutritive value of treated forages for ruminants. J. Anim. Sci. 41: 1742-1751.

Chapter 6.1

WET TREATMENT WITH
SODIUM HYDROXIDE

by

Thor Homb

Department of Animal Nutrition
Agricultural University of Norway
P.O. Box 25, 1432 Ås-NLH
Norway

6.1.1. INTRODUCTION

It is generally accepted that purified cellulose from wood or
straw is highly digestible and has a high net energy value for
ruminant animals. This fact was established in the early work initiated by Henneberg and his co-workers in Göttingen (Lehmann,
1891) and followed up by the Möckern group (Kellner and Köhler,
1900; Kellner, 1905). This is one of several examples demonstrating the value of the balance trial technique, with its clear and
conclusive answer even though the biochemical steps in rumen physiology were not clearly understood at the time.

The researchers in Möckern used straw cellulose produced in a
paper factory. The straw was boiled under pressure (7 atmospheres)
in a solution of NaOH, Na_2CO_3, Na_2S and $Na_2S_2O_3$. Clearly, this was
an expensive method of improving the feeding value of straw, not
only because of the severe processing conditions, but also because
of the low output of feed produced and the large amount of environmental pollution generated. Most of the lignin disappears under
such treatment, while cellulose and part of the hemicellulose remain in the final product.

6.1.2. BOILING METHODS WITH SODIUM HYDROXIDE

All German methods for treatment of straw with sodium hydroxide
during the period 1890-1917 were based on the assumption that boiling was necessary. The following methods have been described in
the literature:

Boiling in open vessels (Lehmann, 1891, 1895, 1904, cited by
Homb, 1948). 100 kg oat straw was boiled in 200 l water and 4 kg
NaOH and the remaining lye was subsequently washed out. The product had the same digestibility as wheat bran, indicating a clear
improvement in energy value. However its digestibility was lower
than straw processed by Kellner (1905) who found that the organic
matter digestibility was 88%. The palatability of Lehmann's product was not good. Lehmann therefore, tried boiling under pressure.

Boiling under pressure. This followed the principle used in
paper factories. The chemical concentration used, however, was
much lower; 40 g NaOH per kg straw. This low dosage was thought to
make washing-out of remaining lye unneccessary. After the boiling
process air was blown into the autoclave, and it was found that
sufficient quantities of organic acids were formed to neutralize
the lye. Lehmann recommended boiling at 5-6 atmospheres pressure
for 6-8 hours.

Another method was introduced by Colsman: Chopped straw was
mixed with a NaOH solution (80 g NaOH per kg straw) in a special
machine. This wet material was stored for 12 hours before being
transported to vats, where a steaming process took place. The temperature in the vat was held at 100°C for the the first four
hours. In the next 1-2 hours more steam was forced in, leading to
a somewhat raised temperature. A washing process followed, either
in the same vat, or in washing drums (see Homb, 1948).

These three methods are probably the more interesting of those
developed during intensive research into finding an economical
straw feeding system during the period from 1890 to the First
World War. Other modifications have been mentioned in the literature (see Fingerling, 1924; Homb, 1948). The starting point for
development and experiment during this period was apparently the
original discovery by Lehmann and Kellner that straw treated in
paper factories had a high nutritive value. This raised the question of whether it was necessary to spend so much energy and lose

so much organic matter in order to obtain a useful fodder. A compromise between the almost pure cellulose and the lignified straw was looked for. All these methods had in common that boiling was necessary to loosen the bond between cellulose and lignin. The results obtained also appeared to be a compromise, as the digestibility figures were somewhere between those for untreated straw and "Kraftstroh" made in paper factories. However, all these methods required relatively advanced facilities, and the feed obtained was not very palatable. The most important drawback was the high cost.

Some chemicals other than NaOH were also tested in the older German studies. Of greatest interest is $Ca(OH)_2$ which did not compete with NaOH (see Homb, 1948). This has largely been verified in subsequent research (see Chapter 8).

6.1.3. THE ORIGINAL BECKMANN METHOD

Fingerling (1924) stated that a deficient supply of coal led to trials on cold treatment with NaOH solution. "Geheimrat" E. Beckmann in Kaiser Wilhelm-Institut, Dahlem near Berlin, has been honoured for developing this method, which has been described by Beckmann (1919), Fingerling (1919, 1924) and Fingerling and Schmidt (1919). Originally it was suggested that the treatment should compensate for the low temperature by lengthening the soaking period to three days. The solution recommended was 1.5-2 per cent NaOH. The remaining lye was washed out to neutrality, indicated by litmus paper (Fingerling, 1924). Beckmann's promising results were verified by Fingerling and Schmidt (1919) and Fingerling et al. (1923). The first of these publications showed an almost linear increase in OM digestibility (%) of rye straw from 45.7 to 71.2 for an increase in the concentration of NaOH from 0 to 1.5%. Higher rates of NaOH were not tested. Figure 6.1.1. illustrates OM digestibility as a function of NaOH added. The time factor was tested by Fingerling et al. (1923). As three days gave only a small increase in digestibility, a 12 hours soaking was recommended for practical purposes.

Compared with the boiling methods this type of treatment led to a lower loss of dry matter, i.e. about 20 per cent. Thomann (1921) in Switzerland found that almost all the cellulose remained in the feed, while 25-30 per cent of the lignin and 8-15 per cent of the pentosans disappeared during the washing process.

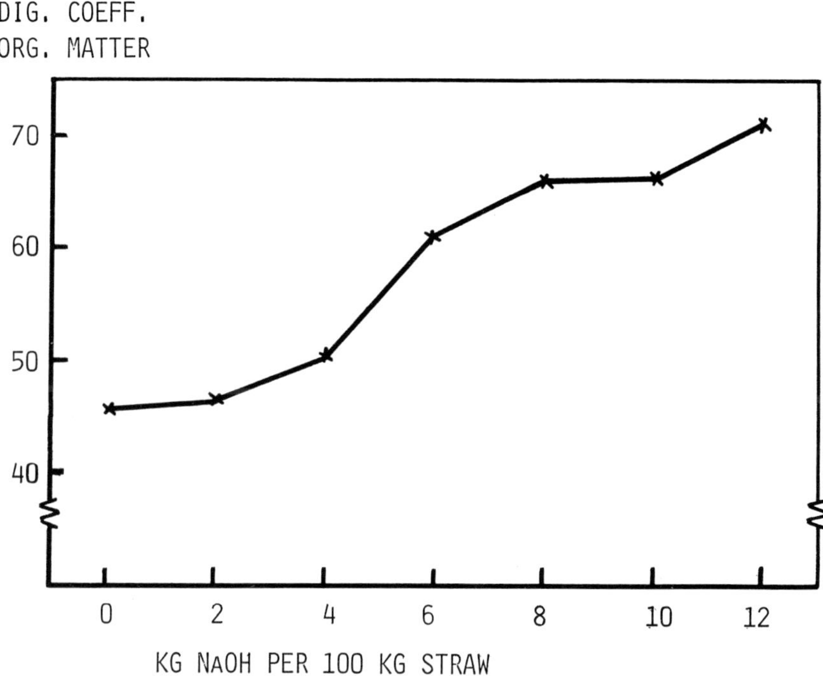

Figure 6.1.1. Digestibility of organic matter in Beckmann-treated straw at different levels of sodium hydroxide application (Fingerling et al., 1923)

Fingerling (1924) mentioned several advantages of using this method instead of the older boiling methods - treated straw remained in its natural physical state, had a good aroma, was more palatable and yield was higher (smaller loss). Most important were the simple and inexpensive facilities needed, combined with a saving of coal. The soaking process took place in shallow wooden barrels (often 0.3 m deep) sealed with asphalt. These were advantageous because it was easier to avoid leakages, and the washing process was more efficient. Usually there were two barrels, one for soaking in NaOH-solution, and one for washing. Chopped straw was added to water containing 120 g NaOH per kg straw. Following soaking the rest lye was reused, some 70-80 g NaOH per kg straw was then needed to recharge the concentration to the necessary 1.5 per cent in the solution.

In the light of present knowledge the whole procedure was very laborious. They had to stir by hand to obtain a homogeneous solution. The straw was moved by hand from one barrel to another. To save some of the hand work the barrels (two or more) were placed stepwise at different heights, enabling lye for reuse to pass from barrel to barrel by gravity. However, the remaining lye still had to be taken by hand from the lowest barrel to the top one.

The heavy work involved was probably one reason why this method had only limited practice on German farms. However, the results had been published and this made it possible to continue the search for improved technology. The biological aspects of the method have subsequently been shown to hold true. This early German work played an important role in setting the foundations for straw treatment methods which were subsequently developed.

6.1.3.1. <u>British experiments 1937-43</u>

As early as 1920 some preliminary experiments on treatment of straw were published by Godden, interest presumably having been aroused by the German findings. Godden used a 1.5 per cent solution of NaOH, with a subsequent short period of steaming, with or without washing. This type of study was continued at Jealott's Hill Research Station during the late 1930's (Slade et al., 1939; Watson, 1941; Ferguson, 1943). As to NaOH concentration and soaking time, their results seemed to confirm the German results. In the digestion studies both sheep and cattle were used, and different types of straw were tested. Treatment at 7°C was as effective as at 30-40°C. No steaming was performed.

The facilities for the straw treatment were somewhat different from the German ones. Plate 6.1.1. is an illustration of the concrete vats used at Jealott's Hill. They were placed outside the barn. The lower vat was used for NaOH-treatment. After the immersion period, the straw (long or chopped) was lifted on to the slanted area between the two vats. From there lye drained to the bottom vat and was used again for the next batch of straw. The washing process took place in the upper vat by letting a continous flow of water pass through.

Plate 6.1.1. Ramp plant for treating straw. Jealotts Hill, England
(Watson, 1941)

This facility was simple and inexpensive. One disadvantage was
frozen fodder during cold weather, but such periods occur relatively seldom in England. Heavy hand work was also a drawback here.
In Scotland digestibility trials with horses fed treated straw
were carried out during the same period (Williamson, 1941).

6.1.3.2. <u>Scandinavian experiments and technology with Beckmann-
          treatment of straw</u>

The initiative for starting straw treatment in Norway was taken
by Breirem (1939) who reviewed the early German work. His conclusion was in favour of the Beckmann method. A few small-scale farm
facilities for NaOH-treatment were built as early as 1939-40. Wooden barrels (400-500 l each) were used in the first year, but by
1941 larger vats made of concrete had appeared (Hesthamar, 1940;

112

1943; Blakstad, 1943). Two of these, at Tjerne and Holstad, produced treated straw for group feeding experiments during the 1943-
44 winter (Homb, 1948). Figure 6.1.2 illustrates the system which
has been used most frequently on Norwegian farms from the 1940's
until the early 1970's.

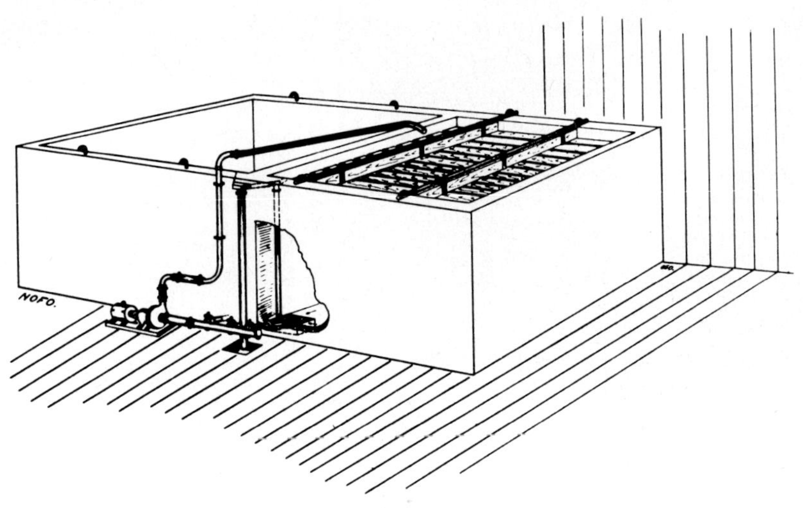

Figure 6.1.2. On-farm facilities for treating straw in Norway,
            produced by Norwegian Feed Conservation Association
            (NOFO)

Feed shortage due to the war accelerated on-farm constructions.
Also those who started during the war, thought treated straw would
make a substantial contribution to cattle and sheep rations under
normal conditions. The reasons for the successful beginning and
later development of Norwegian on-farm constructions could be attributed to:
1) The system chosen in 1941, comprising two vats, one pump and a
    system of pipes for linking the two vats, and for circulating
    the lye, happened to be a successful one.
2) Chopped straw, used in the early years, was normally stored on
    the first floor, while cows were on the ground floor. This

enabled the straw to be easily conveyed by gravity from the store to the treatment vats, which were usually located close to the cow shed.

3) Later on, government agricultural policy made straw treatment economically rewarding.

Chopped straw was considered useful during the 1940's. Although straw transfer from vat to cows was done by hand, the work was not very time-consuming (Homb, 1950). Later, bales of long straw were preferred, together with the use of mechanical equipment for transporting the straw to the cows.

Typically the treatment procedure was as follows: Vat 1 was filled with straw, water and NaOH. From the early 1940's NaOH was bought in 180 kg drums. They had to be divided by hand, using an axe. Later on, flakes came on the market, and handling was easier. Solution was prepared by placing the NaOH flake in a box, fitted with a wire bottom and putting this under the water tap. Some farmers were not satisfied with this type of equipment and built separate tanks for dissolving NaOH. During the 1950's NaOH solution (40-50% NaOH) was introduced for distribution to cooperative plants as well as farmers who had tanks. The lye was allowed to circulate for some time. After 20-22 hours the lye was pumped over to vat 2, which had been filled with straw. The first washing from vat 1 was also pumped to basin 2. Further water and NaOH were then added to vat 2. The remaining lye in vat 1 was washed out. The treated straw in vat 1 was ready for feeding 40-48 h after stacking. In this way one vat was ready daily. The lye was used repeatedly, often over the whole winter. The routine usage of NaOH was 70-80 g NaOH per kg dry straw. Equipment for monitoring NaOH concentration was made available to farmers from the early 1950's (Hvidsten and Simonsen, 1952; Hvidsten and Gloppe, 1953). The dip-treated straw may be enriched by adding, for instance, nitrogen (urea) and sulphur to the alkali solution before treatment.

6.1.3.3. Cooperative plants for straw treatment

Not all farmers had sufficient water resources for treatment of straw according to this method. For efficient washing the demand was great. Each kg of dry straw required 40-50 l water for soaking and washing. Cooperative treatment plants placed near streams sol-

114

ved this problem, although transportation to and from the plants represented some cost. The first cooperative plant was built as early as 1944. It comprised many vats and in principle was operated in the same manner as an on-farm plant. However, the heaviness of the work caused loss of interest in this type of plant.

In 1954-55 NOFO (Norwegian Feed Conservation Association) found a system that solved the problem, and in the next five years some 7,000 farmers became members of such cooperatives (see Homb,1956). The main features of the system were: 1) only two vats, 2) six or eight bales in a chained bundle were transported from the lorry to the vat by means of a traverse crane. The chain was not removed from the bales during soaking. The same equipment was used for moving the treated straw from vat to lorry. 3) An extra tank was constructed for preparing the NaOH solution and incorporated automatic dosing with NaOH and water to bring the reused lye to the correct concentration of NaOH. 4) Washing-out the residual lye by an automatic siphon-system. The vat was filled with water and emptied several times (Figure 6.1.3).

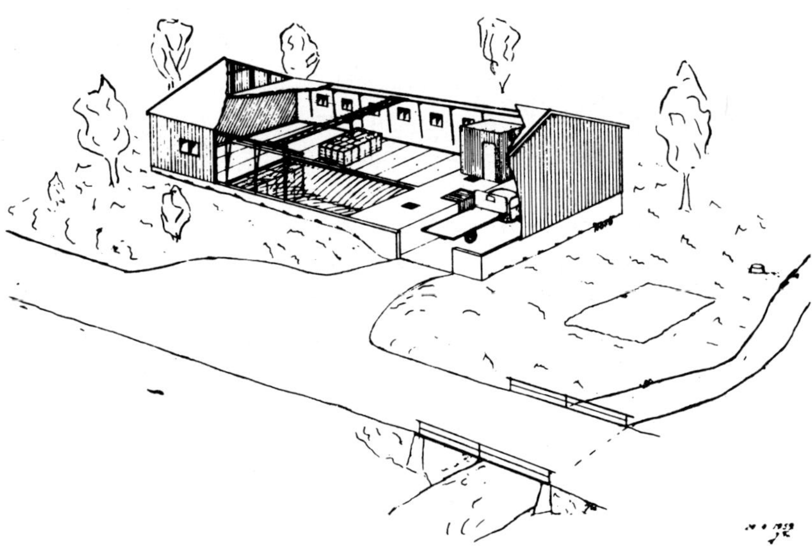

Figure 6.1.3. Cooperative factory for treating straw in Norway, planned by NOFO

6.1.3.4. <u>Feeding experiments and practical experience with Beckmann-treated staw</u>

Digestibility trials with sheep during the 1940's showed almost
the same coefficients as found by Fingerling, i.e. 66-68 per cent
for organic matter (Homb, 1948, 1956). Based on two production experiments with lactating cows an energy value of 0.73 FU per kg
DM was calculated. With an averge dry matter content of 17.5 per
cent, 1 kg treated straw is equal to 0.13 FU (later calculated to
1.7 MJ ME). Growth studies with young cattle and lambs have confirmed this value. Treated straw has been used as a part of the
rations for dairy cattle, beef cattle and sheep. During the war
horses were also fed treated straw (Hvidsten, 1945). Homb (1956),
Homb et al. (1977) and Jackson (1978) have reviewed Norwegian feeding practice with emphasis on the use of Beckmann-treated straw.

During the 1970's several restrictions on the pollution of
streams were enforced. Production of wet-treated straw therefore
declined (Figure 6.1.4). To make up for the deficit several dry
methods were introduced. Of these, treatment with ammonia has had
the most rapid growth, reaching about 100,000 tonnes of straw in
1981-82. Relatively small amounts of straw have been dry-treated
with NaOH.

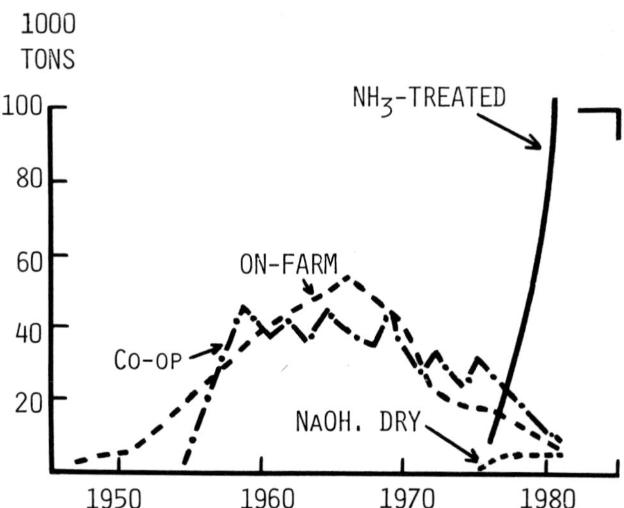

Figure 6.1.4. Yearly amounts of straw treated in Norway in thou-
sand tons (Presthegge, 1982)

6.1.3.5. <u>Water-saving modifications</u>

Looking back one might look at these as preparatory steps towards closed systems (see section 6.1.4). This applies to experimental work by Fyrileiv and Torgrimsby (1954) in Norway, Lampila (1963) in Finland, and Baludan and Piatkowski (1972) in DDR (see also Arnason and Sundstøl, 1978).

6.1.4. CLOSED SYSTEMS

Some of the methods cited in 6.1.3.5 were small modifications of the original Beckmann system. It is perhaps arguable that the closed systems to be discussed here should also be classified as modifications of Beckmann's original procedure. There is an important similarity in that the NaOH concentration is usually set at about 1.5 percent. Treatment at environmental temperature is another similarity. A closed system is one where treatment occurs without leakage to the environment. Of course, this system is also water-saving.

In 1971 Torgrimsby constructed a pilot system consisting of four vats with internal communication. By means of a rather complicated procedure he achieved a product with a high digestibility containing little NaOH (0.75 g NaOH per kg treated straw). Further details of this modification have been described by Arnason and Sundstøl (1978), and Jackson (1978). Wethje made another modification in the system on his farm in Sweden using five vats for straw treatment (see Arnason and Sundstøl, 1978, Theander, 1981).

Both modifications, which have several similarities, might also be considered steps towards simpler closed systems. Both are probably too complicated for modern farmers. However, from resource and biology points of view, they seem well founded.

6.1.4.1. <u>The circulation method (formerly Boliden process)</u>

This process was developed by Boliden A.B. in Helsingborg, Sweden and dates back to 1975. Using a closed treatment chamber straw bales are sprayed with a solution of NaOH and $Ca(OH)_2$ (15-25 g Na and 10-15 g Ca per kg DM straw). The next step is to spray a neutralizing agent, phosphoric acid, over the bales (Figure 6.1.5). After the excess liquid has been allowed to drain off, the bales are ready for feeding. Further details have been described by Mattsson and Lagerström (1976), Homb et al. (1977), Lagerström and Mattsson (1978, 1981), Arnason and Sundstøl (1978) and Arnason (1978, 1980).

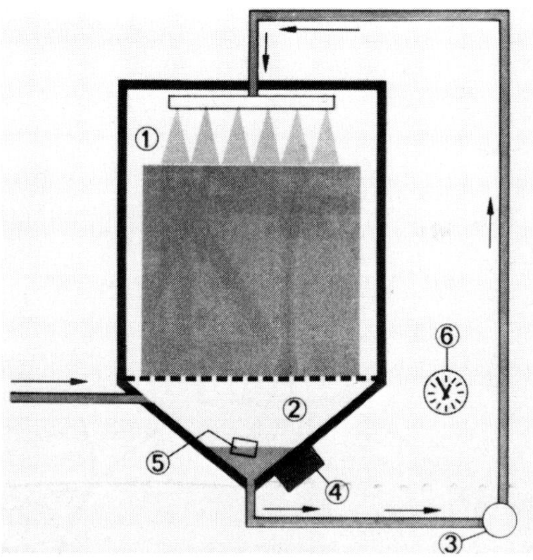

Figure 6.1.5. Facilities for treating straw according to the circulatio method (Lagerström and Mattson, 1981) (Boliden method)
1. Insulated chamber with a tightfitting door, a bottom drain, and a movable spraying device.
2. Container collecting excess solution from the chamber and serving as a sump.
3. A centrifugal pump transporting the solution from the sump to the spraying device.
4. A heater controlled by a thermostat.
5. A device for automatic adjustment of the water intake to variations in straw weight. A maximum concentratio of alkali is assured at any time.
6. A timer to control the circulation and movement of th spraying device.

Several other nutrients might be added to the neutralizing solution, for example trace elements, urea, molasses and vitamins. The method has been used for producing treated straw for digestibility trials and growth experiments with heifers (Arnason, 1980).
OM digestibility using sheep was found to be 66 per cent. In comparison with untreated straw heifers consumed more dry matter after treatment, thus giving a saving in concentrates. Swedish trials confirmed the digestibility found by Arnason (Den Braver and Eriksson, 1980). Metabolizable energy content was calculated to be 9.2 MJ/kg DM.

6.1.4.2. Dip treatment

This method was developed by Sundstøl and coworkers in Norway and Tanzania (Sundstøl et al., 1979; Kategile et al., 1981; Sundstøl, 1981 and Randby, 1982). To avoid excessive amounts of NaOH in the final feed it has been shown that soaking time may be shortened to 2 h, or even less than 1 h (Randby, 1982). Straw is then lifted out of the vat and the lye is drained or pumped back into the vat. Some water may be washed through the straw to fill the vat ready for the next batch. At the same time 60-65 g NaOH per kg straw is added to achieve the desired 1.5% NaOH (titrated value)

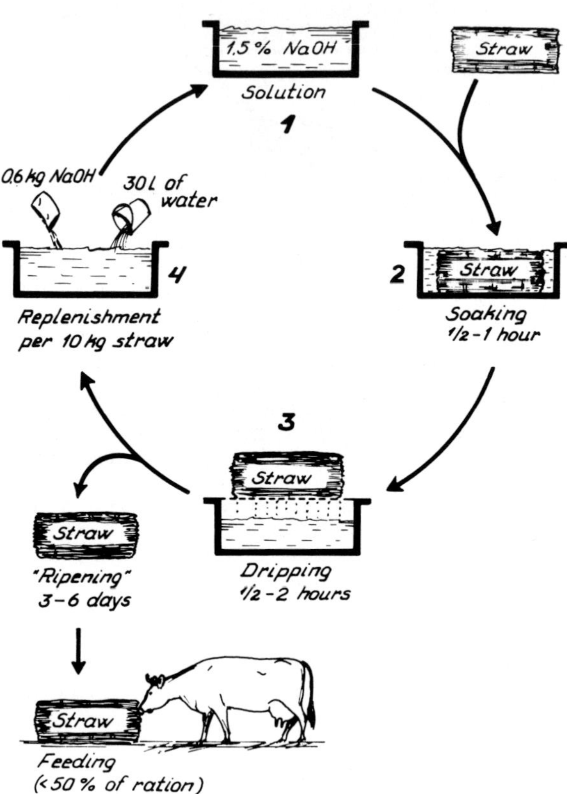

Figure 6.1.6. The principles of dip treatment of straw (from Sundstøl, 1981)

solution. Thereafter, the "ripening" process takes place, and this should last at least 3-6 days depending on temperature. There is experimental evidence for increasing digestibility during the "ripening" process, whereas palatability does not seem to decline during 7 days post-treatment. Figure 6.1.6 illustrates the principles of dip treatment (Sundstøl, 1981).

The Na content of dip-treated straw is relatively high (ca 30-35 g/kg DM) and comparable to dry-treated straw. It is not recommended therefore to feed treated straw as the only roughage. Garmo (1981) refers to a small scale experiment where dip treated straw gave somewhat higher milk yield than dry-treated straw. During the years 1980-82 another two Norwegian experiments were conducted with three groups of dairy cattle (Randby, 1982).

```
  I.   6 FU grass silage + concentrates
 II.   4 FU    "        "    +       "      + 2 FU dip treated straw
III.   2 FU    "        "    +       "      + dip treated straw
                                               ad lib.
```

During the 6 week long experimental period straw consumption increased markedly to reach 11-12 kg DM/d in the last week. Average consumption for the whole period was 8.3 kg DM/d. Concentrates were regulated according to forage intake and milk yield. There were no significant differences between groups in milk yield.

Dip treatment does not require more than two vats. An extra tank for rinse-lye may be useful. Space for storing treated straw for some days must be available. Sundstøl (1981) suggests that the temperature during storage should not be too low. Low temperatures (3-5°C) are known to delay the "ripening" process. In cold climates, such as Scandinavia, treated straw should be "ripened" in barns. Recent data indicate that the ambient temperature should be at least 10°C. The dip-treated straw may be enriched by adding, for instance, nitrogen (urea) and sulphur to the alkali solution before treatment.

6.1.5. MISCELLANEOUS ROUGHAGES

Hitherto only straw treatment has been mentioned, the reason being that most of the wet treatment covered by the literature, has dealt with straw of wheat, rye, barley and oats. Balch (1977) refers to statistics on other widespread types of straw, from maize, rice, millet, sorghum and soybean. He also includes hulls of different origin in his review. On the whole, there are meagre contributions in the literature on wet treatments of such roughages. Archibald (1924) was probably the first one to apply the Beckmann method to hulls. He had studied the German work very thoroughly and used facilities as described from Germany. His digestibility figures are very varied, due to there being small quantities of hulls in the total rations (100-150 g dried material per day for wethers). In conclusion oat hulls and barley hulls had double the feeding value of untreated hulls after treatment in a solution of 1.5% NaOH and washing. The final product was dried gently. Treatment with a one per cent solution gave a considerably lower digestibility. Rice straw was also positively affected by the same treatment, but reached much lower levels of digestibility. The results with cottonseed hulls and flax shives were negative.

More recently El-Shazly and Naga (1981) have published excellent results from wet treatment (modified Beckmann) of rice straw, OM digestibility being raised from 52.5 to 74.9%. Similar figures were obtained with the Boliden method, while dry treatment gave a less positive response.

Said (1981) in his review refers to East-African digestibility trials with maize stover treated by sprinkling with NaOH solution and left to ripen overnight. Judged by growth rate of sheep, treated maize stover is comparable with untreated *Chloris gayana* hay.

Kategile et al. (1981) treated maize stover in a solution of $Ca(OH)_2$ and NaOH. Compared to a solution of NaOH the results were negative. In their estimate dip treatment with an allowance of 100-150 g of NaOH per kg DM and a replenishment of 40-60 g NaOH per kg would be considered optimal.

Urio (1981) treated maize cobs with a 1.5 per cent NaOH-solution for 48 hours. After dripping and partial washing, cobs were ground in a hammer mill, air-dried, and fed to goats and sheep. This treatment led to an increased digestibility of dry matter of the ration (including a fixed amount of concentrates) from 55.6 to 64.1 in goats, and from 56.5 to 59.3 in sheep. Gain in live weight, carcass weight and carcass fat percentage of goats also responded positively due to the "dip and drip" treatment of maize cobs. A third group, fed *Chloris gayana* hay did not differ significantly from the group on treated cobs. However, a similar experiment with sheep showed no significant differences in growth rate etc. between untreated and treated cobs.

Tropical grasses seem to respond less to NaOH-treatment than temperate grasses, according to Owen (1981). This conclusion was drawn on the basis of laboratory studies, and he stated that further studies with animals were needed.

6.1.6. DISCUSSION

Wet treatment methods usually require more and heavier hand work than the modern dry-treatment systems (Jackson, 1978). On the other hand, from the biological and nutritive point of view wet treatment has given higher digestibility figures than dry methods (Homb et al., 1977; Jackson, 1978; Sundstøl et al., 1978; Garmo, 1981; Westgaard, 1981). It is reasonable to believe that moisture itself disrupts lignification. Furthermore, the original Beckmann method, with washing, presupposes higher dosage of NaOH than has been recommended in dry treatment. Several investigations with dry treated straw have shown in vivo digestibility increases up to 40 or 50 g NaOH per kg straw, while in vitro figures continue to increase with higher NaOH levels. If water supply is abundant, and no harm to the surroundings is apparent, the original Beckmann system should be chosen. One of the reasons for this is the low amount of Na in the feed. On-farm treatment is more favourable than treatment in cooperatives, because of the cost of transport.

If any release of material to the environment is prohibited, then different alternatives of closed and wet methods are available, together with a number of dry methods. Of considerable interest is the amount of Na in the final product. The following figures are typical:

	g Na/kg DM
Beckmann-treatment with washing	4-6
Torgrimsby and Wethje-method	20
Boliden-straw	20
Dip treatment	30-35
Dry methods	30-35

All systems except the original Beckmann method result in substantial amounts of Na (or NaOH) remaining in the straw. There are different opinions as to whether Na^+ or OH^- (or both) represent a hazard to the animals. Most probably, both should be taken into consideration. Owen (1981) concludes that treated straw (dry) fed with low-pH feedstuffs such as grass silage or concentrates (acidosis promoting) does not need neutralizing agents like HCl. Heavier kidneys and lower blood-Ca and Mg have been related to Na-load, because Na excretion also drains other minerals (see Kristensen, 1981). Kristensen suggests that extra supplementation of certain minerals may be helpful (K, Cl, Mg). Danish work by Stigsen 1975) disclosed relatively high amounts of net base in the urine when cows were fed dry-treated straw. This was counteracted by adding hydrochloric acid. Without neutralizing with strong acid, a constant compensated alkalosis occurred in cows fed 10 kg per day of straw treated with 5o g NaOH/kg. These experiments lasted only a few weeks. More recent results from Denmark have been reviewed by Rexen and Bach Knudsen (section 12.5.3). However, long term experiments are needed to clarify this point, and studies are currently in progress in Denmark.

It seems advisable that high-straw rations with treated straw containing about 30 g Na/kg DM should be avoided. Kristensen (1981) states that the rations of dry NaOH-treated straw to cows should be limited to 7 kg DM, besides liberal amounts of concentrates. As mentioned earlier Randby (1982) gave somewhat higher amounts of dip-treated straw to one of the groups. All in all, it seems desirable to wait for additional evidence from long-term studies before recommending rations of this size for use in practice. In agreement with the view of Pigden (1981) NaOH-treated roughages might make up 20-40 per cent of production rations. Even at this level considerable amounts of conventional forages can be saved.

In contrast, Beckmann-treated straw, efficiently washed, and NH_3-treated straw may be fed to appetite without causing any harm.

Of the different non-polluting wet methods, the dip treatment requires the least expensive facilities. Only two vats with a simple arrangement for lifting and transporting the straw are needed. Capital cost is highest for the Boliden method. The Torgrimsby and Wethje methods require several basins and are somewhat complicated. Otherwise, they are well suited to practical use. For developing countries the dip treatment seems to hold much promise

for improving the nutritive value of low-quality roughages. This
method is also promising for industrialized countries, if recent
results hold true in long-term experiments.

6.1.7. SUMMARY

An historical review is given of wet treatment of straw from
the old German work, prior to developing the Beckmann method, to
the recently developed closed systems. The use of the Beckmann
method and its modifications in Scandinavian countries covers a
great part of the review. The non-polluting closed systems such as
the Boliden method (Circulation method) and dip treatment are considered. A brief discussion of treating oat hulls, barley hulls,
rice straw, maize stover, maize cobs and tropical grasses is included. The review is concluded by a general discussion of wet and
dry methods.

6.1.8. REFERENCES

Archibald, J., 1924. The effect of sodium hydroxide on the composition, digestibility and feeding value of grain hulls and other fibrous material. J. Agric. Sci. 27: 245-265.

Arnason, J., 1978. Progress of straw treatment in Norway. Paper presented at 4th straw utilization conference, Oxford, 30 Nov. - 1 Dec. 1978.

Arnason, J., 1980. Halm våtlutet og nøytralisert etter Boliden-metoden som fôr til kviger. Meld. Norges Landbr.høgsk. 59: (27).

Arnason, J. and Sundstøl, F., 1978. Våtluting av halm. Beckmann's metode og dens utvikling til lukkede systemer. NJF-halmseminar Denmark 28-31 March 1978.

Balch, C.C., 1977. The potential of poor-quality agricultural roughages for animal feeding. FAO Animal Production & Health Paper 4, Rome.

Baludan, G. and Piatkowski, B., 1972. Untersuchungen zum Aufschluss von Getreidestroh. I. Behandlung mit Natronlauge und die anschliessende Neutralisierung. Arch. Tierernähr. 22: 485-492.

Beckmann, E., 1919. Preussiche Akademie der Wissenschaften, Berlin. Stizungsberichte 275.

Blakstad, L., 1943. Halm-luting i praksis. Tidsskr. f.d. norske landbr. 50: 54-59.

Breirem, K., 1939. Halmproblemet. Tidsskr. f.d. norske Landbr. 46: 157-166.

den Braver, E. and Eriksson, S., 1980. Näringsvärdet hos halm lutad enligt Bolidenmetoden. Rapport 48 Avd. for husdjurens näringsfysiologi, Sveriges lantbruksuniversitet.

El-Shazly, K. and Naga, M.A., 1981. Supplementation of rations based on low quality roughages under tropical conditions. Proc. Workshop at Arusha, Tanzania, 18-22 Jan. 1981: 157-170.

Ferguson, W.S., 1943. The digestibility of straw pulp. J. Agric. Sci. 88: 174-177.

Fingerling, G., 1919. Fütterungsversuche mit aufgeschlossenem Stroh. Landw. Versuchsst. 92: 1-56.

Fingerling, G:, 1924. Die Ernährung der landwirtschaftlichen Nutztiere. Paul Parey, Berlin.

Fingerling, G. and Schmidt, K., 1919. Die Strohaufschliessung nach der Beckmannschen Verfahren. I. Einfluss der Aufschliessungszeit auf den Umfang der Nährwerterschliessung. Landw. Versuchsst. 94: 115-152.

Fingerling, G., Schmidt, K. and Schuster, A., 1923. Strohaufschliessung nach der Beckmannschen Verfahren. II. Einfluss der Laugenmenge auf den Unfang der Nährwerterschliessung. Landw. Versuchsst. 100: 1-19.

Fyrileiv, E. and Torgrimsby, J., 1954. Vannforbruket ved luting av halm kan reduseres. Norsk Landbr. 20: 55-57.

Garmo, T.H., 1981. Milk production of cows after fed on NaOH- and NH$_3$-treated barley straw. Proc. Workshop at Arusha, Tanzania, 18-22 Jan. 1981, 113-117.

Godden, W., 1920. The digestibility of straw after treatment with soda. J. Agric. Sci. 10: 437-459.

Hesthamar, T.B., 1940. Luting av halm for å heve fôrverdien. Norsk landbr. 6: 518-521.

Hesthamar, T.B., 1943. Halmluting frå teoretisk og forsøksmessig synspunkt. Tidsskr. f.d. norske landbr. 50: 113-120.

Homb, T., 1948. Fôringsforsøk med lutet halm. Meld. Norges Landbr. høgsk. 28: 277-365.

Homb, T., 1950. Arbeidsstudier i 8 fjøs på Østlandet. Forskn. forsøk i landbruket 1: 447-471.

Homb, T., 1956. Norwegische Erfahrungen bei der Strohaufschliessung nach dem Beckmannschen Verfahren. Futterkonservierung 2:
129-146.

Homb, T., Sundstøl, F. and Arnason, J., 1977. Chemical treatment
of straw at commercial and farm level. FAO Animal Production &
Health Paper 4, Rome.

Hvidsten, H., 1945. Erfaringer med trecellulose og luta halm til
hester. Meld. Norges Landbr.høgsk. 25: 210-252.

Hvidsten, H. and Simonsen, H., 1952. Undersøkelser over luting og
utvasking av hel halm. Tidsskr. f.d. norske landbr. 59: 85-93.

Hvidsten, H. and Gloppe, K.E., 1953. Halmluting. En enkel metode
til bestemmelse av lutstyrken og utvaskingsgraden av halmen i
praksis. Norsk Landbr. 19: 122-123.

Jackson, M.G., 1978. Treating straw for animal feeding. FAO Animal
Production & Health Paper 10, Rome.

Kategile, J.A., Urio, N.A., Sundstøl, F. and Mzihiriwa, Y.C.,
1981. Simplified method for alkali treatment of low-quality
roughages for use by small-holders in developing countries.
Anim. Feed Sci. Technol. 6: 133-143.

Kellner, O., 1905. Die Ernährung der landwirtschaftlichen Nutztiere. Paul Parey, Berlin.

Kellner, O. and Köhler, A., 1900. Untersuchungen uber den Stoff-
und Energie-Umsats des erwachsenen Rindes bei Erhaltungs- und
Produktionsfutter. Landw. Versuchsst. 53: 1-474.

Kristensen, V.F., 1981. Use of alkali-treated straw in rations for
dairy cows, beef cattle and buffaloes. Proc. Workshop at Arusha, Tanzania, 18-22 Jan. 1981: 91-106.

Lagerström, G. and Mattsson, A., 1978. Boliden-metoden for lutning
och näringsberikning. NJF-halmseminar, Denmark 28-31 March
1978.

Lagerström, G. and Mattsson, A., 1981. Straw treatment according
to the Boliden metod. Translated from Foder-Journalen, No. 1-2,
1981. Mimeographed paper.

Lampila, M., 1963. Experiments with alkali straw and urea. Ann.
Agric. Fenniae, 2: 105-108.

Lehmann, F., 1891. Der Nährwert der Cellulose. Landw. Versuchsst.
38: 337-338.

Matsson, A. and Lagerström, G., 1976. Boliden-metoden for lutning
och näringsberikning av halm. Foderjournalen 15: 77-78, 80-81.

Owen, E., 1981. Use of alkali-treated low quality roughages to
sheep and goats. Proc. Workshop at Arusha, Tanzania 18-22 Jan.
1981: 131-150.

Pigden, W.J., 1981. Use of low quality forages in the future needs
for research and implenation. Proc. Workshop at Arusha, Tanzania 18-22 Jan. 1981: 201-213.

Presthegge, K., 1982. NOFO A/S Norsk Fôrkonservering 50 år, 1932-
82. Landbruksforlaget, Oslo.

Randby, Å., 1982. Dyppeluting - en ny,- lovende våtlutingsmetode
for halm. Kraftfôrnytt 24 (2): 7.

Said, A.N., 1981. Sodium hydroxide and ammonia-treated maize stover as a roughage supplement to sheep and beef feedlot cattle.
Proc. Workshop at Arusha, Tanzania, 18-22 Jan. 1981: 107-112.

Slade, R.E., Watson, S.J. and Ferguson, W.S., 1939. Digestibility
of straw. Nature 143: 942-943.

Stigsen, P., 1975. Kulhydratkildens og neutralisationens betydning
for udnyttelse af natriumhydroxyd-behandlet halm hos malkekøer.
Lic. afhandl. i Kvægets fodring. Den Kgl. Veterinær- og Landbohøjskole, Copenhagen.

Sundstøl, F., 1981. Methods for treatment of low quality roughages. Proc. Workshop at Arusha, Tanzania, 18-22 Jan. 1981: 61-79.
Sundstøl, F., Coxworth, E. and Mowat, D.N., 1978. Improving the nutritive value of straw and other low quality roughages by treatment with ammonia. World Anim. Rev. (FAO) (26) 13-21.
Sundstøl, F., Urio, N.A., Garmo, T.H. and Tubei, S.K., 1979. Dyppeluting av halm. Foredrag i Halmlutingslagenes landsforening. Porsgrunn 1 Nov. 1979. Mimeographed paper 66, Dept. Anim. Nutr., Agric. Univ. Norway.
Theander, O., 1981. Chemical composition of low quality roughages as related to alkali treatment. Proc. Workshop at Arusha, Tanzania 18-22 Jan. 1981: 1-15.
Thomann, W., 1921. Vergleichende Versuche über die Zusammensetzung und Verdaulichkeit von Rohstroh und aufgeschlossenem Stroh. Landw. Jahrb. d. Schweiz 35: 667-723.
Urio, N.A., 1981. Alkali treatment of roughages and energy utilization of treated roughages fed to sheep and goats. Thesis PhD, University of Dar-es-Salaam.
Watson, S.J., 1941. Increasing the feeding value of cereal straws. J. Roy. Agric. Soc. Engl. 101 Part II: 37-43
Westgaard, P., 1981. Factors influencing the effect of alkali treatment of low quality roughages. Proc. Workshop at Arusha, Tanzania, 18-22 Jan. 1981: 29-47.
Williamson, G., 1941. The effect of Beckmann's treatment by sodium hydroxide on the digestibility and feeding value of barley straw for horses. J. Agric. Sci. 31: 488-499.

Chapter 6.2

INDUSTRIAL-SCALE DRY TREATMENT WITH SODIUM HYDROXIDE

by

Finn P. Rexen and Knud E. Bach Knudsen

Carlsberg Research Laboratory
Department of Biotechnology
Gl. Carlsberg vej 10, DK-2500 Valby
Denmark

6.2.1. INDUSTRIAL PROCESSES - HISTORICAL REVIEW

6.2.1.1. Introduction

In this compilation it is demonstrated that straw may be turned by chemical treatment into an energy rich feed for ruminants. This renewable resource may be an important feed resource especially in emergency situations when supplies of roughage and other feeds are scarce.

Such an emergency situation occurred during World War One, when the production of alkali treated straw on an industrial scale was first introduced. Considerable amounts were produced for a number of years and also wood was used as raw material. The straw/wood was cooked under pressure for several hours, then washed and dried. The final product, called fodder cellulose, was an excellent fodder, but due to its high cost, production was stopped after the war when the feed supply situation became normal. During World War Two large amounts of alkali treated straw were also produced on an industrial and farm scale, although production decreased after the war.

In the beginning of the 1970's, a revival of interest in the industrial production of straw feed was observed, and attempts were made to find new and cheaper treatments.

In 1968 a new "semi-dry" industrial process based on the production of pellets (Rexen, 1972) was introduced. Instead of eliminating the excess of NaOH by washing (as in the old Beckmann process) it was neutralized with acidic gasses thus drastically reducing the water demand.

The pelleted product had a digestibility comparable to that obtained from Beckmann-processed straw. Some of the machinery necessary for the process is standard equipment in green crop drying plants. The principle only found limited practical application.
An excellent review of the historical development of alkali treatment methods is given by Homb et al. (1977).

6.2.1.2. <u>Industrial plants for dry alkali treatment</u>

The first commercial plants for dry alkali treatment of straw were built in the beginning of the 1970's in England and Denmark, where the price of traditional energy feed sources such as barley had begun to rise. Alkali-treated straw is a low protein energy feed, which to a great extent can replace grain in feedstuff rations. The price obtained for lye-treated straw will therefore depend upon the grain price. If grain is cheap it will not be economical to buy alternative feeds such as lye-treated straw. The price of fodder grain in England and Denmark which rose constantly in the period 1972-75, gave rise to the building of a number of processing plants in both countries.

During the drought of 1976, which resulted in scarce feed supplies and high prices, interest in alternative feeds such as lye-treated straw was renewed and several plants were built in different European countries.

It was natural for the grass drying industry in Denmark and to some extent in England, to take up the alkali treatment of straw in the beginning of the 1970's. These industries had economical problems due to the increasing oil prices and therefore they were most interested in finding an alternative use for the machinery. A grass drying plant in Northern Europe normally only produces grass pellets for a few months of the year and stands idle the rest of the time. All grass drying plants have pellet presses, so it was natural to look for an application which could take advantage of these presses and some of the other existing process equipment. By taking up the production of straw pellets it would make use of the equipment all year.

Straw treatment plants have also come to be integrated into existing feed mills in order to produce composite feeds (full-feed or concentrated feed) for ruminants, based on alkali treated straw.

At the end of 1977, 11 industrial plants were operating in England with a total capacity of 200,000 to 250,000 tonnes of alkali-treated pellets per year (Wilson and Brigstocke, 1977), while in Denmark the production was approximately 50,000 tonnes per year. Production, however, has decreased ever since in both countries due to price competition from such alternative grain substitutes as imported citrus waste and tapioca meal and to a general trend to decrease the consumption of industrially produced feed.

One of the main producers today is Poland with 20 factories in operation and a total capacity of approximately 200,000 tonnes per year of composite feed containing 30-40% alkali treated straw. Other socialist countries including the USSR, DDR, Czechoslovakia, Hungary, Yugoslavia and Bulgaria have erected alkali treatment plants.

In Western Europe, apart from England and Denmark, plants are operating in Sweden, Norway, Finland, France, Greece, Spain, Portugal, Holland and Austria.

Outside Europe, alkali treated straw is produced industrially in the USA, Australia, Israel, the Phillippines, Thailand and Kenya.

6.2.2. EXPERIMENTAL BACKGROUND

6.2.2.1. Introduction

The chemical reaction between sodium hydroxide and straw depends on reaction-conditions such as temperature, pressure, alkali concentration, and reaction time.

Application of high temperature and pressure enables a reduction in the addition of chemicals, and the reaction time is reduced considerably compared with reactions at low temperature and atmospheric pressure (farm scale processes).

The first industrial plants for production of fodder from straw and wood operated with elevated pressure and temperature in alkaline water suspensions (10 kg of water per kg material). The product obtained had an excellent feeding value, but the process was expensive. The reaction time was 3-4 hours.

The first farm scale processes, such as the Beckmann process operated at atmospheric pressure and ambient temperature and required relatively large amounts of chemicals (10-12% on straw basis) and a comparatively long reaction time (12-14 hours). The process had no effect on wood.

The modern industrial dry treatment methods operate with extremely short reaction times ($\frac{1}{2}$ to 1 min.) under very high pressure. The demand for process-water is minimal in contrast to earlier methods, which were both water consuming (up to 40-50 l of water was used per kg straw) and, consequently, polluting. In the older processes the treated product was thoroughly washed in order to remove excess sodium hydroxide, while there is no washing step in the dry treatment methods. The unreacted alkali and the sodium ion from the reacted alkali remain in the product thus limiting the use of alkali treated straw in practical feeding.

6.2.2.2. Experimental work

Since the late 1960's, emphasis has been placed on developing dry methods which operate with small quantities of water and do not cause pollution problems. In other words, the aim is to add precisely as much water and NaOH as is necessary to avoid excess of chemicals and a washing of the treated product.

6.2.2.2.1. Demand for process water

As early as in 1933, experiments performed by Schwalbe indicated that the large amount of process water used in the Beckmann process is not necessary, provided the NaOH is properly mixed into the material. Schwalbe used wood as raw material and a pan mill (Kollergang) was used as the reaction chamber. The pan mill assured an intimate mixing of the wood and lye as well as subjecting the material to strong mechanical forces. Schwalbe made trials with both sodium hydroxide and calcium hydroxide and in order to increase the swelling ability of the cellulosic fibres he added sugars.

Wilson and Pigden later (1964) carried out experiments which confirmed that a satisfactory effect on the digestibility of straw could be obtained by the addition of a small amount of water. Ground wheat straw was mixed with a solution of 30% water and various amounts of NaOH, and the mixture was allowed to react for 13-21 days. It was found that the in vitro digestibility increased in accordance with the addition of increased amounts of NaOH (up to 9 kg per 100 kg straw). No washing was carried out.The dry matter digestibility increased from 30-40% to a maximum of 70-80%.

Similar experiments were later carried out in farm scale by Chandra and Jackson (1971) and Piatkowski et al. (1974) which showed effects similar to those found by Wilson and Pigden (1964).

On the other hand other experiments have shown that the amount of process water used may influence the effect of treatment. Donefer et al. (1969) found that at a fixed NaOH percentage (based on straw) the digestibility increased with the addition of increasing amount of water (up to 120 ml per 100 g straw). Boludan and Piatkowski (1972) concluded from their experiments that the relationship between liquid (lye solution) and straw should be at least 1:1, as this was the maximal amount which could be absorbed by the straw. Such amounts of water are not theoretically necessary for the chemical reaction, and the explanation for the above mentioned findings may be that the water helped to distribute the NaOH evenly on the surface of the straw. Later findings, (Junker, 1976; Friis Kristensen et al., 1978) have shown that with an efficient distribution of the NaOH on the straw surface the reaction can take place with small amounts of water (10%).

6.2.2.2.2. Temperature, pressure and lye dosage

Reaction speed and the duration of the reaction, are strongly influenced by temperature and pressure. The effect of temperature has been investigated by Ololade et al. (1970). Their results clearly indicate that temperature has a considerable influence on the reaction time and digestibility. A digestibility of 63% was achieved by using 4% NaOH on straw weight basis (distributed in a diluted solution) either at 130°C for 5 min. or at 100°C for 45 min. The same effect could only be obtained at lower temperatures by increasing the amount of NaOH and/or the reaction time.

Laboratory experiments were carried out by Rexen et al. (1975) to investigate the relationship between pressure, temperature, and percentage of NaOH. A concentrated NaOH solution (33%) was used (see Figure 6.2.1). Increasing any of these three parameters (at temperatures above 20°C) produced an increase in digestibility. The reaction time was fixed at 1 min.

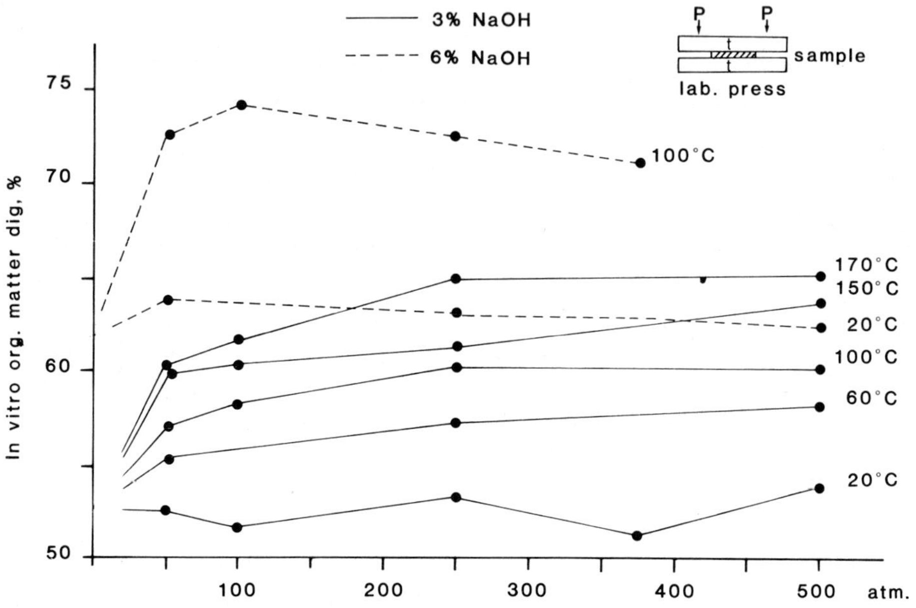

Figure 6.2.1. Effect of pressure and temperature upon digestibility (Rexen, 1975)

Junker (1976) has also investigated the influence of pressure, temperature, and amount of NaOH on in vitro digestibility (Table 6.2.1). Junker found that an increase in temperature from 60°C to 150°C had no positive effect on in vitro digestibility which is not in accordance with the findings by Ololade et al. (1970) and Rexen et al. (1975).

Each figure is an average of the results from all experiments, where a particular parameter was held constant. Thus the figure for digestibility by treatment at 60°C is an average for all samples treated at 60°C irrespective of the variation in pressure and reaction time etc.

Table 6.2.1. Influence of different parameters on the in vitro digestibility (Junker, 1976)

Parameter		OMD
Temperature	60°C	71
	150°C	68
Pressure	400 bar	69
	900 bar	68
Reaction time	1 (2) min	68
	3 (4) min	69
	7 min	69
NaOH added	2%	58
	4%	64
	6%	71
	8%	74
	10%	77

It can be concluded from Figure 6.2.1 and Table 6.2.1, that the amount of NaOH added has a higher impact on digestibility than pressure and temperature. It should be noted, however, that an increase in pressure and temperature reduces the reaction time considerably compared with farm scale processes operating at ambient temperature and pressure.

The industrial dry alkali treatment method, described by Rexen (1972), Junker (1976), Wilson et al. (1976) and Tesic (1977) is based on the use of a ring-die press or a piston press as reaction chamber. The lye impregnated straw is subjected to both high pressure and high temperature in the press, but retention time in the press, and thus the reaction time, is very short, however still adequate for an efficient treatment.

Wilkinson (1978) has compared the relationship between level of NaOH (g/100 g) (x) and digestible organic matter (OMD) in vitro (%) for both factory and farm scale treatment of chopped straw. The following relationships were reported:

Factory scale process:

1. $OMD = 43.9 + 5.28x - 0.23x^2$ Wilson and Brigstocke (1977)

2. $OMD = 50.1 + 4.02 x$ Rexen and Vestergaard Thomsen (1976)

Farm scale process:

3. $OMD = 42.3 + 7.01 x - 0.34x^2$ Wilkinson and Gonzales Santillana (1978)

The three equations are equivalent; for straw treated with 5 g NaOH per 100 g straw DM, OMD is predicted to be 65%, 70% and 69% by equations 1, 2 and 3 respectively. The influence of the level of NaOH added on in vitro and in vivo digestibility is further discussed in section 6.2.5.

6.2.3. THE EFFECT OF DRY ALKALI TREATMENT ON VARIOUS RAW MATERIALS

Most of the experimental work on dry alkali treatment has been carried out on barley and wheat straw, but also other lignocellulosic materials have been studied. Table 6.2.2. shows the results from an experiment carried out with various materials in a semi industrial alkali treatment plant with a capacity of 500 kg/h (Rexen, 1979). The raw materials were chopped and treated with 4-5% NaOH. Pellet diameter was 14 mm.

There are substantial differences in the effect of treatment on various raw materials. The best effect seems to be achieved on ragi and sorghum straws.

Friis Kristensen et al. (1978) investigated the effect of NaOH treatment on 4 of the most common straw sources in Europe (rye, wheat, oat and barley). The results indicated that the in vitro digestibility is increased linearly with increased addition of NaOH up to 6%, and that the effect of treatment is independent of the type of straw used.

Fodder crops such as grass, alfalfa, and maize (stems) decrease in digesibility after the maturity, and therefore a delayed harvest will reduce the fodder value. Such overripe fodder plants could be improved by alkali treatment. Laboratory experiments with dry alkali treatment (5% NaOH) in a hydraulic press carried out by Tesic (1977) showed that grass and maize straw in different stages of maturity responded well to the treatment, while the effect of processing on the value of the whole chopped maize plant was limited. The increase in digestibility seemed to be independent of the maturity stage. The alkali treatment had no effect on alfalfa.

Mwakatundu and Owen (1974) have investigated the effect of alkaline treatment (wet treatment at ambient temperature in 24 hours) on ryegrass cut at weekly intervals over 14 weeks.

Table 6.2.2. The effect of alkali treatment on various raw materials (Rexen, 1979)

Raw material	Density of pellets bulk kg/m^3	Before treatment	After treatment	Increase
		----------- IVOMD, % ---------		
Barley straw	450	45.3	66.3	21.0
Oat straw	416	54.5	73.6	19.1
Wheat straw (Danish)	596	51.1	64.1	13.0
Wheat straw (Indian)	560	39.1	70.1	31.0
Rye straw	530	46.7	74.5	17.8
Rice straw	380	46.3	74.2	27.9
Rye grass	400	49.7	72.9	23.2
Ragi	600	41.1	72.9	31.8
Sorghum straw	456	38.1	68.4	30.3
Maize stems	424	56.1	74.6	18.5
Bagasse	268	32.0	59.1	27.1

Digestibility of material harvested in the first 5 weeks showed little improvement with alkali treatment, but thereafter treatment increased digestibility significantly, and for a given material the response to treatment increased with increasing concentration of sodium hydroxide.

6.2.4. INDUSTRIAL SCALE TREATMENT - TECHNICAL DESCRIPTION

The industrial dry treatment process is simplified in a flow chart (Figure 6.2.2). The process may include the following unit operations:

Disintegration (chopping, grinding) - drying - addition of lye - pressing - cooling - storage.

A plant may be designed to produce from 2 to 5 t/h of pelleted straw. When the moisture content of the raw material is lower than 20%, the drier stage may be deleted, and the chopper may feed the straw either directly to the weigher or through the grinder, if ground straw is preferred.

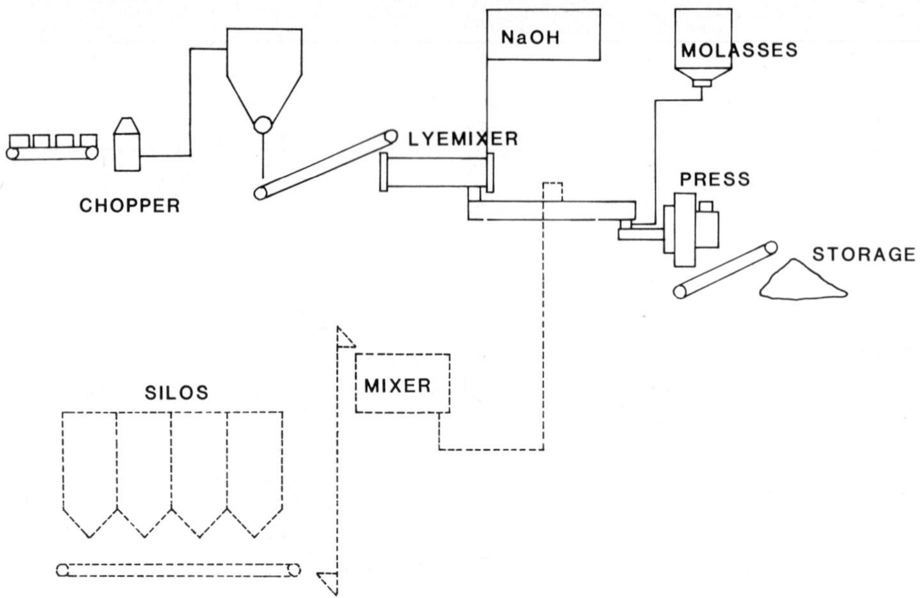

Figure 6.2.2. Industrial dry treatment process, flow chart

Raw material can be delivered to the plant in different forms either loose or as small or large bales.

After disintegration (and drying) the material is transported via a band weigher or volume measurer to the lye mixer, where lye is added at prescribed rate controlled by the straw capacity measurer. When the impregnated straw leaves the mixer, it still looks dry, but the colour is much brighter and more yellow than the raw material.

The impregnated straw is conveyed to the pellet mill and pelleted and cooled. Molasses, fat, and dry feed ingredients may be added before pressing to produce a composite feed directly.

In the following text each operation will be discussed in detail.

6.2.4.1. <u>Disintegration</u>

The raw material is normally delivered to the factory in bales which must be disintegrated before processing.

Bales are usually tied with twine, plastic ties or wire. The wire must be removed, but the sisal twine and plastic ties may be disintegrated and enter the plant together with the straw. According to Wilson et al. (1976) the sisal twine and plastic ties are totally inert and completely harmless to livestock. However, the heavier grades of plastic twines, which are used on larger bales have to be manually cut and removed.

The straw length has to be reduced in order to achieve a uniform flow and efficient lye impregnation. The plant capacity (especially the press capacity) is also dependent upon the rate of disintegration. For some fodder applications (as part of a complete feed) a coarse structure is advantageous, while for other applications (in concentrated feeds) ground straw is preferred.

Friis Kristensen et al. (1978) described the effect of the rate of disintegration upon pellet density, enzyme solubility and the amount of remaining NaOH in the pellets. The straw was ground in a hammer mill with various screen sizes (3.8, 12 and 25 mm) or chopped to a length of 20-100 mm. From these experiments it was concluded that the degree of disintegration had little influence on increase in digestibility and excess of NaOH in the pellet, and there was no effect on pellet density. Technical problems such as dust formation and uneven lye distribution occurred when finely ground straw was used.

Tesic (1977, 1982) has found that pellets produced from short chopped straw and ground straw have a lower density than pellets produced from long straw, provided that the straw has a low moisture content. At higher moisture contents only small differences in pellet density were observed.

Tesic's experiments showed that the pellet strength was higher in pellets from long chopped straw than in pellets from ground straw. He also found that the pellets from ground straw had a lower durability (1-3%, measured by the ASAE standard) than pellets from chopped straw. This resulted in a higher formation (1-3%) of fines during a mechanical treatment (tumbling) under standard conditions.

The pellets from ground straw had a higher digestibility (3%) than pellets from chopped straw. This result is not in accordance with the findings by Friis Kristensen et al. (1978). The reason could be that a concentrated lye solution was used in the latter experiments, while Tesic used NaOH powder which is easier to mix with ground straw than with chopped straw. In contrast, Friis Kristensen et al. (1978) pointed out that it was difficult to distribute a lye solution evenly on ground straw.

The initial moisture content in the straw bales influences the capacity of the disintegrator and the quality of the disintegrated straw. Experience in practice shows that capacity may decrease from 5 t/h to 1.5 t/h, if moisture content is increased from 15% to 25%.

The moisture content of the straw bales should therefore not exceed 17-20%. Disintegration at higher moisture levels is very power consuming and will often result in a heterogeneous product with low volume weight leading to reduced capacity and causing stoppage in conveyers, cyclones, rotating valves etc.

The most commonly used disintegrators in the straw feed factories are hammer mills and tub grinders.

A tub grinder (Plate 6.2.1.) consists of a bottom frame, a tub and the grinding cyclinder. The tub, approximately 3 m in diameter, makes it possible to load the machine with different bale sizes and thus also large round or square bales. The tub rotates and leads the material to the grinding system which is placed in the stationary bottom of the tub. The grinding system consists of a heavy rotor equipped with hammers and a sieve.

Many plants have installed both tub grinders and hammer mills. While the tub grinder is used for chopping long straw, the hammer mill is often placed after the drier and used for grinding chopped straw, when a fine pellet structure is required.

6.2.4.2. <u>Drying</u>

It is not always possible for factories to purchase straw with a moisture content under 17-20% which is the upper limit for a good chaff quality and a "smooth" flow through the system. It is therefore sometimes necessary to dry the straw. This is normally carried out in a semi-pneumatic drum drier, where the material comes in direct contact with oil heated air. Although it is very expensive to dry straw, some factories still prefer to dry all of the straw, in order to get a more homogenous flow of raw material to the lye mixer and the press thus securing an increase in capacity.

Plate 6.2.1. Tub grinder (Desmi)

6.2.4.3. <u>Addition of lye</u>

6.2.4.3.1. <u>NaOH - quality and concentration</u>

NaOH (or caustic soda) is made by electrolysis of brine using both mercury cells and diaphragm cells. Mercury cell caustic soda liquor is a high grade product which is low in chloride and chlorate impurities and is known as rayon grade. It is produced directly by amalgam decomposition at the strength at which it is sold, and usually contains 46-47% NaOH.

Diaphragm cells yield a weak caustic liquor, which is further processed (evaporation) to yield caustic soda liquor which can vary from 27% (winter lye) to 47% in concentration. Caustic soda is also marketed as fused solid or solid flakes although most factories prefer the cheaper NaOH solutions which can be delivered in bulk, avoiding redissolving equipment. As previously mentioned, the optimal addition of NaOH is 4-6% based on straw dry matter, and this ratio is used in most factories. However, the concentration of the lye solution used may vary much from one factory to another depending on the moisture content of the raw material and the concentration of lye available on local markets.

According to Tesic (1977), the optimal moisture content in the straw before pressing is about 20%. The pellet strength is reduced with lower moisture content and in cases where the moisture content is only 11-13% it is not possible to produce cohesive pellets. A moisture content above 20% often leads to a tendency to build up material in the press which may block the press completely.

Friis Kristensen et al. (1978) have tested the effect of lye concentration on digestibility. Four different concentrations were used: 16-19%, 21-23%, 26-28% and 46%, and the lye solutions were added so that the final NaOH content on straw basis corresponded to approximately 4% in each experiment. It was unexpected to find that each of the four concentrations tested gave within statistical limits the same effect on digestibility.

6.2.4.3.2. Mixing and dosing

An even distribution of NaOH on the straw surface is essential in optimizing the treatment.

In practice, screw conveyors with spray nozzles are used (molasses mixers) or mixers specially designed for addition of alkali. The principle is shown in Figure 6.2.3 (Friis Kristensen et al., 1978). The rotating shaft shown is hollow and mounted with paddles and nozzles through which the solution is sprayed into the straw. Plate 6.2.2. shows a commercially produced lye mixer.

In some factories the lye dosage is simply determined by setting the flow rate of lye at a constant figure corresponding to the expected flow rate of the straw. This is not a good method because of unavoidable variations in flow rate of straw. A better system is to continously register the straw flow rate and to use this value to regulate the addition of lye.

The amount of straw may be measured as volume per time unit or as weight per time unit. The latter is the more accurate method, because the straw volume varies with chaff length and straw quantity.

The weight of the straw may be measured continously on a band weigher which consists of a slow moving belt on which a thick bed of straw is evenly built up to ensure accurate weighing.

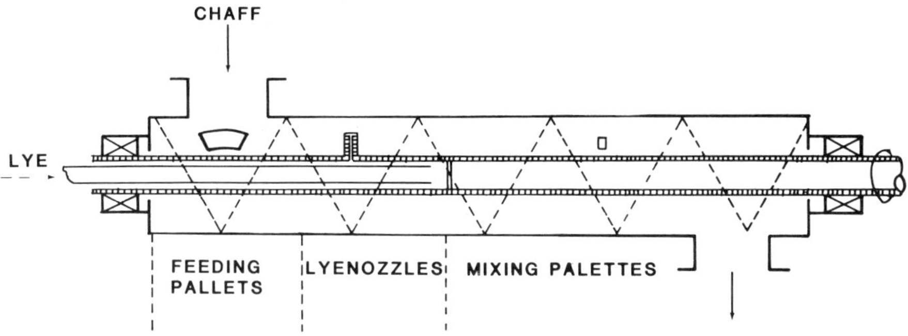

Figure 6.2.3. Principle of lye mixer (Friis Kristensen et al., 1978)

Plate 6.2.2. Commercial lye mixer (Schmidt and Jessen)

The band weigher discharges the material directly into the mixer, where the lye is added at a prescribed rate. The band weigher controls the dosage of alkali in direct proportion to the weight of the straw recorded.

To achieve a better control, some factories have equipped the alkali pump with a moisture-meter which automatically adjusts the alkali dosage relative to the moisture content of straw. The addition of caustic soda will then be regulated on the basis of NaOH per kg of dry straw, regardless of the moisture content of the straw.

6.2.4.4. Pressing

The press is the economic bottleneck in a plant. It is the most expensive machine unit, and its capacity determines the throughput of the plant. The press both compresses the material into pellets or briquettes and it acts as a reaction chamber.

The efficiency of the reaction is dependent on the added amount of NaOH, reaction time, temperature and pressure.

In Figure 6.2.1 it can be seen that an increase in pressure above 100 atm has little effect on digestibility. Commercial presses operate at much higher pressures, and pressure is therefore not a limiting factor.

The temperature has a pronounced effect on the increase in digestibility. The increase of temperature during pressing is usually caused by friction forces, but the exothermal reaction between NaOH and straw also increases the temperature.

The reaction temperature is thus determined by:

a. Friction forces: Depend on type of press and design of the pressing channel. Also type of raw-material and moisture content influence.
b. Heat reaction: Increases with increasing addition of NaOH.
c. As a supplement steam may be added into the press chamber.

In principle there are two main types of presses in use today: The piston press and the roller press.

6.2.4.4.1. Piston press

A piston press (Plate 6.2.3) consists of one or two pressing channels equipped with pistons. The diameter of the press channel may be rather large, and the compressed product is normally characterized as briquettes with a diameter of 40-80 mm. The density is 500-1000 kg/m^3, and the bulk weight is 250-500 kg/m^3 (Plate 6.2.4).

Plate 6.2.3. Commercial alkali treatment plant equipped with a piston press (Desmi)

Straw is only slightly disrupted in this press and the product therefore more or less retains its coarse structure. This is an advantage, when the product is used as roughage or as a component in composite feed for dairy cows.

The capacity of piston presses is rather low (1 to 3 tons per hour). Power consumption is proportionally low (20-25 kwh/ton) compared with roller presses (30-40 kwh/t)

6.2.4.4.2. <u>Roller press</u>

Roller pressed pellets have a smaller diameter than the briquettes: 10-25 mm, and the straw structure is more or less disrupted depending on the press type, conicity of holes and thickness of matrix.

Plate 6.2.4. Briquettes from a piston press

There are two principally different types of roller presses: Presses with a vertical ring die (Plate 6.2.5) and presses with a horizontal disk die (Plate 6.2.6 and 6.2.7). The pellets may have a bulk weight of 400-600 kg/m^3 (Plate 6.2.8).

There are technical differences within both ring die presses and disk presses. Shape of die and holes, number of rollers, and the choice of stationary and rotating parts may vary.

Thus, there are many parameters which have to be taken into consideration. Until today only little research has been carried out to determine the influence of the above mentioned parameters on the efficiency of the process.

Plate 6.2.5. Vertical roller press, opened to show the two rollers
and the ring die (EMM, Matador)

Friis Kristensen et al. (1978) have investigated the effect of
die thickness and hole diameter in a ring die press with one rotating roller. They concluded that some improvement, measured as
enzyme solubility, could be obtained by increasing the thickness
of the die. This parameter also affected both reaction temperature
and capacity. The bulk density was increased from 350 kg/m^3 to
600 kg/m^3 by increasing the thickness from 50 to 135 mm. Enzyme
solubility was also increased by approximately 14%.

Increased die thickness and decreased hole diameter considerably reduced the particle size, an effect which may be counteracted
by introducing steam into the press chamber.

146

Plate 6.2.6. Horizontal disk press (Amandus Kahl)

    Tesic (1977) has made a comparison between a ring die press and
a disk press. He only found small differences regardless of the
fact that the mechanical impact, retention time in press channel,
and die temperature were different in the two types of presses.
The pellets from both machines had almost the same density (aprox-
imately 1000 kg/m^3), and only small differences in pellet strength
could be measured. The ring die press produced pellets with a
slightly better adherence. The most pronounced difference between
the two types of machinery was with regard to particle size, the
disk press being significantly inferior to the ring die press.
With the addition of 5% NaOH, a mean particle size of approximately 4 mm was measured in pellets from the disk press, while the
ring die pellets had a mean particle size of 7 mm. Chopped straw
was used in both experiments.

Plate 6.2.7. Principle of the disk press (Amandus Kahl)

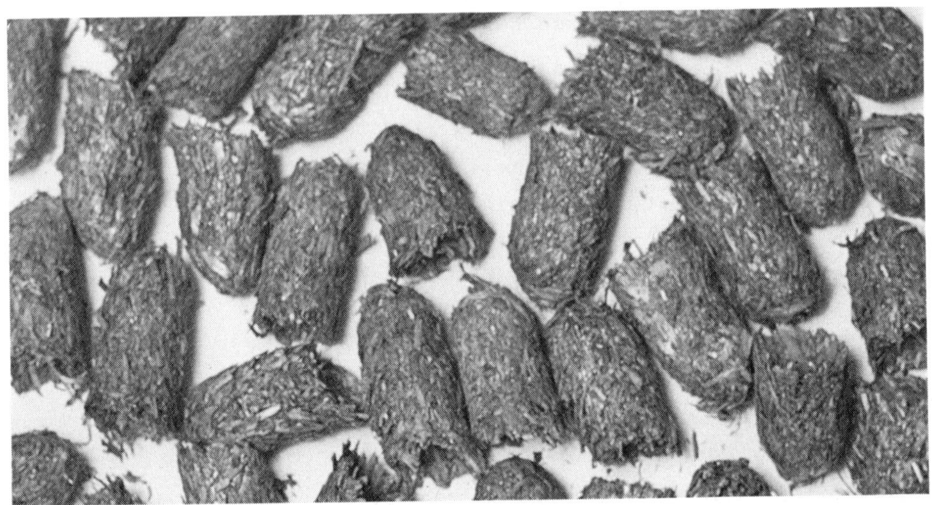

Plate 6.2.8. Pellets from a roller press

In this study it was also observed that the digestibility of organic material was 2-5% higher in pellets from the disk press. This might be due to the higher die temperature.

6.2.4.5. Cooling

The pellets leave the press with a temperature of 70-90°C, and must be cooled before storage.

Horizontal band coolers are the most common, but also vertical coolers are used.

In the band cooler the pellets are transported on a slowly moving, perforated band, through which cool air is injected. In this system both air velocity and pellet retention time can be regulated. 1-5% moisture is evaporated from the pellets during the cooling.

After cooling the pellets pass a screen to remove brokens and dust before they are transported to the storehouse or directly to a customer.

6.2.4.6. Storage

Upon storage the pellets still contain unreacted sodium hydroxide and have a pH of 9-11. Experiments carried out by Friis Kristensen et al. (1978) showed that the pH in pellets could remain constant during storage for as long as a year.

The per cent of unreacted NaOH falls within the first months of storage to a level, which remains constant for a long period. This reduction could indicate both a continued reaction between NaOH and the straw and a neutralization of the alkali by carbon dioxide from the air. A slight increase in digestibility could be observed under certain circumstances which also proves that the NaOH reaction on straw continues under storage.

Industrially produced straw pellets as well as farm scale alkali treated straw may combust spontaneously in spite of the high pH which should preserve against microbiological attack. Kristensen (1978) has investigated some of the accidents in Denmark caused by spontaneous combustion of alkali treated straw and found that even straw pellets with a low average moisture content (straw dried before impregnation with lye) could spontaneously combust. The risk seemed to be greatest when freshly harvested straw containing grass fragments and dust was used. Pellets with an uneven distribution of lye and moisture also present a risk. The critical point is normally reached after 10-20 days storage. The development starts with an increase in moisture on the surface of the heap which is probably due to the diffusion of water from lower levels where the temperature has risen due to microbial activity or from heat, generated from the reaction between the straw and surplus NaOH. If no precaution is taken, steam and heavy smelling smoke will be generated within 2-5 days preceding combustion. When an increase in moisture content of the surface layer is noticed it is simple to stop the development of combustion by repeatedly removing the outer wet layer and dry the whole lot.

6.2.4.7. Production of composite feed with lye treated straw

Two different methods for production of composite feed with alkali treated straw are in use (Friis Kristensen et al., 1978).

In the so-called "2-step technique" the straw is first pressed into pellets, then crushed and mixed with other components, and the mixture is finally pelleted and cooled.

The 2-step technique results in a reduction of the straw particle size but ensures an effective treatment and a product with a high bulk density and durability even from feed mixtures with a high content of fat.

In the "1-step technique" lye impregnated straw is mixed with the other feed ingredients before pelleting. The production costs are lower but the technique has usually some nutritional disadvantages compared with the 2-step technique (Table 6.2.4). The product has a coarser physical structure, however, the effect of the alkali treatment on digestibility tends to be reduced. Also the bulk density and durability of pellets/briquettes are decreased. These disadvantages may be overcome by heating the lye solution before application and by introducing a conditioning step before mixing with the other ingredients (Table 6.2.3).

Table 6.2.3. The effect of different processing techniques on handling properties and in vitro digestibility (Friis Kristensen et al., 1978)

Technique	Bulk density kg/m3	Durabi- lity %	Particle size, mm	IVOMD %
1-step pressing				
Normal treatment: lye: 10°C, no steam, fat before pressing	384	91	1.7	63.7
Heating of lye to 90°C	+1	0	-0.1	+1.0
Addition of 6% steam in press	-55	+1	+0.2	+1.0
Fat added after pressing	+111	+7	+0.1	-0.3
All three treatments	440	98	1.7	66.0
2-step pressing	608	98	1.2	66.9

6.2.5. FEEDING WITH INDUSTRIALLY PRODUCED ALKALI-TREATED STRAW

6.2.5.1. Introduction

It is well established that industrial dry alkali treatment of straw results in a substantial increase in the energy value of processed straw. This has been documented in in vitro and in vivo digestibility trials and in feeding experiments with such polygastric animals as dairy cows, bulls, beef cattle and sheep (Rexen et al., 1975; Junker, 1976; Rexen and Vestergaard-Thomsen, 1976; Wilson and Brigstocke, 1977; Wilkinson, 1978; Friis Kristensen et al., 1978). The in vivo organic matter digestibility increased from 45-50% of untreated straw to 60-70% in straw treated with 5% NaOH (Figure 6.2.4). The feeding value (expressed as Scandinavian feed Units (Sc.f.U)) rose from 30 to 60-64 Sc.f.U./100 kg DM (Friis Kristensen et al., 1978). It has further been shown that the in vitro degradation rate increased considerably by alkaline treatment. This in vitro effect is correlated to the turnover time and the quantity of feeds which can be assimilated in vivo (Vestergaard Thomsen et al., 1973; Coombe et al., 1979). However, due to the disruption of the long fibre structure of the straw and to the unreacted NaOH and sodium ions from the reacted NaOH, the use of lye treated straw in practical feeding is somewhat restricted.

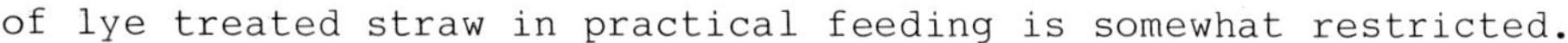

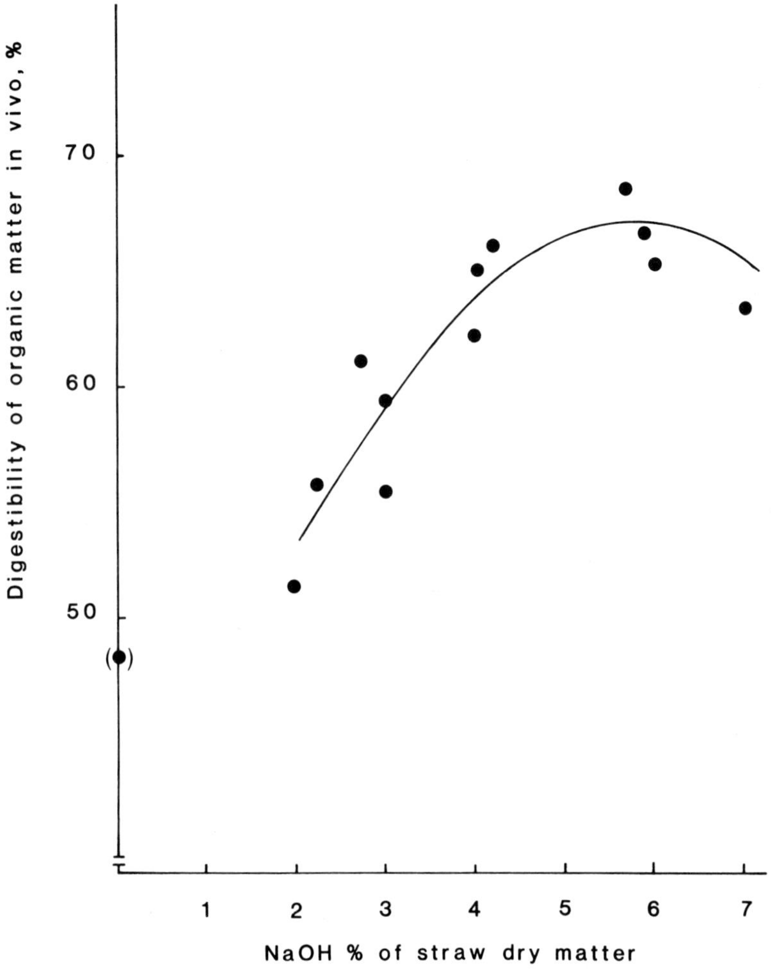

Figure 6.2.4. The relationship between NaOH dosing and the in vivo
digestibility (Friis Kristensen et al., 1978)

6.2.5.2. <u>Physical structure</u>

It has been shown in several experiments that the inclusion of
long fibre in the diet for dairy cows avoids a low content of fat
in the milk (Broster et al., 1978). Straw may provide the fibre,
but the low energy value of untreated straw imposes a severe limitation to its use in the diets of high yielding dairy cows.

On the other hand, the higher energy value of alkali treated straw means that it can be used with less adverse effect on the overall energy value of the total rations. A maximum voluntary straw inta- ke of up to 8.7 kg dry matter per day was obtained in experiments with dairy cows, fed treated straw alone or with small amounts (10-15%) of molasses (Friis Kristensen et al., 1978). In rations where it was necessary to feed straw together with other kinds of roughages to obtain a coarse physical structure, straw intake was reduced to a level of only 3-5 kg per day.

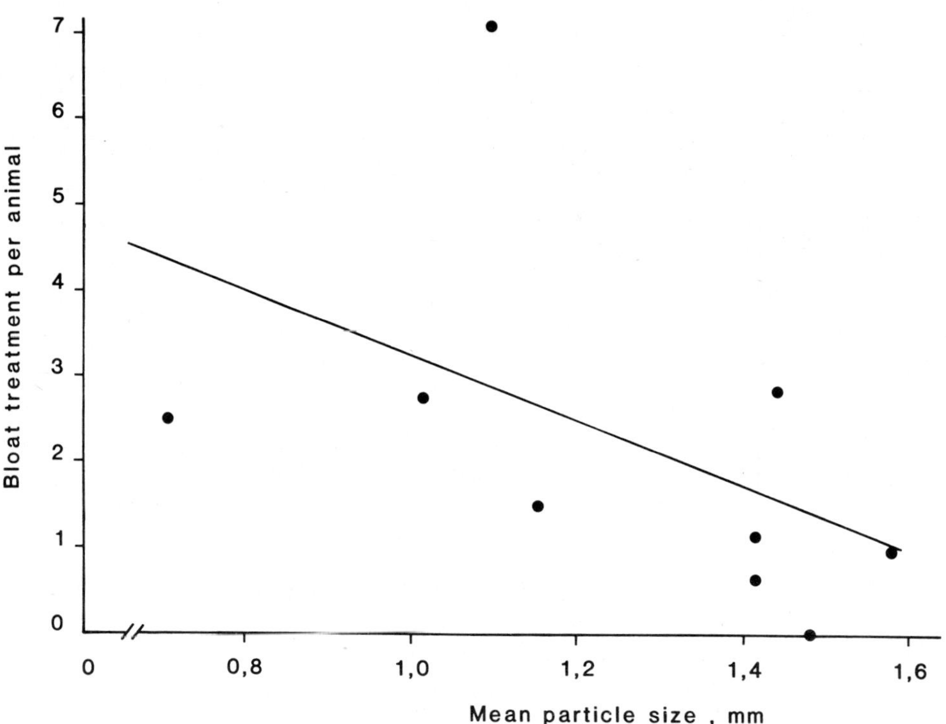

Figure 6.2.5. The relationship between the occasional incidence of bloat and the mean particle size of complete feed (Data from Henriksen, 1978)

These figures demonstrate on the one hand, how cows preferentially select between different roughages when they reach the upper limit of their voluntary feed intake, and on the other the limitation in the intake of alkali treated straw due to the disruption of the long fibre structure. This limitation is further demonstrated in experiments with straw based complete feeds fed to young bulls and dairy cows. Friis Kristensen et al. (1978) found a mean particle size of 1.2-1.6 mm for a NaOH treated fullfeed mixture containing straw. This is considered too low for a complete feed and might be expected to be the main reason for the occasional incidence of bloat reported in investigations where industrially produced complete feeds (Figure 6.2.5) were fed to young bulls and dairy cows (Vind, 1974; Friis Kristensen et al., 1978; Henriksen, 1978).

It is difficult to recommend exact lower limits for coarseness of feed. Henriksen (1978) showed that the frequency of rumen disturbance increased when young growing bulls were fed a complete feed with a mean particle size below 1.8 mm (Figure 6.2.5). The values of Vind (1974) demonstrated the advantage of 16 mm pellets over 6 mm pellets, the former giving rise to coarser particles in the rumen than the latter.

Complete feeds with a coarser structure than those reported by Friis Kristensen et al. (1978) can be produced by using the "1-step technique" described in section 6.2.4.7 instead of the "2-step procedure". Table 6.2.4 shows the in vivo digestibility results for sheep fed feeds produced by these two methods. The digestibility of the feed was lowest when produced in the simple 1-step metod.

Table 6.2.4. The influence of the processing technique on particle size and in vivo digestibility of treated straw in sheep (Friis Kristensen et al., 1978)

Technique	In vivo digestibility, %				Particle size, mm
	Organic matter	Crude protein	Crude fat	Crude fibre	
1-step pressing:					
Normal treatment,					
lye 10°C, no steam	66	74	77	48	1.7
Heating of lye to 90°C	69	74	75	53	1.5
Addition of steam in press	67	74	78	50	1.9
Both treatments	70	74	75	55	1.7
2-step pressing	70	69	77	57	1.2

Heating lye before its application and adding steam to the straw before mixing with the concentrate, improved the digestibility in the one step method to the same level as the 2-step procedure. The improvement in digestibility was obtained without any adverse effect on the coarseness of the feed. The mean particle size was increased from 1.2 to 1.7 mm.

6.2.5.3. <u>Palatability problems with NaOH-treated straw</u>

Sodium hydroxide treated straw is not well accepted by the animals when it is fed alone. Straw treated with 5% NaOH contains approximately 25 g sodium per kg straw and about 10 g of this quantity are left in the straw as NaOH (Figure 6.2.6). This results in a high pH of sodium hydroxide treated straw and leads to a depressed feed intake as indicated by Stigsen (1975). He found that when the pH value of the upper rumen fluid was above 8.0, which was reached a short time after the start of feeding, the feed intake was stopped, and it was not initiated again before the pH fell below 7.0. Neutralization of the straw with HCl stabilized the pH of the rumen fluid and feed consumption was increased by 2.4% (Stigsen, 1975). Other investigations reported positive responses in feed intake in the order of 5-20% due to neutralization (Junker and Pfeffer, 1976; Friis Kristensen et al., 1978).

Experiments carried out by Friis Kristensen et al. (1978) showed that molasses and molassed dried pulp also improved palatability and feed intake.

6.2.5.4. <u>Recommended levels of NaOH-treated straw in composite feeds</u>

The above results regarding physical structure and palatability of NaOH treated straw together with high sodium levels (see Chapter 12) show that it may be necessary to restrict the amount of treated straw in the diets of polygastric animals.

Wilson and Brigstocke (1977) stated that even though acceptability studies showed that addition of only 20% of other feed components improved the acceptability considerably, the use of more than 50% industrially processed sodium hydroxide-treated straw in the feed is not recommended. Moreover, due to the unknown long term effect of high sodium intake, Wilson and Brigstocke (1977) recommended even lower inclusion rates in practical feeding situations (Table 6.2.5).

Table 6.2.5. Conservative inclusion rates of NaOH treated straw (Wilson and Brigstocke, 1977)

Class of stock	% inclusion rates
Dairy cows	10-15
Beef cattle	15
Sheep	10-15
Young stock	10-15
Young calves and lambs	Nil
Rabbits	5-8

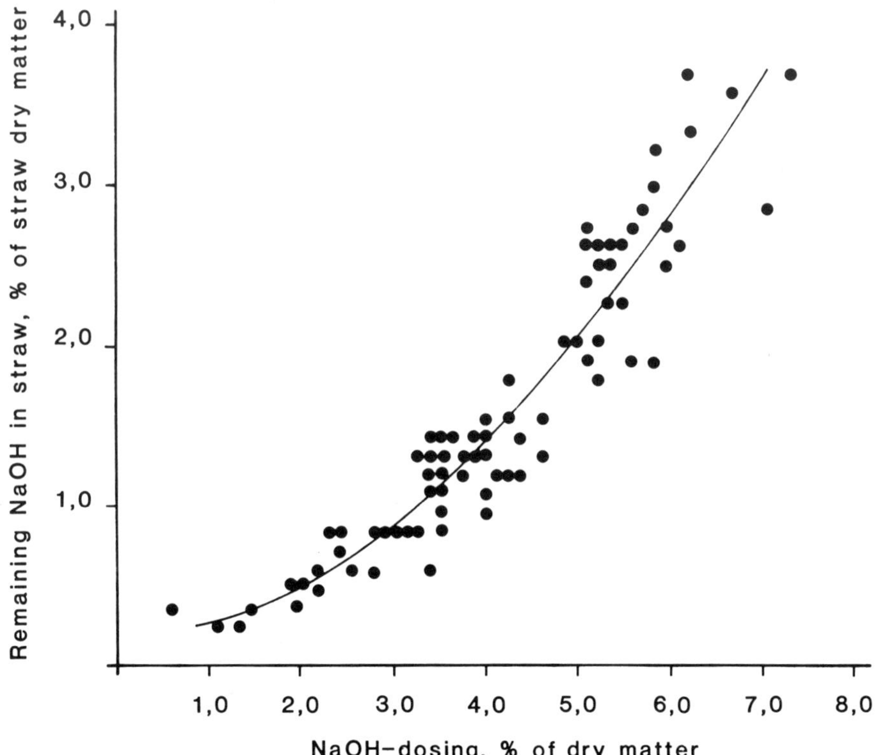

Figure 6.2.6. The relationship between NaOH dosing and the content of remaining NaOH in straw (Friis Kristensen et al., 1978)

156

Friis Kristensen et al. (1978) in a comprehensive investigation
with industrially produced NaOH-treated straw show examples of
rations with high amounts of treated straw (Table 6.2.6). The
rations were formulated so that amounts of other roughages were
limited. It was also assumed that the straw was treated with a
maximum of 4-5% NaOH and had an organic matter digestibility above
65%.

Table 6.2.6. Examples of dairy cow diets containing sodium hyd-
             roxide treated straw (Friis Kristensen et al., 1978)

Ration	1	2	3	4
Beets/molasses/beet pulp, Sc.f.U	3	3	3	3
Grass silage, kg dry matter	0	3	6	8
NaOH treated straw, kg dry matter	6.5	4.0	1.5	–
Barley, kg	1.8	2.5	3.2	3.7
Oil cakes, kg	6.2	5.0	3.8	3.0

These two examples of recommendations for dairy cows differ
considerably ending up with an upper limit for inclusion of treated straw of 15 and 38% respectively (Table 6.2.5 and 6.2.6). However, it is worth mentioning that the recommendation of Wilson and
Brigstocke (1977) are based on finely ground straw, whereas the
composition of the rations of Friis Kristensen et al. (1978) are
based on straw produced according to the process in Figure 6.2.2.
This process gives rise to a much coarser structure which has a
positive effect on rumen physiology. On the other hand, ration 1
(Table 6.2.6) represents the upper limits for inclusion of NaOH
treated straw in diets for lactating dairy cows due to restriction
in physical structure and available energy supply.

6.2.6. FUTURE ASPECTS FOR INDUSTRIAL ALKALINE TREATMENT OF STRAW

6.2.6.1. <u>Industrialized countries</u>

Industrial plants for alkali treatment of straw have mostly
been erected in industrialized countries where the level of production of treated straw has been determined to a great extent by
the price of competitive energy feeds. The extent of roughage production by farmers for their own use also influences the demand

for industrial alkali treated straw pellets.

The product has proved its value both in experiments and in practical feeding for sheep, dairy cows and beef cattle and its potential as a part of composite feeds for ruminants is clearly demonstrated.

Alkali treated straw pellets alone or containing small amounts of molasses are also commercially marketed. Although these products are often more expensive than farm treated straw, the pelleted product has a number of advantages over loose straw treated in farm equipment. First, pellets are produced under controlled conditions which yield a homogenous product with an evenly lye distribution. Second, the pelleted product is free flowing, dense, dust-free and it is easily transported and stored. It can therefore be distributed to animals by modern automatic feeding equipment.

As previously mentioned, many factors determine the demand for alkali treated straw pellets, and it is therefore not easy to predict future production quotas. Nonetheless, the construction of industrial plants for alkali treatment of straw may be seen as a part of a commencing tendency towards industrialization of agriculture.

As pointed out by Munck (1981): "The products from industrialized agriculture will not be cheaper than at the present production based on family farms. The current income level of farmers is much lower than of other groups, even though the farmers work longer hours. The question, which comes to mind in many countries today, is how long the farmers will be willing to endure the present economic and social conditions. The answer to this question may mark the start of a fully economical industrial agriculture including plant husbandry and material handling".

6.2.6.2. <u>Developing countries</u>

In most developing countries the demand for animal protein is rising steadily and these demands are often met by higher imports of both food and feed, which leads to ever increasing trade balance deficits. To counteract such trends it is important to increase domestic animal production from domestic feed resources. Although husbandry methods are not efficient enough to meet demands for meat and milk at a suitable price for the consumption to be raised.

158

A key factor limiting animal production is inadequate animal nutrition, which cannot be improved by increasing grazing areas, because land for food crops is becoming scarcer and food production must have priority over feed production. A possibility for immediate increases in animal production in the developing countries thus lies in an improved utilization of straws and other lignocellulosic wastes from agriculture, which cannot be consumed directly by man.

Both farm scale treatment systems and industrial systems would be applicable, although industrial straw based feed production secures a homogenous quality, and makes it possible to produce composite pellets with a balanced content of nutrients, thus securing an optimal diet for the ruminants.

6.2.6.3. Future reseach

Much research has been done both on alkali treatment techniques and on feeding aspects, and our knowledge of these subjects has increased substantially, however, there still remain a number of unsolved problems with the utilization of alkali treated straw.

First, and most urgently, we must find ways of reducing the negative effects of the sodium ion and the high pH. When this problem is solved, it will be possible to use much higher percentages of alkali treated straw in diets.

Second, although digestibility is greatly improved by alkali treatment, the product must still be classified as a low energy feed with a digestibility comparable to good quality hay. Research should be performed to increase the effect of alkali treatment so that also high energy feed can be produced.

Third, the disruption of the physical structure of straw in the pellet press presents a problem, when the product is used as a structure feed. The problem must be overcome by modifications in equipment and processing conditions.

Fourth, investigation of the interaction between alkali treated straw and other feed components are other important areas of research.

6.2.7. SUMMARY

The industrial alkali treatment technique was introduced in Denmark and Great Britain in the beginning of the nineteen seventies. Since then, plants have been built in many countries throughout the World.

The industrial treatment principle involves the passage of chopped NaOH impregnated straw through a pellet press or briquette press. The reaction takes place at a temperature of 80-100°C under 50-100 atm pressure. The retention time in the press, and thus the reaction time, is very short, less than one minute. The alkali is sprayed on the straw in a concentrated solution (27-46% NaOH), and the amount varies between 3 and 6% based on straw dry matter.

The process does not include any washing step, and therefore there is no loss of dry matter. All the added NaOH remains in the product, mainly in a neutralized form. The amount of free NaOH should not exceed 2% of straw dry matter. The final product appears in a compact storeable form.

Two principally different types of presses are used, viz. pellet presses and piston presses (or briquette presses). The pellets have a diameter between 6 and 25 mm, while the briquettes usually have a diameter of 50 mm.

The industrial alkali treatment results in a substantial increase in the digestibility and the energy value of the straw. The in vivo organic matter digestibility increases from 45-50% to 60-70% and the energy value (expressed as Scandinavian feed Units (Sc. f.U.) rises from approximately 30 to 60-64 Sc.f.U./100 kg DM. Due to the higher energy value, diets to high yielding dairy cows can contain up to 38% of lye treated straw.

The industrial methods have relevance for both industrialized and developing countries throughout the World.

ACKNOWLEDGEMENT

The authors are indebted to Fil.Dr. Lars Munck, head of the Department of Biotechnology, Carlsberg Research Laboaotry, Copenhagen, for valuable suggestions during preparation of the manuscript.

6.2.8. REFERENCES

Bolduan, G. and Piatkowski, B., 1972. Untersuchungen zum Aufschluss von Getreidestroh. I. Mitteilung. Behandlung mit Natronlauge und die anschliessende Neutralizierung. Arch. Tierernähr. 22: 485-492.

Broster, W.M., Sutton, J.D. and Bines, J.A., 1978. Concentrate: forage rations for high yielding dairy cows. In: Proc. of the 12th Nottingham Conference for Feed Manufacturers, 4th-6th January 1978. University of Nottingham, England.

Chandra, S. and Jackson, M.G., 1971. A study of various chemical treatments to remove lignin from coarse roughages and increase their digestibility. J. agric. Sci. (Camb.) 77: 11-17.

Coombe, J.B., Dinius, D.A. and Wheeler, W.E., 1979. Effect of alkali treatment on intake and digestion of barley straw by beef steers. J. Anim. Sci. 49: 169-176.

Donefer, E., Adeleye, I.O.A. and Jones, T.A.O.C., 1969. Effect of urea supplementation on the nutritive value of NaOH treated oat straw. Cellulases and their application. Adv. in Chem. Series No. 95: 328-339.

Friis Kristensen, V., Andersen, P.E., Stigsen, P., Vestergaard Thomsen, K., Refsgaard Andersen, H., Sørensen, M., Ali, C.S., Mason, V.C., Rexen, F., Israelsen, M. and Wolstrup, J., 1978. Natriumhydroxyd-behandlet halm som foder til kvæg og får. 464. Beretning fra Statens Husdyrbrugsforsøg. Beretning 82 fra Bioteknisk Institut, Denmark.

Henriksen, J., 1978. In vitro teknik til bestemmelse af kvægfoders energiværdi. Thesis, Husdyrbrugsinstitutet. Den Kgl. Veterinær- og Landbohøjskole, København. 119 + XVII pp.

Homb, T., Sundstøl, F. and Arnason, J., 1977. Chemical treatment of straw at commercial and farm levels. New Feed Resources. FAO Anim. Prod. & Health Paper 4, 25-37.

Junker, T., 1976. Untersuchungen über die Verbesserung des Futterwertes von Stroh für Wiederkäuer durch trockenen Aufschluss beim Brikketieren. Dissertation, der Georg August Universität zu Göttingen.

Junker, T. and Pfeffer, E., 1976. Untersuchungen über den Einsatz von trocken aufgeschlossenen Stroh in der Fütterung von Wiederkäuern. Kongressband 88. VDLNFA Kongress, Oldenburg, 22nd-25th Sept.: 259-267.

Kristensen, T.P., 1978. Undersøgelser over årsagerne til selvantændelse i ludbehandlet halm. Beretning 83, Bioteknisk Institut, Kolding, Denmark.

Munck, L., 1981. Barley for food feed and industry - In Cereals: A renewable resource. Theory and practice. American Association Cereal Chemists: 427-459.

Mwakatundu, A.G.K. and Owen, E., 1974. In vitro digestibility of sodium hydroxide-treated grass harvested at different stages of growth. East Afric. Agr. For. J. 40: 1-10.

Ololade, B.G., Mowat, D.N. and Winch, J.E., 1970. Effect of processing methods on the in vitro digestibility of sodium hydroxide treated roughages. Can. J. Anim. Sci. 50: 657-662.

Piatkowski, B., Bolduan, G., Zwierz, P. and Lengerken, J.V., 1974. Untersuchungen zum Aufschluss von Getreidestroh mit Natronlauge. Beziehungen zwischen der NaOH Konzentration im behandelten Stroh, dem säurefallbaren Ligninanteil und der Verdaulichkeit in vivo und in vitro. Arch. Tierernähr. 24: 513-522.

Rexen, F.P., 1979. Low quality forages improve with alkali treatment. Feedstuffs 51 (42): 33-34.

Rexen, F.P., 1972. Forøgelse af halmens fordøjelighed ved kemisk behandling. Ugeskr. Agr. Hort. 18: 364-365.
Rexen, F.P., Stigsen, P. and Friis Kristensen, V., 1975. The effect of a new alkali technique on the nutritive value of straw. In Proc. of Ninth Nutrition Conference for Feed Manufacturers, 5th-7th January, 1975. University of Nottingham, England.
Rexen, F.P. and Vestergaard Thomsen, K., 1976. The effect on digestibility of a new technique for alkali treatment of straw. Anim. Feed Sci. Technol. 1: 73-83.
Schwalbe, C.G., 1933. Futterstoff aus Holz. Angewandte Chemie 45: 707-718.
Stigsen, P., 1975. Kulhydratkildens og neutralisationens betydning for udnyttelse af NaOH behandlet halm hos malkekøer. Thesis. Den Kgl. Veterinær- og Landbohøjskole, Copenhagen.
Thesic, M., 1977. Das Verdichten unter Nährstoffaufschluss von Futterpflanzen in Matrizenpresen. Dissertation. Georg August Universität zu Göttingen.
Tesic, M., 1982. Eigenschaften von Strohcobs als Rauhfutter für Wiederkäuern. Die Mühle- und Mischfuttertecknik. Heft 23: 318-321.
Vestergaard Thomsen, K., Rexen, F. and Friis Kristensen, V., 1973. Forsøg med natriumhydroxybehandling af halm. 1: behandlingens indflydelse på halmens fordøjelighed. Ugeskr. Agr. Hort. 25: 436-439.
Vind, R., 1974. Complete feeds for ruminants – an experimental approach. Paper presented at the Annual Convention of the British Association of Green Crop Driers, Chester, 5th-7th Nov., 8 pp.
Wilkinson, J.M., 1978. Recent development in chemical treatment of straw for use as feed for livestock. Report on straw utilization Conference, Oxford, 17-21.
Wilkinson, J.M. and Gonzales Santillana, R., 1978. Ensiled alkali treated straw. Anim. Feed Sci. Technol. 3: 117-132.
Wilson, R.K. and Pigden, W.J., 1964. Effect of a sodium-hydroxide treatment on the utilization of wheat straw and poplar wood by rumen microorganisms. Can J. Anim. Sci. 44: 117-132.
Wilson, P.N., Sangster, I. and Walker, C., 1976. Straw utilization. Experience of running a commercial straw processing plant. MAFF/ADAS Conference, 23rd January, Oxford.
Wilson, P.N. and Brigstocke, T., 1977. The commerical straw process. Process Biochem. 9: 17-20.

Chapter 6.3

FARM-SCALE DRY TREATMENT
WITH SODIUM HYDROXIDE

by

J. Michael Wilkinson

Chief Scientists' Group
Ministry of Agriculture, Fisheries and Food
Great Westminster House, Horseferry Road
London SW1P 2AE, United Kingdom

6.3.1. INTRODUCTION

The fact that straw and other fibrous by-products are produced
on farms which may also have livestock, leads inevitably to consideration of upgrading these materials so that they may more effectively contribute to the nutrition of those livestock. Further, it
is arguable that the very production of such by-products should be
followed by their exploitation as feeds, so that they may contribute (albeit indirectly) to the total yield of human food from the
farming enterprise, and at the same time increase the financial
return to the farmer.

The upgrading of low-quality by-products on the farm of their
origin avoids the need to transport them to a central place, eg.
for industrial-scale treatment, before they are consumed by livestock. But there is an increased probability that, due to variability in technique, on-farm methods might give rise to a greater
variability in feed value in the stored, treated product than
would be the case with a factory-scale process. A further feature
of farm-scale treatment is that there is less likelihood of par-

[1]Present address: "Chalcombe", Highwoods Drive, Marlow Bottom,
Marlow, Bucks SL7 3PU, UNITED KINGDOM

ticle size being reduced as occurs during the pelleting of industrially-treated material.

Treatment with sodium hydroxide (NaOH) is the most cost-effective method for the upgrading of low-quality straws and by-products (Owen, 1981a), though mixtures of NaOH and other chemicals, particularly those such as urea which have a high content of nitrogen, may eventually prove to be more effective. Donefer, Adeleye and Jones (1969) and Ørskov and Grubb (1978) highlighted the interaction between NaOH and urea; in the absence of the other both the response in voluntary intake and in digestibility to either NaOH or to urea was limited. Thus addition of NaOH to straws and fibrous by-products exacerbates the deficiency of N which already exists in these materials. "Dry" treatment implies the use of minimal amounts of water, and generally there is no addition of water above that which is contained in the solution of NaOH, or which might be added to ensure distribution of the chemical or its reaction with the feed. Thus the feature of the treatments to be considered in this section is that the NaOH is added to the crop or by-product, rather than the other way round. There is usually no recirculation or neutralization of the NaOH.

Treatment on the farm may occur at a number of points between harvest (or arrival on the farm) and feeding (see Figure 6.3.1). At its simplest, the crop is harvested manually, and treated by hand before either being stored or being given to the animal. By contrast, specialised processing equipment may be used, comprising purpose-built machines for chopping the crop, mixing it with NaOH and delivering the treated product to the store or the animal. Intermediate technology comprises the use of existing forage harvesters for either harvesting and treating the crop in the field, or for use as a mixer-blower for treating the material after transportation to the farm, prior to storage. These techniques are discussed in the sections which follow.

Whatever process is used, it is important that adequate precautions are taken to protect people from the hazard of direct contact between sodium hydroxide and the body. Particular attention must be given to the protection of the eyes, face and hands by the wearing of a plastic mask over the face and gloves on the hands. A suitable eyewash should be at hand in case of accident, together with an ample supply of clean water. Guidelines for the safe use of sodium hydroxide are available from the supplier, and they should be strictly followed.

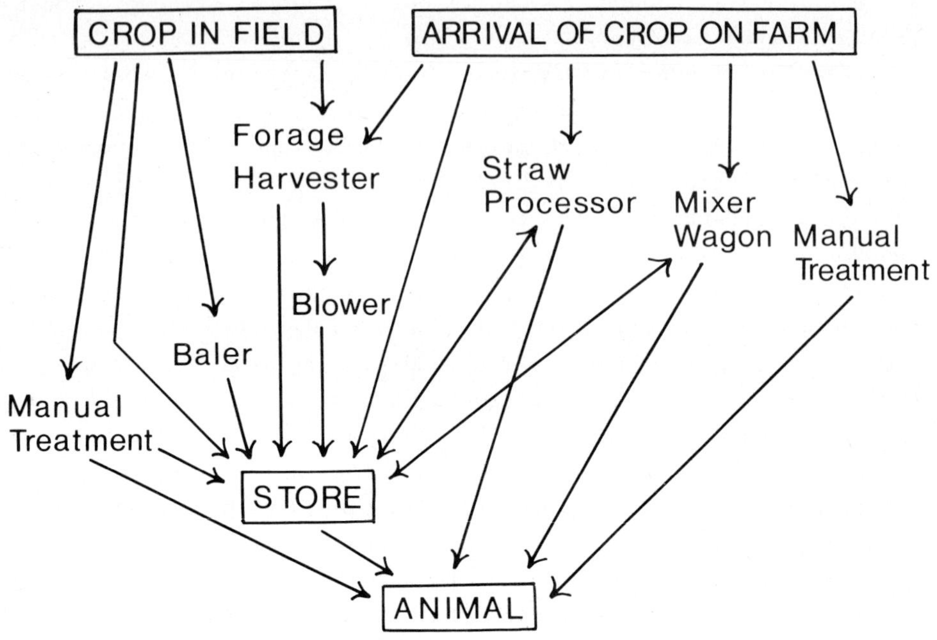

Figure 6.3.1. Possibilities for the on-farm treatment of straws and other low quality feeds to improve their feed value

6.3.2. MANUAL TREATMENT

Jackson (1978) describes procedures for the daily treatment of dry straw with NaOH by hand. They are summarised in Table 6.3.1. The simplest method is to add NaOH in solution to straw via a garden watering can so that it is uniformly wetted. The straw is turned by hand fork during the sprinkling. Batches are prepared the day before being given to the animals. Uniformity of spraying is enhanced if a knapsack or other hand-operated type of garden pressure-sprayer is employed. Using this equipment, the concentration

of solution of NaOH can be increased. If the straw can be chopped
prior to treatment, this will aid distribution of alkali.

Table 6.3.1. Equipment and recommended volumes of solution for
             manual treatment of straws with NaOH (from Jackson,
             1978 and Owen, 1981b)

Equipment	Solution of NaOH
Garden watering can	2 l/kg
Pressure sprayer	1 l/kg
Fork or chopper and horizontal batch mixer (500 kg capacity) and	
Pump/spray nozzles	0.5 l/kg
or cement mixer	1.5 l/kg

Manual treatment may also be accomplished with the aid of a
horizontal batch mixer (if there is an appropriate •source of power
for its operation). This equipment, which is basically a hopper
with an auger at its base, can be fitted with spray nozzles on an
overhead boom. The solution of NaOH is then pumped through the
nozzles on to the straw as it turned in the hopper. The volume of
solution required to treat dry crops can be further reduced with
this technique (Table 6.3.1). Other dietary ingredients may also
be added and mixed, after allowing about 10 minutes for the initial reaction between alkali and straw to occur. Another possibility is to chop the crop prior to treatment with NaOH in a cement
mixer (see Plates 6.3.1 a-d). The recommended quantity of solution
is between 0.5 and 1.5 l/kg (Table 6.3.1), and treatment is done,
as with the other methods, on a daily basis.

The effect of treating straws by the above procedures on the
digestibility of the organic matter is summarised in Table 6.3.2.

The data indicate that substantial improvements in digestibility can be achieved following manual additions of NaOH to straws,
at the rate of about 5% of the DM (range 2.5 to 7.0). Those of
Jayasuriuya and Owen (1975) indicate that the volume of solution
should exceed 0.3 l/kg with hand spraying and hand mixing. The
average increase in digestibility in vivo was 13 units (range 7.8
to 17.9).

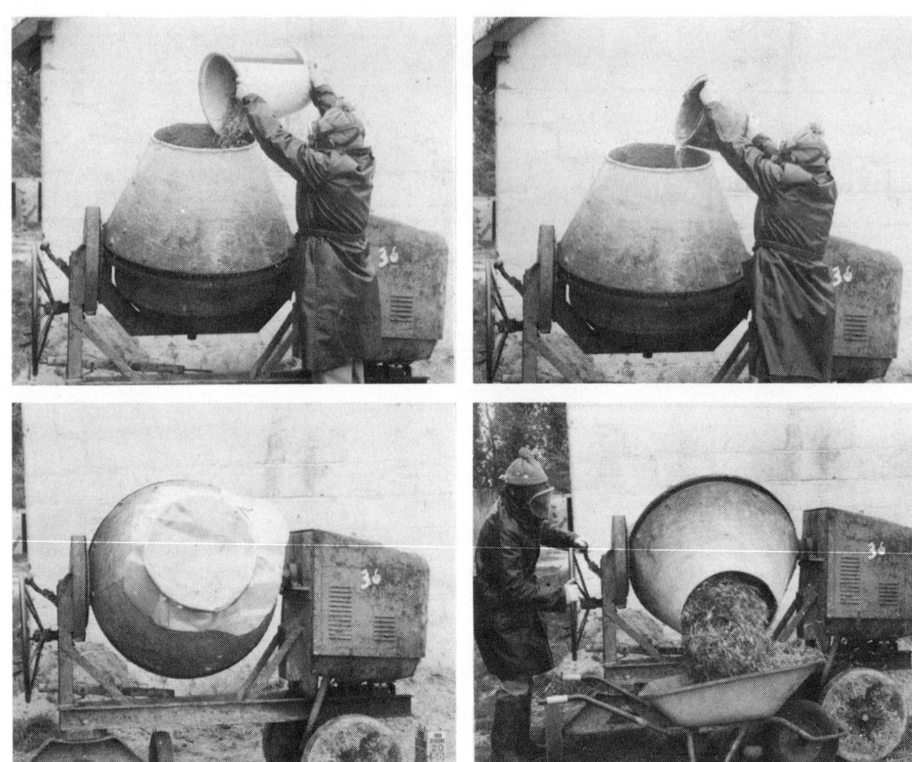

Plate 6.3.1. Small-scale treatment of straw with alkali using a
cement mixer. a) Loading with chopped straw, b) adding
alkali solution, c) mixing, d) discharging (Courtesy
E. Owen)

Plate 6.3.2. Small-scale treatment of straw with alkali; storing in
sealed plastic bags in oil drums (Courtesy E. Owen)

Table 6.3.2. Manual treatment of straw with NaOH: procedures, level of NaOH and digestibility
in vivo of untreated and treated straws

Crop	Procedure	Solution l/kg straw	NaOH g/kg straw DM	Digestibility of OM %		Reference
				untreated	treated	
Barley straw	Garden sprayer	0.3	45	51.6[1]	59.4	
	+ hand mixing	0.3	67.5	51.6	60.6	
		0.6	45	51.6	64.2	Jayasuriya and Owen
		0.6	67.5	51.6	65.5	(1975)
		1.2	45	51.6	64.3	
		1.2	67.5	51.6	65.3	
Barley straw	Watering can	1.5	35	46.9[1]	57.4	Owen (1981b)
	+ cement mixer	1.5	70	46.9	63.9	
Wheat straw	Watering can	1.0	25	55[2]	70	Singh and Jackson
	+ horizontal batch mixer	1.0	52	55	71	(1971)
Rice straw	Spray, sun-dried	0.5	50	51.5[2]	69.4	Mehrez et al. (1981)
Maize stalks				50.2	64.2	
Maize cobs	Spray, sun-dried	0.75	50	59.9[2]	64.6	Kategile and Frederiksen (1979)

[1] Derived values for the straw components of the diet

[2] Whole diet

Table 6.3.5. Mechanical treatment of straw and low quality forages with NaOH in the field: equipment and solutions added in the UK, Australia and Tanzania

UK	Australia	Tanzania
Crops: Barley straw, wheat straw	Crops: Oat straw, wheat straw, rice straw	Crop: *Hyparrhenia*
Equipment: Metered-chop forage harvester Trailer Applicator for pearl NaOH Water pump Oil drum (in trailer, 200 l capacity)	Equipment: Forage harvester + trailer or "Stak hand" Two tanks Two pumps	Equipment: Flail forage harvester Boom sprayer, 6 nozzles Tank (400 l capacity) Pump
Rate of work: (man, minutes per tonne harvested and put into store 75	Rate of work (man min./t) approx. 50	
Additives: Water (1/100 kg straw 19 NaOH (% of straw DM) 5 to 7	Additives[1]: Solution 1 (1/100 kg straw) 5 Solution 2 (1/100 kg straw) 19 NaOH (% of straw DM) 5	Additives: Solution (8% w/v NaOH) (1/100 kg crop fresh wt.) 25 NaOH (% of forage DM) 4
Reference: Tetlow and Wilkinson, (unpublished data, 1977)	Reference: Kellaway et al. (1978) and Kellaway (personal communication, 1978)	Reference: Kategile (1981)

[1] Solution 1 (per 100 litres): 50 l H_2O, 34.7 l black H_3PO_4 + H_2SO_4 (72:28 w/w), 50 kg urea.

Solution 2 (per 100 litres): 66.8 l H_2O, 33.2 l NaOH (50% w/w).

6.3.3. MECHANICAL TREATMENT

6.3.3.1. <u>Via the forage harvester in the field</u>

The possibility that straws and low-quality forage crops may be treated with NaOH at the time of harvest in the field via a forage harvester has been studied in several countries (UK, Australia and Tanzania). The principle object of the excercise is to harvest the crop and treat it with alkali in a single operation, using existing farm equipment with the minimum of modification. Generally, the intention is to store the treated crop by ensilage (see Chapter 6.4). A further possibility is that the moisture in, or on the surface of the crop may be used as the means of solubilising granular NaOH; thus removing the need to move large quantities of solution from farm to field.

Details of some of the procedures which have been tested are in Table 6.3.3. The approach taken in the UK work was to add NaOH in solid form, using an applicator developed for adding powders to grass crops prior to ensilage. With dry straws, it was also necessary to add water from a separate tank and pump. Both alkali and water were added to the crop as it passed through the chopping chamber of the forage harvester. By contrast, the Australian approach concentrated on rectifying not only the low digestibility of the crop, but also the deficiencies in nitrogen, phosphorus, sulphur and, if necessary, trace elements. Thus two solutions were applied simultaneously, one containing urea and acids (and trace elements as an optional extra); the other of NaOH. Both were added to the crop as it passed through the delivery chutes of the forage harvesters.

In Tanzania, a standing mature crop of *Hyparrhenia*, with a DM content of 50% was harvested by a flail forage harvester fitted with a boom sprayer. The six nozzles were connected to spray into the chopping chamber of the harvester. The flow rate of relatively dilute solution of NaOH was adjusted so that the rate of addition of alkali was 2% of the fresh crop weight (4% of the original crop DM).

The digestibility in vivo of the untreated and treated (ensiled) materials is in Table 6.3.4. With the exception of the diets of oat straw, treatment in the field with NaOH was reflected in marked improvements in digestibility. A possible explanation of the lack of effect of NaOH on the oat straw is that, in the presence of adequate supplementary N and minerals, the digestibility of the untreated material was relatively high.

Table 6.3.4. Mechanical treatment of straw and low-quality forages with NaOH in the field: digestibility of OM in vivo of untreated and treated materials (ad libitum)

	UK (Alawa, 1980) Wheat straw[1]	Australia (Thiago et al. 1979) Wheat straw	Oat straw	Tanzania (Kategile, 1981) Hyparrhenia[3]
Untreated	44.7	48.7[2]	60.4[2]	50.9
Treated	62.7	68.3	59.6	57.2

[1] By difference, diet contained 80% straw and 20% concentrates
[2] Supplementary urea and minerals sprayed on to straw at feeding
[3] Simsim meal (200 g/d) comprised 25% and 20% total diet DM for untreated crops, respectively

The response of *Hyparrhenia* to treatment with alkali was lower than with wheat straw, reflecting the somewhat higher digestibility of the untreated crop (field-dried hay). Associated with the improvement in digestibility (6.3 units), the sheep consumed 41% more DM of the treated crop than the untreated material. The increase in digestibility was also lower than those predicted by the equations for tropical grasses derived by Thomas (1978) and Owen (1981b), which were 9.1 and 15.3 units, respectively. Probably the difference reflects the level of feeding (ad libitum), and the fact that the prediction equations were derived from determinations of digestibility in vitro.

6.3.3.2. By mixer-wagon

The use of a mixer-wagon to facilitate uniform distribution of NaOH with crop or by-product is essentially a scaling-up of the horizontal batch mixer described in section 6.3.2. The procedure involves chopping, or grinding straw through a coarse screen (4 cm) to give particles of 1 to 5 cm in length, and spraying a solution of NaOH at low pressure on to the chopped or milled straw as it is mixed in the mixer-wagon. The concentrations of NaOH solution which have been used vary from 16% w/v (Pirie and Greenhalgh,

1978) to 12% (Coombe et al., 1979b) and 6.25% (Wilkinson and Gonzalez Santillana, 1978). Values for the digestibility of barley straws treated by this procedure are in Table 6.3.5.

Plate 6.3.3. Adding prilled sodium hydroxide to straw as it is harvested by forage harvester (Courtesy E. Owen)

In the trial with young calves given 90% of their diet DM as treated straw with a protein supplement, digestibility was also over 70%, and markedly higher than that determined for the untreated straw. With a lower proportion of straw in the diet the improvement, though substantial, was not as great (though it is likely that the digestibility of the straw fraction would have been increased to a greater extent than that indicated by the values for the whole diet (see Chapter 13). The lower (4%) addition of NaOH in the trial of Coombe et al. (1979a) was reflected in a 10-unit improvement in digestibility.

Jackson (1978) concluded from a review of some 36 digestibility trials that under ad libitum feeding regimes, treatment of straw with 4 to 5% NaOH and a 24 h "curing" period was likely to be reflected in a 10 unit increase in digestibility. At higher rates of adding of NaOH (7 to 9% of straw DM), digestibility should be expected to increase by 17 units.

Plate 6.3.4. Addition of sodium hydroxide to straw via a mixer waggon: drum containing solution of NaOH and pump.
(Courtesy Rowett Research Institute)

Plate 6.3.5. Addition of sodium hydroxide to straw via a mixer waggon: spraying boom and cover for mixer wagon .
(Courtesy Rowett Research Institute)

Table 6.3.5. Mechanical treatment of barley and wheat straw with NaOH in a mixer-wagon: digestibility of OM in vivo (%) of untreated and treated materials (given ad libitum to cattle)

Liveweight (kg)	Diet (% of total DM as straw DM)	Level of NaOH (g/kg straw DM)	Digestibility of OM		Reference
			untreated	treated	
		Barley straw			
114	90	75	47.1[1]	71.0	Wilkinson and Gonzalez Santillana (1978)
420	60	80	60.1[2]	72.3[2]	Pirie and Greenhalgh (1978)
268	80	40	57.3	67.0	Coombe et al. (1979a)
		Wheat straw			
277	90	40	54	69	Coombe et al. (1979b)

[1] Digestibility in vitro, straw alone

[2] Digestibility of DM

6.3.3.3. <u>By feed processor</u>

In this section, the treatment of baled straw by feed processing equipment is discussed. Principally, the equipment comprises specialised units for rolling and/or chopping (or grinding), treating and mixing straw with alkali (see Walker, Chapter 5, Rexen and Bach Knudsen, Chapter 6.2).

Some values for the digestibility in vivo (sheep) of the OM of barley and wheat straws treated by three feed processors, with or without addition of NaOH (at the recommended level of 40 to 50   g NaOH/kg straw DM) are in Table 6.3.6.

Table 6.3.6. Digestibility of OM in vivo of straws[1] treated on commercial   farms   by   feed   processing   equipment (Barber et al., 1979)

	Untreated	Treated
Wheat straw	44 (33 to 56)	63 (56 to 79)
Barley straw	51 (46 to 59)	71 (59 to 80)

[1]Values for straws calculated by difference when given to sheep at the maintenance level of feeding with dried grass of known feed value. Values are the mean (and range) of seven paired comparisons for each species of straw.

The average improvement in digestibility of OM was 20 percentage units (approximately 2 MJ metabolizable energy per kg DM), and there was apparently no difference between the three machines. A comparison between two straw processors in Scandinavia also showed no significant difference in the feeding value of the treated products (Arnason, 1978).

At the low volumes of concentrated NaOH solution (27% w/v) added to straw with the feed processing machinery (0.15 to 0.18 l/ kg straw DM), there is a rapid increase in temperature in the heap of treated material. This increase in temperature can reach 80 to 90 $^{\circ}$C (Jackson, 1978) and at these high temperatures there is risk of combustion. It is important to let the heat be dissipated as it is produced, either by allowing air to pass through the straw to dry it out, or, if the straw is to be ensiled, by having sufficient water to absorb the heat. With relatively dry straw, a prudent policy is to add water to reduce the content of DM to between 50 and 60% (i.e. to add about 0.60 to 0.70 litres of liquid per kg of straw DM). This relatively wet material may then be con- solidated in a silo and sealed, to prevent air movement and to inhibit the supply of oxygen to the heated straw. In this way the risk of fire can be minimised.

Plate 6.3.6. Application of sodium hydroxide in solution to straw via a feed processor (Courtesy T. Smith)

6.3.4. FARM-SCALE VERSUS INDUSTRIAL-SCALE

A common feature of the processes described in this section is that the reduction in particle size is less than that which occurs in the industrial process. The data in Tables 6.3.7 and 6.3.8 sug- gest, however, that the difference in feeding value between the two is likely to be relatively small. In these trials wheat straw was treated with 55 g NaOH/kg straw DM either in the field, via a metered-chop forage harvester (see Table 6.3.3) or by the indust- rial process (see Chapter 6.2). Intake of milled, industrially- treated straw diets was higher than that of the farm-treated mate- rial, but digestibility was lower for the milled material. Both these effects can be attributed to the difference in particle size of the treated straws, rather than to the method of treatment with NaOH. Performance of finishing beef cattle (Table 6.3.8) was similar between the two forms of treated straw.

Plate 6.3.7. Application of sodium hydroxide in solution to straw via a feed processor (Courtesy W.R. Butterworth)

Table 6.3.7. Comparison of farm-scale and industrial dry treatment of wheat straw with NaOH: effect on digestibility and intake in sheep (from Owen, 1981a)

	Untreated		Treated	
	Farm	Industrial	Farm	Industrial
Intake of straw OM				
(g/kg LW.day)	14.3	19.2	20.8	28.6
Digestibility of straw OM[1], %	44.7	39.5	62.7	55.3
Intake of DOM (g/kg LW.day)	6.39	7.58	13.0	15.8

[1] By difference: diet contained 80% straw DM, 20% supplement DM

Plate 6.3.8. Addition of alkali to maize cobs as it is ensiled in
a Silopresse (Courtesy E. Owen)

Table 6.3.8. Performance of finishing beef cattle given farm
treated or industrially treated wheat straw
(from Owen et al., 1980)

	Farm-treated	Industrially-treated
Diet, % of DM:		
Treated straw	33	33
Grass silage	33	33
Rolled barley	33	33
Intake of DM, kg/d	9.09	9.71
Liveweight gain, kg/d	1.05	1.14
Feed efficiency, kg/100 kg DM intake	11.6	11.7

6.3.5. SUMMARY

Various techniques for adding sodium hydroxide to straws and other fibrous by-products are described. Manual treatment may involve use of a garden watering can, or a batch mixer such as a cement mixer. The recommended rate of addition of NaOH is 50 g chemical in 0.5 to 2.0 l solution per kg crop dry matter (DM). Mechanical treatment in the field at harvest may be effected via a forage harvester. Alternatively, a complete diet mixer wagon or a proprietary feed processor may be used. In the latter two cases, concentrated (27% w/v) solutions of NaOH are generally added at 40 to 50 g NaOH per kg crop DM. A response in digestibility of OM in vivo to treatment of 10 to 15 units should be anticipated for all methods, provided mixing of alkali and feed is adequate and uniform. The difference in product feed value between farm-scale and industrial-scale treatment is small.

6.3.6. REFERENCES

Alawa, J., 1980. The nutritive value of wheat straw treated with sodium hydroxide using a farm method or an industrial method. MPhil Thesis. University of Reading, UK.

Arnason, J., 1978. Progress of straw treatment in Norway. Proceedings, 4th MAFF/ADAS Straw Utilisation Conference, Oxford, UK. pp. 22-28.

Barber, W.P., Ibbotson, C. and Palmer, F.G., 1979. The nutritional value of untreated and on-farm sodium hydroxide-treated straw. Paper N5.11, 30th Meeting of the European Association of Animal Production, Harrogate, UK.

Coombe, J.B., Dinius, D.A., Goering, H.K. and Oltjen, R.R., 1979a. Wheat straw-urea diets for beef steers: alkali treatment and supplementation with protein, monensin and a feed intake stimulant. J. Anim. Sci. 48: 1223-1233.

Coombe, J.B., Dinius, D.A. and Wheeler, W.A., 1979b. Effect of alkali treatment on intake and digestion of barley straw by beef steers. J. Anim. Sci. 49: 169-176.

Donefer, E., Adeleye, I.O.A. and Jones, T.A.O.C., 1969. Effect of urea supplementation on the nutritive value of NaOH treated oat straw. In: R.F. Gould (ed.): Cellulases and their Applications. Advances in Chemistry Series No 95, American Chemical Society, Washington DC, USA, pp. 328-342.

Jackson, M.G., 1978. Treated straw for animal feeding. FAO Animal Production and Health Paper No 10. FAO, Rome.

Jayasuriya, M.C.N. and Owen, E., 1975. Sodium hydroxide treatment of barley straw: effect of volume and concentration of solution on digestibility and intake by sheep. Anim. Prod. 21: 313-322.

Kategile, J.A., 1981. Simultaneous cutting and alkali treatment in a modified forage harvester for ensiling. In: J.A. Kategile, A.N. Said and F. Sundstøl (eds.): Utilization of Low Quality Roughages in Africa. AUN - Agricultural Development Report No 1, Aas, Norway, pp. 81-84.

Kategile, J.A. and Frederiksen, J.H., 1979. Effect of level of sodium hydroxide and volume of solution on the nutritive value of maize cobs. Anim. Feed Sci. Technol. 4: 1-15.

Kellaway, R.C., Crofts, F.C., Thiago, L.R.L., Redman, R.G. and Leibholz, J.M.L., 1978. A new technique for upgrading the nutritive value of roughages under field conditions. Anim. Feed Sci. Technol. 3: 201-210.

Mehrez, A.Z., El-Shinnawy, M.M., Abou-Raya, A.K. and El-Ayek, M., 1981. A proposed approach for evaluating NaOH-treated roughages. In: J.A. Kategile, A.N. Said and F. Sundstøl (eds.): Utilization of Low Quality Roughages in Africa. AUN - Agricultural Development Report No 1, Aas, Norway, pp. 25-27.

Ørskov, E.R. and Grubb, D.A., 1978. Validation of new systems for protein evaluation in ruminants by testing the effects of urea supplementation on intake and digestibility of straw with or without sodium hydroxide treatment. J. Agric. Sci. (Camb.) 91: 483-486.

Owen, E., 1981a. Straw: research work. In: B.A. Stark and J.M. Wilkinson (eds.): Upgrading of Crops and By-Products by Chemical or Biological Treatments. Ministry of Agriculture, Fisheries and Food, London, UK. pp. 1-9.

Owen, E., 1981b. Use of alkali-treated low quality roughages to sheep and goats. In: J.A. Kategile, A.N. Said and F. Sundstøl (eds.): Utilization of Low Quality Roughages in Africa. AUN - Development Report No 1, Aas, Norway, pp. 131-150.

Owen, E., Alawa, J., Thomas, C., Wilkinson, J.M. and Robb, J., 1980. Sodium hydroxide-treated straw - a comparison of farm-treated and industrially-treated straw as partial replacers of grass silage for beef cattle. Anim. Prod. 30: 388 (Abstract).

Pirie, R. and Greenhalgh, J.F.D., 1978. Alkali treatment of straw for ruminants. 1. Utilization of complete diets containing straw by beef cattle. Anim. Feed Sci. Technol. 3: 148-154.

Singh, M. and Jackson, M.G., 1971. The effect of different levels of sodium hydroxide spray treatment of wheat straw on consumption and digestibility by cattle. J. Agric. Sci. (Camb.) 77: 5-10.

Thiago, L.R.L., Kellaway, R.C. and Leibholz, J.M.L., 1979. Kinetics of forage digestion in the rumen. Annales Recherches Veterinaires 10: 329-331.

Thomas, C., 1978. Effect of sodium hydroxide treatment on the organic matter digestibility of hay from three tropical grasses. Trop. Agric. 55: 325-327.

Wilkinson, J.M. and Gonzalez Santillana, R., 1978. Ensiled alkali-treated straw. II. the nutritive value for young beef cattle of mixtures of ensiled or frozen alkali-treated straw and ryegrass silage. Anim. Feed Sci. Technol. 3: 133-142.

Chapter 6.4

ENSILING WITH SODIUM HYDROXIDE

by

J. Michael Wilkinson

Chief Scientists' Group
Ministry of Agriculture, Fisheries and Food
Great Westminster House, Horseferry Road
London SW1P 2AE, United Kingdom

6.4.1. INTRODUCTION

Ensiling is generally linked with the conservation of grass and forage crops. It implies preservation by bacterial fermentation of carbohydrates (usually glucose and fructose) to short-chain organic acids such as lactic and acetic.

By contrast, straws and other by-products which have been treated with NaOH have a relatively high pH (up to pH 12). In this context, "ensiling" may only mean storage of the treated crop in a silo, though some bacteria can tolerate alkaline conditions, and if sufficient moisture is present fermentation, albeit at a very restricted level, may occur.

In this section, the principles of ensiling are considered in relation to straws and by-products which have been treated with NaOH. Physical and compositional changes during ensilage of treated materials are discussed, and finally the effects of these changes on the nutritive value of the stored product are evaluated.

[1]Present address: "Chalcombe", Highwoods Drive, Marlow Bottom, Marlow, Bucks SL7 3PU, United Kingdom

6.4.2. PRINCIPLES

The main principle of ensiling is to maintain anaerobic conditions within the silo during the period of storage. This means that for relatively dry materials such as cereal straw and by-products of the maize crop, attention should be paid to compaction to minimise the amount of air entrapped in the mass of the material. Thus procedures such as chopping, rolling or treading and putting a weight on top of the silo after filling and sealing are beneficial. In addition, the presence of moisture in the crop is likely to improve compaction.

Prevention of air movement is also essential to the achievement of anaerobic conditons. Heat generated by the treatment of the crop with NaOH can enhance air movement (hot air rises). Complete sealing (by use of plastic, manure or earth) throughout storage is the most important single factor in the successful ensilage of all crops, including those which have been treated with NaOH.

6.4.3. PHYSICAL CHANGES

The most obvious physical changes associated with the ensiling of materials after treatment with NaOH are the production of heat, a loss of structural rigidity and a colour change - usually to a bright or deep yellow or yellow/brown. These changes reflect the effect of NaOH on the straw; in some cases they are more apparent after ensiling than before, particularly if the crop is ensiled at a relatively low level of moisture. The loss of structural rigidity can mean that the density of ensiled treated straw (compacted during filling) may be as high as that achieved by baling (100 kg/DM/m^3).

The production of heat can be very marked, especially when NaOH is added to dry straws in concentrated solution (see Chapter 6.3). If the intention is to store the treated material in a silo, then it is recommended that the dry matter content of the crop is in the range 40 to 70%. At these levels of dry matter there is sufficient moisture to achieve compaction, and there is also adequate moisture to act as a "sink" for the heat produced as a result of the reaction between NaOH and straw. Since storage is likely to be for weeks rather than days, the beneficial effects of heat on the rate (and extent) of reaction between alkali and straw are likely to be of relatively little consequence.

The change in colour can be used as a crude indicator of uniformity of distribution of alkali in the straw. Thus the presence of dark brown patches indicates over-treatment (and possibly over-heating), whilst areas which have apparently not changed colour are unikely to have been treated with the alkali.

6.4.4. COMPOSITIONAL CHANGES

Ensiling implies an extended period of storage; for weeks rather than hours. This storage period might be expected to enhance the extent of the reaction between alkali and straw so that the proportion of free alkali is reduced and the nutritive value of the product increased.

Chandra and Jackson (1971) studied the amount of unreacted alkali in maize cobs and wheat straw treated with 6% NaOH in 1.0 l solution per kg DM. They found (Table 6.4.1) that there was little decrease in residual alkali during the first 18 hours of storage and concluded that it was not worthwhile to allow more than 10 to 15 minutes of."curing" time. By contrast, straw treated with a lower volume of solution by a feed processor contained 46% of unreacted NaOH (added at 5% of the straw DM in 0.19 l solution per kg DM) in samples taken as the straw left the machine. After 15 days storage, this value had fallen to 8%. During this period, the digestibility of the straw in vitro had increased by 7 percentage units (Jackson, 1978).

Possibly, a much longer storage period may result in complete neutralization of excess alkali. This aspect was studied by Gonzalez Santillana (1977) who found in three experiments a consistent reduction in pH as a result of ensiling barley straw for 90 days after treatment with different amounts of NaOH in 1.2 l of solution/kg straw DM. The average reduction in pH was 1.9 units (Figure 6.4.1).

Wilkinson and Gonzalez Santillana (1978a) recorded significant amounts of both lactate and acetate in ensiled barley straw treated with different amounts of NaOH. In addition, butyrate was present in untreated straw, and in that treated with 2.5% of NaOH. The content of acetate most likely reflects liberation of acetyl groups from the xylans of the plant cell walls (Tarkow and Feist, 1969; Theander, 1981), but the presence of lactate and butyrate is indicative of microbial activity during the storage period.

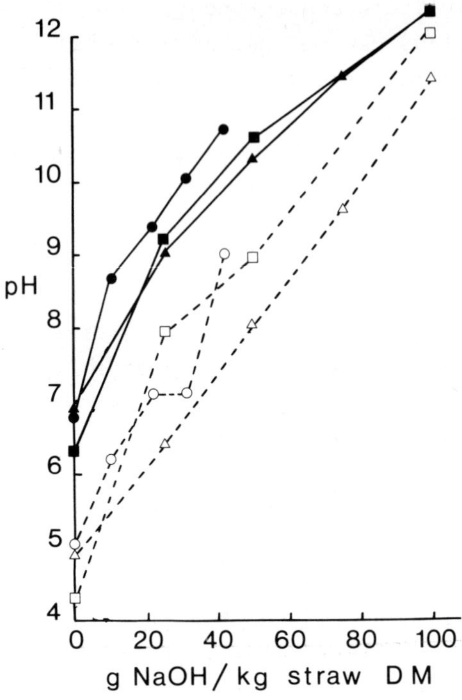

Figure 6.4.1. Changes in pH as a result of ensiling barley straw after treatment with NaOH (1.2 l solution/kg straw DM) (Gonzalez Santillana, 1977). ————— pH at ensiling; ----- pH after 90 days of ensiling; ●○ Experiment 1; ▲△ Experiment 2; ■□ Experiment 3.

Table 6.4.1. Residual NaOH (% of total added) in treated straws (Chandra and Jackson, 1971)

	Maize cobs	Wheat straw
Time after treatment		
5 minutes	23	32
30 minutes	22	30
1 hour	20	29
4 hours	19	29
18 hours	19	27

The effect of ensiling on digestibility was inconsistent (Figure 6.4.2), though at zero NaOH, digestibility of OM in vitro was lower after 90 days of ensiling than in samples taken at ensiling and frozen prior to analysis.

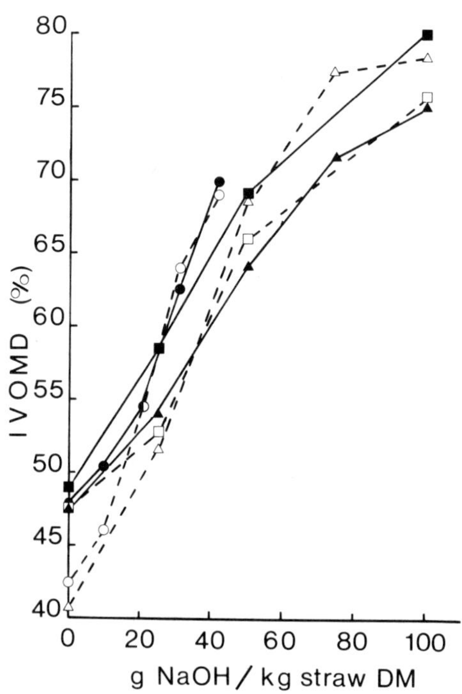

Figure 6.4.2. Changes in digestibility as a result of ensiling barley straw after treatment with NaOH (1.2 l solution/kg straw DM) (Gonzalez Santillana, 1977) ——— pH at ensiling; ----- pH after 90 days of ensiling; ●o Experiment 1; ▲ △ Experiment 2; ■ □ Experiment 3.

These results were confirmed in a subsequent trial with young calves, in which frozen and ensiled alkali-treated straws were given to the animals in different proportions with ensiled grass and a fixed amount of supplement (Wilkinson and Gonzalez Santillana 1978b). Intake of metabolizable energy and liveweight gains were very similar between the two forms of straw (Figure 6.4.3).

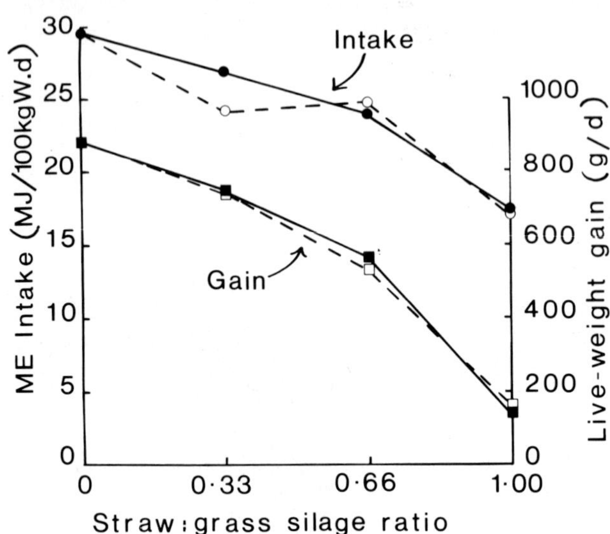

Figure 6.4.3. Intake of metabolizable energy (ME) and live-weight
           gain by calves given ensiled (---) or frozen (——)
           alkali treated straw and grass silage in different
           proportions (Gonzalez Santillana, 1977)

In these trials, the content of DM was relatively low (42 to
45%) as a result of the addition of 1.2 l of solution per kg straw
DM. This is ten times the amount added when straw is treated by
use of feed processing equipment (see Chapter 6.3 and Sundstøl et
al., 1979). The reasons for working with relatively wet material
were a) to facilitate uniform distribution of alkali in the straw,
b) to simulate the harvesting and treatment of moist straw in the
field and c) to aid compaction in the silo.

Sundstøl et al. (1979) found that when relatively dry straw was
treated with only 0.1 l of NaOH solution per kg straw DM (i.e.
similar amount of liquid to that added via feed processing equipment) and ensiled, there was a progressive increase in digestibility in vitro during a 63-day storage period (Figure 6.4.4).

In a further trial, barley straw treated with 40 g NaOH/kg
straw DM was ensiled for 21 days at different contents of DM. Increasing the volume of solution from 0.1 to 0.4 l/kg straw DM
gave only a slight benefit to digestibility. Thus, provided distribution of alkali is good, and provided a storage period of at

least 9 weeks is allowed, the digestibility of ensiled straw treated with 4 to 5% NaOH may be expected to reach 70% (in vitro).

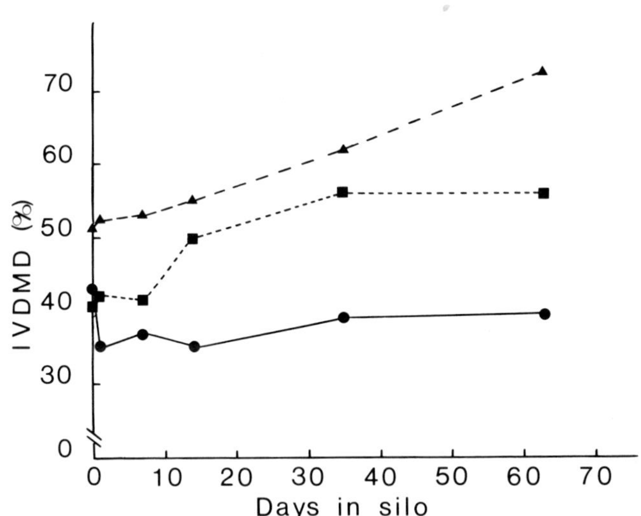

Figure 6.4.4. Effect of time in store on digestibility in vitro of
wheat straw treated with NaOH (0.1 l/kg straw DM)
(Sundstøl et al., 1979). ▲-----▲ 5.4% NaOH;
■···■ 2.7% NaOH; ●——● 0% NaOH.

Further, the alkali may be added in the presence of a wide
range of moisture (0.1 to 1.2 l/kg straw DM). The proviso made in
the previous Chapter, of using higher volumes of liquid with
manual treatment, to aid distribution of alkali, is also relevant
in the context of ensiling.

Straw contains relatively little fermentable carbohydrate compared to conventional grass and forage crops. Thus there is less
opportunity for fermentation, even if there is adequate "free"
water to allow bacterial growth to take place. But with other
forages the addition of NaOH, followed by ensilage, may give rise
to poorly-preserved material. Indeed, the addition of alkali may
buffer the fermentation to the extent that the normal decrease in
pH and dominance of lactic acid bacteria is prevented (Bolsen,
Tetlow and Wilson, 1983). Examples of the effect of treating

lucerne, whole-crop barley, ryegrass, ryegrass straw, whole-crop maize and maize stover with NaOH on composition of the ensiled products are in Table 6.4.2.

With lucerne, addition of 3.0% NaOH exacerbated the problem of secondary fermentation so that the quality of preservation of the alkali-treated materials was even lower than that of the untreated crops. With whole-crop barley, addition of 5% NaOH induced a clostridial fermentation in the crop harvested at 28% DM, but prevented it in later-harvested material (36.5% DM). In another study, the reverse effect was recorded, though here the level of addition of NaOH was lower (3%).

Ryegrass straw, ensiled with the addition of molasses, urea and limestone at a low content of DM (23%) fermented to give a well-preserved product in the absence of alkali, but was poorly-preserved following addition of 4.5% hydroxide, which clearly buffered acid production and induced a secondary fermentation of lactic acid to butyric. Whole-crop maize and maize stover were relatively well-preserved when ensiled with 4% NaOH (Table 6.4.2).

Ryegrass, ensiled at 28 to 30% DM with a short wilting period was poorly-preserved when treated with 3.5% NaOH, but well-preserved when either untreated or treated with 7.0% NaOH (Table 6.4.2). In the latter case, fermentation was inhibited by the relatively high level of addition of alkali. Addition of the lower rate of NaOH gave little improvement in digestibility of OM in vitro, but at the higher rate a 20-unit increase was observed (Herrod-Taylor, 1980).

Ensiled, untreated whole-crop barley gave lower values for digestibility in vitro than that of the crop before ensiling (Table 6.4.3). Addition of NaOH appeared to prevent the decrease in digestibility during storage at least in the case of the crops harvested at 38% and 46% DM, which were well-preserved.

The treatment of straws with NaOH prior to ensilage appears to be reflected in reasonable stability of the product in aerobic conditions once the silo is opened. This stability for several days in air occurs despite pH values at which rapid bacterial and fungal growth would be expected (Wilkinson and Gonzalez Santillana, 1978a). Possibly the relatively low content of "free" water, and the restricted amount of fermentation during storage are contributory factors to stability.

Table 6.4.2. Composition of silages made from crops either untreated or treated with sodium hydroxide

Crop	NaOH (% of DM)	DM (%)	pH	Fermentation acids (% of DM)			NH_3-N (% total N)	Reference
				Lactic	Acetic	Butyric		
Lucerne	0	26.0	5.4	0.60	NR[1]	NR	23.7	
	3	25.7	7.6	0.13	NR	NR	34.9	Flipot et al.(1976)
Lucerne	0	33.2	5.0	0.66	NR	NR	12.4	
	3	34.3	6.0	1.52	NR	NR	15.2	
Whole-crop barley	0	27.2	3.9	2.9	3.5	0	10.9	
	5	27.0	7.0	0	1.7	6.8	3.5	Bolsen et al.(1983)
Whole-crop barley	0	34.4	4.9	6.1	0.7	2.7	11.4	
	5	39.9	10.1	0.2	1.4	0	3.0	
Whole-crop barley	0	39.1	5.2	2.1	0.8	0.9	16.1	
	3	35.8	6.5	0.8	1.9	4.4	33.5	Mowat (1976)
Whole-crop maize	0	40.2	4.0	3.98	8.90	NR	6.5	
	4	41.7	4.4	3.01	14.9	NR	1.9	Flipot et al.(1976)
Maize stover	0	30.7	4.7	3.48	1.63	NR	11.4	
	4	33.2	5.3	5.76	2.39	NR	12.0	
Ryegrass straw[2]	0	22.9	3.8	6.3	5.9	1.4	NR	
	4.5	22.7	4.9	3.0	10.6	11.8	NR	Schultz et al.(1974)
Ryegrass	0	28.6	4.0	6.6	0.0	0.0	8.9	
	3.5	26.7	5.6	6.0	9.2	9.2	4.3	Herrod-Taylor (1980)
	7.0	30.4	10.2	0.6	1.0	0.0	1.3	

[1] Not recorded

[2] Ensiled with either water or 2.25% NaOH + 2.25% KOH in solution + 20% molasses + 1% urea + 0.5% limestone

Table 6.4.3. IVOMD of whole-crop barley treated with NaOH prior to ensiling; effect of alkali and of the storage period (Bolsen, Tetlow and Wilson, unpublished data)

	Untreated		Treated with 5% NaOH	
	Before ensiling	After 60-days ensiling	Before ensiling	After 60-days ensiling
Harvest 1, low DM	69.5	64.7	81.3	79.7
Harvest 2, medium DM	71.3	63.6	84.7	84.6
Harvest 3, high DM	69.9	66.1	81.3	84.7

6.4.5. NUTRITIVE VALUE OF THE ENSILED PRODUCT

The data in Figure 6.4.3 indicate that ensiling straw after treatment with NaOH is unlikely to be reflected in reduced animal performance, compared to freshly-treated material. However, it is important to note that in this trial there was a short-period of storage (in a freezer) of 10 to 14 days before the material was given to the cattle. Gonzalez Santillana and Wilkinson (1978) noted that the reaction between NaOH and straw continued during a 24-day period of storage of the material at -15°C. Thus it appears likely that benefits to digestibility can accrue during the storage period; if the amount of liquid added to the straw is low, ensiling should be for not less than 60 days (see Figure 6.4.4). Garmo (1981) found that the storage of ensiled alkali-t-reated straws for 120 days was reflected in a 3-unit increase in digestibility of OM in vitro, compared to material ensiled for 60 days. This improvement occurred in relatively wet materials which had received 1.0 l of alkali solution per kg straw DM. It may, therefore, be concluded that the longer the period of ensilage the better, provided the silo is completely sealed.

Greenhalgh et al. (1978) compared "ensiled" barley straw, trea-ted with a 16% w/w solution of NaOH at the rate of 500 g/kg straw DM (to give 80 g NaOH/kg straw DM), with the same material treated on a daily basis. The "ensiled" material was, however, stored as an open heap in a barn in a 6-tonne lot for 60 days. Both "en-siled" and freshly-treated straws were mixed with concentrates and immediately given to cattle, initially 385 kg liveweight for a 66-day period. The results are in Table 6.4.4.

Table 6.4.4. Alkali-treated barley straw in complete diets for beef catttle: comparison of stored material (60 days) with straw treated daily (Greenhalgh, Pirie and Shin, 1978)

	"Ensiled"	Treated daily
Level of NaOH, % of DM	8	8
Proportion of straw in diet	0.6	0.6
Intake, kg DM/d	10.23	9.06
DMD in sheep, %	69.7	62.5
Liveweight gain, kg/d	1.08	1.00
Liveweight gain, kg/100 kg DM intake	10.6	11.0

The period of storage was beneficial to the extent that the cattle ate more of the diet which contained the stored straw, which also had a higher digestibility when given to sheep (by 7 units). Liveweight gain was 8% higher, but efficiency of feed use (weight gain per unit of feed intake) was marginally lower than with the freshly-treated straw.

Table 6.4.5. Liveweight gains by cattle given ensiled alkali-treated crops and by-products

Crop	Proportion in diet	Liveweight gain, kg/d		Reference
		Untreated	Treated with NaOH	
Wheat straw	0.90	0.11	0.33	Coombe et al. (1979)
Oat straw	0.95	0.14	0.56	Kellaway et al. (1978)
Ryegrass straw	0.82	0.40	0.45	Schultz and Ralston (1974)
Whole-crop barley	0.50	0.90	1.01	Mowat (1976)

In other trials (Table 6.4.5) higher rates of liveweight gain have been recorded with ensiled, alkali-treated straws compared to untreated straws. However, where preservation was poor (Schultz and Ralston, 1974) or where in addition the proportion of treated crop in the diet was relatively low (Mowat, 1976) the response in animal performance was relatively small.

6.4.6. ENSILING TREATED STRAW BELOW DIRECT-CUT FORAGE CROPS

The production of effluent from direct-cut (unwilted) forage crops during the ensiling period can cause a problem with regard to its collection and disposal, particularly in countries which have introduced regulations designed to prevent pollution of rivers and streams. The high biochemical oxygen demand of the effluent, coupled with the fact that its production is very extensive in the first two to three weeks after the crop is ensiled makes the risk of pollution at this time a significant problem.

One approach to reduce the risk of pollution is to ensile straw below the unwilted forage crop. The juice from the wet crop is collected in the dry straw. A trial with a very wet crop of rape demonstrated that a substantial amount of effluent could be retained in small tower silos by both untreated and NaOH-treated straw (Table 6.4.6). Despite a low content of DM in the treated straw, it absorbed more effluent than the untreated material which was much drier.

Addition of effluent to the straws was reflected in an improved digestibility of the untreated straw (from 54.6% to 61.6%), but there was no increase in the case of the treated straw so that the net energy values of the two straws was estimated to be very similar.

Table 6.4.6. Production of effluent, and its absorption in the silo by straw ensiled under a wet crop of rape[1] (Sundstøl, Arnason and Kjus, 1980, and Sundstøl (personal communication)

	Rape + un-treated straw	Rape + NaOH treated straw[2]
Effluent produced from silo (l/tonne)	200	200
Effluent absorbed by straw (l/kg)	3.5	3.6
Digestibility of straw OM (%)	61.6	67.7

[1]Rape was 10% dry matter content, dry untreated straw 87.4% DM
[2]6% NaOH (DM basis). Dry matter content of treated straw was 31%

6.4.7. SUMMARY

Ensiling straws and other fibous by-products after treatment with sodium hydroxide essentially involves preservation by anaerobic storage, rather than by fermentation. Complete sealing of the treated product is essential to prevent oxidative loss of dry matter. The storage period is usually associated with a reduction in pH, and where concentrated solutions of NaOH have been added to the straw there is also a reduction in the amount of unreacted alkali in the stored product. At relatively high levels of moisture (70% or more), and in crops where the content of fermentable carbohydrate is also relatively high, addition of NaOH may buffer the fermentation to the extent that a secondary (clostridial) pattern of fermentation is induced. Ensiling treated crops and by-products is unlikely to be reflected in a marked change in feed value compared to similar material offered to animals directly after treatment. In some cases, however, small improvements in digestibility and voluntary intake associated with an extended period of storage have been recorded.

6.4.8. REFERENCES

Bolsen, K.K., Tetlow, R.M. and Wilson, R.F., 1983. The effect of calcium and sodium acrylate on the fermentation and digestibility of ensiled whole-crop wheat and barley. Anim. Feed Sci. Technol. (in press).

Chandra, S. and Jackson, M.G., 1971. A study of various chemical treatments to remove lignin from coarse roughages and increase their digestibility. J. agric. Sci. (Camb.) 77: 11-17.

Coombe, J.B., Dinius, D.A., Goering, H.K. and Oltjen, R.R., 1979. Wheat straw - urea diets for beef steers: alkali treatment and supplementation with protein, monensin and a feed intake stimulant. J. Anim. Sci. 48: 1223-1233.

Flipot, P., Mowat, D.N., Parkins, J.J. and Buchanan-Smith, J.G., 1976. Ensiling characteristics of silages treated with sodium hydroxide. Can. J. Plant Sci. 56: 935-940.

Garmo, T.H., 1981. Ensiling of alkali-treated straw. In: J.A. Kategile, A.N. Said and F. Sundstøl (eds.): Utilization of Low Quality Roughages in Africa. AUN - Agricultural Development Report No 1, Aas, Norway, pp. 55-59.

Gonzalez Santillana , R., 1977. The ensiling of alkali-treated straw. MPhil thesis, University of Reading, UK.

Gonzalez Santillana, P. and Wilkinson, J.M., 1978. Effect of time of storage and method of drying on the digestibility in vitro of alkali-treated barley straw. Anim. Feed Sci. Technol. 3: 381-384.

Greenhalgh, J.F.D., Pirie, R. and Shin, H.T., 1978. Nutritive value of barley straw ensiled after alkali treatment. Anim. Prod. 26: 400-401 (Abstract).

Herrod-Taylor, M.M., 1980. The ensiling of alkali treated grass. MPhil Thesis, University of Reading, UK.

Jackson, M.G., 1978. Treating straw for animal feeding. FAO Animal Production & Health Paper No 10. FAO, Rome.

Kellaway, R.C., Crofts, F.C., Thiago, L.R.L., Redman, R.G. and Leibholz, J.M.L., 1978. A new technique for upgrading the nutritive value of roughages under field conditions. Anim. Feed. Sci. Technol. 3: 201-210.

Mowat, D.N., 1976. Sodium hydroxide treated barley silage. Research Report, Department of Animal Science, University of Guelph, Ontario, Canada.

Shultz, T.A. and Ralston, A.T., 1974. Effect of various additives on nutritive value of ryegrass straw silage. II. Animal metabolism and performance observations. J. Anim. Sci. 39: 926-930.

Shultz, T.A., Ralson, A.T. and Shultz, E., 1974. Effect of various additives of ryegrass straw silage. I. Laboratory silo and in vitro dry matter digestion observations. J. Anim. Sci. 39: 920-925.

Sundstøl, F., Arnason, J. and Kjus, O., 1980. Alkali treatment of straw in the baler, in the bottom of forage silos and experiments with pelleted straw added fat. Scientific Report no 202, Department of Animal Nutrition, Agricultural University of Norway.

Sundstøl, F., Said, A.N. and Arnason, J., 1979. Factors influencing the effect of chemical treatment on the nutritive vlue of straw. Acta Agric. Scand. 29: 179-190.

Tarkow, H. and Feist, W.C., 1969. A mechanism for improving digestibility of ligno-cellulose materials with dilute alkali and liquid ammonia. In: R.F. Gould (ed.): Cellulases and their Applications. Advances in Chemistry Series 95: American Chemical Society, Washington, D.C., USA, pp. 197-218.

Theander, O., 1981. Chemical composition of low quality roughages as related to alkali treatment. In: J.A. Kategile, A.N. Said and F. Sundstøl (eds.): AUN - Agricultural Development Report No 1, Aas, Norway, pp. 1-15.
Wilkinson, J.M. and Gonzalez Santillana, R., 1978a. Ensiled alkali-treated straw. I. Effect of level and type of alkali on the composition and digestibility in vitro of ensiled barley straw. Anim. Feed Sci. Technol. 3: 117-132.
Wilkinson, J.M. and Gonzalez Santillana, R., 1978b. Ensiled alkali-treated straw. II. The nutritive value for young beef cattle of mixtures of ensiled or frozen alkali-treated straw and ryegrass silage. Anim. Feed Sci. Technol. 3: 133-142.

Chapter 7

AMMONIA TREATMENT

by

Frik Sundstøl[1] and Ewen M. Coxworth[2]

[1] Department of Animal Nutrition
Agricultural University of Norway
P.O. Box 25, N-1432 Aas-NLH
Norway

[2] Resources Sector
Saskatchewan Research Council
20 Campus Drive, Saskatoon
Saskatchewan S7N OX1
Canada

7.1. INTRODUCTION

The marked improvement in the feeding value of straw obtained when treated with caustic soda (NaOH) encouraged a number of people to study other chemicals and methods for upgrading of straw as feed. As can be seen from Chapter 8 a great number of other chemicals have been tried throughout the century. One of the first systematic studies of the effect of ammonia on straw was carried out in Germany about 50 years ago (Kronberger, 1933). In a laboratory experiment he studied the effect of increasing amounts of ammonia on the colour and water solubility of wheat straw. Treatment temperatures were 10, 30 and 45°C and time of treatment 1 week, 4 weeks or 5 weeks. The intensity of the colour increased with increasing NH_3 dosage, increasing temperature and increasing length of treatment. The solubility of the straw increased with increasing ammonia treatment as shown below:

Treatment	Water soluble organic matter, %
Distilled water	1.6
" " + 0.34% NH_3	3.4
" " + 1.70% "	4.7
" " + 3.40% "	10.7

In the USSR Nikolaeva (1938) used aqueous ammonia and obtained a marked increase in digestibility of the straw. Thus NH_4OH solutions were recommended for treatment of straw to be fed to farm animals.

In 1954 F. Juncker in Denmark applied for a patent (2141/54) for ammonia treatment of straw for animal feed. Juncker reported that ammonia-treated straw was used as feed for his cattle and that satisfactory results were obtained. Inspired by the good experience in Denmark the Norwegian companies Norwegian Feed Conservation Assocation (NOFO) and Norsk Hydro, in cooperation with the Department of Animal Nutrition, Agricultural University of Norway, carried out a series of laboratory scale and "semi-practical" experiments with ammonia treatment of straw. In an unpublished report from Norsk Hydro it was shown that the NH_3-content of ammoniated wheat straw increased with increasing moisture content (Gunnes, 1957). The effect of temperatures between 21°C and 74°C was also studied and it was concluded that increasing temperature had a negative effect upon the NH_3-content, probably because of decreased NH_3-vapour pressure. The treatment time in these experiments varied from 1/2 h to 23 hours and no clearcut effect of time was observed. Neither had increasing NH_3-vapour pressure above 1 atmosphere any significant effect on the N-content of the straw. When ammonia-treated straw with a high moisture/high NH_3-content was dried at 100-105°C most of the ammonia evaporated off.

The general trend of the experiments was that after such drying the N content of the straw was about 8 g/kg DM higher than that of untreated straw, regardless of treatment conditions.

Digestibility experiments with sheep fed straw treated with steam and thereafter ammonia showed that the digestion coefficient

(%) for organic matter was increased up to 60.5 (Fyrileiv and Ulvesli, 1958). The nitrogen content of the straw was increased up to 13.4 g/kg DM.

Based on these experiments it was concluded that, in spite of the positive effect obtained, the ammonia treatment could not compete with the Beckmann method for treatment of straw which was widely adopted by the Norwegian farmers at that time.

In 1959 Zafren (USSR) published results from a production trial with growing bulls fed untreated or ammonia-treated rye straw. Based on the daily gain of the animals it was estimated that the energy value of the treated straw was more than twice that of un-treated straw. The extra nitrogen supplied could replace 20-25% of the protein-N in the ration.

In Poland Chomyszyn et al. (1960) soaked chopped straw in an NH_3 solution (40 g NH_3/kg straw), mixed with hay and field-peas and stored for three days before feeding it to growing wet-hers. The control diet was treated similarly except that the straw was soaked in water. From the results obtained the authors conclu-ded that ammoniation of straw

- improved the palatability of the ration
- increased the digestibility, particularly that of crude pro-tein and crude fiber
- increased the rate of gain and the efficiency of feed utiliza-tion
- ammonia produced NPN in the straw could replace 30-40% of the protein in the diet of fattening lambs

Zafren (1961) described a method whereby straw was treated with 3% ammonia, or 120 l of a 25% solution per tonne. The treatment could be carried out in pits, trenches or stacks covered with plastic film. It was necessary to expose the straw to air for 4 or 5 days to get rid of excess ammonia before feeding it to animals. In an experiment with yearling cattle the daily gain was signifi-cantly greater for ammonia-treated straw than for untreated straw.

Other scientists have tried aqueous as well as anhydrous ammo-nia for treatment of straw with varying effect. In many of these studies the intention has been to increase the digestibility and energy utilization of the straw, combined with a supplementation of non-protein nitrogen for microbial protein synthesis in the forestomachs of ruminants.

7.2. AMMONIA AS A CHEMICAL FOR TREATMENT OF STRAW

At normal pressure and temperature ammonia (NH_3) is a colourless gas with a penetrating odour. The gas is easily liquefied under pressure and dissolves readily in water. At 20°C the vapour pressure is 8.5 atmospheres and the specific gravity at 0°C is 0.63. Boiling point at atmospheric pressure is -33.4^{O}C and freezing point is -77.7^{O}C. Ammonia is normally available at a high degree of purity, >99.8 per cent NH_3 (weight basis). It is packed in steel cylinders of varying size or in tank-waggons and tankers.

Ammonia is extensively used as a raw material in the fertilizer industry or as fertilizer directly. Its use as a fertilizer is of great importance in the discussion of ammonia as a chemical for treatment of straw. The advantage of an existing distribution system will allow the dealer to sell the ammonia at a much lower price than otherwise possible.

Urea is a crystalline solid produced technically from ammonia and carbon dioxide. It is easily dissolved in water and is widely used as a fertilizer or fed directly to ruminants as a source of NPN (46.6% N).

It is inherently safer to use than ammonia. Many countries which do not have access to anhydrous or aqueous ammonia, will usually have urea available since it is one of the commonest nitrogen fertilizer sources. Hence methods of using urea to generate ammonia in situ in straw are very important for many places. Urease enzymes which decompose urea to ammonia are present in some crop residues, and are also present in certain seed meals (e.g. raw soybean meal) and in urine. Various methods to use these urease sources and urea to generate ammonia in mixtures with straw are discussed in section 7.4.3.2.

7.3. FACTORS INFLUENCING THE EFFECT OF AMMONIA TREATMENT

Like other chemical reactions there are a number of factors that influence the effect of ammonia treatment of straw (Westgaard, 1981).

7.3.1. Amount of ammonia

One of the first factors to be studied was the effect of dosage of ammonia to straw. Nikolaeva (1938) reported that satisfactory results were obtained with 5% NH_4OH. Waiss et al. (1972) found that the enzyme solubility of rice straw treated with 5.2% ammonia was markedly higher than when 2.6% was applied. A further increase from 5.2 to 7.8% of added ammonia had only a marginal effect when the straw was treated at ambient temperature (22°C). Waagepetersen and Vestergaard Thomsen (1977) compared 3.4, 4.4 and 5.9% of ammonia when treating barley straw in a laboraory scale for up to 14 days, at temperatures ranging from 15 to 55°C. At 15°C and partly at 30°C there was a positive effect of increasing the dosage of ammonia from 3.4 to 5.9%. However, since an increasing amount of water was added along with the ammonia, the results represent a combined effect of the two. By treating oat straw with from 1.0 to 5.5 g NH_3 per 100 g straw DM, Sundstøl et al. (1978) showed a great improvement in the in vitro digestibility by increasing the ammonia dosage from 1.0 to 2.5 and a slightly higher digestibility at 4.0 g NH_3 added. The straw was treated for four weeks at 4, 17 and 24°C (Figure 7.1).

In a systematic study of the factors influencing the effect of ammonia treatment, Kernan and Spurr (1978) found an optimum level of 3.0-4.0% NH_3 above which the effect of increasing dosage was negative.

Based on a number of experiments Sundstøl et al. (1979) concluded that even if some beneficial effects from increasing the NH_3 level up to 5.5 and 7.0 are obtained, it can hardly be justified to exceed 4.0% NH_3 on straw DM basis. The economical optimum for dosage of NH_3 lies probably between 2.5 and 3.5% of DM.

7.3.2. Temperature

In general chemical reactions run faster at high rather than at low temperatures. When anhydrous ammonia is injected into a stack of straw the temperature rises rapidly reaching a maximum 2-6 hours after injection. According to Waagepetersen, Vestergaard Thomsen and Kristensen (1976) the increase varies between 40 and 60°C, depending on the temperature at the start, dosage of ammonia, moisture content, and other factors. The highest temperature

is at the top of the stack (Kristensen et al., 1977). The decline
in temperature also depends on a number of factors but ambient
temperature is normally reached within 1-2 weeks after injection
(Waagepetersen, Vestergaard Thomsen and Ubbesen, 1976; Kristensen
et al., 1977).

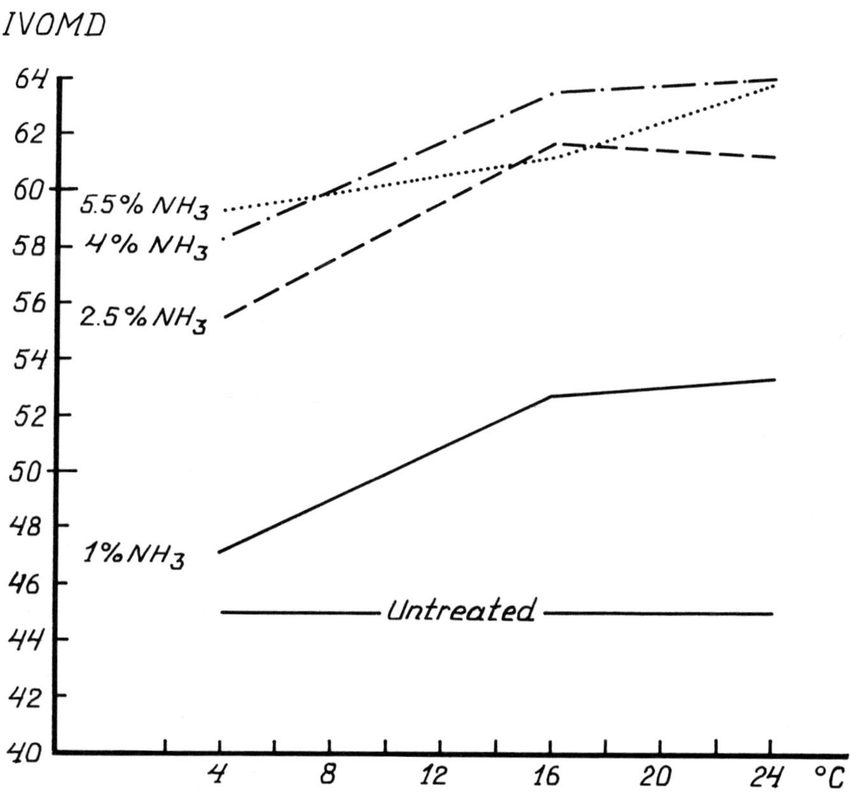

Figure 7.1. The effect of varying the NH$_3$ level in straw DM, and
treatment temperature, on the in vitro digestibility
of organic matter (IVOMD) in dry oat straw treated for
4 weeks (Sundstøl et al., 1978)

Liquid anhydrous ammonia vapourizes gradually from the bottom
of a stack rendering a drastic cooling of the surroundings. Danish

studies revealed that the temperature at the bottom of such stacks stayed below 0°C for five days after injection, and reached the ambient temperature of 13-14°C after another 6-7 days (Kristensen et al., 1977).

The immediate rise in temperature after injection is considered to be of limited significance. The ambient temperature will have a major effect on the speed of reaction between the chemical and the straw. At temperatures near 100°C the reaction is almost immediate whereas the reaction is extremely slow near 0°C. Therefore, in countries with cold climates it is important that the treatment is done as soon as possible after harvest of the grain while the temperature is still relatively high. Waagepetersen and Vestergaard Thomsen (1977) showed in short term experiments (3-7 days) that the effect of temperature increased up to 45°C. Sundstøl et al. (1979) also found a positive effect of increasing temperatures from -20°C to +25°C, but the effect was less marked at 8 weeks than at 4 weeks of treatment (Figure 7.2). This indicated that at least part of the temperature effect can be compensated for by increasing the time of treatment. Similar results were obtained by Becker and Pfeffer (1977) and Richter et al. (1980).

At short treatment times (6-15 days) Kernan and Spurr (1978) found a positive effect of increasing the treatment temperature from -20°C to +20°C. At longer treatment times (30-90 days) increasing temperature had a negative effect both on nitrogen content and in vitro OMD of ammonia-treated wheat straw. The explanation for these conflicting results is not yet found.

From Spain Alibes et al. (1983) reported that the N-content as well as the digestibility of straw was higher when treated with ammonia in summer (38°C) than when treated in winter (7°C).

7.3.3. Time of treatment

Compared with sodium hydroxide, ammonia is a slow reacting chemical. As mentioned already, the time necessary to treat straw and other crop residues depends on the temperature. According to Sundstøl et al. (1978) the effect of treatment increased up to 4 weeks at 17 and 25°C whereas at lower temperatures (4 and -20°C) the increase continued at least up to 8 weeks of treatment. Thus, below 0°C a very long treatment time is required, whereas treat-

ment in ovens may take 24 hours or less. At intermediate temperatures the treatment time required is also intermediate. In the
beginning of on-farm ammonia-treatment studies in Norway, three
weeks of treatment was applied, but this proved to be too short
under the prevailing conditions. In the study of Kernan and Spurr
(1978) there was a positive effect of increasing the time of
treatment, particularly at low temperatures.

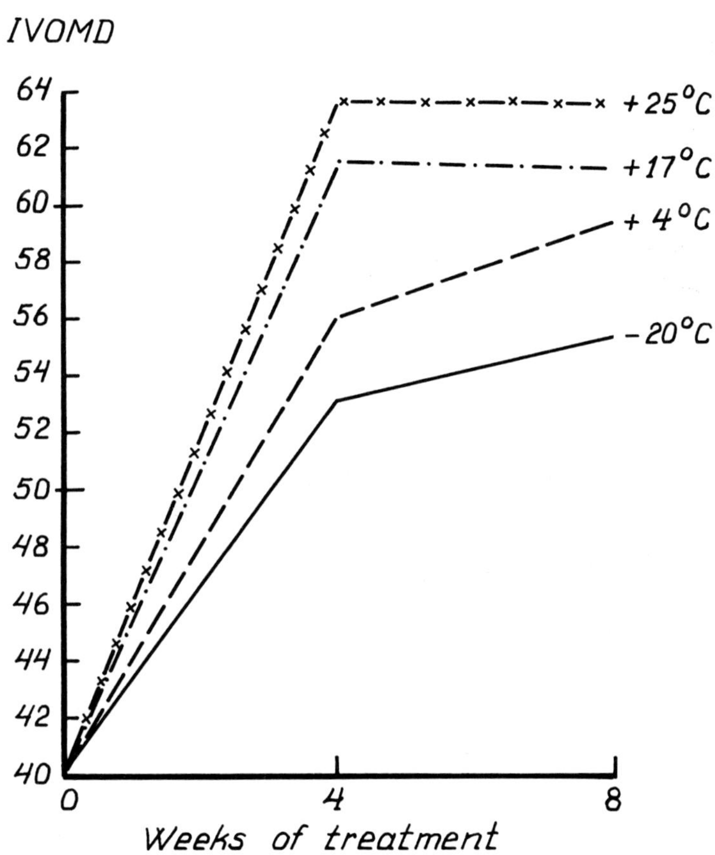

Figure 7.2. The effect of treatment temperature and time on the in
vitro organic matter digestibility (IVOMD) of oat
straw containing 25% moisture and treated with 3.4%
NH_3 in DM (Sundstøl et al., 1978)

Under temperate conditions it is therefore recommended to keep the
stack sealed for at least 8 weeks and it is advocated not to open
the stack before the feed (straw) is needed. According to Tohrai
et al. (1978) 5 d of treatment was sufficient when rice straw was
treated with 2.5% ammonia at 45°C. When treating barley straw
with aqueous ammonia at 20°C Hartley and Jones (1978) obtained
no improvement in the in vitro digestibility by extending the
period of treatment from 1 week to 4 or 13 weeks.

As a general rule for length of treatment time, the following
suggestions have been given:

Temperature	Kernan et al. (1977)	Sundstøl et al. (1978)
0°C	60 days	
Below 5°C		more than 8 weeks
10°C	30 "	
5-15°C		4-8 weeks
20°C	15 "	
15-30°C		1-4 "
Above 30°C		less than 1 week

7.3.4. Moisture content

Moisture content is another important factor determining the
effect of ammonia treatment. Waiss et al. (1972) concluded that an
optimal effect of the ammonia treatment was obtained at a moisture
content of about 30%. Sundstøl et al. (1979) found that increasing
the moisture content of the straw from 12 to 50% had a positive
effect on the in vitro organic matter digestibility of ammonia-
treated straw (Figure 7.3). These results were confirmed by Solaiman et al. (1979).

Experiments conducted by Musimba and Sundstøl (see Sundstøl et
al., 1979, p. 186) indicated that anhydrous ammonia had no positive effect on straw with extremely low moisture content (3.3%).
Since the range of moisture content between 3 and 12% was not
covered properly before, Borhami and Sundstøl (1982) dried straw
to a moisture content of 2.5, 5.0, 7.5 and 10.0%. The corresponding in vitro OMD was 52.1, 58.5, 59.1 and 66.0% respectively when
treated with 2.0% anhydrous NH_3 for 6 weeks. For untreated material the OMD was 49.1%. When aqueous ammonia was used, considerably greater effect was obtained even at 10% moisture indicating

that the optimal level of moisture probably lies between 15 and 20%. Even if higher moisture content may give a better effect of the treatment, this advantage is outweighed by the greater risk of storage damage e.g. mould. Another disadvantage of high moisture content is that the handling problems are enhanced.

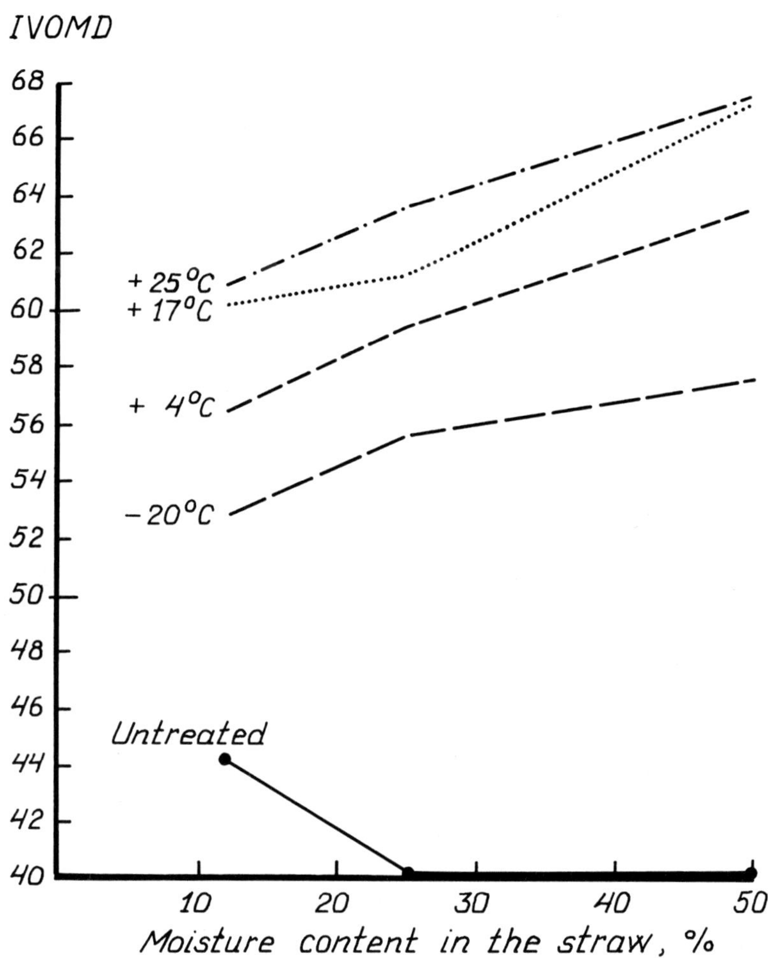

Figure 7.3. The effect of varying moisture content and treatment temperature on the in vitro organic matter digestibility (IVOMD) of straw treated with 3.4% NH_3 in DM for 8 weeks (Sundstøl et al., 1978)

In arid and tropical areas with very dry straw the advantage of using aqueous ammonia seems obvious. If anhydrous ammonia is the only alternative a wetting of the dry straw might be considered (Said et al., 1982).

7.3.5. Type and quality of material being treated

At an early stage of the ammonia treatment it was observed ·that different materials responded differently to the treatment. Work by Waiss et al. (1972) indicated that the effect of ammonia was more pronounced for plant materials with an initially low digestibility. This is in accordance with the work of Mwakatundu and Owen (1974) who treated grass cut at varying stages of maturity using caustic soda.

Kernan et al. (1979) studied the effect of ammonia treatment of 14 cultivars of wheat, oats and barley grown at four locations. They found that the improvements in crude protein were 8.1, 4.7 and 5.3 percentage units for wheat, oats and barley straw respectively. The corresponding improvements in digestible organic matter were 8.6, 6.1 and 6.6 percentage units. They concluded from the experiments that the resultant digestibility after ammoniation was highly dependant on the quality of the starting material. Horton and Steacy (1979) and Horton (1981) also found greater improvement in the digestibility of wheat straw than in barley and oat straw, although the digestibility after treatment was still highest for the latter two.

Chomyszyn et al. (1972) treated chopped rape straw with 2.5% NH_3 in aqueous form for 4 days at room temperature in polyethylene bags. The treatment increased the N-content of the straw but had apparently no positive effect on fiber digestibility or N balance. A plausible explanation might be that the time of treatment was relatively short.

Recent experiments by Ibrahim and Pearce (1983) demonstrated the positive effect of ammonia treatment on the in vitro OM digestibility of barley straw, pea straw and, to some extent, sugar cane bagasse. The digestibility of sunflower hulls was very low (16.6%) and was not improved by treatment with aqueous ammonia or urea-ammonia. A slight increase was obtained when using anhydrous ammonia.

Ammonia treatment of maize plant residues has been studied by a number of people. Rounds et al. (1976) obtained disappointing results when treating maize cobs with 4% NH_4OH. The time of treatment was relatively short in these experiments. Oji et al. (1977) treated maize stover for at least 30 days with either 3% NH_3 or 5% NH_3 plus 30% water. Both treatments increased the digestibility of organic matter by 8-9 percentage units. One interesting feature was that there was a significant increase in the content of acetic acid, from 2.1% in the control to 3.1 and 3.5% in the ammoniated maize stover on a DM basis.

Oji and Mowat (1979) compared ammonia treatment of maize stover at 21°C (30 days) and at 90°C (12 hours) and found a statistically significant increase in the NDF digestibility with increasing treatment temperature. The effect of the temperature on organic matter digestibility was not statistically significant, although significantly higher for the ammoniated maize stover compared to the control. The amount of acetic acid in the ammoniated material reached about 5% of DM. Mean digestion retention time was lower for the ammoniated than for the untreated maize stover (Oji et al., 1979).

Kiangi et al. (1981) studied the effect of ammonia treatment under varying conditions on the in vitro digestibility of maize stover, wheat straw and rice straw. The greatest effect of the ammonia was obtained with rice straw, which had the lowest initial digestibility. In spite of this, maize stover showed the highest digestibility also after treatment. In vivo experiments showed even greater improvements in the digestibility than did in vitro measurements. When feeding ammonia-treated maize stover ad libitum the digestion coefficients were consistently lower than when feeding restricted amounts. Said (1981) reported experiments done in Kenya showing that ammonia-treated maize stover came out slightly better than *Chloris gayana* hay when fed to growing steers as 26% of the diet. In an experiment with sheep Tubei and Said (1981) found that the daily gain was higher for ground and ammonia-treated maize cobs than for ammonia-treated maize stover and untreated stover and cobs. In the USA, Paterson et al. (1981) obtained a significant increase in intake and digestibility in lambs when treating maize stover with 3% ammonia for three weeks. In growth experiments with steers, ammonia treatment of maize stover showed no positive effect on the performance of the animals, whereas ammonia treatment of maize cobs improved the daily gain from 390 g/d (untreated) to 720 g/d.

Rice hulls are known for their low digestibility. Studies by White (1966), Fahmy et al. (1968) and Itoh et al. (1979) indicated that it is possible to improve the nutritive value of rice hulls by treatment with ammonia but not to the extent that it can be considered as a valuable feed for farm animals.

The effect of ammonia treatment on the N content and in vitro digestibility of awns and off-fall (chaff) from the combine was studied by Kjus and Torgrimsby (1980). The N-content and in vitro digestibility (DM) increased by 64 and 12 percentage units respec- tively for oat and 119 and 12 percentage units for wheat off- falls. A feeding test with sheep showed that the material was relatively palatable.

Ammonia treatment of wood residues e.g. sawdust, has revealed a marked increase in the in vitro digestibility (Millett et al., 1970). According to Tarkow and Feist (1969) this can, at least partly, be accounted for by the increase in swelling capacity of the material resulting from the treatment.

Studies by Kernan et al. (1981) with ammonia treatment of cereal straws and other crop residues have rendered more evidence for the assumption that the effect of ammonia is relatively more pronounced for materials of low digestibility.

Some materials, e.g. aspen wood, did respond quite well to the ammonia treatment, but treatment increased the in vitro OMD only to the level of untreated wheat straw. It may be questioned whet- her it is worthwhile treating materials with an organic matter di- gestibility between 55 and 65%. However, if the N content of the material is low and the ration has to be supplemented with NPN anyway, this may justify ammoniation of materials with relatively high digestibility. If the intake of treated material by the ani- mal is improved (as it often is) then this might also help justify treatment. For materials with an OM digestibility higher than 65- 70% ammoniation is hardly economic unless the ammonia is used as a preservative (Winther, 1978) and/or source of NPN.

7.4. METHODS FOR AMMONIA TREATMENT OF STRAW

The technique applicable for treating low quality roughages with ammonia may vary considerably with:
- form of ammonia
- level of technology
- type of material to be treated - system of production
- weather conditions
  etc.

7.4.1. Anhydrous ammonia

Anhydrous ammonia is the most concentrated form of the chemical (100%) and therefore small amounts are needed for treatment of straw. Another advantage is its gaseous nature which ensures a rapid penetration of the straw.

Use of anhydrous ammonia has also some disadvantages in ammoniation of low quality roughages. Pressure containers are required to keep the anhydrous ammonia as a liquid. A distribution network is necessary to ensure a regular and cheap supply of the chemical. If anhydrous ammonia is applied as fertilizer in an area, both these conditions are normally satisfied.

If anhydrous (gaseous) ammonia is to be applied, sealable containers for the roughage are needed to reduce the loss of ammonia. This is another prerequisite for economical application which, in addition to those mentioned above, makes this way of ammonia treatment almost impossible in many developing countries.

Anhydrous ammonia is a potentially dangerous and toxic material, even though it is widely used in certain countries as a fertilizer. Stringent safety precautions need to be observed in using this material. Norwegian and Canadian information bulletins on straw ammoniation have been careful to stress the safety measures needed to be taken when using anhydrous ammonia (see section 7.7.1).

7.4.1.1. Stack methods

In 1969 the Norwegian Feed Conservation Association (NOFO) and Norsk Hydro again took up the development project mentioned before, aiming to find a practical method for utilizing the solubilizing effect of ammonia on straw (Lie, 1975). The work was started mainly because of the new laws prohibiting outlet of rinsing water from sodium hydroxide straw treatment plants (Beckmann system, see Chapter 6.1). The work was based on the studies carried out earlier (see section 7.1). Closed chambers, varying temperatures and treatment time were tried, and in 1974-75 a stack method for ammonia treatment of straw at ambient temperature was introduced to the farmers.

During the harvesting season 1975 this new technique was tried on about 1000 farms and a brochure describing the method was printed (NOFO, 1976). Training courses for ammonia truck drivers and technical staff at supply units were organized. Special emphasis was laid on safety precautions. Demonstrations were made for the farmers who showed a growing interest. Due to a serious drought, with reduced forage crops in 1975 and 1976, the ammonia treatment method had a "flying start" in Norway (Homb et al., 1975; Arnason, 1976; Homb et al., 1977; Arnason and Mo, 1977; Sundstøl et al., 1978). There are many reasons for the rapid increase in ammoniation of straw in Norway:

1. Norwegian farmers are accustomed to the use of treated straw as feed (Beckmann procedure)
2. Regulations concerning the disposal of rinsing water from the Beckmann straw treatment plants (pollution)
3. High price of purchased feed
4. Ammonia (and NaOH) used for straw treatment subsidized by the government
5. Ammonia acts as an efficient preservative for high moisture straw
6. Ammonia treatment method is simple and relatively inexpensive (no storage room is required)
7. The treated straw is easy to handle

Almost parallel to the work in Norway a series of experiments with ammonia treatment of straw was carried out in Canada (Kernan et al., 1977). The method described was in principle the same for the two countries. One minor difference was that a black plastic cover was applied in Canada whereas clear plastic is used in Norway.

In Denmark Kristensen et al. (1977) obtained very satisfactory results with chopped barley straw treated with 4% anhydrous ammonia in a silage bag of polyethylene. There was very little difference in the enzyme solubility between the 8 layers above the bottom layer, being on an average 38.5% (untreated 23.1%). The enzyme solubility of the bottom layer of straw, which was wetted by condensed water, was significantly higher (53.2%).

The total amount of straw treated with ammonia in Norway since 1975 is shown in Table 7.1.

Table 7.1. Amount of straw treated with ammonia in Norway

Year	tonnes	% of total straw produced
1975	(tried by some farmers)	
1976	9000	1.2
1977	15000	2.0
1978	32000	4.3
1979	65500	8.7
1980	85000	11.3
1981	103000	13.7
1982	125000	16.7

In Denmark the total amount of straw treated with ammonia is greater than that treated in Norway (Fodgaard, 1982), but as percentage of total straw yield, it is less.

In Eastern Europe a stack method for straw treatment with anhydrous ammonia was also developed. Martynov (1972) treated stacks of 20 tonnes of straw with gaseous ammonia for 6 days after which the airtight cover was removed to get rid of the excess NH_3. The nitrogen content increased by 91% and the digestibility of organic matter, protein, crude fiber and N-free extracts of the straw was increased, because of the treatment.

According to Latvietis and Ruwalds (1980) the amount of straw treated with ammonia in Latvia increased from 17500 t in 1975 to 74500 t in 1977.

In Bulgaria, Sandev (1980) has described a pipe system for injection of anhydrous ammonia in the bottom of a stack of maize stover. The organic matter digestibility, determined in wethers, rose from 61.8% for untreated stover to 73.2% for maize stover treated with 3% NH_3. The corresponding figure for stover treated with 5% NaOH was 74.2%.

<u>Description of a stack method used in Norway</u>:

The stack should be placed away from buildings in case of fire. It should not be more than 10 m from a road so that it is accessible to the truck. A standard size of a stack is 4.6 x 4.6 x 2.1 m which will contain 2-4 tonnes of dry straw depending on the density of the bales. This stack will require

    1 bottom sheet of polyethylene 6 x 6 m

    1 covering sheet of 10 x 10 m

    4 wooden laths, 2.5 x 5 cm, 4.5 m long

    20-25 sand bags + 20 m rope (belt)

or 8-20 sand bags + 80 m rope (belt)

In the latter case, 4 ropes are crossed under the bottom and tied around the stack when finished. A 0.2 mm thick polyethylene sheet, stabilized against ultraviolet light, is used.

The length of the stack can be increased to for example, 11 m, which requires a 12.5 x 6 m bottom plastic and a 17 x 10 m covering sheet. This stack will contain about 10 tonnes and reduce the plastic costs by about 25% per tonne. However, inexperienced farmers are recommended to start with a small stack. Check the site for the stack and remove sharp items, e.g. stones, which can puncture the plastic. Loose straw may be spread over the ground to protect the bottom sheet. Place the ropes (belts) before the bottom sheet and then the first layer of straw bales (see Figure 7.4). Make sure that there is at least 70 cm free plastic at the sides of the stack. The next layer of bales should be placed at right angles to the bales below. The top layer should be smaller than those below to avoid water ponds on top of the stack (Figure 7.5).

Weighing of some bales will allow an estimation of the weight of the whole stack. The total weight is necessary for calculation of the correct amount of ammonia to be injected.

When sealing the stack special care should be taken in folding the corners (see Figures 7.4, 7.5 and 7.6). A minimum of two people are required. When rolling the two plastic sheets over the lath, it is recommended that all four sides of the stacks are sealed before injection. The ammonia is injected through a perforated metal pipe (spear) which is pressed through the plastic and into the middle of the stack before injection (Plate 7.1). The amount of ammonia normally used is 30-35 kg per ton of straw. The hole left when taking the pipe out of the stack is sealed with good quality tape.

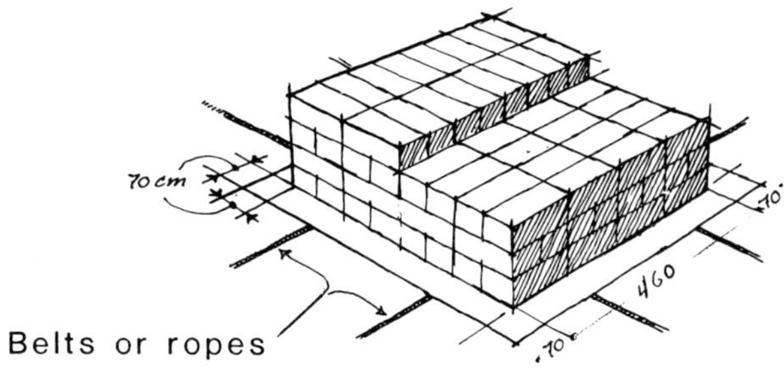

Figure 7.4. Stacking of straw

Figure 7.5. Folding of the plastic sheet is important

Figure 7.6. Rolling of the bottom sheet and the cover
over a pole or lath to seal the stack

As shown before, the time needed to treat the straw depends upon the temperature. It is advocated not to open the stack before it is necessary. In cold climates this may not be before 6-8 weeks; in warmer climates a shorter time is possible. If the moisture content of the straw is high the stack should be opened and the straw used while the temperature is low (winter). If such a stack is opened in the spring it may be attacked by mould within a short time. If not used, a stack of straw may be stored unopened over the following summer. When opening a stack of ammonia treated straw it should be aerated for some time before used as feed.

Plate 7.1. Injection of ammonia into the stack of straw. A net and worn-out tyres are used to prevent the stack from being torn apart by wind (Courtesy Hargreaves Fertilizers Ltd., UK)

The time required for airing depends on a number of factors such
as:

a.	Temperature	Low temperature - longer time of airing
b.	Moisture content in the straw	High moisture - longer time of airing
c.	Density of the straw	Big bales of high density - longer time of airing
d.	Wind	Strong wind - shorter time of airing

Plate 7.2. Stack of ammonia treated straw (Courtesy NOFO, Norway)

Airing may take as little as 2-3 days, but under unfavourable
conditions it may take weeks. Farmers are advised, if in doubt,
tear a bale apart in front of the animals and notice the smell.

Treatment of big bales can be done in stacks or each bale can
be treated individually. For individual ammonia treatment of big
bales special plastic bags are made. The bales are lifted by means

of a tractor equipped with two spears and the plastic bag is easily pulled over the bale. Since such bales can be handled easily by tractors and with other equipment, they are often used when treated straw is made for sale and transport. Airing of such bales may be a problem because the bales are very compact, especially when the moisture content of the straw is high. O'Shea et al. (1981) treated large cylindrical bales with ammonia in stacks containing 12 bales (3 x 2 in two layers). An increase in in vitro digestibility (DMD) of 12.3 units and a 250 g increase in daily gain of steers was obtained when compared with untreated barley straw.

Plate 7.3. "Sausages" of ammonia treated round bales (Courtesy Hargreaves Fertilizers Ltd., UK)

In Denmark and Norway a vacuum system has been developed where the stacks are made as above but the covering plastic is a cap tailored to fit a stack of a particular size and shape. The purpose is to reduce the plastic costs. The stack is sealed by folding the sleeves of the plastic in an open pipe (see Figure 7.7). Then a tube is put into the pipe and pumped up to press the plastic sheets against the inner wall of the pipe.

The air is partly evacuated out of the stack before the ammonia is injected. Three percent of anhydrous ammonia is used and the recommended time of treatment is 1-3 weeks (Agrozym, European patent No 82900503.2)

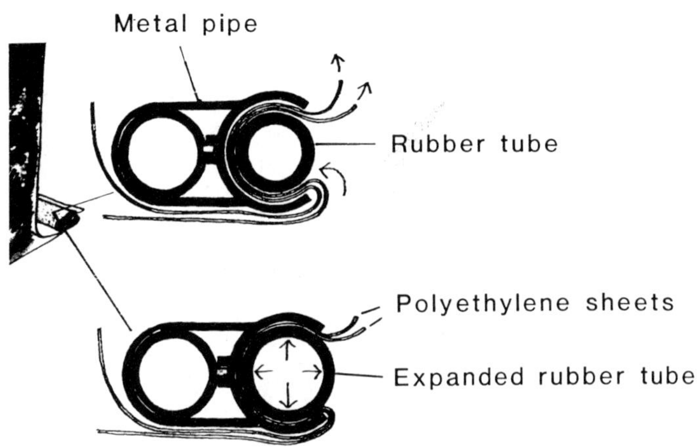

Figure 7.7. Device for sealing a "vacuum" stack

7.4.1.2. Ammoniation in ovens (chambers)

The long time needed to treat straw with ammonia at low temperatures has stimulated the development of ovens for ammonia treatment (Busk and Kristensen, 1977; Lundsby, 1978). When hot ammoniated air is circulated through straw in a chamber, the treatment may take less than 24 hours, including 3-4 hours for ventilation of the treated straw (see Figure 7.8). Such ovens are now in use in Denmark, Norway, United Kingdom and some other countries. The equipment can be fully automated so that the manual work is mainly loading and unloading of the chamber. This work can also be done to a great extent by a tractor/fork, using a spike for big bales, or pallets for conventional small bales. The cost of the equipment is relatively high. Compared with the stack method the plastic costs can be saved, provided that a store for dry straw is at hand. A clear disadvantage with the oven method, as with most other methods for straw treatment, is that the straw must be dry before baling and storing.

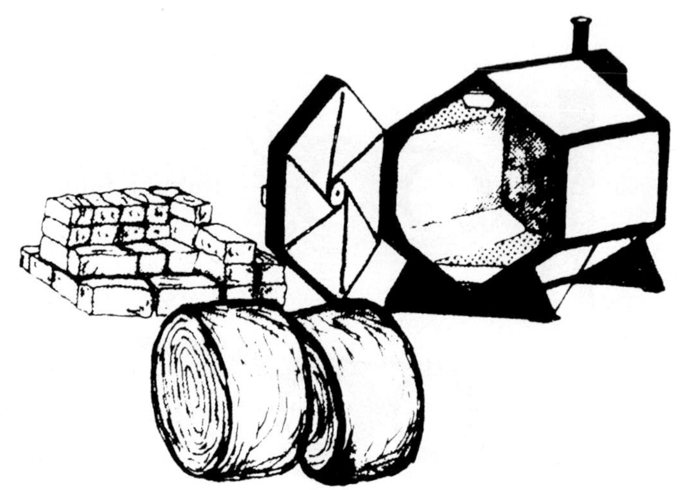

Figure 7.8. Oven for treatment of straw with ammonia (Fma)

7.4.1.3. Ammonia freeze-explosion process (AFEX)

Another fast industrial process for industrial treatment of straws and hays with ammonia has been developed by Dale and his associates at Colorado State University (Dale and Moreira, 1982). Liquid ammonia and rice straw (1:1 ratio by weight) are mixed in a reactor capable of withstanding high pressures. At room temperature, but in the sealed reactor, high pressures (12 kg/cm^2) of ammonia vapour are reached in a few minutes. The pressure is then suddenly released through a rotating ball valve. The temperature of the ammonia and straw drops well below 0°C, at the same time as the liquid ammonia trapped within cell structures rapidly expands into the vapor state and physically tears apart the embrittled cell walls. Costs of processing can in theory be kept low by recompressing and recycling most of the ammonia, since only a small amount is retained by the treated straw.

There is a substantial improvement in cellulase enzyme digestibility of the straw after this treatment. Both chemical and physical mechanisms may be involved since the treated material is finely shredded and has a greatly increased surface area.

While this process was primarily developed for the production of ethanol fuel and high protein content feed from alfalfa and other forages, it may have possibilities for improved feed from straws in some circumstances.

7.4.2. Aqueous ammonia

Aqueous ammonia is ammonia dissolved in water. The maximum content of ammonia that water can hold at normal pressure is determined by the temperature as shown in Table 7.2.

Table 7.2. Solubility of NH_3 in water at a pressure of 760 mm Hg

Temperature, $^\circ$C:	10	20	30	40	50
Ammonia in the solution, g/kg	400	342	285	237	185

A common solution of aqueous ammonia contains 25% NH_3 by weight and this can be marketed in plastic containers. This may undoubtedly be an advantage if ammonia is not readily available in a short and hectic straw treatment season. If the moisture content of the straw is high, the aqueous ammonia stored on the farm can be used immediately after the stack is finished and before the straw is damaged by mould. Also, in a very dry climate, the aqueous ammonia has an advantage over anhydrous ammonia as mentioned before. Addition of water to the dried straw before treating with anhydrous ammonia can have the same effect as aqueous ammonia (Sundstøl et al., 1979). Kiangi (1981) found no clear positive effect of increasing the moisture content from 20 to 40% before treatment with anhydrous ammonia.

Aqueous ammonia can be added to the material in different ways. Most common is to inject it into the stack through a pipe by use of a pump. Opened cans (containers) may also be placed on top of the stack and pushed over allowing the solution to flow through the straw when the stack is covered and sealed. The solution will then flow down through the straw and the ammonia evaporate gradually and penetrate the stack.

In 1971 Bergner and Marienburg described a method by which 100 kg chopped straw was mixed with 10 kg of a 25% ammonia solution before pelleting. Digestibility experiments with sheep revealed a marked increase in the digestibility of the straw. In later experiments with dairy cows these results were confirmed (Bergner and Marienburg, 1972a). Muller and Bergner (1975) showed that when oat straw was treated with 15% aqueous ammonia at room temperature for 5 days or with 10% aqueous ammonia at 150°C for 5 hours a great proportion of the lignin was dissolved into ammonia-lignin which could be precipitated with HCl.

According to Hartfiel and Ali (1979) aqueous ammonia has proved more beneficial than anhydrous ammonia because it is more easily transported and stored and can be at hand when needed. One disadvantage is, however, that the weight and volume of aqueous ammonia is about four times that of anhydrous ammonia. In some cases the extra water added to the straw may increase the risk of post-treatment moulding. A compromise may be to use a 33% solution of NH_3 as is presently done in the United Kingdom.

7.4.3. Urea treatment

When urea is decomposed it forms ammonia according to the formula:

$$NH_2\overset{\overset{\textstyle O}{\|}}{C}NH_2 + H_2O \longrightarrow 2\ NH_3 + CO_2$$

Molecular weight:	60	18	17	44
Weight produced	60	18	34	44

Thus adding 6.2% (wt/wt) of urea is equivalent to adding 3.5% ammonia, assuming 100% conversion. The use of urea as a source of ammonia for straw treatment has been known for some time and the application of this chemical has developed in two main directions: 1) in industrial processing of straw combined with grinding and pelleting (Marienburg and Bergner, 1973, Bergner, Zimmer and Münchow, 1974), 2) at a farm scale by mixing solutions of urea with straw or other low quality roughages and leaving it for some time to allow the enzyme urease to decompose urea and form ammonia (Van der Merwe, 1976; Ahmed and Dolberg, 1980). In some cases the straw contains sufficient enzyme; in other cases an independent source of urease has to be added, e.g. jackbean meal.

Since in these treatment methods, ammonia is only released after the straw has been mixed with urea and is safely within the confines of a pelleter, silo or bag, these are inherently much safer methods of treatment than those requiring handling of anhydrous or aqueous ammonia.

Cost and availability of different reagents may also dictate methods of treatment. In certain countries (e.g. Bangladesh) urea is considerably cheaper than either sodium hydroxide or ammonia, and is much more readily available.

7.4.3.1. Urea as a source of ammonia for straw treatment on an industrial scale.

Pure urea is decomposed completely at a temperature of $>133^{\circ}C$ When mixed with other chemical components the melting point and temperature at which urea is decomposed is markedly lowered (Bergner, 1980). When 3% urea is included in a feed mixture and pelleted the decomposition of urea in the pellet press begins already at $75^{\circ}C$ (Muller et al., 1976). The ammonia released is, however, insufficient for ammoniation of straw. If the temperature measured in the pellets is $90^{\circ}C$ the temperature in the ring dye must have been $110^{\circ}C$ which is sufficient to release about 1 kg urea for upgrading of 100 kg straw pellets (Bergner, 1980).

7.4.3.2. Urea as a source of ammonia for straw treatment on farm scale

Ammonia in the form of urea is pleasant to handle and transport. In roadless areas where ammonia cannot be supplied from truck tanks urea is a good alternative for ammoniation of straw. Urea has also the advantage compared to aqueous or anhydrous ammonia that no health risks are involved in handling and using the chemical.

Oji and Mowat (1977) treated maize stover with urea solution in polyethylene bags at room temperature. After 2 days 70% of the urea was decomposed and virtually no urea was detected by day 20 after treatment.

Coxworth and Kullman (1978) compared urea with aqueous and anhydrous ammonia for treatment of wheat straw. In contast to the results of Oji and Mowat, the wheat straw used did not contain significant levels of endogenous urease enzymes. Jackbean meal had to be added as a source of urease before decomposition of urea took place, and an improvement in organic matter digestibility was observed. (Unless urea decomposition is ensured by one means or another, very high levels of urea will remain in the straw, and be potentially dangerous if this straw is fed to animals.) Anhydrous ammonia was slightly more effective than the other two reagents in these experiments. The authors hypothesized that the better effect was related to the heat developed when adding anhydrous ammonia to the straw. The favourable effect of anhydrous ammonia over urea and aqueous ammonia was not confirmed in the work by Kiangi et al. (1981). The most extensive work on the use of urea for farm scale treatment of straw with urea is probably that done in Bangladesh (Ahmed and Dolberg, 1980; Dolberg et al., 1981). Rice straw was "ensiled" with a urea solution either in earthen pits (Plate 7.4) or bamboo bags. The walls of the pit were lined with banana leaves etc. to avoid contamination of the straw with soil. The top of the "silo" was also covered with local material, e.g. jute bags, soil and cow dung to avoid ammonia losses. If the rice straw was dry, it was mixed with equal weights of a solution containing 5% urea. Recommended treatment time is about 3 weeks. Saadullah et al. (1981) measured the intake and digestibility of rice straw treated with urea in various ways and the results are presented in Table 7.3.

Table 7.3. Protein content, intake (OMI) and digestibility (OMD) of organic matter of urea treated rice straw (Saadullah et al., 1981)

	Crude protein g/kg DM	OMI g/kg $W^{0.75}$.d	OMD %
Untreated rice straw	33	46.2	45
Treated with 3% urea in earthen pit (20 days)	74	51.7	54
Treated with 5% urea in earthen pit (20 days)	80	60.9	56
Treated with 5% urea in earthen pit (40 days) +10% molasses	78	63.4	57
Treated with 5% urea in bamboo basket (20 days)	83	57.5	56

The results obtained indicated that under the Bangladeshi conditions urea increased the protein content and the digestibility of the straw to the same extent as that normally found in experiments with anhydrous or aqueous ammonia. Bamboo baskets seemed as good "silos" for urea treatment of straw as did the earthen

pit. The advantage of using 5% urea instead of 3% was marginal in
terms of digestibility, although intakes were higher.

Plate 7.4. Urea-ammonia-treatment of rice straw in Bangladesh
        (Courtesy F. Sundstøl)

In a 84 day experiment with rice straw fed to calves, Saadullah
et al. (1982) found a 6 unit increase in the DM digestibility when
untreated straw was supplemented with urea at feeding. If the
straw was ensiled with urea for 10 days prior to feeding there was
an 11% unit increase in the digestibility. The daily gain was also
higher when the straw was ensiled with urea than when urea was
just added to the straw before feeding. The intake of straw in
this experiment was remarkably high, but the effect of the treatments on the intake was not very clear.

Jayasuriya and Perera (1982) found that 4% urea and 3-4 weeks
of treatment gave optimal results when rice straw was ensiled in
polyethylene bags.

In spite of the good results obtained by Saadullah et al. (1981) with urea-ensiling rice straw in bamboo bags, the importance of an airtight container has been a matter of debate. Ibrahim, Wijeratne and Costa (1983) compared various types of silos for urea-ammonia treatment of straw and based on the results they may be ranged in the order given in Table 7.4.

Kumarasuntharam et al. (1983) found that the treatment time for rice straw ensiled with urea could be reduced to 3 days by adding 8.5% soybean-powder (a source of urease enzymes).

Ibrahim, Fernando and Fernando (1983) studied a number of urease enzyme sources and their influence on the effect of urea-ammonia treatment of straw. Inclusion of a urease source reduced the treatment time to less than 5 days compared with 21 days without urease.

Table 7.4. The effect of silo type on mould attack, digestibility and intake of urea-ensiled rice straw (Ibrahim, Wijeratne and Costa, 1983)

	Moulded %	DM digestibility %	DM intake g/kg.d
1. Earthen pit	3	61	55
2. Polyethylene bag	18	60	51
3. Coconut leaves	31	60	47
4. Urea bags	37	57	46
5. Big bag	41	58	38
6. Open stack	36	53	40

In Thailand ensiling of rice straw in big bamboo baskets with 5% urea and 0.2% salt gave good results in growing steers (Wanapat et al., 1982) and in growing water buffaloes (Wanapat et al, 1983). Schmidt et al. (1982) in the German Democratic Republic improved the energy concentration of high moisture winter wheat straw 30% by ensiling with 2.5-6.0% urea.

In Scotland Ørskov et al. (1981) found no effect of treating straw with urea. The authors concluded that the nutritive value of straw was not improved by treatment with urea, whereas ammonia-treated straw could be used for low level production systems such as over wintering of store or replacement stock. Similar results were obtained by Mira et al. (1983). Williams and Innes (1983) showed that the moisture content of straw was of major importance for the decomposition of urea added to the straw. Results indicating that ammonia was released from urea through microbial activity were also obtained. Other factors that may inhibit the breakdown of urea are low temperature and low urease activity. In Norway, M. Wanapat, F. Sundstøl and J.M.R. Hall (unpublished results 1983) increased the OM digestibility of barley straw from 52.4 to 56.4 by ensiling with 5% urea for 8 weeks. A small amount of soybean meal added to the urea solution as a source of urease increased the digestibility further to 59.0.

7.4.4. Urine as a source of ammonia for straw treatment

Urine is a source of ammonia which is available everywere that animal production is taking place. According to Lauer (1975) the loss of ammonia from manure produced in USA is 1.6×10^6 t/year equivalent to 12 kg N/ha cropland or 7.6 kg N per person per year. The total animal protein consumed yearly was equivalent to about 5.0 kg N per person. The content of nitrogen in the urine varies normally from 2 to 20 g/l.

Coxworth and Kullman (1978) compared cattle urine with urea plus jackbean meal and with ammonia. Urine gave nearly as good improvements in OMD and nitrogen content as the other methods, provided sufficient urine was added. Urine is a good source of urease enzymes, but adds large amounts of soluble ash if used as the sole source of urea (see Table 7.5). In Bangladesh rice straw was treated with one litre of animal urine per kg straw and ensiled for 20 days (Saadullah et al., 1980). The nitrogen content increased from 0.53% to 0.90% and the organic matter digestibility from 45% to 55%. The dry matter intake in sheep increased by 70%. In an experiment with young calves (Haque et al., 1983) the daily gain was 110 g, 166 g and 181 g for untreated, urine-treated and urea-treated rice straw respectively.

Mahyuddin (1982) found that the nitrogen content of urine varied from 5.3 g/l on a low protein diet (8% CP) to 14.3 g/l on a high protein diet (15.5% CP). Of the total N 76-82% was in the form of ammonia-N and very little in the form of urea (1.8-1.9%). Urine (l) and rice straw (kg) were mixed in the ratios 1:1, 0.75:1, 0.5:1 and 0.25:1. The moisture content of urine-treated rice straw was adjusted to 50%. Treatment with urine for eight weeks increased the in vitro DMD up to 14 per cent units (35.5% -

49.4%) compared to 19 units increase for NaOH treated rice straw. Highest increase was obtained with urine from animals on a high protein diet (Mahyuddin, 1982). In Norway, Hall (unpublished results, 1983) was able to increase the in vivo OM digestibility of barley straw from 52.4 to 56.3 by means of treatment for 8 weeks with animal urine at about $10^{o}C$. Addition of soybean meal to the straw at the time of urea treatment did not improve the digestibility of the straw further (see also Table 7.5).

When considering urine as a source of ammonia the difficulty in collecting the urine even from stall fed animals should be emphasized. The possibility of collecting human urine and using it for treatment of straw and other low quality roughages should be investigated. No doubt this represents an enormous source of nitrogen which is largely underutilized today.

7.4.5. Other N-containing compounds

Semjakina (1968) treated rye straw with solutions (15%) of NH_3, $(NH_4)_2CO_3$, NH_4HCO_3, $(NH_4)_2SO_4$ and urea (cited by Bergner, 1981). The straw was treated with 1.5 l/kg in drums for five days. All chemicals reduced the content of crude fiber of the straw whereas the lignin content was reduced by NH_3, NH_4HCO_3 and urea only. Bergner and Marienburg (1972b) proposed NH_4HCO_3 as a chemical for straw treatment in pellet presses.

At temperatures above $60^{o}C$ ammonium hydrogen carbonate is decomposed to NH_3, CO_2 and H_2O. By means of an isotope technique Bergner et al. (1974) showed that NH_4HCO_3 was completely decomposed in a pellet press at $110^{o}C$. About 20% of the nitrogen was detected in the pellets and half of this was extractable by cold water.

In long term experiments with dairy cows (Marienburg and Bergner, 1975) the animals performed reasonably well (6310 kg in 546 days) on a diet of straw pellets treated with NH_4HCO_3 as the only roughage and supplemented with concentrate. The positive effect of ammonium bicarbonate was also demonstrated by Hasselmann et al. (1978). In experiments by Schiemann et al. (1978) and Jentsch et al. (1978), however, no positive effect on the digestibility of straw pellets was obtained when ammonium-bicarbonate was added during processing.

7.5. THE VALUE OF AMMONIATED MATERIALS RELATIVE TO UNTREATED STRAW, NaOH-TREATED STRAW AND OTHER FEEDSTUFFS

The effect of alkali treatment on the chemical and physical properties of fibrous materials and its consequence for the ruminal degradation has been discussed under Chapters 4, 11 and 12 and will therefore not be dealt with in depth here. A brief discussion of the energy value of ammonia-treated low quality roughages relative to untreated straw, NaOH treated straw and other feedstuffs is relevant, however. The role of ammonia-treated material in practical feeding is dealt with in Chapters 14.1, 14.2, 15 and 16 (see also Sundstøl, 1983).

When comparing with sodium hydroxide-treated straw the digestibility of ammonia-treated straw seems to be slightly inferior. Experiments with wethers by Frenzel and Pfeffer (1979) showed a digestibility of organic matter of 57.3% for ammonia-treated straw versus 66.3% for sodium hydroxide-treated straw.

In Norway M. Wanapat, F. Sundstøl and J.M.R. Hall (unpublished results, 1983) compared 15 different ways of treating barley straw before feeding it to wethers at maintenance with a supplement of 75 g of herring meal per day. The results (Table 7.5) showed that wet treatment with NaOH was superior to all other treatments. The digestibility of the straw treated with anhydrous (or aqueous) ammonia was more or less the same regardless of whether it was treated in a stack or in an oven and similar to spray (NaOH)-treated straw. Other comparisons have shown results different from these (Sundstøl, 1982b). Urea or urine treatment resulted in very small improvements in OM digestibility. In this study there was no effect on the digestibility of adding urea to the straw at the time of feeding. This means that the effect of ammonia and urea treatment is not an "NPN-effect" but a real effect on the degradability of the fibrous material.

While in vitro and in vivo digestibility trials often indicate that sodium hydroxide treatment is superior to ammonia treatment, a review of performance trials up to 1978 (Sundstøl et al., 1978) suggested that rates of gain are frequently more similar than digestibility trial differences would lead one to expect.

When comparing ammonia-treated and sodium hydroxide-treated straw the higher ash content of the latter should be taken into account.

228

Table 7.5. Ash and nitrogen content and in vitro, in sacco (cows) and in vivo (sheep) digestibility of barley straw (M. Wanapat, F. Sundstøl and J.M.R. Hall, unpublished results, 1983)

Treatment (Abbreviations in parenthesis refer to company or method)	Contents g/kg DM Ash	N	Digestibility, % in vitro ------ 48 h ------ DM	in sacco DM	in vivo OM
Untreated straw	40	3.4	42.2	41.4	52.4
" " + urea at feeding					52.0
Urine treated straw	97	17.1	58.5	56.2	56.3
" " "					
+ soya (urease)	98	19.5	58.6	56.2	57.1
Urea-NH$_3$-treated straw	46	22.2	46.3	46.8	56.4
" " " "					
+ soya (urease)	46	21.0	49.6	50.0	59.0
Anhydrous ammonia:					
Stack (NOFO)	42	14.1	56.1	56.7	67.8
Stack, vacuum (Agrozym)	41	16.0	57.1	57.6	66.7
Oven (Fma)	35	17.0	55.9	59.7	63.6
Aqueous ammonia:					
Stack	47	15.8	58.0	61.6	59.0
Sodium hydroxide:					
Dry treated (JF)	92	3.4	65.9	67.3	67.8
Dry " , pellets (FK)	91	3.8	65.5	67.4	64.7
Beckmann treated	37	1.9	69.2	78.1	75.7
Wet treated, chamber (CLM)	99	3.4	69.0	74.9	72.8
Dip treated	154	3.2	73.3	88.4	73.6
" " + urea	143	14.9	65.5	86.6	74.8

There is one matter that has been discussed frequently, viz., the stability of the ammoniated material. Research in East Germany indicated that the energy value of ammoniated straw pellets was reduced by about 20% when the feed was stored for one year (Bergner et al., 1978). Later studies in which the treated straw pellets was stored for 32 weeks confirmed this observation (Bergner et al., 1979). Hartfiel and Ali (1979) found no reduction in the enzyme solubility of ammonia treated straw after storage at 21°C for 170 days. Gordon and Chesson (1983) measured the digestibility in sacco of DM and cellulose of ammonia treated straw for 112 days subsequent to opening of a stack of barley straw. The improvement in digestibility achieved during the 54 days of treatment was 15.4 and 21.8 percentage units for DM and cellulose respectively and it remained constant throughout the 112 day period.

Reduced digestibility of ammonia-treated straw may occur when combined with readily available carbohydrates as shown by Hvelplund et al. (1978) and Williams and MacDearmid (1982).

7.6. THE PROTEIN VALUE OF AMMONIA TREATED MATERIALS

Straw and other fibrous materials are normally low in protein. Therefore the added nitrogen is considered as one of the three main advantages of ammonia treatment - together with increased intake and digestibility (Potthast and Hartfiel, 1977).

In an early study Obradovic (1973) found that the protein equivalent of barley straw increased from 2.75 to 9.88% when treated with 3% NH_3 (aqueous). The digestibility and retention of nitrogen was increased when diets containing ammonia-treated straw were fed. After ammoniation at high temperature and pressure Hamad and El-Saied (1982) found highest N-content in sugar cane bagasse (2.94%) and lowest in rice hulls (1.30%) with maize stover (2.63%) and rice straw (2.24%) inbetween . There was a very strong correlation between the contents of N and pentosans in this study.

Gordon and Chesson (1983) divided the nitrogen in ammonia-treated straw into three fractions (see Table 7.6):

        a) Water soluble ammonia-N

        b) Water soluble non-ammonia-N

        c) Water insoluble non-ammonia-N

The major reduction in N content after opening of the stack occurred in the water soluble ammonia N fraction.

The utilization of the ammonia added to straw during treatment depends on a number of factors such as:
- Total amount of N in the straw
- Speed at which the N is released
- Amount of available energy (carbohydrates) in the rumen
- Degradability of dietary protein

In other words the ammonia-treated material may be considered as an ordinary NPN source.

Table 7.6. Nitrogen content of untreated and ammonia-treated barley straw (Gordon and Chesson, 1983)

Sample	Nitrogen, g/kg straw DM			
	Ammonia N	Non-ammonia N		Total
	Water soluble	Water soluble	Water insoluble	
Untreated straw	0.2	0.2	4.6	6.8
Ammonia treated straw (at opening)	7.6	5.1	8.0	20.7
" " " (112 d after opening)	4.6	4.6	7.2	16.4

Danish experiments with fistulated cows showed no positive
effect on the microbial protein synthesis of supplementing an
ammonia-treated straw diet (5% molasses) with urea or soybean meal
(Møller and Hvelplund, 1978 and 1982). The N content of the ammoniated straw diet was 1.53% of DM. The microbial protein synthesis
was 21.4 and 29.8 g/100 g OM digested in the forestomachs which is
equivalent to an average protein synthesis on a normal ration. The
authors concluded that ammonia-treated straw is used as a substrate for the rumen microorganisms and that the bound N can be used
for protein synthesis in the forestomachs of ruminants. Results in
support of this were obtained by Borhami and Johnsen (1981) although they concluded that a proportion of the ammonia was tightly
bound to the straw and not released during the passage through the
alimentary tract. Solaiman et al. (1979) showed that 12.6% of the
N retained in the straw after ammonia treatment was bound to the
ADF fraction of the straw. Al-Rabbat and Heaney (1978) did not
find higher microbial protein synthesis in sheep fed ammonia-treated straw than when fed untreated straw at 64% of the ration. The
N contents of the diets were relatively high, 2.9 and 1.8% respectively, which may have influenced the results obtained. Males and
Gaskins (1982) reported higher N-balance in lambs fed wheat straw
treated with anhydrous ammonia than when feeding diets based on a)
bromegrass-lucerne hay, or b) chopped untreated wheat straw. Work
by Herrera-Saldana et al. (1982) indicated that even if the nutritive value of wheat straw was improved by ammonia treatment supplemental energy might be needed for maximal utilization of the
added nitrogen.

Borhami et al. (1983) compared the N metabolism in sheep fed
sodium-hydroxide-treated straw plus urea versus ammonia-treated

straw. The results obtained were somewhat conflicting, but the flow of microbial protein through the duodenum tended to be higher for the ammonia-treated straw diet.

In an experiment with growing bulls Mo (1977) found that animals fed 5 kg ammonia-treated straw per day had a nitrogen balance similar to that of bulls fed 5 kg untreated straw + 150 g herring meal. Sundstøl (1982a) fed ammonia-treated straw as the only roughage supplemented with either a low-protein concentrate (LP) or a medium protein concentrate (MP) with soybean meal as the main source of protein to growing bulls. The average daily gain for the 18 bulls in each group was 893 and 953 g respectively. Taking the higher price of the MP concentrate into account the two groups came out equal economically.

The potential for protein synthesis in ruminants fed ammonia treated straw was demonstrated in an experiment with steers by Sundstøl and Matre (1980). On a ration of ammonia-treated straw, 250 g barley plus mineral and vitamin supplements, the steers gained more than 400 g/d for a period of more than five months.

7.7. APPLICATION OF AMMONIA TREATMENT UNDER VARYING CONDITIONS

The applicability of ammonia treatment varies greatly from one country to another, making it very difficult to generalize. When discussing the potential for upgrading of straw and other fibrous by-products in an area there are many factors to be considered. No attempt will be made to cover all situations, but the factors may be of technological, economical or sociological nature.

7.7.1. Technological factors

The first prerequisite when considering chemical treatment of straw and other low quality roughages is the availability of the chemical. It seems obvious that for use of anhydrous and aqueous ammonia to treat straw a certain level of technological development is required. Relatively sophisticated equipment is needed for handling and transport of the ammonia. Infrastructure is also a necessity for distribution of ammonia gas in pressure containers. Most favourable are probably the conditions where the use of anhydrous ammonia as fertilizer is already established and a distribution network exists. For application of the stack method the farmers should be trained in making stacks properly and the straw should preferably be baled in one way or another. Polyethylene sheets or other suitable covers should be available together with sand bags and/or belts (ropes) or nets to protect the stack from being torn apart by the wind.

Anhydrous as well as aqueous ammonia can be dangerous chemicals to handle for untrained personnel. Anhydrous ammonia is stored and transported as a liquid under a pressure of 115 psi at 20°c. It is very toxic and a mixture of 15-28% anhydrous ammonia in air is explosive. When handling anhydrous or aqueous ammonia rubber gloves and eye goggles should be worn (see Kernan et al., 1977).

To protect personnel from injuries due to ammonia, restrictions have been given for maximum concentrations allowable for exposure (Table 7.7).

Table 7.7. Physiological response to various concentrations of ammonia

Physiological response	Ammonia, ppm in air
Least detectable odor	53
Least amount causing immediate irritation to the throat	408
Least amount causing immediate irritation to the eyes	698
Least amount causing coughing	1720
Maximum concentration allowable for prolonged exposure	100
(In Norway	25)
Maximum concentration allowable for short exposure (1/2-2 h)	300-500
Dangerous for even short exposure (1/2 h)	2500-4500

Adapted from: Encyclopedia of Chemical Technology, Vol. 2, 2nd Ed.
Editor: Kirk-Othmer, Interscience Publishers
New York, London, p. 291, 1963

If stacks of ammoniated straw with high moisture content are opened in a cold climate and the straw is brought into a warm barn the excess ammonia is released. Under such conditions concentrations of up to 200 ppm have been measured, i.e., 8 times the upper limit set by the health authorities in Norway.

Norway has also very strict laws to prevent propagation of weeds e.g. wild oats. All feed grains that are traded as concentrate in Norway have to be finely ground. Straw is not allowed to be transported in open vessels unless the farm from which it originates is declared free of wild oats.

Trade of straw from farms on which wild oats has been found is prohibited. If such straw is transported by road or rail it must be thoroughly covered. In 1975-76 nylon bags containing wild oats were placed at various points in stacks of straw before treatment with anhydrous ammonia. Repeatedly it was found that the viability of the wild oat seeds was destroyed by the treatment (Norwegian Plant Protection Institute). These results were confirmed by Coxworth (1978) who treated 18 weed species with 3.5% anhydrous ammonia for 14, 30 or 60 days. Even after 14 days of treatment the germinating capacity of all weed seeds was gone. This is considered as a great advantage and consequently there are no restrictions on sale of ammonia-treated straw in Norway.

Another great advantage with the ammonia treatment is the preservative effect of the ammonia (Knapp et al., 1974; Winter, 1978; Schmidt et al., 1982). In laboratory scale experiments oat straw containing 15, 30 or 45% moisture was treated with 1, 4 or 7% anhydrous ammonia at 4, 17 or 25°C for 8 weeks (F. Sundstøl, unpublished results, 1978). The only sample damaged by mould was that containing 45% moisture and treated with 1% NH_3 at 25°C. In countries with humid climate this is probably one of the greatest advantages of the ammonia treatment. It means that straw with a relatively high moisture content can be baled and stored without being spoilt. If such a stack is opened in the winter when the temperature is low the straw will keep for weeks without reduction in quality. Ammonia-treated straw with high moisture content should not be opened in late spring. At high temperatures the straw will soon get spoilt. If the stack is tight it can normally be kept over the following summer without any adverse effects.

High moisture straw should be ammoniated soon, 2-4 days after stacking, otherwise the straw may be spoilt. On the other hand high moisture will enhance the upgrading effect of the ammonia (see section 7.3). Practical experience in Norway indicates that high moisture content also improves the voluntary intake of ammonia-treated straw (Westgaard and Torgrimsby, 1977).

The fact that 2/3 of the added ammonia is wasted when treated straw is uncovered may be considered a serious drawback of the method. Experiments have shown that it is possible to trap most of the straw with acid when opening a stack (Borhami et al., 1982).

7.7.2. Economical and socio-economical factors

Economic considerations are dealt with in Chapter 18. Therefore only a few points related to ammonia treatment will be dealt with here. One aspect of considerable interest is the amount of energy required for production of one feeding unit (FU, ME or TDN) of ammonia treated straw. In Norway a working group with Professor K. Breirem as chairman studied the use of commercial energy (mainly oil and electricity) in feed production (NLVF, 1982). Table 7.8 shows the amount of commercial energy used in the production of one MJ metabolizable energy in various feedstuffs.

Table 7.8. The use of commercial energy in feed production (NLVF, 1982)

	MJ per MJ metabolizable energy (ME)
Barley	0.45
Ammonia treated straw	0.37
Hay, wire fences + pasture	0.51
" , field cured + "	0.54
Pasture	0.64
Grass silage	0.67
Artificially dried grass	2.10

The figures given demonstrate that the total amount of commercial energy required per feeding unit (ME) of ammonia-treated straw is relatively low in spite of the energy involved in the production of ammonia and polyethylene sheets (see also Breirem, 1982).

The economic feasibility of ammonia treatment of straw and other fibrous by-products is determined by a number of factors of which a few are mentioned below:

a) Cost of ammonia or urea. This is to a great extent determined by alternative uses of the NH_3-source e.g. as fertilizer.
b) Cost of concentrate. If cheap concentrate can be bought there is no incentive to take the trouble to treat low quality roughages with NH_3. This may be the case in many countries that

have surpluses of grain e.g. Canada, USA, Sweden, Argentina, Australia.

c) Cost of straw etc. Up to now straw and other by-products have been cheap because they have been considered as wastes and therefore of no alternative value. High energy prices and new technologies have made straw more attractive as a source of fuel, fiber for paper, building materials etc. (see Lockeretz, 1981). In some countries the price of straw is even higher than that of concentrate (e.g. Egypt).

d) Cost of other roughages. In some areas the conditions for production of high quality forages, mainly pasture, are favourable (New Zealand, some states in the USA and some European countries). High quality forages at low prices make straw treatment less attractive.

e) The improvement achieved by ammoniation. The response achieved in higher digestibility, increased intake and higher protein value is the benefit of the treatment and therefore to a great extent determines whether the process is economical. This is demonstrated in Figure 7.9 which shows how the maximum price of ammonia that can be paid depends on the price of energy (concentrate) and the gain achieved by the treatment. In this example the energy value only is considered.

An assessment of the technical and economic feasibility of treatment of straw for animal feeding was made by Jackson (1978). He concluded that alkali treatment is clearly profitable in countries where straw is traditionally fed to livestock.

In Norway the treatment of straw is encouraged by the government through subsidizing the price of ammonia and sodium hydroxide. The main idea behind this is to stimulate the use of national resources in food production, since the animal production today is in part based on imported feedstuffs. One could also argue that straw treatment in such situations has a currency saving effect and that land otherwise used for forage production may be used for food (grain) production. A simplified example of how to calculate processing costs for ammonia treatment of straw in Norway (1983) is given next page:

	US$/t straw (85% DM)
Cost of straw	31
Cost of plastic	17
Cost of ammonia (1 US$/kg NH_3)	30
Total cost	78[1]
Subsidy	26
Net cost of treatment	52

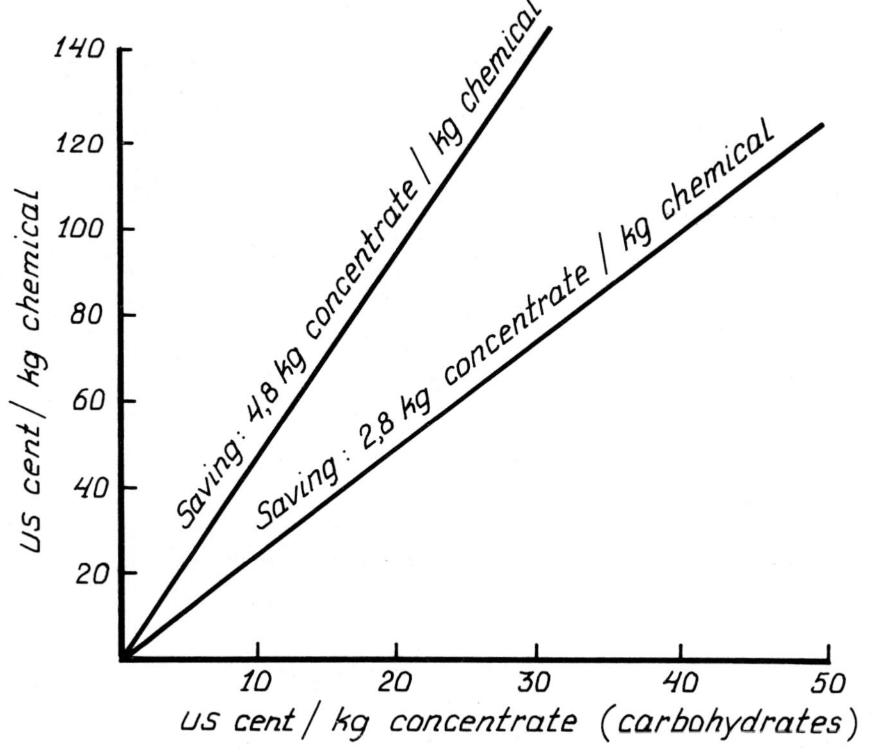

Figure 7.9. Cost/benefit relationship for alkali treatment of low quality roughages with two chemicals, one being more effective than the other

[1]Comparable costs in Canada (in U.S.$) have been calculated to be $43/t (no subsidy) based on a 38 tonne stack (to reduce plastic costs), ammonia delivered on farm at $0.38/kg, and polyethylene sheeting costing $3.30 per tonne of straw covered. Saenger, Lemenager and Hendrix (1983) were able to place 88.5 tonnes of

straw (large round bales) under one 30 m x 12 m piece of polyethylene sheeting, by placing the straw bales in a concrete bunker silo and thus needing the polyethylene only for sealing the top and ends of the stack. Costs of plastic would be reduced to $1.42 (U.S.) per tonne of straw ammoniated by this stratagem.

A detailed example of ammonia treatment costs in Canada is given by Potts (1982) who concludes that economic evaluation should be included in research programs at an early stage to eliminate processes which are not economically feasible. It may, however, be difficult in some instances to decide what will be economical in 5 or 15 years time and what will not. In a recent study from USA Saenger et al. (1983) concluded that "The ammoniation procedure used in this study was effective and economically practical in improving the nutritive value of wheat straw". On the contrary, Horton (1979) concluded that the process of ammoniation could not be justified in economic terms when feeding steers 4 kg concentrate per day plus ammonia treated wheat straw ad libitum.

In many parts of the world ammonia and urea are made from natural gas. Their price may therefore increase substantially in the next decade. One strategy to counteract this trend would be to develop economically attractive ways to use the 2/3's of the ammonia which is vented into the air by present methods. Borhami et al. (1982) have shown that various acids, including acetic, can be used to absorb the excess ammonia. One possible source of cheap organic acids might be silage. Combinations of treated straw with silage (e.g. grass silage) might thus provide a method for capture of the excess unreacted ammonia. Since silages often contain appreciable amounts of NPN such a combination should be supplemented with a low nitrogen source of available energy for the ruminal synthesis.

In some areas of the World all the farmed land has to be used for human food production and no-one can afford to produce forage for animal feed (e.g. Bangladesh, India). Under such conditions the animals normally have to rely on by-products and what grass they can find on the road-side, etc. Untreated rice straw is normally the staple feed in these areas. Often the amount of straw has to be restricted, otherwise some animals will have nothing to eat. The potential for an improvement in energy as well as protein value of the straw in such a situation seems obvious (see also

Chapter 18).

In some areas straw is abundant but not used to a great extent because the animals rely entirely on the pasture, e.g., Africa and Latin America. In periods of drought straw is transported over long distances to save animals from starving to death (N.K.R. Musimba, personal communication, 1978) Ammonia-treated straw will provide more net energy and protein than will untreated straw and will thus be a more efficient emergency feed.

Ammoniation of straw should presumably be viewed always in the context of other options which may be available to improve the straw quality and effective utilization or indeed whether straw should be used at all. Mosi and Lambourne (1982) suggested a strategy for Africa of intercropping forage legumes with cereal crops so that cereal crop residues would be grazed with the forage legume. Several authors have suggested that there may be possibilities to breed more digestible straw without sacrificing the agronomic qualitites of the crop. White et al. (1981) in the Norther United States found high yielding wheat cultivars with straw digestibilities up to 12 units higher than that of the poorest cultivars. This difference is comparable to what would likely be achieved by ammoniation of the poor quality straw. Studies in Canada have come to a similar conclusion (Coxworth et al., 1981), with the additional observation that selection of grain cultivar and crop residue components can be combined with ammoniation to increase feed values even further.

7.8. SUMMARY

The effect of ammonia on breakdown of straw has been known for more than 50 years. Systematic studies of the effect of ammonia in upgrading low quality roughages has not been undertaken until recently. Increasing dosage of ammonia up to 3-4% of the straw, temperature, moisture content of the straw and time of treatment all seem to have positive effects on the treatment response. The effect of high temperature over long treatment times seems rather unclear. The response to ammonia treatment depends to a great extent on the material in question. In some instances the treatment is relatively more effective when the intial value of the material is low. For very poor materials such as rice hulls and bagasse the improvement achieved by ammonia treatment is marginal.

Methods for treatment of straw etc. in stacks, pits, baskets and ovens are described. The sources of ammonia may be anhydrous ammonia, aqueous ammonia, urea, animal urine, ammoniumbicarbonate and others.

If the treatment of straw with ammonia is done properly the increase in organic matter digestibility and nitrogen content is normally 10-12 and 0.8-1.0 percentage units respectively. There are, however, several examples in the literature of both higher and lower values.

The energy value of ammonia-treated straw is considered lower than that of wet-treated and dry-treated NaOH straw. The value of the added nitrogen will depend on the composition of the whole diet. As a roughage, ammonia-treated straw may be compared with medium to low quality grass hay. Most reports in the literature show improved intake of the materials after ammonia treatment.

In addition to improved energy and protein value and improved intake, ammonia treatment of straw and other fibrous materials has other advantages such as:
- Most methods for ammoniation are simple and relatively inexpensive, depending on the level of technology applied.
- Ammonia has a preservative effect i.e. high moisture materials can be stored.
- Weed seeds, e.g. wild oats, are killed by effective ammonia treatment.
- No extra store for the straw is needed.
- The amount of commercial energy spent in the process of ammonia treatment is small relative to the production of other feedstuffs.
- Ammonia treatment causes no pollution of soil or water.

Disadvantages:
- Excess ammonia (2/3) is lost into the air.
- Under certain conditions excess ammonia may cause air pollution in animal barns.
- The energy value of the end product is lower than that obtained with some other methods (NaOH-treatments).
- Most ammonia is manufactured from fossil fuels and may increase considerably in price over the next decade.
- Great care must be taken in handling anhydrous and aqueous ammonia. Methods employing in situ decomposition of urea are in this respect more desirable.

7.9. REFERENCES

Ahmed, R. and Dolberg, F., 1980. Practical ways of improving utilization of straw. ADAB NEWS, (Dacca), 7: 12-14.

Alibes, X., Munoz, F. and Faci, R., 1983. Treated straw for animal feeding. Some results from the mediterranean area. OECD workshop, G.R.I. Hurley, U.K. 15 -17 February.

Al-Rabbat, M.F. and Heaney, D.P., 1978. The effects of anhydrous ammonia treatment of wheat straw and steam cooking of aspen wood on their feeding value and on ruminal microbial activity. 2. Fermentable energy and microbial growth derived from ammonia nitrogen in the ovine rumen. Can. J. Anim. Sci. 58: 453-463.

Arnason, J., 1976. Feeding experiments with ammonia treated straw. Norsk landbruk 15.

Arnason, J. and Mo, M., 1977. Ammonia treatment of straw. Report on Straw Utilization Conference, Oxford 24-25 February. Ministry of Agric. Fisheries & Food.

Becker, K. and Pfeffer, E., 1977. Treatment of straw with ammonia. Das Wirtschaftseig. Futter 23: 83-87.

Bergner, H., 1980. Chemische Grundlagen des Strohaufschlusses in der Pelletier-presse. Arch. Tierernähr. 30: 239-256.

Bergner, H. 1981. Chemical treatment of straw. Plant Research Development 14: 61-81.

Bergner, H., Hasselmann, L., Marienburg, J. and Müller, J., 1979. Untersuchungen zur Prüfung des Futterwertes von Strohmaterialien im in vitro-System. 7. Untersuchung zum Rückgang des Aufschlusseffektes von ammonisierten Strohpellets. Arch. Tierernähr. 29: 797-803.

Bergner, H. and Marienburg, J., 1971. Herstellung und Futterwert von ammonisierten Roggenstrohpellets. Arch. Tierernähr. 21: 557-566.

Bergner, H. and Marienburg, J., 1972a. Ammoniated straw pellets for feeding ruminants. Monatshefte für Veterinärmedizin 27 (15): 565-568. (Nutr. Abstr. Rev. 44: 3538, 1974).

Bergner, H. and Marienburg, J., 1972b. Quoted by Bergner (1981).

Bergner, H., Müller, J., Marienburg, J., Adam, K. and Zimmer, H.-J., 1974. Untersuchungen zur Charakterisierung von Strohpellets. 3. Herstellung von Strohpellets mit ^{14}C- und ^{15}N-markierten Aufschlussmitteln und ihre chemische Untersuchung. Arch. Tierernähr. 24: 567-576.

Bergner, H., Püschner, A., Müller, J. and Marienburg, J., 1978. Ergebnisse von Vergleichsuntersuchungen an ammonisierten Strohpellets in zwei Forschungseinrichtungen der Tierernährung. Arch. Tierernähr. 28: 417-425.

Bergner, H., Zimmer, H.-J. and Münchow, H. 1974. Untersuchungen zur Charakterisierung von Strohpellets. 6. Verdaulichkeitsuntersuchungen an Weizenstrohpellets. Arch. Tierernähr. 24: 689-700.

Borhami, B.E.A. and Johnsen, F., 1981. Digestion and duodenal flow of ammonia-treated straw, and sodium hydroxide-treated straw supplemented with urea, soybean meal, or fish viscera silage. Acta Agric. Scand. 31: 245-250.

Borhami, B.E.A. and Sundstøl, F., 1982. Studies on ammonia-treated straw. I. The effects of type and level of ammonia, moisture content and treatment time on the digestibility in vitro and enzyme soluble organic matter of oat straw. Anim. Feed Sci. Technol. 7: 45-51.

Borhami, B.E.A., Sundstøl, F. and Garmo, T.H., 1982. Studies on ammonia-treated straw. II. Fixation of ammonia treated straw by spraying with acids. Anim. Feed Sci. Technol. 7: 53-59.

Borhami, B.E.A., Sundstøl, F. and Harstad, O.M., 1983. Nitrogen utilization in sheep when feeding either sodium hydroxide treated straw plus urea or ammonia treated straw. Acta Agric. Scand. 33: 3-8.

Breirem, K., 1982. The use of commercial energy in Norwegian agriculture. In A. Ekern and F. Sundstøl (eds.): Energy Metabolism of Farm Animals. EAAP publication no 29: 314-324.

Busk, J. and Kristensen, T.P., 1977. NH_3-behandling av halm i isolert gårdanlæg. Ugeskr. Agron. Hort. Forst. Lic. 24: 486-490.

Chomyszyn, M., Bielinski, K. and Slabon, W., 1960. The use of ammoniated feed in the feeding of ruminants. 7. The use of ammoniated straw in fattening growing wethers. Roczniki Nauk Rolniczych (Polish Agric. Annual) TOM 75: B-4: 531-540.

Chomyszyn, M. Zak, Z. and Kowalczyk, J., 1972. Feeding value of rape straw treated chemically. Roczniki Nauk Rolniczyh B-94 (2): 89-108. (Nutr. Abstr. Rev. 44: 3537, 1974).

Coxworth, E., 1978. The effect of treatment with ammonia on the germination of various weed seeds. Saskatchewan Research Council C 78-14, Volume 2, section 8.

Coxworth, E., Kernan, J., Knipfel, J., Thorlacius, O. and Crowle, L., 1981. Review: Crop residues and forages in western Canada; potential for feed use either with or without chemical or physical processing. Agric. Environm. 6: 245-256.

Coxworth, E. and Kullman, P., 1978. Improving the feeding value of straw and other forages by the use of ammonia released from urea by the action of an urease enzyme. Saskatchewan Research Council C 78-14, Volume 2, section 6.

Dale, B.E. and Moreira, M.J., 1982. A freeze-explosion technique for increasing cellulose hydrolysis. Biotechnol. Bioeng. Symp. 12: 31-43.

Dolberg, F., Saadullah, M., Haque, M. and Ahmed, R., 1981. Storage of urea-treated straw using indigenous material. World Anim. Rev. 38: 37-41.

Fahmy, S.T.M., El-Shazly, K. and Badr, M.F., 1968. The effect of treating rice hulls with ammonia on its nutritive value. J. Anim. Prod. U.A.R. 8: 11-24.

Fodgaard, S., 1982. Halmoverskud på lands-, amts- og kommuneniveau. Jordbrukgsøkonomisk Institut, Denmark, Report no 11.

Frenzel, E. and Pfeffer, E., 1979. Vergleich des Futterwertes von Ammoniak- und Lauge-behandeltem Stroh. Das Wirtschaftseigene Futter 25: 193-197.

Fyrileiv, E. and Ulvesli, O., 1958. The digestibility of ammonia treated straw 1956-58. Unpublished report.

Gordon, A.H. and Chesson, A., 1983. The effect of prolonged storage on the digestibility and nitrogen content of ammonia-treated barley straw. Anim. Feed Sci. Technol. 8: 147-153.

Gunnes, S., 1957. Luting av halm med ammoniakk. Norsk Hydro, Forskningslaboratoriet, RP 263. Unpublished report.

Hamad, M.A. and El-Saied, H., 1982. The ammoniation of agricultural residues. J. Sci. Food. Agric. 33: 253-254.

Haque, M., Davis, C., Saadullah, M. and Dolberg, F. (1983). Performance of cattle fed treated paddy straw with animal urine as a source of ammonia. Trop. Anim. Prod. (in press).

Hartfiel, W. and Ali, A., 1979. Untersuchungen uber den Strohaufschluss mittels Ammoniak-Starkwasser. Landwirtsch. Forschung, Sonderheft 36: 285-291.

Hartley, R.D. and Jones, E.C., 1978. Effect of aqueous ammonia and other alkalis on the in vitro digestibility of barley straw. J. Sci. Food Agric. 29: 92-98.

Hasselmann, L., Bergner, H. and Müller, J., 1979. Untersuchungen zur Prüfung des Futterwertes von Strohmaterialien im in vitro-System. 6. Strohaufschluss mit Ammoniumhydrogenkarbonat, NaOH sowie NH_3-Wasser und Prüfung der in vitro-Eiweissynthese. Arch. Tierernähr. 29: 743-749.

Herrera-Saldana, R., Church, D.C. and Kellems, R.O., 1982. The effect of ammoniation treatment on intake and nutritive value of wheat straw. J. Anim. Sci. 54: 603-608.

Homb, T., Saue, O., Matre, T. and Berg, N., 1975. Halmbehandling og fôring med halm. Husdyrforsøksmøtet, Agric Univ. Norway.

Homb, T., Sundstøl, F. and Arnason, J., 1977. Chemical treatment of straw at commercial and farm level. FAO Anim. Prod. & Health Paper 4: 25-37.

Horton, G.M.J., 1979. Feeding value of rations containing nonprotein nitrogen or natural protein and of ammoniated straw for beef cattle. J. Anim. Sci. 48: 38-44.

Horton, G.M.J., 1981. Composition and digestibility of cell wall components in cereal straws after treatment with anhydrous ammonia. Can. J. Anim. Sci. 61: 1059-1062.

Horton, G.M.J. and Steacy, G.M., 1979. Effect of anhydrous ammonia treatment on the intake and digestibility of cereal straws by steers. J. Anim. Sci. 48: 1239-1249.

Hvelplund, T., Møller, P.D., Thomsen, K.V. and Kragelund, Z., 1978. Stigende mængder melasse i rationer med ammoniak-behandlet halm. Statens Husdyrbrugsforsøg, Danmark, Medd. 238.

Ibrahim, M.N.M., Fernando, D.N.S. and Fernando, S.N.F.M., 1983. Evaluation of methods of urea-ammonia treatment application at village level. Australian-Asian Fibrous Agric. Residues Res. Network, Univ. Peradeniya, Sri Lanka, 18-22 April.

Ibrahim, M.N.M. and Pearce, G.R., 1983. Effects of chemical pre-treatments on the composition and in vitro digestibility of crop by-products. Agric. Wastes 5: 135-156.

Ibrahim, M.N.M., Wijeratne, A.M.U. and Costa, M.J.I., 1983. Evaluation of different sources of urease for reducing the treatment time required to treat with ammonium hydroxide generated from urea. Australian-Asian Fibrous Agric. Residues Res. Network, Univ. Peradeniya, Sri Lanka, 18-22 April.

Itoh, H., Terashima, Y. and Tohrai, N., 1979. Evaluation of ammonia treatment for improving the utilization of fibrous materials in low quality roughages. Jap. J. Zootechn. Sci. 50: 54-61.

Jackson, M.G., 1978. Treating straw for animal feeding - an assessment of its technical and economic feasibility. Wld. Anim. Rev. 28: 38-43.

Jayasuriya, M.C.N. and Perera, H.G.D., 1982. Urea-ammonia treatment of rice straw to improve its nutritive value for ruminants. Agric. Wastes 4: 143-150.

Jentsch, W., Schiemann, R., Wittenburg, H. and Hoffmann, L., 1978. Untersuchungen zur Verdaulichkeit und Verwertung von Rationen mit Stroh unterschiedlicher Behandlung. 2. Untersuchungen zur energetischen Verwertung von Rationen mit unterschiedlich behandelten Stroh durch das Schaf. Arch. Tierernähr. 28: 397-406.

Kernan, J.A., Coxworth, E.C. and Spurr, D.T., 1981. New crop residues and forages for Western Canada: assessment of feeding value in vitro and response to ammonia treatment. Anim. Feed Sci. Technol. 6: 257-271.

Kernan, J., Coxworth, E.C., Nicholson, H. and Chaplin, R., 1977. Ammoniation of straw to improve its nutritional value as a feed for ruminant animals. Agric. Sci. Bulletin, Univ. Saskatchewan, Coll. Agric., Extension Public. 329.

Kernan, J.A., Crowle, W.L., Spurr, D.T. and Coxworth, E.C., 1979. Straw quality of cereal cultivars before and after treatment with anhydrous ammonia. Can. J. Anim. Sci. 59: 511-517.

Kernan, J. and Spurr, D., 1978. The effect of reaction conditions during ammoniation on the in vitro organic matter digestibility and the crude protein content of Neepawa wheat straw. Saskatchewan Research Council, C-78-14. Volume II, section 2C.

Kiangi, E.M.I., 1981. Ammonia treatment of low quality roughages to improve their nutritive value. In J.A. Kategile, A.N. Said and F. Sundstøl (eds.): Utilization of Low Quality Roughages in Africa. Agric. Univ. Norway, Agricultural Development Report 1: 49-54.

Kiangi, E.M.I., Kategile, J.A. and Sundstøl, F., 1981. Different sources of ammonia for improving the nutritive value of low quality roughages. Anim. Feed Sci. Technol. 6: 377-386.

Kjus, O. and Torgrimsby, J., 1980. Oppsamling av bøss og agner ved skurtresking. Institute of Agricultural Engineering, Norway, Mimeographed paper, Series A, No 667.

Knapp, W.R., Holt, D.A. and Lechtenberg, V.L., 1974. Anhydrous ammonia and propionic acid as hay preservatives. Agron. J. 66: 823-834.

Kristensen, T.P., Busk, J. and Danielsen, F., 1977. NH_3-behandling af halm i silopose eller stak. Ugeskr. Agron. Hort. Forst. Lic. (31): 652-655.

Kronberger, M., 1933. Zur Aufschliessung des Strohes durch Ammoniak im Stalldünger. Prakt. Bl. Planzenbau 10: 255.

Kumarasuntharam, V.R., Jayasuriya, M.C.N., Joubert, M. and Perdok, H.B., 1983. The effect of method of urea ammonia treatment on the subsequent utilization of rice straw by draught cattle. Austalian-Asian Fibrous Agric. Residues Res. Network, Univ. Peradeniya, Sri Lanka, 18-22 April.

Latvietis, J. and Ruvalds, I., 1980. Verfahrens des Strohaufschlusses in der Lettischen SSR. Arch. Tierernähr. 30: 267-271.

Lauer, D.A., 1975. Limitations of animal waste replacement for inorganic fertilizers. In W.J. Jewell (ed.): Energy, Agriculture, and Waste Management, Ann Arbor Science Publishers Inc., Ann Arbor, Mich.

Lie, O.H., 1975. Internordisk halmmøte, Sundvollen, April. Report on straw utilization, Agric. Research Council, Norway.

Lockeretz, W., 1981. Crop residues for energy: Comparative costs and benefits for the farmer, the energy facility and the public. Energy Agric. 1: 71-89.

Lundsby, A., 1978. NH_3-halmanlæg. Scand. Assoc. Agric. Scientists, Straw Seminar, Middelfart, Denmark, March.

Mahyuddin, P., 1982. Increasing digestibility in rice straw by urine treatment. Australian-Asian Agric. Fibrous Residue Res. Network, Serdang, Malaysia, May 3-8.

Males, J.R. and Gaskins, C.T., 1982. Growth, nitrogen retention, dry matter digestibility and ruminal characteristics associated with ammoniated wheat straw diets. J. Anim. Sci. 55: 505-515.

Marienburg, J. and Bergner, H., 1973. Untersuchungen zur Charakterisierung von Strohpellets. 2. Fütterungsversuche mit aufgeschlossenen Harnstoff-Strohpellets. Arch. Tierernähr. 23: 521-529.

Marienburg, J. and Bergner, H., 1975. Untersuchungen zum Einsatz von ammonisierten Strohpellets als alleinige Grundration in der Wiederkaufütterung. 1. Fütterungsversuche an Milchkühen mit ammonisierten Strohpellets als alleinige Grundration. Arch. Tierernähr. 25: 393-403.

Martynov, S.V., 1972. Treatment of straw with dry ammonia. Nutr. Abstr. Rev. 43: 2012, 1973.

Millett, M.A., Baker, A.J., Feist, W.C., Mellenberger, R.W. and Satter, L.D., 1970. Modifying wood to increase its in vitro digestibility. J. Anim. Sci. 31: 781-788.

Mira, J.J.F., Kay, M. and Hunter, E.A., 1983. Treatment of barley straw with urea or anhydrous ammonia for growing cattle. Anim. Prod. 36: 271-275.

Mo, M., 1977. Ammoniakk-behandlet halm brukt i N-balanseforsøk med okser. Mimeographed paper, Dept. Anim. Nutr., Agric. Univ. Norway.

Møller, P.D. and Hvelplund, T., 1978. Omsætningen av protein i mave-tarmkanalen hos kvæg ved fodring med ammoniakbehandlet halm. Statens Husdyrbrugsforsøg, Denmark, Beretn. 239.

Møller, P.D. and Hvelplund, T., 1982. Nitrogen metabolism in the forestomachs of cows fed ammonia treated barley straw supplemented with increasing amounts of urea or soya bean meal. Z. Tierphysiol., Tierernähr. Futtermittelk. 48: 46-57.

Mosi, A.K. and Lambourne, L.J., 1982. Research experiences in the African Research Network on Agricultural By-products (ARNAB). In B. Kiflewahid, G.R. Potts and R.M. Drysdale (eds.): By-product utilization for animal production. IDRC - 206e, Ottawa, Canada.

Müller, J. and Bergner, H., 1975. Untersuchungen zur Charakterisierung von Strohpellets. 8. Veranderungen von Strohlignin durch Ammoniakeinwirkung. Arch. Tierernähr. 25: 37-45.

Müller, J., Bergner, H. and Marienburg, J., 1976. Untersuchungen zur Charakterisierung von Strohpellets. 9. Herstellung und Untersuchung von ^{14}C-^{15}N-Harnstoff-Strohpellets, ^{14}C-Saccharose-Strohpellets, ^{32}P-Phosphat-Strohpellets und ^{3}H-Strohpellets. Arch. Tierernähr. 26: 221-229.

Mwakatundu, A.G.K. and Owen, E., 1974. In vitro digestibility of sodium hydroxide-treated grass harvested at different stages of growth. East Afric. Agr. For. J. 40: 1-10.

Nikolaeva, L.I., 1938. Ammonium hydroxide treatment of straw. Problems Animal Husbandry (USSR), 7 (3): 175-178. (Chem. Abstr. 35, 817, 1941).

NLVF (Norwegian Agric. Research Council), 1982. Energibruk ved ulike driftsformer i jordbruket, og muligheter for å redusere bruk av energi. Report.

NOFO, 1976. Ammoniakkbehandling av halm (Brochure). Spesialtrykk A.S. 12 February.

Obradovic, M., 1973. Estimation of the efficiency of non-protein nitrogen fixation in carob meal and barley straw treated with ammonia solution. Acta Veterinaria, Beograd, 23: 211-216.

Oji, U.I. and Mowat, D.M., 1977. Breakdown of urea to ammonia for treating corn stover. Can. J. Anim. Sci. 57:828 (Abstract).

Oji, U.I. and Mowat, D.N., 1979. Nutritive value of thermoammoniated and steam-treated maize stover. I. Intake, digestibility and nitrogen retention. Anim. Feed Sci. Technol. 4: 177-186.

Oji, U.I., Mowat, D.N. and Buchanan-Smith, J.G., 1979. Nutritive value of thermoammoniated and steam-treated maize stover. II. Rumen metabolites and rate of passage. Anim. Feed Sci. Technol. 4: 187-197.

Oji, U.I., Mowat, D.N. and Winch, J.E., 1977. Alkali treatments of corn stover to increase nutritive value. J. Anim. Sci. 44: 798-802.

Ørskov, E.R., Tait, C.A.G. and Reid, G.W., 1981. Utilization of ammonia- or urea-treated barley straw as the only feed for dairy heifers. Anim. Prod. 32: 388 (Abstract).

O'Shea, J., Lawlor, M.J. and Hopkins, J.P., 1981. A note on the ammoniation of large cylindrical straw bales. Irish J. Agric. Res. 20: 101-103.

Paterson, J.A., Klopfenstein, T.J. and Britton, R.A., 1981. Ammonia treatment of corn plant residues: Digestibilities and growth rates. J. Anim. Sci. 53: 1592-1600.

Potthast, V. and Hartfiel, W., 1977. Verdaulichkeitsversuche an Hammeln mit NH_3-behandeltem Stroh. Kraftfutter (12).

Potts, G.R., 1982. Application of research results on by-product utilization: Economic aspects to be considered. In B. Kiflewahid, G.R. Potts and R.M. Drysdale (eds.): By-product utilization for animal production. IDRC-206e, Ottawa, Canada.

Richter, W.I.F., Baranowski, A. and Koch, G., 1980. Zum Futterwert von aufgeschlossenem Stroh. 2. Untersuchungen zur Aufschlusswirkung von Ammoniak. Das Wirtschaftseigene Futter 26: 165-172.

Rounds, W., Klopfenstein, T., Waller, J. and Messersmith, T., 1976. Influence of alkali treatments of corn cobs on in vitro dry matter disappearance and lamb performance. J. Anim. Sci. 43: 478-482.

Saadullah, M., Haque, M. and Dolberg, F., 1980. Treating rice straw with animal urine. Trop. Anim. Prod. 5: 273-277.

Saadullah, M., Haque, M. and Dolberg, F., 1981. Effectiveness of ammonification through urea in improving the feeding value of rice straw in ruminants. Trop. Anim.Prod. 6: 30-36.

Saadullah, M., Haque, M. and Dolberg, F., 1982. Treated and untreated rice straw for growing cattle. Trop. Anim. Prod. 7: 20-25.

Saenger, P.F., Lemenager, R.P. and Hendrix, K.S., 1983. Effects of anhydrous ammonia treatment of wheat straw upon in vitro digestion, performance and intake by beef cattle. J. Anim. Sci. 56: 15-20.

Said, A.N., 1981. Sodium hydroxide and ammonia-treated maize stover as a roughage supplement to sheep and beef feedlot cattle. In J.A. Kategile, A.N. Said and F. Sundstøl (eds.): Utilization of low quality roughages in Africa, Agric. Univ. Norway, Agricultural Development Report 1.

Said, A.N., Sundstøl, F., Tubei, S.K., Musimba, N.K.R. and Ndegwa, F.C., 1982. Use of by-products for ruminant feeding in Kenya. In: B. Kiflewahid, G.R. Potts and R.M. Drysdale (eds.): By-product utilization for animal production. IDRC-206e, Ottawa, Canada.

Sandev, S., 1980. Über die Perspektive der Bearbeitung und Verwendung des Stroh's als Futtermittel in der VR Bulgarien. Arch. Tierernähr. 30: 293-297.

Schiemann, R., Jentsch, W., Wittenburg, H. and Hoffmann, L., 1978. Untersuchungen zur Verdaulichkeit und Verwertung von Rationen mit Stroh unterschiedlicher Behandlung. 1. Untersuchungen zur Verdaulichkeit mit unterschiedlich behandeltem Stroh beim Schaf. Arch. Tierernähr. 28: 387-396.

Schmidt, L., Weissbach, F., Block, H.J. and Haacker, K., 1982). Harnstoff als Konservierungsmittel bei der Lagerung feuchter Futterstoffe. 2. Konservierung und Aufschluss von Stroh durch Harnstoffzusatze. Arch. Tierernähr. 32: 57-67.

Semjakina, (1968). Quoted by Bergner (1981).

Solaiman, S.G., Horn, G.W. and Owens, F.N., 1979. Ammonium hydroxide treatment of wheat straw. J. Anim. Sci. 49: 802-808.
Sundstøl, F., 1982a. Proteininnholdet i kraftfôret til okser fôr med ammniakkbehandla halm. Husdyrforsøksmøtet, Agric. Univ. Norway 204-209.
Sundstøl, F., 1982b. Energy utilization in sheep fed untreated straw, ammonia-treated straw or sodium hydoxide-treated straw. In A. Ekern and F. Sundstøl (eds.): Energy metabolism of farm animals. EAAP publication no 29: 120-123.
Sundstøl, F., 1983. Ammonia treatment of straw - Methods for treatment and feeding experience in Norway. Anim. Feed Sci. Technol. (in press).
Sundstøl, F., Coxworth, E. and Mowat, D.N., 1978. Improving the nutritive value of straw and other low-quality roughages by treatment with ammonia. Wld Anim. Rev. 26: 13-21.
Sundstøl, F. and Matre, T., 1980. Bruk av ammoniakkbehandla halm i kjøttproduksjonen. Husdyrforsøksmøtet, Agric. Univ. Norway 399-404.
Sundstøl, F., Said, A.N. and Arnason, J., 1979. Factors influencing the effect of chemical treatment on the nutritive value of straw. Acta Agric. Scand. 29: 179-190.
Tarkow, H. and Feist, W.C., 1969. A mechanism for improving the digestibility of ligno-cellulosic materials with dilute alkali and liquid ammonia. In: Cellulases and Their Applications, Advances in Chemistry Series 95, Amer. Chem. Soc., Washington, D.C.
Tohrai, N., Terashima, Y. and Itoh, H., 1978. Effect of processing conditions on the nutritive values of ammonia treated rice hulls and rice straw. Jap. J. Zootechn. Sci. 49: 69-74.
Tubei, S.K. and Said, A.N., 1981. The utilization of ammonia-treated maize cobs and maize stover by sheep in Kenya. In J.A. Kategile, A.N. Said and F. Sundstøl (eds.): Utilization of low quality roughages in Africa, Agric. Univ. Norway, Agricultural Development Report 1: 151-156.
Van der Merwe, P.K., 1976. Animal feed material. Chem. Abstr. 87: 20898y, 1977.
Waagepetersen, J. and Vestergaard Thomsen, K., 1976. Metoder til NH$_3$-behandling af halm belyst ved laboratorieforsøg. Ugeskr. Agron. Hort. Forst. Lic. 36: 714-716.
Waagepetersen, J. and Vestergaard Thomsen, K. 1977. Effect on digestibility and nitrogen content of barley straw of different ammonia treatments. Anim. Feed Sci. Technol. 2: 131-142.
Waagepetersen, J., Vestergaard Thomsen, K. and Kristensen, T., 1976. Korttidsbehandling af halm med NH$_3$ i forsøgsanlæg. Ugeskr. Agron. Hort. Forst. Lic. 39, 776-782.
Waagepetersen, J., Vestergaard Thomsen, K. and Ubbesen, H., 1976. NH$_3$-halm fremstillet ved langtidsbehandling. Ugeskr. Agron. Hort. Forst. Lic. 42: 867-869.
Waiss, A.C.Jr., Guggolz, J., Kohler, G.O., Walker, H.G.Jr. and Garrett, W.N., 1972. Improving digestibility of straws for ruminant feed by aqueous ammonia. J. Anim. Sci. 35: 109-112.
Wanapat, M., Praserdsuk, S. and Chanthai, S., 1983. Supplementation of dried cassava leaves to urea-ensiled rice straw for water buffaloes. Trop. Anim. Prod. (in press).
Wanapat, M., Praserdsuk, S., Chanthai, S. and Sivapraphagon, A., 1982. Improvement of rice straw utilization by urea-ensiling and/or supplementation with cassava chip for cattle during the dry season. In P.T. Doyle (ed.): The Utilization of Fibrous Residues as Animal Feeds, University of Melbourne Printing Services.

Westgaard, P., 1981. Factors influencing the effect of alkali treatment of low quality roughages. In J.A. Kategile, A.N. Said and F. Sundstøl (eds.): Utilization of low quality roughages in Africa. Agric. Univ. Norway, Agricultural Development Report 1: 29-47.

Westgaard, P. and Torgrimsby, J., 1977. Praktiske erfaringer med ammoniakkbehandling av halm. Norsk Landbruk (15): 8-9, 29, 32, 42.

White, T.W., 1966. Utilization of ammoniated rice hulls by beef cattle. J. Anim. Sci. 25: 25-28.

White, L.M., Hartman, G.P. and Bergman, J.W., 1981. In vitro digestibility, crude protein, and phosphorus content of straw of winter wheat, spring wheat, barley, and oat cultivars in eastern Montana. Agron. J. 73: 117-121.

Williams, P.E.V. and Innes, G.M., 1983. Factors affecting the hydrolysis of urea when applied to barley straw. Br. Soc. Anim. Prod., Winter Meeting, Paper no 36.

Williams, P.E.V. and MacDearmid, A., 1982. Treated straws and turnips in rations for beef steers. Proc. Nutr. Soc. 41 (1), 22A.

Winther, P., 1978. Anvendelse af ammoniak som hø-konserveringsmiddel. Statens Planteavlsforsøg, Denmark. Medd. 1440.

Zafren, S.Ja., 1959. Increasing the nutritive value of straw and at the same time adding digestible nitrogen. Nutr. Abstr. Rev. 31: 252. 1961.

Zafren, S.Ja., 1961. Increasing the feeding value of straw by treatment with ammonia. Nutr. Abstr. Rev. 32: 1309. 1962.

Chapter 8

TREATMENT WITH OTHER CHEMICALS

by

Emyr Owen[1], Terry Klopfenstein[2] and Ndelilio A. Urio[3]

[1]Department of Agriculture and Horticulture
University of Reading, Earley Gate
Reading, Berks RG6 2AT, United Kingdom

[2]   Department of Animal Science
University of Nebraska, Lincoln 68583
Nebraska, USA

[3]Department of Animal Science and Production
Faculty of Agriculture, Forestry and
Veterinary Science
P.O. Chuo Kikuu, Morogoro, Tanzania

8.1.   INTRODUCTION

8.1.1. Characteristics of "ideal" chemical

A large number of chemicals are known to react with lignocellu-lose materials and might therefore be considered potential upgra-ding agents for poor quality roughages. In order to rationalise what might otherwise be a long catalogue, it is useful to recol-lect characteristics of the "ideal" chemical from the standpoint of upgrading roughages for animal feed. Clearly it must be effec-tive in improving digestibility and/or intake. The cost of treat-ment (chemical and application) in relation to improved nutritive

value must be economic. The chemical should be readily available
in a given locality and should remain available, notwithstanding
increased usage. This points to a naturally-occurring chemical or
one that is easily manufactured. The high costs of energy and its
future shortage calls for a chemical that is energy-sparing in its
production and application. Chemical residue in treated roughages
should be non-toxic to animals and the faeces and urine voided
should be non-polluting to soils and water courses. In addition to
improving roughage digestibility and/or intake the "ideal" chemical should itself be a nutrient required by the animal or have a
fertilizer-value. Treatment by "ensiling" a mixture of roughage
and chemical, or simply conservation of treated roughage frequently requires the chemical to have a preservative action also. The
"ideal" chemical should be non-hazardous to handle by man and
should be non-corrosive to machinery; these are characteristics of
importance on farms, especially in less developed countries. In
some circumstances a rapid reaction of the chemical on the roughage is required. This is why sodium hydroxide has proved so successful for industrial processing of straw (Chapter 6.2).

8.1.2. The case for alternatives to sodium hydroxide and ammonia

Chapters 6 and 7 show that sodium hydroxide and ammonia have
been extensively researched and developed for improving the nutritive value of straw and other by-products. Nevertheless, both chemicals fall short of the "ideal" defined in 8.1.1. Both are energetically expensive to manufacture (NaOH, 51 MJ/kg; NH_3, 76 MJ/kg,
Leach, 1976) and are potentially hazardous for on-farm handling.
During the past five years sodium hydroxide has been particularly
expensive and less available in many countries. Anhydrous ammonia
is only widely available for on-farm application in countries where it is already used as a fertilizer and therefore infrastructure
for its distribution already established. For example ammonia has
not been widely used as a fertilizer in the United Kingdom, although one company (Hargreaves Limited, York) has marketed aqueous-
ammonia fertilizer and has recently launched a system ("Nutramon")
for treating straw on farms with aqueous ammonia (330 g NH_3/kg
solution). The cost of transporting an aqueous solution, though
high, is still less than for setting-up and operating a method of
distributing anhydrous ammonia (C. Dawson, personal communication,

1983).The high residual sodium in caustic soda-treated roughages is accompanied by disadvantages of accelerated passage-rate, increased urinary output and (in some countries) the risk of soil salinity problems. Ammonia-treated roughages do not present the same disadvantages, but ammonia is less effective than sodium hydroxide in improving digestibility and the usefulness for rumen microorganisms of nitrogen added during ammoniation is in question.

8.2. CHEMICALS INVESTIGATED

8.2.1. Categories

Table 8.1 emphasises the wide range of chemicals that have been investigated. By far, the alkalies, especially NaOH, have received most study.

At current economics and technical knowledge many can be dismissed because of falling short of the desired characteristics discussed in 8.1.1. For example, peracetic acid is a powerful oxidising agent and therefore potentially effective (Streeter and Horn, 1982). But it is explosive on heating and is strongly irritating to skin and eyes. Furthermore there is no information regarding its effectiveness using animal feeding trials. Chlorine dioxide is also a potent delignifier and was shown by Sullivan and Hershberger (1959) to increase cellulose digestibility in vitro in wheat straw. However, chlorine dioxide is dangerously explosive. This disadvantage may be overcome by future research eg. generation of nascent ClO_2 from "safe" precursors applied to straw.

8.2.2. Apparently-ineffective chemicals

Of the acids listed in Table 8.1 neither volatile fatty acids, formic acid nor orthophosphoric acid appear to be effective upgrading reagents. Stone et al. (1965) used a mixture of acids (3 CH_3COOH : 6 CH_3CH_2COOH : 1 $CH_3CH_2CH_2COOH$) at 20 or 40 g/kg bagasse before or after 24 h treatments with similar amounts of NaOH, KOH and NH_4OH mixtures. Acid treatment did not improve cellulose digestibility under these conditions. Owen et al. (1977) ensiled chopped wheat straw at 500 g DM/kg for 78 days with 12.5 to 50

Table 8.1. Chemicals investigated for upgrading straws etc.

Category	Chemical	Formula	Reference
Alkalis	Ammonium hydroxide (ammonia, urea)	NH_4OH (NH_3, $CO(NH_2)_2$)	Chapter 7
	Calcium hydroxide (quick lime)	$Ca(OH)_2$ (CaO)	See 8.3.1
	Potassium hydroxide	KOH	See 8.3.2
	Sodium hydroxide	$NaOH$	Chapter 6
Acids	Acetic + propionic + butyric	$CH_3COOH, C_2H_5COOH, C_3H_7COOH$	Stone et al. 1965
	Formic	$HCOOH$	Owen et al. 1977
	Hydrochloric	HCl	Klopfenstein et al. 1967
	Orthophosphoric	H_3PO_4	Owen et al. 1977
	Peracetic	CH_3COOOH	Streeter & Horn 1982
	Propionic	CH_3CH_2COOH	Owen et al. 1977
	Sulphuric	H_2SO_4	Holzer et al. 1980, Fahmy & Ørskov 1983
Salts	Ammonium bicarbonate	NH_4HCO_3	Bergner 1981, Chapter 7
	Sodium bicarbonate	$NaHCO_3$	Owen et al. 1977
	Sodium carbonate	Na_2CO_3	See 8.3.3, 8.3.4
	Sodium chloride	$NaCl$	Owen et al. 1977
	Calcium carbide	CaC_2	Zaharjan 1960 (cited by Bergner 1981)
Oxidising agents			
Chlorine com- pounds	Bleaching powder	$CaCl_2O$	Chandra & Jackson 1971, Owen et al. 1977
	Clacium hypochlorite	$Ca(OCl)_2$	Yu et al. 1975
	Chlorine	Cl_2	Ibrahim & Pearce 1983, Yu et al. 1975
	Chlorine dioxide	ClO_2	Sullivan & Hershberger 1959
	Potassium chlorate	$KClO_3$	Yu et al. 1975
	Sodium chlorite	$NaClO_2$	Owen et al. 1977, Yu et al. 1975, Ibrahim & Pearce 1983
	Sodium hypochlorite	$NaClO$	Owen et al. 1977, Yu et al. 1975, Gharib et al. 1975
Other compounds	Hydrogen peroxide	H_2O_2	Chandra & Jackson 1971, Halliwell 1965
	Ozone	O_3	Ben-Ghedalia & Miron 1981
	Sodium peroxide	Na_2O_2	Klopfenstein et al. 1972
Sulphur compounds	Sodium bisulphite	$NaHSO_3$	Owen et al. 1977
	Sodium sulphide	Na_2S	Chandra & Jackson 1971, Owen et al. 1977, Gharib et al. 1975
	Sodium sulphite	Na_2SO_3	Chandra & Jackson 1971, Owen et al. 1977
	Sulphur dioxide	SO_2	Ben-Ghedalia & Miron 1981, Ibrahim & Pearce 1983
Surfactants/	Ethylenediaminetetracetic acid (EDTA)		Owen et al. 1977
	Sodium lauryl sulphate	$NaC_{12}H_{25}SO_4$	Owen et al. 1977
	Water	H_2O	Chaturverdi et al. 1973, Holzer et al. 1975

g/kg straw DM of formic, orthophosphoric or propionic acids. None of the treatments changed the in vitro OM digestibility compared with untreated straw.

In the same study, Owen et al. (1977) also found sodium chloride (25 to 75 g/kg DM), sodium bicarbonate (25 to 100 g/kg DM) and sodium bisulphite (25 to 100 g/kg DM) to be ineffective.

It is notable that chlorine-containing oxidising agents (Table 8.1) with the exception of chlorine dioxide and possibly sodium chlorite (eg. Terashimi et al., 1981) have been shown to be without effect on straw. This is despite frequently being shown to decrease acid detergent lignin (eg. Yu et al., 1975). It appears that treatment with these chemicals produces a residue toxic to rumen microorganisms. Yu et al. (1975) showed that washing treated straw could alleviate the problem.

Based on the one "ensilage" study by Owen et al. (1977), surfactants EDTA (12.5 to 50 g/kg straw DM) and sodium lauryl sulphate (25 to 75 g/kg straw DM), do not appear to hold promise as upgrading agents. The study has involved commercial products (12.5-50.0 g "Teepol"/kg DM, 6.3-25.0 g "Brij 35"/kg DM) but these were also digestibility-depressants.

8.2.3. Effective on laboratory-scale

Sulphuric acid treatment of barley straw or NH_3-treated barley straw improved DM digestibility, as measured by the nylon bag technique, in studies by Fahmy and Ørskov (1983) (Figure 8.1). The acid was added as a 20% solution to give 20, 40 or 60 g H_2SO_4/kg straw DM. Treated straw was kept in closed polyethylene bags for 14 days before assessment. The low pH of treated material (20 g H_2SO_4, 4.4; 40 g H_2SO_4, 2.9: 60 g H_2SO_4, 2.2), even after NH_3 treatment, would suggest intake problems in practice (L'Estrange and Murphy, 1972). Furthermore, H_2SO_4 is hazardous to handle under farm conditions (cf. A.I.V. silage additive). Fahmy and Ørskov (1983) conclude that further research is needed before acid treatment can be recommended, but point to the study of Holzer et al. (1980) where cattle were successfully fed a mixture of NaOH-treated and H_2SO_4-treated straw. In the study by Streeter and Horn (1982) CH_3COOOH was as effective as NH_3 in improving DM digestibility in vitro. However, this acid is expensive as well as dangerous.

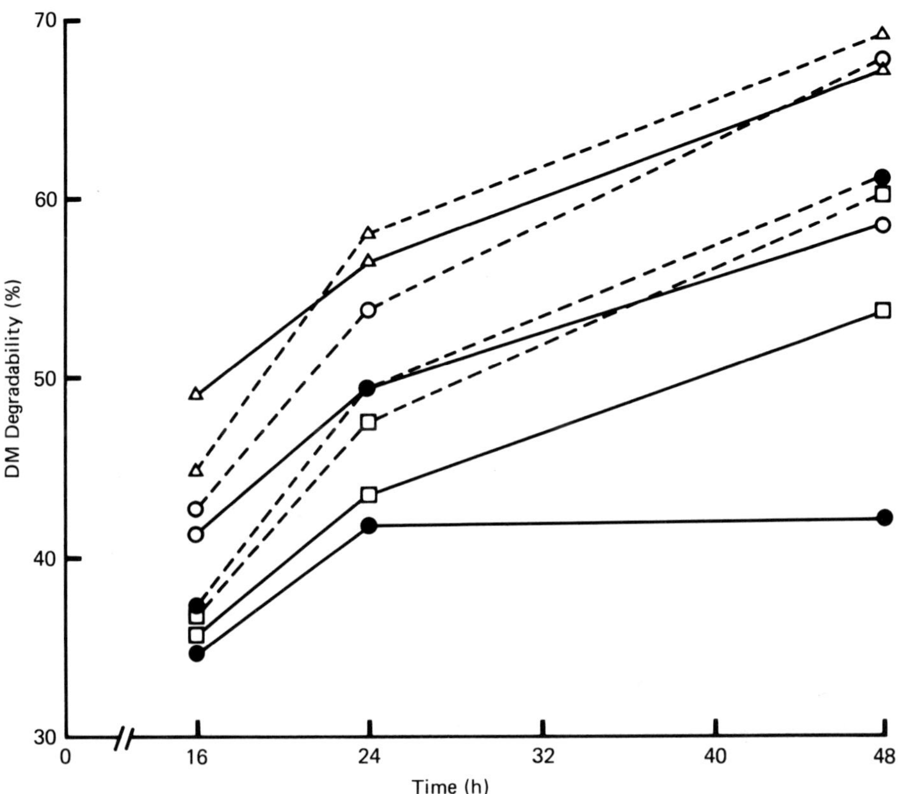

Figure 8.1. Effect of H_2SO_4 treatment of barley straw or NH_3-trea-
ted straw upon rumen degradability in nylon bags (Fahmy
and Ørskov 1983).
————— barley straw; --------- NH_3-barley straw;
● no acid; ☐ 20 g H_2SO_4/kg DM;   o 40 g H_2SO_4/kg DM;
△ 60 g H_2SO_4/kg DM

Of the oxidising agents shown in Table 8.1 hydrogen peroxide,
ozone and sodium peroxide have been shown to have a positive
effect on digestibility under laboratory conditions.

Chandra and Jackson (1971) treated ground maize cobs with 10-60
g H_2O_2/kg cobs.   Treatment was for 1 h using 1.0 l H_2O_2 solution
/kg cobs and 1.0 l of 0.049 g/kg H_2SO_4 followed by drying at 50°C.
Rumen-degradability of DM over 72 h was increased from 40% (unt-

reated) to 55% by 60 g H_2O_2/kg cobs. Earlier, Halliwell (1965) had shown that cotton fibres were completely solubilised in about 7 days after incubating with dilute H_2O_2 and ferrous sulphate. The extent to which the latter would occur, and its implication in a practical situation, is not known. It is notable that in the same study by Chandra and Jackson, NaOH at 60 g/kg cobs increased DM degradability to almost 70%.

Ozone increased wheat straw digestibility in vitro to the ·same extent as NaOH in the study by Ben-Ghedalia and Miron (1981). Treatment involved milled (1 mm) straw moistened to 40% in glass columns, through which O_3 was passed until complete decolorisation of straw. 200 g O_3/kg treated material was required. Treatment with NaOH involved 50 g alkali/kg straw using a 50 g/kg solution, and standing for 7 d. Both treatments produced straw of almost 70% OM digestibility, with untreated material at 44%. O_3 treated straw produced a water extract of pH 2.3 in contrast to NaOH-straw extract at pH 9.7. The acidic nature of O_3-treated straw might create problems of low intakes but this requires investigation together with the economic feasibility of scaling-up the technique.

Sodium peroxide was as effective as NaOH for increasing maize cob digestibility in research by Klopfenstein et al. (1972). Ground cobs were treated at 50% moisture with 40 g/kg cob DM of NaOH or Na_2O_2, for at least 48 h. Ration (77% straw, 23% concentrate) DM digestibilities in sheep were 54.2, 62.8 and 66.4% for control, NaOH and Na_2O_2 treatments respectively. DM digestibilities in vitro for the three straws were 56.1, 70.2 and 70.9%. There have been no further reports involving Na_2O_2, presumably on account of its high cost. This reagent would also have similar disadvantages to NaOH for practical application.

Sodium sulphide was more effective than sodium sulphite in improving in vitro digestibility in the study by Owen et al. (1977) (Figure 8.2). This involved "ensiling" chopped wheat straw at 50% moisture in sealed polythene bags for 78 days. There was little mould in any of the "silages" despite wide ranges in pH (low-high levels, Na_2S 4.5, 5.5, 6.9; Na_2SO_3, 4.6, 5.3, 6.3, 7.3; NaOH 5.7, 7.7, 8.6, 11.1). Treating maize cobs, Chandra and Jackson (1971) also found similar improvements in nylon bag digestibilities with these chemicals, but using a reaction period of only one hour. Sodium sulphite and sodium sulphide however are currently about 50 and 100% respectively more expensive per tonne than sodium hydroxide and are therefore unlikely to be used commercially for upgrading roughages.

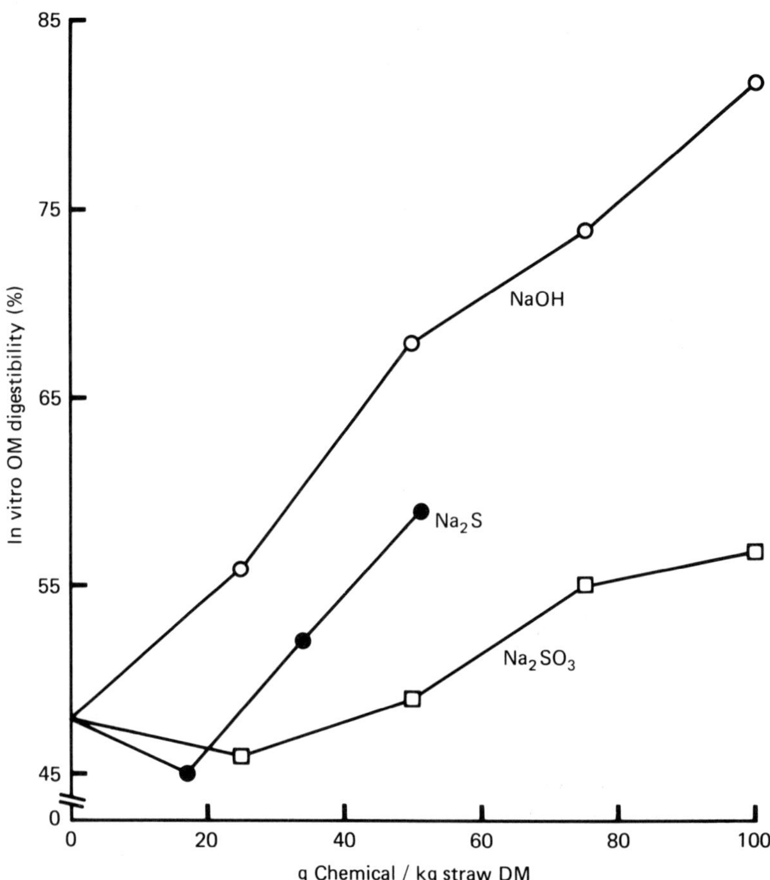

Figure 8.2. Effect of "ensiling" wheat straw with different chemicals upon digestibility in vitro (Owen et al., 1977)

Sulphur dioxide was more effective than NaOH in increasing wheat straw digestibility in vitro, in a study by Ben-Ghedalia and Miron (1981). For treatment with SO_2, straw was ground (1 mm), moistened to 40% and placed in hermetically sealed glass columns. 50 g/kg straw DM of gaseous SO_2 was added and sealed columns were then placed in an oven at 70°C for 72 h, after which the material was aerated and freeze-dried before assessment. NaOH-treatment involved 50 g alkali/kg DM straw at ca 50% moisture for 7 d. After 48 h incubation with rumen liquor, untreated straw OM digestibility was 44%, NaOH-straw was almost 70%, but SO_2-straw was 80%. Digestibility, even after 24 h incubation, was 70% for SO_2-straw. The low pH (3.09) of water extract from SO_2-straw suggests that its acidic nature could present intake problems. SO_2 would have the advantage of increasing S content for optimising microbial utilization of non-protein nitrogen. The treatment conditions used by Ben-Ghedalia and Miron indicate that ammonia ovens (Chapter 7) might be suitable for SO_2 treatment on farms. SO_2 has the further advantage of being a preservative agent. Treatment conditions used by Ben-Ghedalia and Miron (1981) were fairly extreme. Using less drastic conditions ($60^{\circ}C$ for 2 h) Ibrahaim and Pearce (1983) found OM digestibility in vitro to increase from 37.9% (untreated) to 44.1% with SO_2 treatment of moist (70%) barley straw.

8.3. ALTERNATIVES TO SODIUM HYDROXIDE AND AMMONIA STUDIED WITH ANIMALS

8.3.1. <u>Calcium oxide/hydroxide</u>

8.3.1.1. <u>Conditions for optimising effectiveness</u>

$Ca(OH)_2$ has been substituted for NaOH for the previously mentioned reasons. It is less expensive, safer to handle and leaves no Na residue, the Ca residue being little or no problem. Because it has been substituted for NaOH, optimum treatment conditions have not always been used and probably still present some problems in practical application. $Ca(OH)_2$ is a much weaker base than NaOH. NaOH reacts in minutes but it takes 10 to 14 d (Klopfenstein and Owen, 1981; Asadpour, 1979) for reaction of $Ca(OH)_2$ at room temperature. Nwadukwe (1979) even found small responses up to 90 d. In fact, many reports suggest a poor reaction of $Ca(OH)_2$ probably because of insufficient reaction time or other problems which will be discussed.

The problem with the use of $Ca(OH)_2$ is illustrated in the work reported by Paterson et al. (1980). Maize cobs, maize stalks and wheat straw were treated with 50 g $Ca(OH)_2$/kg DM. Treatment was conducted at 20, 40 and 60% moisture for at least 15 days.

Highest digestibility and intake by lambs was at 40% moisture for maize cobs and wheat straw. A subsequent trial showed 45% moisture to be the optimum treatment level for the maize stalks. Observation of the treated residues produced the following conclusions. Moisture of less than 40% was insufficient to wet the residue and provide a medium in which the chemical could react. Some dry chemical was observed after 15 days of reaction. Residues ensiled with $Ca(OH)_2$ above 40-45% moisture fermented giving a smell of good quality fermented silage. It is likely that organic acids produced in the fermentation neutralized the $Ca(OH)_2$ before it could react with the residue. Therefore, the $Ca(OH)_2$ was acting as as fermentation buffer rather than a delignifier.

Residues treated at 40-45% moisture did not appear to ferment and no dry chemical was observed. Therefore, the conditions were optimum for delignification with the $Ca(OH)_2$. However, one problem still remained. Because the residues at 40-45% moisture did not ferment, they tended to mould with time. In fact, one lamb was lost due to aflatoxin in the mould. Moulding has also been reported elsewhere (Owen and Nwadukwe, 1980; Bass et al., 1982).

A possible solution to this problem is to heat the $Ca(OH)_2$-treated residues to increase the rate of reaction similar to the system used for NH_3 treatment (Ørskov, 1981). This has been demonstrated (Klopfenstein, unpublished data) with maize cobs. Maize cobs were mixed with 50 g $Ca(OH)_2$/kg DM and water at 38 ^{O}C to raise the moisture to 60%. The mixture was heated in a well insulated 200 l container to 45°C. Without further heat, presumably because of the exothermic reaction of $Ca(OH)_2$ and cobs, the temperature rose to 70°C within 4 hours and then declined slowly after about 2 h at 70°C. Presumably the reaction was complete when the temperature began to decline. This is a very similar pattern to that with NH_3.

The $Ca(OH)_2$ treatment increased cob digestibility in vitro from 43 to 54%. Lamb intake increased from 1000 g/d to 1455 g/d. After heat treatment the cobs fermented and produced an appealing silage as evidenced by visual observation and lamb intake. There was no mould growth. While heat treatment appears biologically feasible it may not be economical.

Ørskov (1981) suggests that the ambient temperature in some countries may be sufficiently high to create a more rapid reaction of $Ca(OH)_2$ with the crop residue. This would be especially useful in areas where winter wheat or barley is harvested in the heat of the summer. The straw could be treated at that time and take advantage of the high temperature.

Many residues have been treated with $Ca(OH)_2$ including maize (Klopfenstein and Owen, 1981), wheat (Paterson et al., 1980), barley (Owen and Nwadukwe, 1980; Wilkinson and Gonzalez-Santillana, 1978) and rice (Saadullah et al., 1981). Often, treatment has not been as effective as NaOH (eg. Table 8.2) but that might relate to treatment conditions rather than the residue per se. It is assumed that the mode of action of $Ca(OH)_2$ is similar to NaOH treatment if conditions were optimized for $Ca(OH)_2$ treatment. Admittedly, calcium is a divalent cation and this could change its effectiveness.

Level of treatment has usually been 40-50 g $Ca(OH)_2$/kg DM. Because it may not be as effective as sodium hydroxide, higher levels might be used (eg. 90 g/kg DM, Nwadukwe, 1979). It may have less deleterious affects on the animals as well and its cost is less.

It appears that $Ca(OH)_2$ treatment may increase rate of digestion (Brandt and Paterson, 1981). Rate of digestion is very important because of: (1) the slow rate of digestion of crop residues, (2) the high levels of intake achieved with chemically treated residues which may lead to a rapid rate of passage. $Ca(OH)_2$ probably does not increase rate of passage or at least not as much as NaOH (Owen et al., 1982; Brandt and Paterson, 1981).

8.3.1.2. <u>Practical methods of treatment</u>

The optimum conditions to consistently obtain a response to $Ca(OH)_2$ are not yet known. However, the method of practical treatment would require grinding, mixing with water and storing for 2 weeks, probably in a silo. Moistures above 50-60% will likely lead to fermentation, perhaps before complete chemical reaction but, lower moisture would likely lead to moulding. In developed countries, farmers are likely to have mixing equipment for treatment. If straw was sufficiently moist, use of a forage harvester with a powder applicator might work (cf Chapter 6.3). In developing countries it could be mixed by hand. $Ca(OH)_2$ is relatively safe to handle. Covering with plastic or other suitable cover would likely encourage more desirable fermentation and reduce chances of moulding.

In developing countries wishing to minimise capital expenditure on machinery, soaking straw in lime water would be possible. Abou-Raya et al. (1964) used the Beckmann approach. Verma (1981) used 9 l/kg straw in India and investigated various concentrations and soaking times. A concentration of 12.5 g $Ca(OH)_2$/kg and soaking for 3 d seemed optimal for wheat straw. Haque et al. (1981) in Bangladesh used a similar approach, with rice straw, soaking for 2 d. Nylon bag DM degradabilities were increased by 14 units in Verma's study and 4-14 units, depending on straw variety, in Haque et al.'s research. Soaking would necessitate rapid feeding of treated straw to avoid moulding.

8.3.1.3. Application in practice

Conditions where $Ca(OH)_2$ is most likely to be used would be those where residues are too wet for good ammonia treatment, ammonia is not economically available, a gas-tight system for ammonia treatment is not feasible or the residue is in the ground form regardless of chemical used. This would be the case where maize or sorghum residues are being treated and it is impossible or undesirable to allow them to dry. Harvesting these residues with a field chopper, followed by mixing and ensiling seems appropriate. Successful methods of treating roughages with $Ca(OH)_2$ would have wide appeal in developing countries where limestone is locally available.

8.3.1.4. Further research

It is possible that $Ca(OH)_2$ is as effective as NaOH or NH_3 in increasing digestibility of crop residues. The limitation is in defining the treatment conditions so as to assure maximum reactivity. Conditions such as temperature, moisture and time interact to affect treatment and ensiling which influence the overall digestibility and palatability of the product. Mixtures of chemicals may be a solution and will be discussed later. The implications of high-Ca diets on requirements for other minerals need consideration too.

8.3.2. Potassium hydroxide

8.3.2.1. Conditions for optimising effectiveness

KOH is a strong alkali similar to NaOH. Because the atomic

weight of K is greater than Na, KOH supplies only 70% as many hydroxyl ions per unit weight as does NaOH. Research (Garmo, 1981; Rounds et al., 1974; Wilkinson and Gonzalez-Santillana, 1978) shows that KOH is essentially equal in value to NaOH for treating residues on an equal molar basis and 30% more weight would be required to give equal results. Treatment conditions appear to be similar for NaOH and KOH.

Because more weight of KOH is required and because it is more expensive, the only obvious reason for using it compared to NaOH is for mineral balance in the animal. Paterson et al. (1978) have shown that K and Cl are likely to be limiting in animals fed high levels of NaOH-treated residues. Adding K as the hydroxide might be the most economical way to supplement with K without adding to the total mineral load of the animals. Animal performance was improved by using a combination of KOH and NaOH compared to NaOH alone (Lesoing et al., 1980). Potassium has the advantage of being a plant nutrient and thus a fertilizer.

8.3.2.2. Practical methods of treatment

Systems of using KOH would be similar to those for NaOH The two hydroxides would likely be used only in combination to provide K to balance Na.

Wood ash is a crude form of KOH. Bergner (1981) describes a technique used by Kormanovskaya in USSR. Chopped straw was soaked for 10-15 min in 5 g/kg wood ash solution, using 40 g ash/kg straw. Treated straw was then spread out for 10 h before feeding to cows at 20-30 kg/d.

8.3.2.3. Application in practice

There is little likelihood of significant amounts of pure KOH being used because of the cost. Unless NaOH-treated residues are fed above 50% of the ration, there would be little need for the K. Other means of lowering the Na level are likely to be more economical.

8.3.2.4. Further research

The lack of potential for greater use of KOH is based more on economics than lack of biological information. The importance of maintaining mineral balance in the animal is probably the only research area which could be productive. The use of wood ash

merits study in situations where it is plentiful eg. some developing countries.

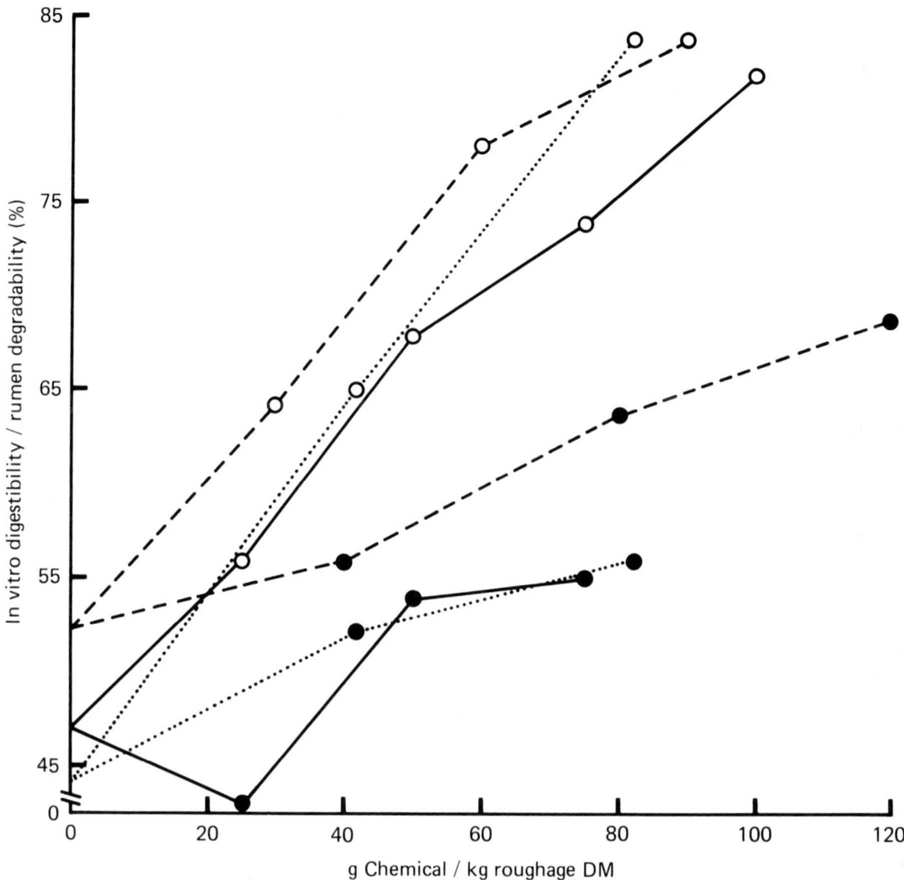

Figure 8.3. Treatment of roughages with Na_2CO_3 (●) or NaOH (o). Effect
on wheat straw OM digestibility, Owen et al. (1977),
———— ; effect on barley straw OM digestibility, Nwadukwe (1979), ------; effect on maize cob DM degradability,
Chandra and Jackson (1971), ...

8.3.3. Sodium carbonate

Sodium carbonate merits special attention because of its safe
handling properties, relatively low cost and natural occurrence as
a major constituent of trona or Magadi (see 8.3.4).

8.3.3.1. <u>Conditions for optimising effectiveness</u>

Several laboratory evaluations (Figure 8.3) show Na_2CO_3 to be less effective than NaOH in improving in vitro digestibility or rumen degradability. For the same amount of chemical, Na_2CO_3 appears to be about half as effective as NaOH. The study by Chandra and Jackson (1971) involved treatment for one hour only, followed by drying. Owen et al. (1977) "ensiled" treated material for 78 d. Nwadukwe (1979) compared treatment durations of 1, 30, 60 and 90 d and found a slight increase in digestibility (2 units) in extending NaOH treatment beyond 1 d, but a slight decrease (6 and 3 units) beyond 1 d with Na_2CO_3 at 40 and 80 g/kg DM. Moisture contents during treatment in the studies by Owen et al. and Nwadukwe were 50% and around 70% in those of Chandra and Jackson. Varying the moisture levels during treatment with Na_2CO_3 has not been studied, but it seems likely that effects would be similar to those for NaOH treatment. Similarly there is no information on temperature effects during Na_2CO_3 treatment.

Urio (1981) found a marked improvement in digestibility in vitro on supplementing maize stover or *hyparrhenia* grass with 10% herring meal after treatment with a range of Na_2CO_3 levels. Treatment was for ca 24 h at 50% moisture followed by sun drying. Without herring meal, OM digestibilities of untreated and 70 g Na_2CO_3 /kg DM-treated roughages were 40.5 and 38.8% respectively for maize stover and 37.1 and 39.6% for *hyparrhenia*. With herring meal, digestibilties were 50.0 and 56.3% for untreated and treated maize stover and 41.5 and 48.6% for *hyparrhenia*.

The disappointing in vitro digestibility improvements with Na_2CO_3 discussed earlier were confirmed in vivo with sheep by Owen and Nwadukwe (1980) and Owen and Kategile (1975). Table 8.2, involving several chemicals, illustrates the overestimation of digestibility in vivo using the in vitro (Tilley and Terry, 1963) technique and the difference in sheep digestibility between restricted and ad libitum feeding. However, the results do show improvements due to Na_2CO_3 treatment, although these are less than with NaOH or NH_3. It is notable that no mould occurred in Na_2CO_3-treated straw, even after 90 d storage. Except for $Ca(OH)_2$-straw, no mould appeared in any of the other straws either. Owen and Kategile (1975) treated ground maize cobs at 50% moisture for 48 h with 60 g NaOH/kg DM or 80 g Na_2CO_3/kg DM. After sun drying, untreated or treated cobs were restricted-fed to sheep with an

equal weight of concentrates. Diet DM digestibilities were 55.3%
for untreated cobs, 61.5% for Na_2CO_3-cobs and 67.3% for NaOH-cobs.
Like much of the literature on chemical upgrading of staws etc.
there is no information concerning the usefulness of treating with
Na_2CO_3in terms of animal production.

Table 8.2. Effect of treatments on barley straw intake and digestibility by wether sheep of 66 kg live weight (Owen and
Nwadukwe, 1980)

Treatment[1] g/kg straw DM	Straw pH	Straw OMD in vitro %	Restricted feeding		Ad libitum feeding	
			Straw intake[2] g OM/kgW.d	Straw OMD[4] %	Straw intake[3] g OM/kgW.d	Straw OMD[4] %
60, NaOH	9.7	77.9	9.3	67.4	16.7	60.4
56, Ca(OH)$_2$	7.9	64.9	9.3	56.9	17.1	53.8
80, Na_2CO_3	9.7	68.9	9.8	61.1	15.9	54.2
40, Na_2CO_3 + 28, Ca(OH)$_2$	9.1	64.6	9.6	58.1	16.1	54.5
35, NH_3	8.2	71.0	10.0	67.3	16.7	59.6
Untreated	-	54.5	9.8	54.8	12.5	49.6

[1]Ensiled for 90 days; 500 g DM/kg treated chopped straw, except
for NH_3-straw (855 g DM/kg straw) and untreated straw (850 g DM/
kg)
[2]Straw 71.6% of diet OM, concentrates 28.4%
[3]Straw 69.6% of diet OM, concentrates 30.4%
[4]Concentrate OMD assumed to be 87%

8.3.3.2. Practical methods of treatment

Except for its non-caustic property, Na_2CO_3 has similar handling characteristics to NaOH. Methods developed for NaOH treatment
would therefore apply to Na_2CO_3. Being non-caustic it would have
considerable advantage over NaOH in developing countries where
hand treatment is more likely.

8.3.3.3. Application in practice

As yet Na_2CO_3 is not used for roughage treatment. Its relative
ineffectiveness and content of sodium are doubtless contributory
factors.

8.3.3.4. Further research

For developing countries with access to inexpensive Na_2CO_3, research is needed to develop simple treatment methods eg. dip treatment (Chapter 6.1). Use of chemical "catalysts" to increase the effectiveness of Na_2CO_3 merit study, although Urio (1977) found disappointing responses to using mixtures of NaOH and Na_2CO_3.

Table 8.3. Chemical composition of *Magadi* soda produced by the Magadi Soda Company in Kenya (Muindi et al., 1981)

Component	g/kg
Sodium	320.00
Potassium	0.27
Calcium	0.56
Magnesium	0.08
Aluminium	1.10
Iron	1.00
Fluorine	0.04
Chlorine	1.40
Sulphate	0.60
Hydrogen carbonate	305.00
Carbonate	246.00
Carbon (total)	105.00
Residue insoluble in 10% HCl	36.00
pH (1 g dissolved in 100 ml H_2O)	10.00

8.3.4. Magadi

8.3.4.1. Chemical composition and availability

"Magadi" as used in East Africa refers to naturally occurring compounds with a wide range of chemical composition. The name is probably associated with Lake Magadi in Kenya, from whose natural trona various sodium salts are manufactured. The product leftover in the purification process is used as an animal salt and is generally referred to as Magadi soda. The unpurified soda ash from the natural trona consists mainly of sodium sesquicarbonate (Na_2CO_3. $NaHCO_3$.2 H_2O) with varying traces of sodium fluoride. However the main chemical compound of Magadi soda is Na_2CO_3 and constitutes over 97% of the soda ash (Musimba, 1980). The chemical composition of the standard Lake Magadi product (Table 8.3) can differ considerably from other unpurified products (Table 8.4) which are also referred to by the common name Magadi. Some of the unpurified forms can have excessively high fluorine contents. These have been reported to adversely affect both humans and livestock in East Africa (McQuilan, 1944; Walker and Milne, 1955; Said et al., 1977).

Table 8.4. Chemical composition (g/kg) of local samples of Magadi Soda in Tanzania (Urio, 1981)

	Lake Manyara	Ngarenanyuki (Arumeru)	Ilongero (Singida)	Kingo'ri (Arumeru)
Phosphorus	2.30	1.30	0.20	1.40
Sodium	304.00	179.00	360.00	320.00
Calcium	7.70	9.30	0.01	8.40
Chlorine	88.00	131.00	423.00	111.00
Fluorine (water soluble)	0.60	2.06	0.17	5.24
Carbon	72.00	45.00	15.00	60.00
pH	9.8	9.9	10.1	9.9

A number of trona lakes are found in East Africa, apart from Lake Magadi. The large quantities found in Lake Magadi (Mathiot, 1982) are probably exceeded by those of Lake Natron and Lake Eyasi in Tanzania. The trona in these two lakes and others has not been exploited to date. There is little chance of the trona being exhausted even at Lake Magadi, where processing has been going on for over 70 years, as the mineral constantly renews itself. Smaller quantities of trona also occur in a number of other lakes and river basins scattered over East Africa.

Large deposits of trona were found in Wyoming in 1938 and most of the Na_2CO_3 in the USA derives from this (Bailar et al., 1978).

8.3.4.2. Conditions for optimising effectiveness

Magadi soda has been used traditionally in East Africa and elsewhere for treatment of both human and animal foods (Muindi et al., 1981). In the treatment of human food it is used mainly for softening the hard seed coats of cereal grains, vegetable leaves and various forms of pulses. In the latter Magadi is also used to buffer acidity associated with such foods. In the case of livestock, the salt has been used mainly for seasoning to increase palatability and feed intake. Also it would have served as a source of sodium.

Boiling or soaking in Magadi as well as sprinkling of Magadi on to dry feeds are some of the methods used for treatment. Literature on optimising the effectiveness of treatment is scarce, but conditions such as moisture, temperature and chemical concentration are likely to have influence. If a solution of Magadi is used for treating straw, boiling strengthens the solution by increasing solubility (Massae, 1977) but it is doubtful whether it has any influence on eventual treatment effectiveness.

Table 8.5. Influence of NaOH and Magadi soda treatment[1] on in vitro digestibility of cereal straws in Kenya (Musimba, 1980)

	Maize stover		Rice straw		Wheat straw	
			Digestibility (%)			
	DM	OM	DM	OM	DM	OM
NaOH solution (% w/v)						
0.625	55.1	49.4	49.4	55.6	43.3	50.3
1.250	61.4	58.0	57.4	65.0	53.6	58.8
1.625	69.0	67.8	64.5	75.0	60.4	66.1
2.000	68.8	70.6	65.0	76.2	59.4	63.8
Magadi soda solution (% w/v)						
0.625	52.6	55.3	55.5	60.3	36.6	42.7
1.250	52.1	57.7	53.3	61.7	39.6	46.5
1.625	52.9	62.4	60.0	60.4	34.9	39.4
2.000	55.7	56.9	54.0	59.2	41.8	46.3
2.500	56.8	61.6	54.7	62.0	46.5	49.9
3.000	66.5	68.5	53.3	59.7	41.3	44.9
3.750	76.8	78.2	59.7	61.1	47.2	52.2
4.500	78.7	80.3	64.7	74.4	44.2	47.5
Control	50.3	48.8	50.2	49.4	39.4	47.7

[1] 4.0 l solution/kg straw DM

Although traditionally Magadi is sprinkled on dry roughages, it is likely that some sort of wetting is necessary if treatment effects other than increased feed intake are to be realised.

Musimba (1980) treated with 4 l Magadi soda solution/kg straw DM and observed a marked increase in digestibility, but no comparison was made of different volumes of solution. In the same studies treatment rates were varied by increasing the concentration of Magadi soda solution (Table 8.5). This resulted in increased digestibility in vitro for various cereal straws although wheat straw response was poor.

8.3.4.3. Practical methods of treatment and application in practice

Interest in Magadi has been aroused by the advent of research on upgrading low quality roughages. Although hardly any research has been done on suitable practical methods of treatment, it is likely that the wet methods used in NaOH treatment will be the most effective with Magadi. Na_2CO_3 (the main constituent) is a weaker base than NaOH, consequently some form of wetting straw with a solution of Magadi is likely to be more effective than dry sprinkling. Apart from the use of Magadi by small farmers, especially for fattening sheep and goats, the product has not been used on any large scale.

8.3.4.4. Further research

Conditions for optimising the effect of Magadi need to be defined, and assessed using animal feeding trials and practical methods of treatment. Treatment of crop residues produced near Magadi deposits deserve priority investigation.

8.3.5. Mixtures: calcium hydroxide and sodium hydroxide

8.3.5.1. Conditions for optimising effectiveness

Several groups have searched for mixtures of $Ca(OH)_2$ and NaOH which would optimize the increase in digestibility and minimize the residual Na (Rounds et al., 1974; Wilkinson and Gonzalez-Santillana, 1978; Klopfenstein, 1978). This information was somewhat inconclusive until the report of Paterson et al., 1980 (cited by Klopfenstein and Owen, 1981). Their results (Figure 8.4) show that the addition of 10 g NaOH to 40 g $Ca(OH)_2$ per kg residue reduced the time of reaction from 10 to 5 days. This together with the previous discussion of fermentation and moulding with $Ca(OH)_2$ treatment indicated good potential for this mixture of chemicals. The question which remains unanswered is whether the chemical reaction proceeds to near completion prior to the onset of fermentation and further, what conditions enhance the speed of reaction without also increasing the onset of fermentation.

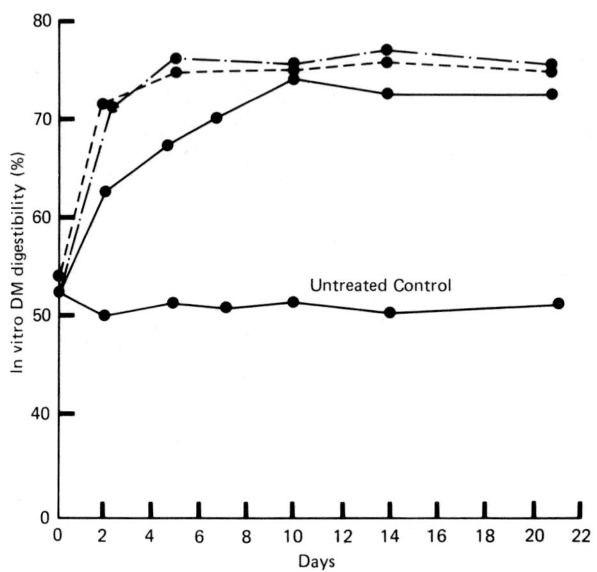

Figure 8.4. In vitro dry matter disappearance for cobs with different combinations of NaOH and $Ca(OH)_2$. NaOH : $Ca(OH)_2$ (g/kg cob DM), ————, 0 : 50; ------, 10 : 40, ·—·—·—, 20 : 30 (Klopfenstein and Owen, 1981)

Table 8.6. Performance of 290 kg heifers fed chemically-treated cob diets[1] (Waller and Klopfenstein, 1975)

NaOH:$Ca(OH)_2$ g/kg DM	Daily gain, kg	Daily feed intake, kg DM	Feed:gain
40:0	1.17	10.10	8.6
30:10	1.34	10.32	7.7
20:20	1.14	9.87	8.7
10:30	1.27	9.97	7.8
0:40	0.97	9.04	9.4

[1]Diet DM: 75% maize cobs, 25% concentrates

Waller and Klopfenstein (1975) showed that a mixture of 30 g $Ca(OH)_2$ and 10 g NaOH was as effective as 40 g NaOH per kg DM (Table 8.6). Corn cobs were treated with mixtures of two chemicals by adding the chemicals in a mixer, raising the moisture to 60%, mixing and ensiling for at least 14 days before feeding over 84 d to yearling heifers. The same results were obtained with sheep. Cook et al. (1982) obtained an excellent response from treating with 30 g $Ca(OH)_2$ and 10 g NaOH/kg cob DM. Rate of gain of calves increased from 0.25 kg/d to 0.55 kg/d and feed efficiency was essentially doubled. This increase was comparable to NH_3 treatment yet in vitro digestibilities were not increased as much as with NH_3. Cobs were treated at 50% moisture and stored in gas-tight silos.

Owen et al. (1982) demonstrated that treatment with a mixture of NaOH and $Ca(OH)_2$ was as good as NaOH alone or NH_3 (Table 8.7). The straw was treated at 60% DM and allowed to react at least 14 days before feeding to lambs in a digestion study. Oji et al. (1977) found 20 g $Ca(OH)_2$ and 20 g NaOH/kg maize stover to be an effective treatment as well. Lamm (1976) found a 3:1 ratio of $Ca(OH)_2$ to NaOH to be as effective as a 1:3 ratio in treating maize cobs for feeding growing calves. However, the same was not true for ensiled treated maize stalks. While a 3:1 ratio of $Ca(OH)_2$ to NaOH markedly improved gain and efficiency compared to the control, ratios of 2:3, 1:3, 1:4 and 0:4 were somewhat superior (Klopfenstein, 1978). In this study ensilage was at 60% moisture which may have encouraged early fermentation.

8.3.5.2. Practical methods of treatment

The same systems of treatment would hold for mixtures of $Ca(OH)_2$ and NaOH as those described previously for $Ca(OH)_2$ alone. The primary advantage of the mixtures is the increased rate of reaction. Probably a moisture content of 50% is optimum, but more research is needed to verify this.

8.3.5.3. Application in practice

Application would be as in the use of $Ca(OH)_2$ alone but more care in handling is necessary to avoid injury to people. A marketing system for the mixture of chemicals would be necessary as well.

Table 8.7. Intake, digestibility and rate of passage of alkali-treated wheat straw diets fed to 36 kg wethers (Owen et al., 1982)

Chemical g/kg DM	Diet DM[1] intake g/d	Diet DM digestibility %	Rate of particulate passage %/h
Control	868(507)[2]	52.1(55.4)	3.24(1.67)
NaOH, 60	1227(508)	62.3(68.3)	3.50(1.08)
$Ca(OH)_2$, 60	1006(508)	50.7(55.5)	3.35(1.43)
NaOH, 45; $Ca(OH)_2$, 15	1331(531)	60.7(67.8)	3.34(1.22)
NH_4OH, 70[3]	1049(534)	59.3(65.5)	3.02(1.33)

[1]Wheat straw comprised 84% of the diet DM, concentrates 16%
[2]Values in parenthesis are at restricted intake
[3]Oven treatment, 24 h at 40% moisture

8.3.5.4. Further research

Mixtures of $Ca(OH)_2$ and NaOH have considerable potential in situations previously described. The problem which remains somewhat unresolved is the optimum treatment condition to obtain delignification without mould. These conditions include moisture, temperature and time, and perhaps other factors which might influence onset of fermentation.

8.3.6. Mixtures: calcium hydroxide and ammonia

8.3.6.1. Conditions for optimising effectiveness

A mixture of NH_3 and $Ca(OH)_2$ has the advatanges that mould would be inhibited and neither the level of residual NH_3 nor Ca is as high as if the chemical were used alone. The disadvantage is the necessity of a closed, gas-tight system. One series of experiments at Nebraska has been conducted to study this mixture ensiled for 21 d (Klopfenstein and Owen, 1981). The mixture was 15 g NH_3 and 20 g $Ca(OH)_2$/kg wheat straw DM and was compared to 30 g NH_3 or 40 g $Ca(OH)_2$ alone or the same mixture but mixed at feeding time. Moisture content of the straw was 60% which may be too high for optimum treatment with this mixture. Fermentation might be encouraged by the high moisture content. Probably a moisture content of 45-50% would be more acceptable. The mixture of NH_3 and $Ca(OH)_2$

increased digestibility, also rates and efficiencies of gains in lambs. While there is some variation in the results in these experiments it does indicate that this mixture at least has biological potential.

Urea and $Ca(OH)_2$ would seem an attractive mixture, but Verma (1981) found rather disappointing results. Wheat straw, ensiled at 65% moisture with 40 g urea and 40 g $Ca(OH)_2$/kg DM for 28 d, was surprisingly less digestible than straw treated with either chemical alone. Heifer growth rates on straw ad libitum (plus minerals and vitamins) over 140 d were poor and least on $Ca(OH)_2$-straw ($Ca(OH)_2$-straw, - 22 g/d; mix-straw, - 14.5 g/d; urea-straw, + 67 g/d; NaOH-straw, + 78 g/d).

8.3.6.2. Practical methods of treatment

This system would require a chopped or ground residue at 45-50% moisture, a mixer and presumably a gas-tight storage facility. $Ca(OH)_2$ and NH_3 (urea) would need to be applied to the mixer separately in that $Ca(OH)_2$ is not readily soluble in water. The $Ca(OH)_2$ would be added dry and the NH_3 in anhydrous or aqueous form.

8.3.6.3. Application in practice

The possibility of this mixture in practice is largely unexplored. The potential looks promising from a biological standpoint but the economics remain to be worked out.

8.3.6.4. Further research

Conditions for optimising this treatment need considerable study. In addition, systems for on-farm tratment need to be developed and the economics of such treatment need careful assessment. Urea or urine and $Ca(OH)_2$ would seem particularly appropriate for developing countries and therefore merit comprehensive research.

8.4. CONCLUSIONS

Based on the evidence reviewed there is no obvious alternative to NaOH or NH_3 which has been evaluated by feeding trial. Based on laboratory evaluation, H_2SO_4 , O_3, SO_2 and ClO_2 look promising. Whether this would be confirmed in vivo remains unanswered along with the question of how to use these chemicals in practice. It is understandable that laboratory evaluation must precede any feeding trial, but the general lack of in vivo evaluation is regrettable. This is especially so in this subject, as laboratory evaluations may not accurately reflect animal utilization.

Despite much research, a common feature of the investigations is inadequate definition of treatment conditions (including type of roughage effect) for optimising the effect of a given chemical. Another feature is the limited consideration given to practical methods of treatment, especially on farms. Economics are also rarely considered.

The obvious conclusion is the need for more research, combining laboratory and animal evaluation, and a greater awareness of practical application.

The "ideal" chemical discussed in 8.1.1 remains elusive. It will probably remain so as long as we think in terms of one chemical only. Much more emphasis on chemical mixtures is required. The possibility of catalysing the effectiveness of desirable chemicals such as $Ca(OH)_2$ and Magadi merits special attention.

8.5. SUMMARY

Characteristics of the "ideal" chemical for upgrading straws etc. are discussed and reasons given for seeking alternatives to sodium hydroxide and ammonia. More than 30 chemicals have been investigated including acids, alkalis, salts, oxidising agents, sulphur compounds and surfactants. Several are ineffective or even depress digestibility in vitro eg. surfactants and most chlorine-based oxidisers. Laboratory-evaluated digestibility enhancers include sodium chlorite, hydrogen peroxide and potassium hydroxide. Sodium carbonate (Magadi) is about half as effective as sodium hydroxide, but calcium hydroxide is even less so. It is concluded that there is no obvious alternative to sodium hydroxide or ammonia at present, but much further research is needed, especially on chemical mixtures. More research emphasis on practical application is also needed.

8.6. REFERENCES

Abou-Raya, A.K., Abou-Hussein, E.R.M., Ghoneim, A., Raafat, M.A. and Mohamed, A.A., 1964. Effect of Ca(OH)$_2$ and NaOH treatments on the nutritive value of maize stalks, sorghum stalks and dry sweet potato vines. J. Anim. Prod. UAR, 4: 55-65.

Asadapour, P., 1979. Calcium and Ammonium Hydroxide Treatment of Wheat Straw. M.S. Thesis. University of Nebraska.

Bailar, J.C., Moeller, T., Kleinberg, J., Guss, C.O., Castellion, M.E. and Metz, C., 1978. Chemistry. Academic Press, New York.

Ben-Ghedalia, D. and Miron, J., 1981. Effect of sodium hydroxide, ozone and sulphur dioxide on the composition and in vitro digestibility of wheat straw. J. Sci. Food Agric. 32: 224-228.

Bass, J.M., Parkins, J.J. and Fishwick, G., 1982. The effect of calcium hydroxide treatment on the digestibility of chopped oat straw supplemented with a solution containing urea, calcium, phosphorus, sodium, trace elements and vitamins. Anim. Feed Sci. Technol. 7: 93-100.

Bergner, H., 1981. Chemical treatment of straw. In: Institute for Scientific Co-operation (ed.): Plant Research and Development. A Biannual Collection of Recent German Contributions. Landhausstrasse 18, 7400 Tubingen, pp. 61-81.

Brandt, R.T. and Paterson, J.A., 1981. Digestibility and rates of digestion of chemically treated wheat straw. American Soc. of Anim. Sci. Abstract 580.

Chandra, S. and Jackson, M.G., 1971. A study of various chemical treatments to remove lignin from coarse roughages and increase their digestibility. J agric. Sci. (Camb.) 77: 11-17.

Chaturverdi, M.L., Singh, U.B. and Ronjahn, S.K., 1973. Effect of feeding water-soaked and dry wheat straw on feed intake, digestibility of nutrients and VFA production in growing Zebu and buffalo calves. J. agric. Sci. (Camb.) 80: 393-397.

Cook, F., Merrill, J., Brink, D. and Klopfenstein, T., 1982. Alfalfa with chemically treated residues. American Soc. of Anim. Sci. Midwest Section. Abstract 92.

Fahmy, S.T.M. and Ørskov, E.R., 1983. Digestion and utilization of straw. 1. The effect of different chemical treatments on degradability and digestibility of barley straw by sheep. Anim. Prod. (in press).

Garmo, T.H., 1981. Ensiling of alkali-treated straw. In: J.A. Kategile, A.N. Said and F. Sundstøl (eds.): Utilization of Low Quality Roughages in Africa. AUN-Agricultural Development Report 1, Aas, Norway. pp. 55-59.

Gharib, F.H., Meiske, J.C., Goodrich, R.D. and El Serafy, A.M., 1975. In vitro evaluation of chemically treated poplar bark. J. Anim. Sci. 40: 734-742.

Halliwell, G., 1965. Catalytic decomposition of cellulose under biological conditions. Biochem. J. 95: 35-40.

Haque, M., Saadullah, M., Zaman, M.S. and Dolberg, F., 1981. Results of a preliminary investigation of the chemical composition and the digestibility of straw from high-yielding varieties of paddy treated with lime. In: M.G. Jackson, F. Dolberg, C.H. Davis, M. Haque and M. Saadullah (eds.): Maximum Livestock Production from Minimum Land. Bangladesh Agricultural University, Mymensingh. pp. 125-131.

Holzer, Z., Levy, D. and Folman, Y., 1980. Performance of fattening cattle and lactating beef cows on diets high in chemically treated wheat straw and poultry litter. Anim. Feed Sci. Technol. 5(4): 299-307.

Holzer, Z., Levy, D., Tagari, H. and Volcani, R., 1975. Soaking of complete fattening rations high in poor roughage. 1. The effect of moisture content and spontaneous fermentation on nutritional value. Anim. Prod. 21: 323-335.

Ibrahim, M.N.M. and Pearce, G.R., 1983. Effects of chemical pre-treatments on the composition and in vitro digestibility of crop by-products. Agricultural Wastes 5: 135-156.

Klopfenstein, T.J., 1978. Chemical treatment of crop residues. J. Anim. Sci. 46: 841-848.

Klopfenstein, T.J., Bartling, R.R. and Woods, W.R., 1967. Treatments for increasing roughage digestion. J. Anim. Sci. 26: 1492 (Abstr.).

Klopfenstein, T.J., Krause, V.E., Jones, M.J. and Woods, W., 1972. Chemical treatment of low quality roughages. J. Anim. Sci. 35: 418-422.

Klopfenstein, T.J. and Owen, F.G., 1981. Value and potential of crop residues and byproducts in dairy rations. J. Dairy Sci. 64: 1250-1268.

Lamm, W.D., 1976. Influence of Nitrogen Supplementation or Hydroxide Treatment upon the Utilization of Corn Crop Residues by Ruminants. PhD. Dissertation. University of Nebraska.

Leach, G., 1976. Energy and Food Production. IPC Business Press, Guildford, UK.

Lesoing, G., Rush, I., Klopfenstein, T. and Ward, J., 1980. Wheat straw in growing cattle diets. J. Anim. Sci. 51: 257-262.

L'Estrange, J.L. and Murphy, F., 1972. Effects of dietary mineral acid on voluntary food intake, digestion, mineral metabolism and acid-base balance of sheep. Br. J. Nutr. 28: 1-17.

Massae, E.E., 1977. The digestibility of maize cob based diets as affected by varying concentrations of Na_2CO_3 solution. B.Sc. Special Projects reports, University of Dar-es-Salaam, Morogoro, Tanzania.

Mathiot, P., 1982. Salts of the earth. The IDRC Reports 11 (2): 19.

McQuilan, C.J., 1944. Chronic fluorine poisoning in the Arusha District, Tanganyika Territory. E.A. Med. J. 21: 131-134.

Muindi, P.J., Thomke, S. and Ekman, R., 1981. Effect of Magadi soda treatment on the tannin content and in vitro nutritive value of grain sorghums. J. Sci. Food Agric. 32: 25-34.

Musimba, N.K.R., 1980. Digestibility and Nutritive Value of Maize Stover, Rice Straw, and Wheat Straw and Oat Straw Treated with Sodium Hydoxide and Magadi Soda. M.Sc. Thesis. University of Nairobi.

Nwadukwe, B.S., 1979. The Nutritive Value for Sheep of Chemically Treated and Ensiled Straw. MPhil Thesis, University of Reading.

Oji, U.I., Mowat, D.N. and Winch, J.E., 1977. Alkali treatments of corn stover to increase nutritive value. J. Anim. Sci. 44: 798-802.

Ørskov, E.R., 1981. Recent overseas and British developments. In: B.A. Stark and J.M. Wilkinson (eds.): Upgrading of Crops and ByProducts by Chemical or Biological Treatments. Ministry of Agriculture, Fisheries and Food. London.

Owen, E. and Kategile, J.A., 1975. Unpublished data, University of Dar-es-Salaam.

Owen, E., Klopfenstein, T.J., Britton, R.A., Rump, K. and McDonnell, M.L., 1982. Treatment of wheat straw with different alkalies. American Soc. of Anim. Sci. Abstract 806.

Owen, E. and Nwadukwe, B.S., 1980. Alkali treatment of barley straw: effect of treatment with different chemicals on digestibility and intake by sheep. Anim. Prod. 30: 489 (Abstr.).

Owen, E., Perry, F.G. and Rees, P., 1977. The effect of various chemicals on the in vitro digestibility of wheat straw under ensilage conditions. Report on research study by University of Reading and BP Nutrition (UK) Limited.

Patterson, J.A., Klopfenstein, T.J. and Britton, R.A., 1978. Effect of sodium hydroxide treatments of roughage on mineral balance and digestibility. American Soc. of Anim. Sci. Abstract 548.

Patterson, J.A., Stock, R. and Klopfenstein, T., 1980. Calcium hydroxide treatment. Nebraska Beef Cattle Report EC 80-218. p. 21.

Rounds, W. and Klopfenstein, T.J., 1974. Chemicals for treating crop residues. J. Anim. Sci. 39: 251 (Abstract).

Saadullah, M., Haque, M. and Dolberg, F., 1981. Practical methods for chemical treatment of rice straw for ruminant feeding in Bangladesh. In: J.A. Kategile, A.N. Said and F. Sundstøl (eds.): Utilization of Low Quality Roughages in Africa. AUN-Agricultural Development Report 1, Aas, Norway.

Said, A.N., Slagsvold, P., Bergh, H. and Laksesvela, B., 1977. High fluorine water to wether sheep maintained in pens. Aluminium chloride as a possible alleviator of fluorosis. Nord. Vet. Med. 29: 172-180.

Stone, C.T., Homan, E.S. Jr., Morris, H.F. Jr. and Frye, J.B. Jr., 1965. The chemical pretreatment of roughages. J. Anim. Sci. 24: 910 (Abstract).

Streeter, C.L. and Horn, G.W., 1982. Effect of treatment of wheat straw with ammonia and peracetic acid on digestibility in vitro and cell wall composition. Anim. Feed Sci. Technol. 7: 325-329.

Sullivan, J.T. and Hershberger, T.V., 1959. Effect of chlorine dioxide on lignin content and cellulose digestibility of forages. Science 130: 1252.

Terashimi, Y., Harada, H., Tarui, M. and Itoh, H., 1981. Nutritive value of ammoniated and oxidised-ammoniated rice straw for sheep. Jpn. J. Zootechn. Sci. 52 (4): 269-274.

Tilley, J.M.A. and Terry, R.A., 1963. A two-stage technique for the in vitro digestion of forage crops. J. Br. Grassld. Soc. 18: 104-111.

Urio, N.A., 1977. Improvement of low quality roughages by alkali treatment under Tanzanian conditions. MSc. Thesis, University of Dar-es-Salaam.

Urio, N.A., 1981. Alkali treatment of roughages and energy utilization of treated roughages fed to sheep and goats. PhD. Thesis, University of Dar-es-Salaam.

Verma, M.L., 1981. New methods of straw treatment using lime. In: M.G. Jackson, F. Dolberg, C.H. Davis, M. Haque and M. Saadullah (eds.): Maximum Livestock Production from Minimum Land. Bangladesh Agricultural University, Mymensingh, pp. 103-124.

Walker, G.W. and Milne, A.H., 1955. Fluorosis in cattle in the northern province of Tanganyika. E. Afr. agric. J. 21: 2-5.

Waller, J.C. and Klopfenstein, T.J., 1975. Hydroxides for treating crop residues. J. Anim. Sci. 41: 424 (Abstract).

Wilkinson, J.M. and Gonzalez Santillana, R., 1978. Ensiled alkali-treated straw. I. Effect of level and type of alkali on the composition and digestibility in vitro of ensiled barley straw. Anim. Feed Sci. Technol. 3: 117.

Yu, Y., Thomas, J.W. and Emery, R.S., 1975. Estimated nutritive value of treated forages for ruminants. J. Anim. Sci. 41: 1743-1751.

Chapter 9

MICROBIAL CONVERSION OF LIGNOCELLULOSE INTO FEED

by

František Zadražil

Institut für Bodenbiologie
Bundesforschungsanstalt für Landwirtschaft
Bundesallee 50, 3300 Braunschweig, FRG

9.1. INTRODUCTION TO THE ECOLOGY OF LIGNOCELLULOSE-DECOMPOSING MICROORGANISMS

An "ideal" fermentor, digesting lignin and cellulose containing plant residues (lignocellulose) modified during evolution, can be found in the rumen of both domestic and wild animals (Hungate, 1975; Prins and Clarke, 1979; Stewart, 1981). Digestion proceeds discontinuously by anaerobic bacteria and protozoa. Several species of anaerobic fungi have also been found in the rumen. Cellulose and hemicellulose is partially metabolized during a relatively short time, 2-3 days. The accumulation of lignin in plant materials used as feed results in a rapid decrease of digestibility for rumen microorganisms.

In order to increase the digestibility of lignocellulose, physical, chemical, and biological methods can be used (Baker et al., 1975; Millett and Baker, 1975; Han, 1978; Jackson, 1978; Kaufmann et al., 1978). The principle of these methods is the splitting of the cellulose-lignin complex by extraction or decomposition of lignin. In this review microbial activities on lignocelluloses are briefly summarized.

The main problem of biological upgrading of lignocelluloses into feed is to find suitable microorganisms, with metabolic patterns different from those of the rumen flora and fauna. "Ideal microorganisms" should have a strong lignin metabolism with a low degradation of cellulose and hemicellulose. The growth properties of cellulase-less mutants of three white rot fungi, with the aim of selective lignin degradation, were studied by Eriksson et al. (1980).

It is very difficult to differentiate the influence of mixed populations of bacteria, yeasts, and fungi on the amount of lignocelluloses degraded in natural ecosystems. Growth activity, synergism and antagonism in microbial associations have been described in detail elsewhere (Rypáček, 1966; Sundman and Nase, 1972; Swift, 1977; Blanchette et al.,1978; Haider and Trojanowski, 1981). Wood, straw, leaves, and other plant residues in nature are mainly mineralized by fungi. Different ecological groups of rot can be distinguished: "sweet rot", "brown rot", "white rot".

Sweet rot fungi, similar to bacteria and yeasts, only metabolize soluble sugars or other easily digestible cell constituents of plant residues. The bacterial degradation of wood (Schmidt, 1978, 1980a, b; Holt and Jones, 1978; Swift, 1977) or the splitting of isolated wood lignin, or synthesized lignin-like substances (Haider et al.,1978; Haider and Trojanowski, 1981; Ellwardt et al., 1981) have been studied under laboratory conditions. The results of individual authors are different. Synthetic lignin can be degraded by bacteria (Haider et al., 1978; Haider and Trojanowski, 1981), but untreated lignocelluloses can only be decomposed under special conditions and with a few microbial species (Crawford et al., 1982). During the degradation of wood some bacteria are able to fix nitrogen (Larsen et al., 1978; Aho et al., 1974, King et al., 1980). In this way, the decomposition of organic matter can be stimulated.

More resistant plant polymers, cellulose and hemicellulose, are primarily decomposed by brown rot fungi at extremely low pH (3.0-2.0). As a result of lignin accumulation, the decomposed material becomes brown.

The white rot fungi are able to decompose and mineralize all plant cell constituents (Rypáček, 1966; Kirk et al., 1980). During the decomposition at pH 6.0-3.5, the colour of the substrate turns white.

At the end of the 19th century decomposed wood was used as an animal feed in southern Chile (Phillippi, 1893; Knoche et al., 1929). It was called "palo podrido". In the etymological dictionary of Chiloe (Tangol, 1976) "palo podrido" is defined as "white coloured, decomposed wood, used as animal feed" (Madera podrida de color blanco, que se utiliza como forraje).

Nikitin (1955) mentioned that the cellulose content of "palo podrido" is up to 80%. Zadražil et al. (1982) analyzed different samples of "palo podrido" and found that average digestibility in vitro (Tilley and Terry, 1963) fluctuated between 30 and 60%. The highest digestibility amounted to 77% in contrast to non-decomposed wood, with a digestibility between 0 and 3%. Digestible samples of "palo podrido" were decomposed by white rot fungi - *Ganoderma applanatum* and *Armillariella spp.* Wood decomposed under the same conditions by a brown rot fungus was practically undigestible and showed the highest lignin content (74%). The lowest lig-lignin content (1%) was found in the samples with the highest in vitro digestibility (77%).

The exact classification of the mineralization processes and the microbial populations in nature is very difficult and practically impossible. The relationship between the different microbial populations and the changes of growth conditions caused by the addition of nitrogen to the substrate (Kirk and Chang, 1981; Reid, 1979; Zadražil and Brunnert, 1980), change in oxygen content in the gaseous phase of the substrate (Reid and Seifert, 1982) or different temperature (Zadražil, 1977; Zadražil and Brunnert, 1981), influence not only the rate of decomposition but also the sequence and quantity of mineralized substrate components. Generally white rot fungi can be divided into three groups (Rypáček, 1966). The first one decomposes hemicellulose and cellulose initially and then lignin. The second one metabolizes more lignin initialy than cellulose and hemicellulose, and the third group is able to degrade all kinds of plant cell polymers simultaneously.

These three groups of fungi are shown in Figure 9.1. Cereal straw was incubated with different fungi and the resulting decomposition of organic matter (loss of OM), loss of lignin and in vitro digestibility of OM was determined.

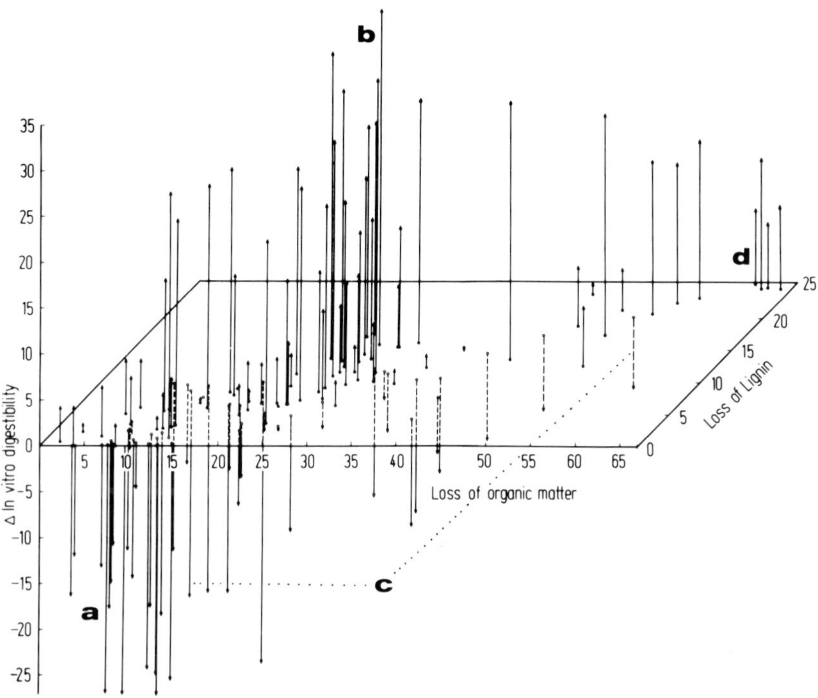

Figure 9.1. Relationship between the loss of organic matter, loss of lignin, and differences in the in vitro digestibility of wheat straw after fungal treatment (60 days at 22, 25 and 30°C; in vitro digestibility of untreated straw = 0; increased digestibility +, decreased -). (a b c d - see text for explanation)

Fungi in group (a) decompose the substrate without lignin degradation (brown rot fungi). In vitro digestibility is negative in comparison with untreated straw. Examples are *Agrocybe aegerita* and *Flammulina velutipes* . Similar results can be obtained by the cultivation of bacteria or yeasts on cereal straw.

The second group (b) includes fungi, which decompose lignin very well, but other substrate components are decomposed only partially. In vitro digestibility increases. Examples are *Dichomitus squalens*, *Abortiporus biennis*, *Stropharia rugosoannulata*, *Pleurotus eryngii* and *Pleurotus sajor caju*.

The third group (c) includes fungi, which decompose lignin and other substrate components, but in vitro digestibility decreases.

This may be due to toxicity of substrate extracts for the rumen microorganisms used for determining digestibility.

The fourth group (d) consists of fungi which decompose lignin and other components very rapidly and change in vitro digestibility only partially. An example is *Sporotrichum pulverulentum* which decomposes ca. 3% of the organic matter daily but causes non significant changes in digestibility of cereal straw (Zadražil and Brunnert, 1982; Zadražil, 1983).

9.2. BIOLOGICAL UPGRADING OF LIGNOCELLULOSES - PROPOSED PROCESSES

Figure 9.2. shows three principal biotechnical possibilities for upgrading plant wastes into feed. Distinction must be made between feed for ruminants, which can contain cellulose and hemicellulose and feed for monogastrics, mainly containing microbial proteins and sugars.

The first system is based on the utilization of lignocelluloses, without pretreatment, for the production of ruminant feed or edible fungi (mushrooms). Lignolytic microorganisms are mainly wood inhabiting fungi. They are able to colonize different plant residues (Zadražil, 1976a, 1979) and increase the digestibility of the substrate (Schánel et al., 1966; Herzig et al., 1968; Kirk and Moore, 1972). This process must be cheap and simple with low cost technology. Similar processes are suitable to convert plant residues into human food (Plate 9.1.) by the cultivation of edible fungi (*Pleurotus spp*, *Volvariella volvacea*, *Stropharia rugoso annulata*). For literature see Singer (1961), Lelley et al. (1976), Chang and Hayes (1978) and Chang and Quimio (1982).

Microbial upgrading of low quality lignocelluloses into feed has been discussed for a long time. For example, Kühler (1848) described a simple process for cultivation of moulds. Further developments of this system by Rhode (1955) and Columbus et al. (1958/1959) were never realized in practice. Seal and Eggins (1976) described a similar process with thermophillic fungi for the fermentation of cereal straw with pig manure. The disadvantages of this system are as follows:

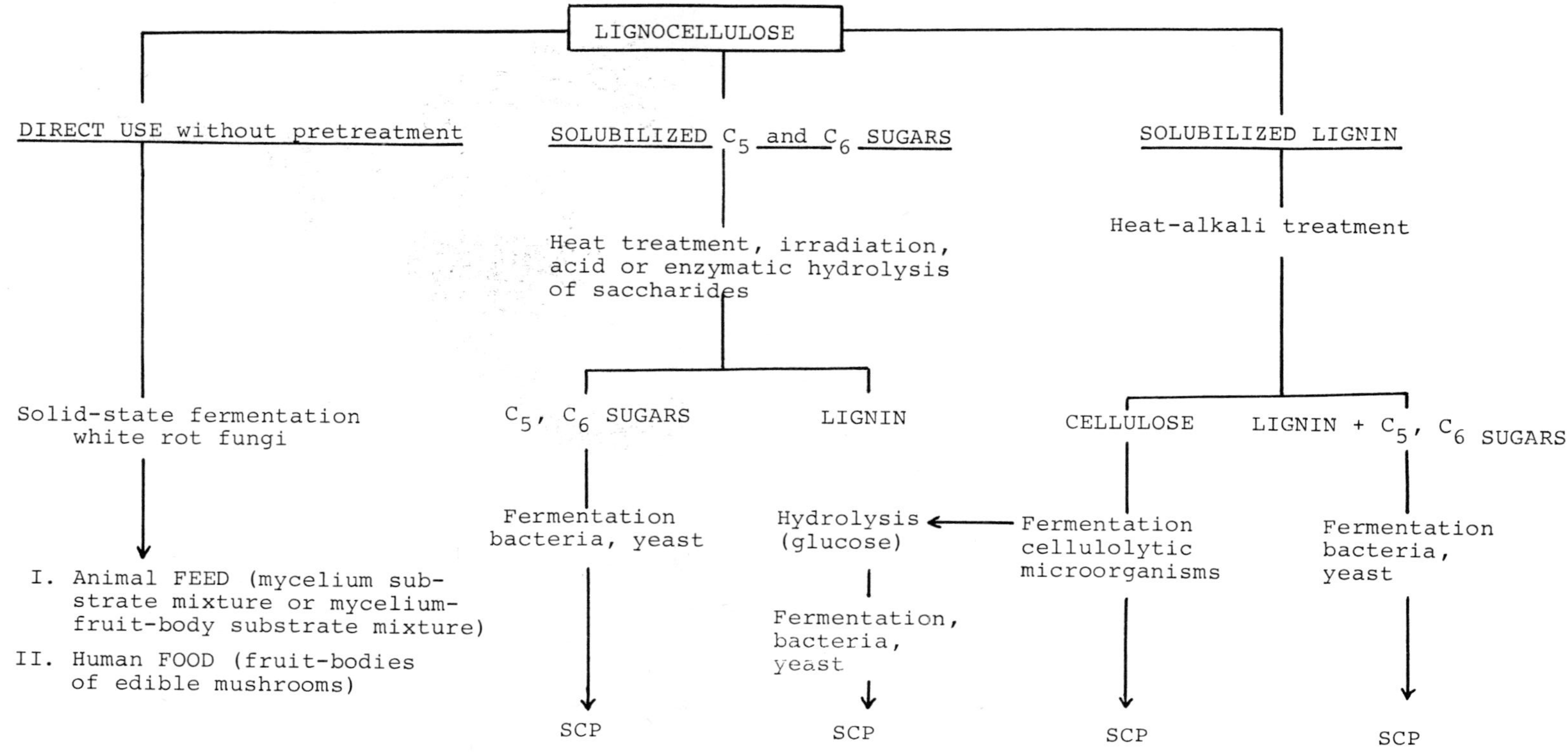

Figure 9.2. Different ways of microbial treatment of fibrous by-products (lignocellulose) for feed/food production

1. The microflora of "fermented" substrate cannot be controlled (production of mycotoxins).
2. The "fermentation" process cannot be controlled and also the quality cannot be guaranteed.
3. The activity of lower fungi reduces the digestibility of lignocellulose-matrix for rumen microorganisms.

Plate 9.1. Edible fungi cultivated on cereal straw

a. Pleurotus sajor-caju
b. Pleurotus eryngii
c. Flammulina velutipes
d. Pleurotus ostreatus

The coprophilous fungus *Coprinus sp.* has been used by Jilek et al. (1979) and Burrows et al. (1979) for the conversion of straw into ruminant feed. Mycelial growth is stimulated by the addition of inorganic nitrogen. Shortly after inoculation, a large quantity of fungal biomass is produced (mycelium and fruit bodies). However, lignin is not decomposed and in vitro digestibility increases for a short time only, i.e. maximum 3 days (Burrows et al., 1979).

Many workers have investigated the use of white rot fungi to improve the digestibility of cereal straw (Kirk and Moore, 1972; Hartley et al., 1974; Zadražil, 1977, 1979, 1980; Kaneshiro, 1977; Latham, 1979; Lindenfelser et al., 1979; Langar et al., 1980; Streeter et al., 1982 and Ibrahim and Pearce, 1980). Cultivation on maple or poplar wood was performed by Matteau and Bone (1980) and McQueen et al. (1980), and on poultry droppings by Virk et al. (1980). The nutritive value of fungal substrate (*Pleurotus ostreatus*) for ruminants was studied by Skultetyova et al. (1978) and Sommer et al. (1978).

The above investigations can be summarized as follows:
1. In vitro digestibility of fungal substrates decreases at the beginning of colonization by white rot fungi and increases afterwards (Zadražil, 1977; Zadražil and Brunnert, 1982). During incubation the contents of soluble substances (partly sugars) increases (Zadražil, 1976b, 1977; Lindenfelser et al., 1979).
2. The increase in digestibility depends on the fungal species (Zadrazil, 1979, 1983), cultivation time, temperature, water/ air ratio in the substrate, and on the preparation, bulk density and composition of the substrate (Zadražil and Brunnert, 1981, 1982)
3. In vitro digestibility of lignocelluloses by white fungi is decreased by addition of inorganic nitrogen (Zadražil and Brunnert, 1980).

Above examples and the analysis of "palo podrido" samples (Phillippi, 1893; Knoche et al., 1929; Zadražil et al., 1982) clearly show that use of white rot fungi for upgrading lignocelluloses into feed is possible, at least on a laboraory scale. On the other hand, only a little is known about the large scale process. Solid-state fermentation of lignocelluloses in deep layers might be proposed (Schuchardt and Zadražil, 1982). A similar process is used for the production of substrates for white mushrooms (Francescutti, 1972; Bech, 1977; Overstijns, 1981).

The chemical pretreatment of lignocelluloses for biological upgrading basically combines two different procedures (Figure 9.2). In the first one, C_5 and C_6 saccharides, di-, tri- and some polysaccharides are liberated from plant tissue by inorganic acids (Gilbert et al., 1952; Hall et al., 1956), steam/hot water treatment (Kaufmann et al., 1978; Sarkanen, 1981) or radiation (Han, 1978; Nehring and Friedel, 1982). After saccharification, lignin is mainly obtained as a solid residue. The extracted solution of sugars can be used for fermentation processes either directly or after removal of the solubilized phenols.

Saccharification of lignocelluloses by microorganisms (*Trichoderma viride*) or their enzymes is very slow, but on pure cellulose it is rapid (Mandels et al., 1974; Spano et al., 1975; Grant et al., 1977; Jurasek, 1979; Linko, 1977; Ghose, 1977). Enzymatic upgrading of wheat bran into monogastric feed supplements was studied by Neudoerffer and Smith (1969).

Wood destroying fungi liberate saccharides during growth in lignocellulosic substrates (Zadražil, 1976b; 1977), but this process is connected with a high loss of organic matter caused by fungal metabolism.

Basic studies of the chemical saccharification of wood (Gilbert et al., 1952; Millett et al., 1976; Sarkanen, 1981) and other lignocelluloses (Detroy et al., 1980) have been carried out. After the First and again after the Second World War saccharification of wood started experimentally in Europe, but this process was not economic and never realized on a large scale (Hall et al., 1956).

Microorganisms used for the fermentation of plant waste hydrolysates are as follows: *Trichoderma viride* , (Han and Anderson, 1974; Kaneshiro et al., 1975), *Candida utilis* and *Cellulomonas sp.* (Han, 1974; Han and Anderson, 1975; Kristensen, 1978), *Torula utilis* and *Oidium lactis* (Leopold, 1965a, b), *Paecilomyces variottii* (Schmidt et al, 1979), *Chaetomium celluloliticum* (Moo-Young et al., 1978), *Phanerochaete chrysosporium, Polyporus anceps, Pleurotus sp.* and other higher fungi (Daugluis and Bone, 1977; Jauhri, 1978; Jauhri and Sen, 1978) and others.

In the third process (Figure 9.2), lignin is solubilized in a manner similar to that used in the pulp and paper industry (Baker et al., 1973); NaOH-treatment may also be used. The cellulose obtained can be hydrolyzed by enzymes (Mandels et al., 1974; Spano et al., 1975; Jurasek, 1979; Fujio and Moo-Young, 1980) or by sulphuric or hydrochloric acid (Hall et al., 1956). Glucose can be used for fermentation processes after neutralization of the hydrolysate. A similar process is available for the conversion of waste paper into single cell protein (SCP) (Brown and Fitzpatrick, 1976). Single cell protein from cellulose can be used for direct human consumption (Von Hofsten, 1976, Humphrey, 1974, and others), because pure compounds (cellulose, glucose and nutrient additives) are used.

Solubilized lignin fractions of wood in the pulp and paper industry contain more than two per cent of saccharides, which can be used for SCP production (see Pekilo-process: Romantschuk, 1974; Alaviuhkola et al., 1975; Farstad et al., 1975; Fors and Passinen, 1976) or fungi (*Sporotrichum pulverulentum* : Thomke et al.,1980). The use of sulphite waste liquors for SCP production for animal feed is discussed by Millett et al. (1973).

In the latter systems, the products cannot be directly fermented because neutralization and sterilization is needed. The addition of nutrients and minerals is necessary to increase biomass (SCP) yield (Han, 1974; Han and Anderson, 1975; Moo-Young et al., 1978; Chahal et al., 1979). The insoluble residues of substrate after hydrolysis have a lower digestibility than the original substrate. Therefore it is necessary to reduce this fraction to a minimum.

At present, with the exception of edible fungi production, none of the described systems have been realized in practice. Further developments and application of the proposed technologies depend on acceptability of the products by animals and on economics.

9.3. SUMMARY

Poorly digested plant residues can be converted into food or feed by usage of different microorganisms and techniques. Such technologies are attractive in being non-polluting to the environment.

Wood-decomposing fungi (*Pleurotus spp.*, *Lentinus edodes*, *Flammulina velutipes* etc). are able to grow on different plant residues and produce edible fruit-bodies. The conversion rate is relatively high (1.0 kg fruit-bodies/kg straw DM).

White rot fungi, after selective lignin decomposition, can increase the in vitro digestibility of plant residue to 77%, without chemical and physical pretreatment and addition of nutrients to the substrate. For example, hard wood after decomposition with *Ganoderma applanatum* had a lignin content of 1% and a digestibility of 74%. This decomposed wood (Palo podrido) was used as feed in Chile. With exact knowledge of the biology of higher fungi and solid-state fermentation low-cost technology for the conversion of lignocelluloses could be realized.

Protein content of plant residues can be increased by fermentation with different species of bacteria, yeast and fungi. Chemical and physical pretreatment of the substrate and addition of nutrients are proposed for this process. Bacteria, yeasts and moulds used during such fermentation with easily digestible substances, produce new, protein-rich, biomass (SCP). Digestibility of the plant-residue matrix for ruminants after such biological treatment is reduced. Soluble sugars, after hydrolysis of plant residues with inorganic acids or heat treatment extraction, can be used following addition of nutrients for fermentation and SCP-production.

Further development and technical realization of these technologies depend on acceptability of the feeds by animals and on economics.

9.4. REFERENCES

Aho, P.E., Seidler, R.J., Evans, H.J. and Raju, P.N., 1974. Distribution, enumeration, and identification of nitrogen-fixing bacteria associated with decay in living white fir trees. Phytopathology 64: 1413-1420.

Alaviuhkola, T., Korhonen, I., Partanen, J. and Lampila, M., 1975. Pekilo protein in the nutrition of growing-finishing pigs. Acta Agric. Scand. 25: 301-305.

Baker, A.J., Mohaupt, A.A. and Spino, D.F., 1973. Evaluating wood pulp as feedstuff for ruminants and substrate for *Aspergillus fumigatus*. J. Anim. Sci. 37: 179-182.

Baker, A.J., Millett, M.A. and Satter, L.D., 1975. Wood and wood-based residues in animal feeds. In: F. Turbak (ed.): Cellulose Technology Research, ACS Symposium Series 10, American Chem. Soc., Washington, D.C.

Bech, K., 1977. Utersuchung von Massenpasteurisierung und Durchwachsen in der Masse im Vergleich zur traditionellen Anbautechnik, wie sie im dänischen Champignon-Laboratorium verwendet wird. Der Champignon 193: 17-27.

Blanchette, R.A., Shaw, C.G. and Cohen, A.L., 1978. A SEM study of the effects of bacteria and yeasts on wood decay by brown- and white-rot fungi. Scanning Electron Microscopy 11: 61-68.

Brown, D.E. and Fitzpatrick, S.W., 1976. Food from waste paper. In: G.G. Birch, K.J. Parker and J.T. Worgan (eds.): Food from waste. Applied Science Publishers, London, pp. 139-155.

Burrows, I., Seal, K.J. and Eggins, H.O.W., 1979. The biodegradation of barley straw by *Coprinus cinereus* for the production of ruminant feed. In: E. Grossbard (ed.): Straw decay and its effect on disposal and utilization. John Wiley & Sons, Chichester, New York, Brisbane, Toronto, pp. 147-154.

Chahal, D.S., Moo-Young, M. and Dhilon, D.S., 1979. Bioconversion of wheat straw and wheat straw components into single-cell protein. Can. J. Microbiol. 25: 793-797.

Chang, S.T. and Hayes, W.A., 1978. The biology and cultivation of edible mushrooms. Academic Press, New York, San Fransisco, London.

Chang, S.T. and Quimio, T.H., 1982. Tropical mushrooms. Biological nature and cultivation methods. The Chinese University Press, Hong Kong.

Columbus, A., Harnisch, W., Göbel, G. and Rojahn, J., 1958/59. Partielle Futterverpilzung unter Anwendung von Pilzkulturen und Stickstoffzusatz. Jahrbuch der Arbeitsgemeinschaft für Fütterungsberatung 2: 309-325.

Crawford, D.L., Barder, M.J., Pometto III, A.L. and Crawford, R.L., 1982. Chemistry of softwood lignin degradation by *Streptomyces viridosporus*. Arch. Microbiol. 131: 140-145.

Daugluis, A.J. and Bone, D.H., 1977. Submerged cultivation of edible white-rot fungi on tree bark. Europ. J. Appl. Microbiol. 4: 159-166.

Detroy, R.W., Lindenfelser, L.A., Julian, G.St. and Orton, W.L., 1980. Saccharification of wheat straw cellulose by enzymatic hydrolysis following fermentative and chemical pretreatment. Biotechnol. Bioeng. 10: 135-148.

Ellwardt, P.-Chr., Haider, K. and Ernst, L., 1981. Untersuchungen des mikrobiellen Ligninaubbaues durch ^{13}-NMR-Spektroskopie an speziphisch ^{13}C-angereichertem DHP-Lignin aus Coniferylalkohol. Holzforschung 35: 103-109.

Eriksson, K.E., Grünewald, A. and Vallander, L., 1980. Studies of the growth conditions in wood for three white-rot fungi and their cellulase-less mutans. Biotechnol. Bioeng. XXII: 363-376.

Farstad, L., Liven, E., Flatlandsmo, K. and Næss, B., 1975. Effect of feeding "Pekilo" single cell protein in various concentrations to growing pigs. Acta Agric. Scand. 25: 292-300.

Fors, K. and Passinen, K., 1976. Utilization of spent sulphite liquor components in the Pekilo protein process and the influence of the process upon the environmental problems of a sulphite mill. Paperi ja Pun. 9: 608-618.

Francescutti, B., 1972. Offenlegungsschrift 2251906, Deutsches Patentamt.

Fujio, Y. and Moo-Young, M., 1980. Isolation of cellulolytic fungi and some properties of isolated fungi for cellulose biodegradation. J. Gen. Appl. Microbiol. 26: 37-44.

Ghose, T.K., 1977. Cellulase biosynthesis and hydrolysis of cellulosic substances. In: T.L. Ghose et al. (eds.): Adv. Biochem. Engineering 6. Springer Verlag, Berlin, Heidelberg, New York 39-76.

Gilbert, N., Hobbs, I.A. and Levine, J.D., 1952. Hydrolysis of wood. Using dilute sulfuric acid. Industrial and engineering chemistry 44: 1712-1720.

Grant, G.A., Han, Y.W., Anderson, A.W. and Frey, K.L., 1977. Kinetics of straw hydrolysis. Dev. Ind. Microbiol. 18: 599-611.

Haider, K. and Trojanowski, J., 1981. Mikrobieller Abbau von ^{14}C- und ^{35}S-markierten Ligninsulfonsäuren durch Pilze, Bakterien oder Mischkulturen. Holzforschung 35: 33-38.

Haider, K., Trojanowski, J. and Sundman, V., 1978. Screening for lignin degrading bacteria by means of ^{14}C-labelled lignins. Arch. Microbiol. 119: 103-106.

Hall, J.A., Saeman, J.F. and Harris, J.F., 1956. Wood saccharification. Unasylva 10: 10-16.

Han, Y.W., 1974. Microbial fermentation of rice straw: Nutritive composition and in vitro digestibility of the fermentation products. Appl. Microbiol. 29: 510-514.

Han, Y.W., 1978. Microbial utilization of straw. In: D. Perlman (ed.): Advances in applied Microbiology. Vol. 23, Academic Press, New York, San Fransisco, London, pp. 119-153.

Han, Y.W. and Anderson, A.W., 1974. The problem of rice straw waste. A possible feed through fermentation. Economic Botany 28: 338-344.

Han, Y.W. and Anderson, A.W., 1975. Semisolid fermentation of ryegrass straw. Appl. Microbiol. 30: 930-934.

Hartley, R.D., Jones, E.C., King, N.J. and Smith, G.A., 1974. Modified wood waste and straw as potential components of animal feed. J. Sci. Food Agric. 25: 433-437.

Herzig, I., Dvořák, M. and Věžník, Z., 1968. Treatment of litter straw by application of the genus *Pleurotus ostreatus* (Jacq.) Fr. Biol. chem. výž. hospodárských zvířat. 3: 249-253.

Hofsten, B. von, 1976. Cultivation of a thermotolerant basidomycete on various carbohydrates. In: G.G. Birch, K.J. Parker and J.T. Worgan (eds.): Food from Waste. Appl. Sci. Publishers, London, pp. 156.

Holt, D.M. and Jones, E.B., 1978. Bacterial cavity formation in delignified wood. Material und Organismen 13: 13-30.

Humphrey, A.E., 1974. Production of food and feed by fermentation. 5NC Symposium paper, The Royal Australian Chemical Institute 285-290.

Hungate, R.E., 1975. The rumen microbial ecosystem. Ann. Rev. Ecol. Syst. 6: 39-66.

Ibrahim, M.N.M. and Pearce, G.R., 1980. Effects of white rot fungi on the composition and in vitro digestibility of crop by-products. Agric. Wastes 2: 199-205.

Jackson, M.G., 1978. Treating of straw for animal feeding. FAO Animal Production & Health Paper 10, pp. 81.

Jauhri, K.S., 1978. Production of protein by fungi from agricultural wastes. Zbl. Bakt. II, 133: 619-622.

Jauhri, K.S. and Sen, A., 1978. Production of protein by fungi from agricultural wastes. Zbl. Bakt. II, 133: 604-608.

Jílek, R., Zezula, V. and Vodičková, M., 1979. Biological transformation of straw cellulose. Mushroom Sci. IX. 303-309.

Jurasek, L., 1979. Enzymatic hydrolysis of pretreated aspen wood. Dev. Ind. Microbiol. 20: 176-183.

Kaneshiro, T., 1977. Lignocellulosic agricultural waste degraded by *Pleurotus ostreatus*. Dev. Ind. Microbiol. 18: 591-597.

Kaneshiro, T., Kelson, B.F. and Sloneker, J.H., 1975. Fibrous material in feedlot waste fermented by *Trichoderma viride*. Appl. Microbiol. 30: 876-878.

Kaufmann, W., Sinner, M. and Dietrichs, H.H., 1978. Zur Verdaulichkeit von Stroh und Holz nach Aufschluss mit gesättigtem Wasserdrampf bei höheren Temperturen sowie Extraktion mit Wasser und verdünnter Natronlauge. Z. Tierphysiol., Tierernähr. Futtermittelk. 40: 91-96.

King, B., Henderson, W.J. and Murphy, M.E., 1980. A bacterial contribution to wood nitrogen. Intern. Biodeterioration Bulletin 16: 79-84.

Kirk, T.K. and Chang, H., 1981. Potential applications of bio-lignolytic systems. Enzyme Microb. Technol. 3: 189-196.

Kirk, T.K., Higuchi, T. and Chang, H., 1980. Lignin Biodegradation: Microbiology, Chemistry, and Potential Applications. Vol. I and II. CRC Press, Inc., Boca Raton, Florida.

Kirk, T.K. and Moore, W.E., 1972. Removing lignin from wood with white-rot fungi and digestibility of resulting wood. Wood Fiber. 4: 72-79.

Knoche, W., Cruz-Koke, E., Connors, W.J. and Lorenz, L.F., 1929. Der "Palo Podrido" auf Chiloe. Ein Beitrag zur Kenntnis der natürlichen Umwandlung des Holzes durch Pilze in ein Futtermittel. Zbl. Bact. II, /9: 427.

Kristensen, T.P., 1978. Continous single cell protein production from *Cellulomonas sp.* and *Candida utilis* grown in mixture on barley straw. Eur. J. Appl. Microbiol. Biotechnol. 5: 155-163.

Langar, P.N., Seghal, J.P. and Garcha, H.S., 1980. Chemical changes in wheat and paddy straw after fungal cultivation. Indian J. Anim. Sci. 50: 942-946.

Larsen, M.J., Jurgensen, M.F., Harvey, A.E. and Ward, J.C., 1978. Dinitrogen fixation associated with sporophores of *Fomitopsis pinicola*, *Fomes fomentarius* , and *Echinodontium tinctorium*. Mycologia LXX: 1217-1222.

Latham, M.J., 1979. Pretreatment of barley straw with white rot fungi to improve digestion in the rumen. In: E. Grossbard (ed.): Straw decay and its effect on disposal and utilization. John Wiley & Sons, Chichester, New York, Brisbane, Toronto, pp. 131-137.

Lelley, J., Schmaus, F. and Musil, V., 1976. Pilzanbau, Verlag Eugen Ulmer, Stuttgart.

Leopold, J., 1965 . Submersní pěstování Oidium lactis v předhydrolysátech slámy. Ceskoslovenská Biologie 3: 170-180.

Leopold, J., 1965 . Složení bílkovin Torula utilis a Oidium lactis, pěstovaných na předhydrolysátech slámy. Československá Biologie 3: 235-242.

290

Lindenfelser, L.A., Detroy, R.W., Ramstack, J.M. and Worden, K.A., 1979. Biological modification of the lignin and cellulose components of wheat straw by *Pleurotus ostreatus*. Dev. Ind. Microbiol. 20: 541-551.

Linko, M., 1977. An evaluation of enzymatic hyrolysis of cellulosic materials. In: T.K. Ghose (ed.): Adv. Biochem. Engineering, Springer Verlag, Berlin, Heidelberg, New York, pp. 25-48.

Mandels, M., Hontz, L. and Nystrom, J., 1974. Enzymatic hydrolysis of waste cellulose. Biotechnol. Bioeng. XVI: 1471-1493.

Matteau, P.P. and Bone, D.H., 1980. Solid-state fermentation of maple wood. Biotechnol. Letters 2: 127-132.

McQueen, R.E. and Reade, A.E., 1980. Changes in composition and digestibility of poplar by fungal fermentation. Can. J. Anim. Sci. 60: 571-572.

Millett, M.A. and Baker, A.J., 1975. Pretreatments to enhance chemical, enzymatic, and microbial attack of cellulosic materials. Biotechnol. Bioeng. 5: 193-219.

Millett, M.A., Baker, A.J. and Satter, L.D., 1976. Physical and chemical pretreatments for enhancing cellulose saccharification. Biotechnol. Bioeng. 6: 125-153.

Millett, M.A., Baker, A.J., Satter, L.D., McGovern, J.N. and Dinius, D.A., 1973. Pulp and papermaking residues as feedstuffs for ruminants. J. Anim. Sci. 37: 599-607.

Moo-Young, M., Chahal, D.S. and Vlach, D., 1978. Single cell protein from various chemically pretreated wood substrates using Chaetomium cellulolyticum. Biotechnol. Bioeng. 20: 107-118.

Nehring, K. and Friedel, K., 1982. Der Einsatz von Enzymen zur Bewertung von Aufschlusstroh unter besonderer Berücksichtigung des Einflusses der Bestrahlung mit ɣ-Strahlen. Arch. Tierernähr. 32: 235-265.

Neudoerffer, T.S. and Smith, R.E., 1969. Enzymatic degradation of wheat bran to improve its nutritional value for monogastrics. Can. J. Anim. Sci. 49: 205-214.

Nikitin, N.C., 1955. Die Chemie des Holzes. Akademie-Verlag, Berlin.

Overstijns, A., 1981. Die Massenpasteurisierung oder das Pasterurisieren in der Masse. Der Champignon 236: 9-15.

Phillippi, F., 1893. Die Pilze Chiles, soweit dieselben als Nahrungsmittel gebraucht werden. Hedwigia : 115-118.

Prins, R.A. and Clarke, R.T.J., 1979. Microbial ecology of the rumen. In: Y. Ruckebusch and P. Thivend (eds.): Nutritive physiology and metabolism in ruminants. MTP, Press Limited, Medical Publishers: 179-204.

Reid, I.D., 1979. The influence of nutrient balance on lignin degradation by the white-rot fungus *Phaneorchaete chrysosporium*. Can. J. Bot. 57: 2050-2058.

Reid, I.D. and Seifert, K.A., 1982. Effect of an atmosphere of oxygen on growth, respiration, and lignin degradation by white-rot fungi. Can. J. Bot. 60: 252-260.

Rhode, G., 1955. Futterverpilzung durch Selbsterhitzung - eine Metode zur Leistungssteigerung im Viehstall. Schulungsbeilage der Zeitschrift"Mitschurin-Bewegung", 7/55 und "Zoottechniker" 4.

Romantschuk, H., 1974. Feeding cattle at the pulp mill. Unasylva. 26: 15-17.

Rypáček, V., 1966. Biologie holzerstörender Pilze. VEB, G. Fischer, Jena.

Sarkanen, K.V., 1981. Principles and practical approaches to chemical and hydrothermal delignification. In: K.H. Domsch, M.P. Ferranti and O. Theander(eds.): OECD/Cost Workshop, Improved utilization of lignocellulosic materials for animal feed. Braunschweig, Commission of the European Communities.

Schänel, L., Herzig, I., Dvořák, M. and Věžník, Z., 1966. Způsob využití méně hodontných druhů slámy ke krmným účelům. Zlepšovací návhr. C. 13: 1-4 (Patent CSSR).

Schmidt, O., 1978. On the bacterial decay of the lignified cell wall. Holzforschung 32: 214-215.

Schmidt, O., 1980a. Über den bateriellen Abbau der chemisch behandelten verholzten Zellwand. Material und Organismen 15: 207-224.

Schmidt, O., 1980b. Mikrobiologische Umsetzung von Holz zu Enzymen und Eiweiss. Mitt. der BFA für Forst- und Holzwirtschaft Hamburg-Reinbeck 130: 105-118.

Schmidt, O., Puls, J., Sinner, M. and Dietrich, H.H., 1979. Concurrent yield of mycelium and xylanolytic enzymes from extracts of steamed birchwood, oat husks and wheat straw. Holzforschung 33: 192-196.

Schuchardt, F. and Zadrazil, F., 1982. Aufschluss von Lignocellulose durch Hohere Pilze. - Entwicklung eines Feststoff-Fermenters. In: H. Delweg (ed.): 5. Symp. Techn. Mikrobiologie, "Energie duch Biotechnologie", Berlin, Institut fur Garungsgewerbe und Biotechnolgie, 422-428.

Seal, K.J. and Eggins, H.O.W., 1976. The upgrading of agricultural wastes by thermophilic fungi. In: G.G. Birch, K.J. Parker and J.T. Worgan (eds.): Food from waste. Appl. Sci. Publishers, London, pp. 58-74.

Singer, R., 1961. Mushroom and truffles. Leonard Hill, London.

Škultétyová, N., Sommer, A. and Ginterová, A., 1978. Investigation into the nutritive value of the harvested substrate of *Pleurotus ostreatus*. I. Nutrient content in the substrate. Polnohospodarstvo, 24: 65-73.

Sommer, A., Škultétyová, N. and Ginterová, A., 1978. Investigation into the nutritive value of the harvested substrate of *Pleurotus ostreatus*. II. Digestibiilty of the nutrients of the harvested substrate of *Pleurotus ostreatus*. Polnohospodarstvo. 24: 152-158.

Spano, L.A., Medeiros, J. and Mandels, M., 1975. Enzymatic hydrolysis of cellulosic wastes to glucose. United States Army, Natick Laboratories, Natick, Mass., USA.

Stewart, C.S., 1981. Rumen Microbiology. In: K.H. Domsch, M.P. Ferranti and O. Theander (eds.): OECD/COST Workshop, Improved utilization of lignocellulosic materials for animal feed. Braunschweig Commission of the European Communities.

Streeter, C.L., Conway, K.E., Horn, G.W. and Mader, T.L., 1982. Nutritional evaluation of wheat straw incubated with the edible muschroom, *Pleurotus ostreatus* . J. Anim. Sci. 54: 183-188.

Sundman, V. and Nase, L., 1972. The synergistic ability of some wood-degrading fungi to transform lignins and lignosulfonates on various media. Arch. Microbiol. 86: 339-348.

Swift, M.J., 1977. The ecology of wood decomposition. Sci. Prog. Oxf. 64: 1175-1199.

Tangol, R., 1976. Diccionario etimologico Chiloe, Chiloe, Chile.

Tilley, J.M.A. and Terry, R.A., 1963. A two stage technique for in vitro digestion of forage crops. J. Br. Grassl. Soc. 18: 104-111.

Thomke, S., Rundgren, M. and Eriksson, S., 1980. Nutritional evaluation of the white rot fungus -   -
as a feedstuff to rats, pigs and sheep. Biotechnol. Bioeng. 22: 2285-2303.

Virk, R.S., Sethi, R.P. and Garcha, H.S., 1980. Note on the conversion of poultry droppings by *Pleurotus ostreatus* into feed. Indian J. Anim. Sci. 50: 293-295.

Zadražil, F., 1976a. Ein Beitrag zur Strohzerzetzung durch hohere Pilze (Basidiomycetes) und Nutzung für Ernährungs- und Düngungszwecke. Landw. Forschung 32/II, 153-167.

Zadražil, F., 1976b. Freisetzung wasserlöslicher Verbindungen während der Strohzersetzung durch Basidiomyceten als Grundlage für eine biologische Strohaufwertung. Z. Acker- Pflanzenbau 142: 44-53.

Zadražil, F., 1977. The conversion of straw into feed by Basidiomycetes. Eur. J. Appl. Microbiol. 4: 291-294.

Zadražil, F., 1979. Umwandlung von Pflanzenabfall in Tierfutter durch Pilze. Mushroom Sci. (Part I) X: 231-241.

Zadražil, F., 1980. Conversion of different plant waste into feed by basidiomycetes. Eur. J. Appl. Microbiol. Biotechnol. 9: 243-248.

Zadražil, F., 1983. Screening of fungi for lignin-decomposition and conversion of straw into feed. Eur. J. Appl. Microbiol. Biotechnol. (in press).

Zadražil, F. and Brunnert, H., 1980. The influence of ammonium nitrate supplementation on degradation and in vitro digestibility of straw colonized by higher fungi. Eur. J. Appl. Microbiol. Biotechnol. 9: 37-44.

Zadražil, F. and Brunnert, H., 1981. Investigation of physical parameters important for the solid state fermentation of straw by white rot fungi. Eur. J. Appl. Microbiol. Biotechnol. 11: 183-188.

Zadražil, F. and Brunnert, H., 1982. Solid state fermentation of lignocellulose containing plant residues with *Sporotrichum pulverulentum Nov.* and *Dichomitus squalens (Karst)*. Eur. J. Appl. Microbiol. Biotechnol. 16: 45-51.

Zadražil, F., Grinbergs, J. and Gonzalez, A., 1982. "Palo podrido" - decomposed wood which was used as feed. Eur. J. Appl. Microbiol. Biotechnol. 15: 167-171.

Chapter 10

WHOLE CROP HARVESTING, SEPARATION AND UTILIZATION

by

David N. Mowat[1] and Brian Wilton[2]

[1]Department of Animal and Poultry Science
University of Guelph
Guelph, Ontario N1G 2W1, Canada

[2]Department of Agriculture and Horticulture
University of Nottingham, School of Agriculture
Sutton Bonington, Loughborough, LE12 5RD
United Kingdom

10.1. INTRODUCTION

One of the problems with cereal straw is that it is a by-product, an inevitable consequence of our decision to feed ourselves on grain, either by direct consumption or indirectly through farm animals. Little thought is given to the quantity, nutrient quality or condition of straw when the primary aim is to produce as much grain of a particular specification as possible.

In Canada the relative values of grain and straw in general range from perhaps 3:1 to almost infinity to one. The position is similar in the UK; at one extreme good oat straw is worth perhaps one third the value of feed oats while at the other winter wheat straw is seldom worth the cost of collection, so much of it is burned on the stubble to ease the establishment of the subsequent crop.

There are, of course, regions in which straw is regarded in a different light and for various reasons these tend to be regions that have not chosen to use the combine as the harvester. In these areas harvesting is still whole crop harvesting - it is only in the countries that have more-or-less completely changed over to the combine that whole crop harvesting can perhaps be regarded as a "new" technique that may have something to offer.

Whole crop harvesting preceded what we have today so why is there any point in going back? Three points are worth mentioning: i) the combine is a weather-dependent machine and some grain losses are inevitable when it is used, particularly in maritime climates; ii) equipment that was not available in pre-combine days is now well established; and iii) there is now a new attitude to straw and a realisation that it may have a role to play in increasing overall levels of food production. Taken together these points justify a second look at whole crop harvesting.

Harvesting may or indeed may not be followed by separation. With the latter it follows that the crop will be utilised by ruminants, and as one would expect there are several options as far as treatment and storage are concerned: the choice will be largely determined by crop moisture content and maturity. If separation is carried out it will almost certainly be on mature or nearly-mature crops and there will be even more options available than with unseparated crops.

A further complication in the consideration of whole crop harvesting is that the specification of the "straw" part of the crop after separation can be much wider than that discharged by combines: in general it will be "better" nutritionally but part of it may be deliberately made "worse". Manipulation of straw quality - or material other than grain (MOG) as it is more correctly called - is an aspect that has to date received little attention, but it may eventually prove to be of considerable importance as the technology of whole crop harvesting advances.

10.2.   EARLY DEVELOPMENT

Prior to introduction of the combine cereal harvesting was an intensely laborious operation. Crops were handled several times and at each stage some grain was shed, a proportion of which could not be retrieved. When the combine appeared it was inevitable that it would be adopted almost universally in areas where the topography and scale of farming were appropriate.

Developments in forage harvesting equipment lagged behind those in cereals, but once basic forage harvesters began to appear on the market their acceptance became almost as rapid as combines. Several types of chopping mechanisms were employed and attachments were developed to enable the basic machines to harvest fresh and pre-cut crops of relatively short species (grasses, clovers, lucerne) and to direct-cut maize and other tall crops.

It was in the 1950's, not long after effective forage harvesters first became available, that they began to be used on cereals in a process that was known at the time as crop threshing. This technique seemed to make most headway in Eastern Europe and it was really regarded as a halfway house to combining. Crops were almost ripe when cut; they were windrowed, picked up and chopped by the harvester (which also almost totally threshed them) and finally fed into a stationary thresher/separator - usually a modified machine from the binder era. Grain driers were not required and investment costs were lower than that for the switch to combines. Eventually, however, combines replaced the crop harvesting in which separation was an integral part.

During the following 20 year period there was sporadic interest in whole crop cereal silage, and indeed vast quantities of immature maize silage are now made. Numerous studies have investigated the yield and feeding value of whole crop cereal silages at differing stages of maturity (Garmo, 1983; Mowat and Slumskie, 1971; Oltjen and Bolsen, 1978). However, several problems face those who wish to ensile small grain crops. First there is the cost of growing them compared with that of growing perennial forages, which also give several cuts per year. Then there is the difficulty of knowing when to cut crops with rapidly changing moisture material. Late harvesting gives rise to a whole range of problems - the hollow stems pack poorly and there is a big risk of mould infestation, fermentation may be unsatisfactory and grain digestibility is frequently low due to the presence of tough seed coats that resist digestion and are almost impossible to treat mechanically.

Developments in the area that could change the situation are not difficult to visualise - for example the addition of, say, ammonia could transform the outlook for ensiled cereals. Alternatively a process such as separation after ensiling (referred to later as a means of increasing the versatility of maize silage) could make ensiling a more attractive way of conserving whole crop cereals than it is today.

In the last few years, however, the most attention has been given to systems designed to deal with relatively mature crops and in particular to those that involve crop separation before storage. This does not mean that other approaches are not being followed - in Denmark a case has been made for studies on a system of ambient or low-temperature storage drying followed by separation immediately prior to use. Surely this must be as close to the binder, field cure, store-in-stack and thresh system as one can get.

Apart from this Danish approach, most of the research and development work has taken place in Sweden and the UK, and two distinct lines have been followed - one aimed at drying before separation (a high temperature, high throughput operation most appropriate for large scale industrial-type operations) and the other at separation before drying (probably more suitable for the on-farm situation).

10.3. THE INDUSTRIAL APPROACH TO DRYING AND SEPARATION

During the 1970's research and development work was proceding simultaneously and along broadly similar lines in Sweden and the UK. The Swedish work was initiated and largely carried out by an engineering company, Kockums AB; their aim was to market plants that could be used by groups of farms and run either by the farmers themselves, by independent organisations or by concerns already handling agricultural commodities - for example it was suggested that the operation might fit in well with the processing of sugar beet. Outside the cereal harvesting season it was thought that the drier and pelleting plant could be used on sugar beet pulp, or perhaps on green forage crops. For commercial reasons only a limited amount of technical information has been published by Kockums.

The UK work was carried out in a University and was modestly financed from public funds. A summary of the work has been published (Wilton, 1978) and more detail is available in reports to the funding bodies.

In both countries crops were direct cut with precision crop forage harvesters and were fed into rotary drum driers. In Sweden it is understood that most of the grain was discharged from the drying drum shortly after entry - a design feature claimed to produce viable grain. A standard single pass machine was used in the UK and at the temperature normally used for crops of the moisture contents encountered (400°C at 30-35%) germination after drying was nil.

Little is known about further separation in the Kockums plants, but it was demonstrated by the provision of samples that the MOG could be split into various components to meet a range of requirements. It is perhaps worth pointing out here that one of the inherent advantages of a whole crop system is that the crop is taken to a central plant and is fed through it at a steady and controllable rate. Within reason space is not limiting in such a plant so machines can be as large and complex as necessary to meet the product specification. Adjustment, control and recirculation are all easy and the materials are dry - the opposite is the case in the combine.

In the UK it was thought that whole crop cereal harvesting and separation could be usefully incorporated into existing greencrop drying programmes in late summer when capacity frequently exceeds the supply of raw material. Minimal plant modification was envisaged and initially separation was carried out in an inclined rotary drum screen to give grain, "short" straw and "long" straw (the dividing line between the two being approximately 50 mm). The final step in the UK work was to build a furnace in which the "long" straw was burned, enabling the oil-fired furnace to be turned off once processing was underway.

Kockums manufactured and installed at least two plants but it is thought that they have not yet commissioned any for use on a strictly commercial basis: this does not mean, however, that the system lacks potential. Unfortunately marketable plants became available as conditions favouring investment disappeared and apparently the demand for separated MOG components was not as strong as anticipated. So for the time being the industrial approach appears to be at a standstill and this situation seems unlikely to change rapidly. One further factor in this situation is the decline of the greencrop drying industry in Europe during the last few years with high energy costs.

10.4. ON-FARM SEPARATION

A farmer who only grows arable crops is never likely to be interested in whole crop harvesting unless there is some dramatic change in the relative values of grain and MOG. Where mixed farming is practised, however, and particularly where the livestock are cattle, forage crops will be grown and whole crop harvesting may have something to offer.

The first possibility is that in many countries mixed farms above a certain minimum size will either be equipped with forage harvesters or will have access to them via contractors because there appears to be a widespread swing from hay towards silage. One type of forage harvester widely used for forage - the precision or metered chop machine - is the one used for cereals. It seems reasonable to assume that there is considerable potential for saving money by using one machine on the two types of crop - forage and cereals - on one farm.

In addition to cutting the combine also separates, hence if a combine is to be replaced a separator is also required. Combines employ sieves and air-assisted sieves, but unfortunately they are restricted to working in relatively dry crops otherwise the sieves block with damp straw. The question asked at Nottingham early in the research programme was: can separation be achieved without sieves? Some basic work in a wind tunnel suggested that it can especially if space is not limited (Wilton et al., 1980).

A tapered wind tunnel was constructed and tested. Separation was not greatly influenced by crop moisture content and four commodities could be produced at a total rate of up to 8 t/h per metre width of tunnel intake. The four were: i) grain; ii) a mixture of light and broken grain, nodes and some heavily lignified straw; iii) "heavy" straw; iv) "light" straw (chaff, leaf and weeds).

One suggestion for dealing with such a range is that iii) could be burned to dry i), and that iv) could be treated with alkali and fed to cattle. About 15% of the crop appeared as mixture ii) and this could either be further separated in some more elaborate equipment or be treated and stored either on its own or mixed with iv).

The current work at Nottingham, which will not be published until 1983 at the earliest, involves the use of a largely pneumatic separator that has been built for use on a commercial farm where approximately 500 cattle (dairy and beef; young and mature) are kept. They are fed on mixtures of grass silage, caustic soda treated grain which is soaked in water before feeding, and ammonia treated straw. At the time of writing (summer 1982) the separator is being used on wheat and barley crops.

10.5. SEPARATION OF LUCERNE AND WHOLE PLANT MAIZE

Although small grain cereal crops have been the main focus of whole crop harvesting, the system has potential for increasing the utilisation of lucerne, possibly of maize and even of oil-seed crops. During the late 1960's researchers in the United States developed a technique for separating whole dehydrated lucerne (Chrisman et al., 1971). This uses controlled velocity air currents in a closed system to separate the light from the heavy particles. The lighter or leafy fraction is high in protein and xanthophyll but low in fibre content; the heavier or stem fraction contains less protein and more fibre. This technique of air separation makes it possible to increase the utilisation of lucerne or other legumes in poultry and pig rations and also in protein supplements for ruminants. Potential also exists to produce various lucerne products to meet the needs of specific rations.

An economic evaluation of air separation of dehydrated lucerne has been conducted (Vosloh et al., 1974). One problem with this system is that the dehydration process is very energy intensive and costly. Secondly, the value of the stem fraction is low and costs of transporting this bulky material back to farms for use as a roughage source in finishing rations or for maintaining beef cows would be high. As an alternative Ganesh (1981) has considered the economic attractiveness of on-farm fractionating low moisture lucerne silage or lucerne hay for use on integrated ruminant farms (eg. growing-finishing cattle and cow-calf). In this system drying costs would be eliminated and transportation costs reduced. However low-cost machines for on-farm fractionating need to be developed, possibly similar to those mentioned earlier for whole plant cereals. Another alternative might be to transport field cured lucerne hay to a local processing plant for gradual separation into leaf and stem fractions. The leaf portion could be used as a high quality protein source as mentioned previously while the low quality stem lignocellulose fraction might be hydrolyzed and fermented to produce ethanol (Dale, 1981). A considerable amount of research is currently in progress in Europe and North America on biomass fuel production. Furthermore, after hydrolysis and fermentation "alfalfa distillers solids" remain, having a crude protein content of approximately 25% (Dale, 1981). This could be used as a liquid protein supplement for local beef or dairy cattle. With dry biomass material (lucerne hay), such a facility could operate throughout the year to reduce economic pressures. Separating the valuable high quality protein from lucerne or other legumes could provide a significant processing credit to improve the economic position of biomass-derived fuels.

Only a very limited amount of the vast quantities of maize crop residue (stover) is utilized for feed - mainly as a maintenance ration for beef cows. Harvest and storage problems as well as low nutritive value have hindered its use. However, numerous attempts have been made to improve the nutritive value of maize stover through processing. An interesting alternative approach to increasing the utilization of maize stover is the possibility of separating mature whole plant maize silage into grain and stover fractions at feeding time. The high moisture grain could be fed on-farm or in the locality to pigs or ruminants with high levels of production. The stover fraction could be fed to growing cattle or ruminants with low to moderate levels of production.

At 50 to 55% moisture in the mature whole plant grain maize (at approximately 35% moisture) is near maximum yield (Daynard and Hunter, 1975). In order to minimize losses during storage, particularly in a horizontal silo, or even in a conventional upright silo at this low moisture content, the crop must be harvested and stored rapidly, tightly packed and sealed with plastic sheets. At feeding, a separator similar to the one previously mentioned for small grain cereals might be used to gradually separate the silage into stover and grain fractions. The stover would be of exceptionally high quality due to the presence of at least some residual grain (fines) and to the very early harvest.

Early harvest maximizes feeding value as well as yield of stover; its digestibility decreases by about 1.5 (Leask and Daynard, 1973) to 1.9 (Berger et al., 1979) units per week following physiological grain maturity. In fact, performance of cattle fed early harvested stover has been shown to be comparable with stover harvested 3 weeks later and treated with NaOH (Berger et al., 1979).

A systems analysis study has recently been completed (Kemp and Mowat, 1982) comparing a conventional system of feeding normal whole plant maize silage and an alternative system of harvesting mature whole plant maize silage and later separating at feeding time into grain and stover fractions. This latter system proved economically attractive even at relatively moderate maize grain prices. However, certain assumptions had to be made based on little data. So further research is needed (and warranted) to fill gaps in knowledge and understanding of the separated stover system (eg. factors affecting proportion of whole intact or separable kernels) as well as further development of a cheap, effective separator for use with mature whole plant corn silage.

10.6. FEEDING VALUE OF SEPARATED COMPONENTS

With cereal crop residue, leaf and chaff components are much higher in feeding value than stems (Table 10.1). The quality of chaff can vary considerably depending on how much cracked grain or weed seed it contains. Coxworth et al. (1981) reported that weight of chaff is a surprisingly high proportion of total wheat residue (26 to 37%); whole crop harvesting would reduce field losses of this high quality fraction as well as permit later separation if desired.

Table 10.1. Digestibility and protein content of different fractions of wheat residue (Coxworth et al., 1981)

Fraction	In vitro DM digestibility	Crude protein g/kg
Chaff	41	64
Leaf	43	39
Stem	26	18
Leaf and stem	33	27
Total residue	36	40

The leaf (plus bud) fraction of first cut lucerne contains over 70% of the plant crude protein while the stem (plus petiole) fraction contains over 75% of the plant crude fibre (Table 10.2).

Table 10.2. Nutritive value and proportions of different fractions of lucerne (Kohler and Chrisman, 1968)

Fraction	Crude protein	Crude fibre	Proportion of total		
			Dry matter	Crude protein	Crude fibre
	----g/kg DM ----		------ g/kg (fresh) ------		
Leaf	322	108	453	653	198
Bud	296	194	48	63	38
Petiole	184	205	61	51	50
Stem	121	407	438	233	714
Average/total	226	252	1000	1000	1000

The concentration of protein in pure lucerne leaves decreases gradually with advancing maturity (Table 10.3). However leaves of
lucerne contain 2 to 3 times more crude protein than stems of
lucerne. In addition, lucerne leaves are markedly higher than
lucerne stems in in vitro dry matter digestibility (Mowat et al.,
1965; Robles et al., 1981). In contrast differences in feeding
value between leaves and stems of many grasses are relatively
small (Mowat et al., 1965) - so for fractionating forages it is
important to have relatively pure stands of lucerne or other legumes.

Table 10.3. Nutritive value of different fractions of first-cut
forages with advancing maturity (Mowat et al., 1965)

	Lucerne		Timothy	
	Crude protein	In vitro	Crude protein	In vitro
Date	g/kg DM	DMD	g/kg DM	DMD
Leaf				
June 4	318	79.0	144	74.4
18	291	77.4	124	68.7
July 3	260	76.3	99	62.3
16	240	74.8	80	56.3
Stem				
June 4	140	66.6	91	76.2
18	106	55.8	61	64.7
July 3	93	50.4	56	54.5
16	92	49.9	48	51.8

As mentioned previously the feeding value of stover separated from mature whole plant maize silage would be relatively high in feeding value due to the earlier harvest and the presence of at least some unseparable grain (fines). The increased feeding value of earlier harvested material could be due to less leaching losses of soluble constituents as well as to reduced field losses of the higher quality (leaf and husk) fractions (Table 10.4).

Table 10.4. Approximate dry weight and nutritional composition of maize stover fractions (Leask and Daynard, 1973)

Fraction	Dry weight g/kg	In vitro DMD %	Crude protein g/kg
Leaf	120	55	70
Stalk	170	50	30
Husk	90	65	30
Cob	120	45	30
Stover	500	54	45

10.7. SUMMARY

As with so many other processes and techniques that are still the subject of research and development it is extremely difficult to predict which way any one will eventually go, if indeed they "go" at all. Whole cereal crop harvesting is no exception, but if the view that it will be cheaper than conventional systems is accepted and to this is added its reduced dependence on the weather and the fact that both harvested yield and feeding value are higher than those from combine/baler or combine/forage harvester systems, then surely it must have a chance.

To some extent the separation of lucerne and whole plant maize silages is easier to visualise since it would merely involve the use of a relatively simple and cheap process prior to feeding. In circumstances where there is an opportunity to use "better" and "worse" fractions from a "normal" silage, this must have an even higher chance of being taken up by the farming community.

10.7. REFERENCES

Berger, L.L., Paterson, J.A., Klopfenstein, T.J. and Britton, R.A., 1979. Effect of harvest date and chemical treatment on the feeding value of corn stalkage. J. Anim. Sci. 49: 1312-1316.

Chrisman, J., Kohler, G.O., Mottola, A.C. and Nelson, S.W., 1971. High and low protein fractions by separation milling of alfalfa. U.S. Dept. Agric., ARS, 74-57.

Coxworth, E., Kernan, J., Knipfel, S., Thorlacius, O.T. and Crawle, L., 1981. Review: Crop residues and forages in western Canada: potential for feed use either with or without chemical or physical processing. Agric. Environm. 6: 245-256.

Dale, E., 1981. Food and fuel from biomass. Proc. Second World Congr. Chem. Eng.

Daynard, T.B. and Hunter, R.B., 1975. Relationships among whole-plant moisture, grain moisture, dry matter yield and quality of whole-plant corn silage. Can. J. Plant Sci. 55: 77-84.

Ganesh, D., 1981. Systems for harvesting, separating and utilizing alfalfa. M.Sc. Thesis. University of Guelph, Guelph, Ont. Canada.

Garmo, T.H., 1983. Avling og kvalitet av byggheilgrøde. Sci. Rep., Agric. Univ. Norway (in press).

Kemp, R.A. and Mowat, D.N., 1983. Systems approach to utilizing corn stover for growing cattle. J. Anim. Sci. (submitted).

Kohler, G.O. and Chrisman, J., 1968. Separation milling of alfalfa. Proc. American Soc. Agr. Eng. Ann. Mtg. pp. 1-23.

Leask, W.C. and Daynard, T.B., 1973. Dry matter yield, in vitro digestibility, percent protein and moisture of corn stover following grain maturity. Can. J. Plant Sci. 53: 515-522.

Mowat, D.N., Fulkerson, R.S., Tossell, W.E. and Winch, J.E., 1965. The in vitro digestibility and protein content of leaf and stem portions of forages. Can. J. Plant Sci. 45: 321-331.

Mowat, D.N. and Slumskie, R.A., 1971. Barley silage, ground whole plant barley and corn silage for finishing beef cattle. Can. J. Anim. Sci. 51: 201-207.

Oltjen, J.W. and Bolsen, K.K., 1978. Wheat, barley and oat silages for beef cattle. Bull. 613. Agr. Exp. Sta., Kansas State University, Manhattan, Kansas, USA.

Robles, A.Y., Martz, F.A., Belyea, R.L. and Warren, W.P., 1981. Preparation and digestibility of alfalfa leaves and stems marked with gold or chromium. J. Anim. Sci. 52: 1417-1420.

Vosloh, C.J., Kuzmicky, D.D., Kohler, G.O. and Enochian, R.V., 1974. Air separation of alfalfa into high and low-protein fractions - an economic evaluation. Agr. Econ. Rep. No 259, U.S. Dept. of Agr. Econ. Res. Ser.

Wilton, B., 1978. Whole crop cereals: harvesting, drying and separation. The Agricultural Engineer 33(1): 4-5.

Wilton, B., Amini, F. and Randjibar, I., 1980. Whole crop cereals: a low-cost approach. The Agricultural Engineer 35 (1): 7-10.

Chapter 11

MICROBIAL DEGRADATION IN THE DIGESTIVE TRACT

by

Andrew Chesson and Egil R. Ørskov

Rowett Research Institute
Bucksburn, Aberdeen AB2 9SB
United Kingdom

11.1. INTRODUCTION. SITES OF FERMENTATION

Microorganisms are essential to the life of functioning ruminants and other herbivores, and alone are capable of utilising the structural polysaccharides which form the major part of plant cell walls. As a result herbivorous animals are totally dependent for energy and growth on the absorption of the end products of the action of gastrointestinal microorganisms when fed diets high in fibre and low in simple sugars or storage polysaccharides.

In ruminants, conditions for the microbial fermentation of plant residues are optimised in the reticulo-rumen in which the bulk of fibre degradation occurs. A high potential for fibre-degradation also exists lower in the ruminant digestive tract although post-ruminal feeding experiments indicate that this is of limited importance in animals fed high-fibre diets. In contrast, in ruminants fed readily-degraded material, particularly legumes, and in some large non-ruminant herbivores, the extent of plant cell wall degradation in the colon/caecum may approach that found in the rumen of sheep and cattle.

The fraction of the total gut devoted to fermentation provides an indication of the importance of this means of digestion to an animal. In ruminants three-quarters or more of the gut is devoted to this function, and in some large hindgut fermenters, such as the horse, the enlarged caecum or colon also represent two-thirds or more of the gut (Parra, 1973).

A limited capacity, at least, to ferment structural polysaccharides of plants is widespread amongst vertebrate species including man. However, this ability is readily overloaded in many species and fibre digestion suppressed by a high-fibre intake: the pig is capable of degrading 40% of cellulose fed as 2% of the diet, but proportionally less when cellulose is included at values > 10% due, presumably, to a reduced retention time.

Herbivores of agricultural importance with the potential to utilize the fibrous residues of processed crops are relatively few in number and are restricted to species with substantial modifications to the hind- or foregut. That ruminants can make use of high-fibre residues is well known. In many non-industrial nations such residues (straws, bagasse etc.) are already the principal source of animal feedstuff.

In this chapter we examine requirements for the efficient utilisation of plant residues in the digestive tract and consider factors which determine the rate and extent of fermentation of plant structural polysaccharides. Principally:

- the need for a substantial and stable population of fibre-degrading microorganisms;
- the duration of contact between plant fragments and microorganisms under conditions which encourage microbial activity;
- the chemical and physical nature of the substrate and the extent to which this determines the availability of its component polysaccharides;
- the rate and extent of degradation, the rate of diminution of large to small particles and the size of the gut.

11.2. MICROBIOLOGY OF FIBRE DEGRADATION

11.2.1. Colonisation of plant fragments

The bacteria of the gastro-intestinal tract form a diverse range of obligate and facultatively anaerobic organisms reaching population densities of 10^{10} - 10^{11} organisms/ml in the rumen and hindgut of herbivores. Relatively few of these strains are capable of attacking plant cell walls and utilizing structural polysaccharides as a sole carbon source (Table 11.1).

As is evident from Table 11.1, most attention has been paid to the microbiology of ruminants (and of the rumen!). The numbers and identity of fibre-degrading microorganisms in many other herbivores are unknown. Nonetheless, despite the considerable difference between foregut and hindgut fermenters, the general principles underlying the microbial colonisation and breakdown of plant cells are a common feature. In this respect the rumen can be considered a model for all major sites of fermentation.

Plant fragments entering the rumen or hindgut become extensively colonised by bacteria within an hour (Cheng et al., 1981). Treating plant fragments to detach bacteria shows that at least as many organisams associate with particles as remain free in the rumen liquor (Minato and Suto, 1978). Preference is always shown for damaged areas during primary colonisation. Chewing by the animal and methods of feed preparation which maximise damage thus promote colonisation and have a marked effect on rates, if not on the ultimate extent, of degradation.

Electron microscopy shows that degradation of non-mesophyll cells is limited in the absence of bacterial attachment to cell walls (Akin, 1979; Cheng et al., 1977). This close association of fibre-degrading microorganisms and their substrate has considerable ecological significance. As Van Soest (1981) has pointed out, attachment allows an organism to attain dominance over substrate and environment and avoids washout at the faster rate of liquid flow.

Adhesion of the cellulolytic bacterium *Ruminococcus flavefaciens* to the epidermal and schlerenchyma walls of ryegrass occurs rapidly. This is followed by colonisation of phloem and mesophyll walls, but not by attachment to the cuticle or vessel elements (Latham et al., 1978a). Adhesion appears to be mediated by a gly-

Table 11.1. The major cell wall degrading bacteria of large herbivores. Organisms of uncertain status or which have been isolated only on limited occasions are omitted. Also omitted are the cellulolytic bacteria known to be present in many other herbivores, including camels, hippotoamuses and the macropod marsupials which have yet to be identified (see Bauchop, 1977)

Strain	Substrate	Host	Site	Reference
Bacteroides ruminicola	Hemicellulose	Sheep, cattle	Rumen and hindgut	Bryant et al. (1958)
	Pectic substances	Goat	Rumen	Dehority and Grubb (1977)
		Pig	Caecum	Robinson et al. (1981)
Bacteroides succino- *genes*	Cellulose	Sheep, Cattle	Rumen and hindgut	Hungate (1966)
	(Hemicellulose)*	Antelope, Camel	Rumen	Hungate et al. (1959)
	Pectic substances	Buffalo	Rumen	Waghmare and Nambudripad (1978)
		Horse, Pig	Caecum, colon	Davies (1965)
Butyrivibrio *fibrisolvens*	(Cellulose)	Sheep, Cattle	Rumen	Hungate (1966)
	(Hemicellulose)	Goat, Reindeer	Rumen	Dehority and Grubb (1977)
	Pectic substances	Buffalo	Rumen	Waghmare and Nambudripad (1978)
		Deer	Rumen	Prins (1976)
		Pig	Caecum, colon	Robinson et al. (1981), Russell (1979)
		Horse	Colon (faeces)	Brown and Moore (1960)
Eubacterium cellulo- *solvens (= Cillobac-* *terium cellulosolvens)*	(Cellulose)	Sheep, Cattle	Rumen	Bryant et al. (1958),Van Gylswyk and
	Hemicellulose	Deer	Rumen	Prins et al. (1972) Hoffman (1970)
	Pectic substances			
Lachnospira multiparus	Pectic substances	Sheep, Cattle	Rumen	Clarke et al. (1969)
Ruminococcus albus	Cellulose	Sheep, Cattle	Rumen	Hungate (1966)
	Hemicellulose	Horse, Pig	Caecum, colon	Davies (1965)
	Pectic substances*			
Ruminococcus *flavefaciens*	Cellulose	Sheep, Cattle	Rumen	Sijpesteijin (1951), Hungate (1966)
	Hemicellulose	Buffalo	Rumen	Waghmare and Nambudripad (1978)
	Pectic substances*	Horse, Pig	Rumen, colon	Davies (1965)
Streptococcus bovis	Pectic substances*	Sheep, Cattle	Rumen	Ziolecki et al. (1972)

( ) Some strains only   *Substrate degraded but end products not assimilated coprotein coat or capsule, the modification by addition of specific antibodies or chemical removal of which, prevents adhesion (Latham, 1978).*Bacteroides succinogenes* also strongly adheres to plant fragments and, although much less frequently isolated by conventional culture techniques, may be the major fibredegrading species in the rumen. Examination of mixed cultures of rumen bacteria on barley straw showed that *B. succinogenes* -like bacteria were considerably more numerous than ruminococci after 48 h incubation (Stewart et al., 1979). Like *R. flavefaciens*, *B. succinogenes* also shows a preference for damaged areas, adhering to the cut edges of most plant cell walls except those of the xylem (Latham et al., 1978b). *B. succinogenes* has a relatively thin capsular layer and, unlike the ruminococci which retain their shape, adjusts its morphology to the contours of the substrate to which it is attached (Plate 11.1). Cavities formed in cell walls by microbial action are often seen to be lined by bacteria morphologically similar to *B. succinogenes* (Akin, 1980).

Axenic cultures of ruminococci on barley straw show that *R. flavefaciens* is more closely associated with areas of cell-wall digestion than the third of the major cellulolytic bacteria of the rumen, *Ruminococcus albus* . Although *R. albus* possesses a polysaccharide-rich capsule similar to that of *R. flavefaciens*(Patterson et al., 1975), it is generally found to be a weakly-adherent species.

Some species of *Butyrivibrio fibrisolvens* also show limited ability to degrade structural polysaccharides, although growth is generally better on oligosaccharides or simple sugars. *B. fibrisolvens* is consistently seen in large numbers associated with, but not adhering to, plant particles in the rumen of sheep and cows. In the rumen of deer, it is the predominant cellulolytic bacterium isolated (Prins, 1976). The rumen of sheep and cattle also harbour weakly-cellulolytic strains of *Eubacterium cellulosolvens*,a species which shares many morphological and biochemical characeristics with *Butyrivibrio*.

The cellulolytic rumen bacteria are not equally effective in their ability to hydrolyse cellulose. While *B. succinogenes* and *R. flavefaciens*, the two organisms which form the closest association with their substrate, are able to degrade highly ordered cellulose such as cotton fibre, *R. albus* can utilise only disordered forms of cellulose or its soluble derivatives (Stewart et al.,

1981; Wood, 1981). Although, curiously, a strain of *R. albus* (SY3) which showed no activity towards cotton fibre but was able to ferment the highly crystalline Avicel has been reported recently (Wood et al., 1982).

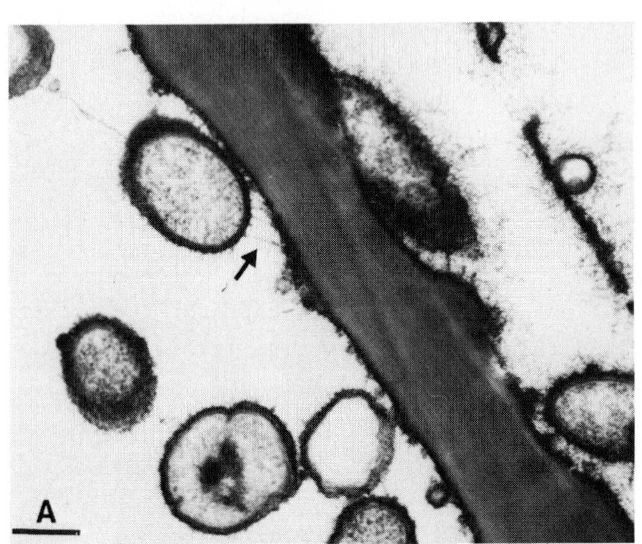

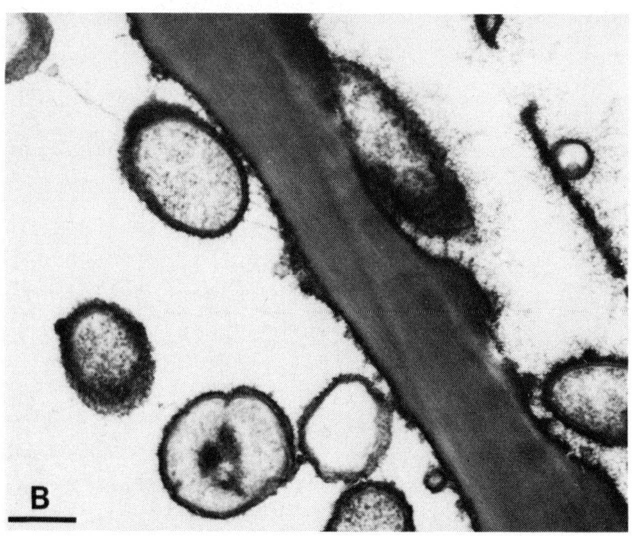

Plate 11.1. Electron micrograph of sections of barley straw incubated with pure cultures of *B. succinogenes* showing the close association of organisms with regions of cell wall degradation. A. Barley leaf, arrow indicates apparent attachment to the substrate by extensions to the glycocalyx. B. Barley stem. Bars, 500 nm.

This differential ability to degrade ordered cellulose is reflected in the preponderance of particular cellulolytic bacteria in the rumen of animals fed different diets. There seems little doubt that the ruminococci predominate in animals fed good quality feeds (Mackie et al., 1978; Akin, 1980) and there is increasing evidence that *B. succinogenes* is of greater importance when less-digestible residues are fed. Inoculation of barley straw with laboratory cultures of rumen bacteria has shown that *B.succinogenes* causes a greater weight loss (42%) after 48 h incubation than do the other cellulolytic rumen bacteria *R. flavefaciens* (37%), *R. albus* (26%) and *B. fibrisolvens* (19%) (Stewart et al., 1979). *B. succinogenes* may be of particular importance in the later stages of digestion when more readily accessible polysaccharides have been degraded.

Microscopic studies have consistently shown a low propensity for bacterial adhesion to highly-lignified surfaces (Akin et al., 1974; Akin, 1979). These observations are consistent with the general relationship found between the lignin content and digestibility of feeds. Yet the effects on bacterial colonisation and adhesion of chemical pretreatments have rarely been considered. In the only report of this kind Latham et al. (1979) examined the effect of NaOH treatment of barley straw and stems of *Lolium termulentum* on the adhesion of *R. flavefaciens* and *B. succinogenes*. In untreated material, both species adhered to unlignified parenchymatous walls but few, if any, colonised other wall types. After treatment most cell walls, including those of the xylem vessels, were extensively colonised and the numbers of adherent bacteria increased ten-fold. There is an evident need for further microbiological studies of the relative effects on colonisation of the many chemical and physical pretreatments known to enhance digestion of plant residues.

Until recently the cellulolytic bacteria have been considered as the major agents of cell wall degradation. This view may have to be reconsidered with the recognition of a substantial anaerobic fungal flora in the rumen of sheep and cattle (Orpin, 1975; Bauchop, 1981), red deer, reindeer and impala, the rumen-like stomach of some marsupials and the hindgut of the horse and African elephant (Bauchop 1979a, 1980; Orpin, 1981). Such fungi are found associated with mainly vascular tissue of leaves and stems and with the more refractory materials such as wheat straw. Like the bacteria the major route of entry is via damaged areas or natural openings such as stomata. The rumen fungi demonstrate cellulolytic and hemicellulolytic activities (Bauchop, 1979a; Orpin and Letcher, 1979; Bauchop and Mountfort, 1981), and their potential importance in the degradation of high-fibre residues is indicated by their ability to colonise and degrade prepared plant fibres such as those from jute or sisal (Bauchop, 1979b).

11.2.2. <u>Enzymology of fibre degradation</u>

The cell wall polysaccharides of poorly-digested crop residues such as the cereal straws differ little in structure from those of more highly digestible herbage, although, of course, they form a far higher proportion of plant dry matter. Methylation analysis has shown that the structural polysaccharides of graminaceous plants, dicytoledons and residues from hardwood are similarly composed of cellulose, in the form of discrete microfibrils, embedded in a matrix of hemicellulose and other polymers. As shown in Table 11.2 the matrix hemicelluloses are (4-0-methyl) glucurono-arabinoxylans with a small proportion of mixed-linked glucans (see also Chapter 4.4). Pectic substances, important constituents of some dicotyledons, are absent or much reduced in proportion in the cell walls of graminaceous plants and heavily lignified residues. As a consequence of this similarity in polysaccharide structure, essentially the same mix of enzymes is required to hydrolyse wall polysaccharides regardless of their phylogenic origins.

In marked contrast to aerobic fungi and bacteria, the anaerobic bacteria of the rumen fail to release substantial quantities of soluble extracellular enzymes. Activities in particle-free rumen liquor are found only in trace amounts. Only *R.albus* , the cellulolytic bacterium which forms the weakest association with its substrate, can be grown to yield reasonable quantities of extracellular cellulase (endo-β-1-4-glucanase) in laboratory culture (Leatherwood, 1965; Smith et al., 1973, Wood et al., 1982). Even with this organism much activity remains associated with the cell wall, although this may be released in part by washing with buffer (Wood et al., 1982). Strains of *R. flavefaciens*, with the exception of that described by Pettipher and Latham (1979), and *B.succinogenes* do not release soluble enzymes in any quantity. Some explanation is offered by recent observations that, in part at least, hydrolytic enzymes are secreted by *B. succinogenes* in sedimentable subcellular

membrane vesicles. These are released from the outer membrane of
the bacterial cell by bleb formation. Such vesicles can be clearly
seen coating and attached to isolated cellulose or to straw recovered from the rumen (Forsberg et al., 1981). Similar particles
derived from *B. succinogenes* were earlier observed by Stewart et
al. (1979) to attach to plant fibres.

Table 11.2. Cell wall and total carbohydrate content and major
            glycosyl linkage composition of some crop residues
            compared to that found in the cell walls of a typical
            forage species - perennial ryegrass (data from
            Chesson, 1983 and Chesson et al., 1983a)

Major glycosyl residue	Deduced linkage	Ryegrass cell walls	Barley straw	Wheat straw	Rice straw Stem	Rice straw Leaf
		--------- % of total sugars ---------				
Arabinosyl	Terminal	4.5	2.6	3.3	2.9	3.8
	1, 3	0.6	0.5	0.3	0.3	0.5
	1, 2	0.6	0.3	0.2	0.2	0.6
Xylosyl	Terminal	0.7	0.2	0.3	0.2	0.3
	1, 4	13.6	24.6	24.5	20.4	16.6
	1, 2, 4 or 1, 3, 4	7.8	6.1	7.9	6.4	8.6
	1, 2, 3, 4	0.4	0.6	0.1	0	0.4
Glucosyl	Terminal	0.8	1.6	1.8	1.7	2.0
	1, 3	1.8	0.9	1.1	0	1.3
	1, 4	62.6	57.4	53.3	59.3	58.8
Uronosyl	–	6.1	4.9	4.9	2.6	1.9
Total carbohydrate, % of DM		65.5	58.7	58.1	53.7	59.8
Cell walls, % of DM		100.0	90.0	88.8	83.5	89.3

Inevitably the cellulase activity of *R. albus* has received clo-
ser attention than that of the other cellulolytic rumen species.
Yu and Hungate (1979) found four cellulases differing in molecular
weight in cell-free extracts of *R.albus*. All released soluble oligosaccharides from ball-milled cotton, Avicel and lucerne cell
walls although rates of digestion differed. Wood and his associates (Wood et al., 1982) also found that *R.albus* produced multiple
forms of endoglucanase but suggested that these arise from a different degree of association of one or more low molecular weight
(25-30,000) enzymes. Wall-bound enzyme was found to be of high

molecular weight and could be released from the wall with little
dissociation. Its stability was enhanced by the presence of rumen
fluid. Treatment with dissociating agents, or some culture conditions, however, led to the production of low molecular weight
forms. High and low molecular weight forms of endo-glucanase are
also elaborated by *B. succinogenes* and *R. flavefaciens* (Pettipher
and Latham, 1979; Groleau and Forsberg, 1981; Forsberg et al.,
1981).

The mechanism of cellulase degradation in the rumen appears to
differ from the now classical picture of ordered cellulose degradation by aerobic fungi which depends on the synergistic action of
endo- and exo-glucanases (cellobiohydrolases) (see Eriksson, 1979;
Wood and McCrae, 1979). Enzymes with activities equivalent to exoglucanase have not been found in rumen bacteria and a mixture of
endo-glucanase from *R. albus* with the exoglucanase of the fungus
Trichoderma koningii failed to enhance the activity of the *R. albus*
enzyme towards native cotton (Wood, 1981).

Endo-xylanases are also mostly of intracellular origin and
hemicellulase preparations can be readily obtained by rupturing
the cell wall of rumen bacteria (Bailey and Gaillard, 1969;
Beveridge and Richards, 1975). Crude hemicellulase preparations of
this nature have been widely used for the study of forage hemicellulose degradation in vitro (see Bailey et al., 1976). Endo-xyla-
nase production has been reported for all of the major cellulolytic bacteria of the rumen (Dehority, 1967; Pettipher and Latham,
1979; Forsberg et al., 1981; Williams and Withers, 1982a), for
Bacteroides ruminicola (Dehority, 1967), for *Butyrivibrio fibrisolvens* (Clarke et al., 1969) and for some rumen protozoa (Bailey
et al., 1962; Bailey and Clarke, 1963). Arabinofuranosidase and
xylosidase activities have also been widely reported (Dekker,
1976; Williams and Withers, 1982b).

To date there are no reports of investigations based on highly
purified enzymes (such as are beginning to be available for the
rumen endoglucanases) and studies have been limited to the range
and extent of substrate utilization. Extracellular hemicellulases
have been found to readily degrade isolated hemicellulose or lignin-hemicellulose complexes, but often to show limited action on
intact cell walls. A far greater loss of hemicellulose from intact
cells occurs in the presence of cellulase activity. A combination
of a hemicellulolytic strain of *B. ruminicola* with *R. flavefaciens*

produced a greater loss of dry weight from intact grass cell walls than did either organism individually. This synergism did not extend to isolated hemicellulose substrates (Dehority, 1973).

The less xylans are substituted with arabinose or glucuronic acid side chains, the more readily they are attacked by rumen hemicellulase. As the extent of substitution decreases with the biological age of the plant the hemicellulose of straw and other mature plant species are potentially more readily degraded than that of immature forage plants (Morrison, 1979; Brice and Morrison, 1982). However, the extent of substitution with arabinose is a minor factor in influencing cell wall degradation compared with other features of the wall, notably the extent of lignification, considered below (Brice and Morrison, 1982).

11.3. LIMITATIONS TO DEGRADATION IMPOSED BY THE SUBSTRATE

11.3.1. Isolated cell wall polysaccharides

Cellulose, whether isolated from cereal straw, from wood or derived from cotton bolls is capable of being fully degraded by rumen microorganisms, albeit at different rates (Figure 11.1). Loss of cereal straw cellulose incubated in nylon bags in the rumen is detectable within 1.5 h and degradation complete within 36 h. The duration of the lag period, which may be interpreted as the time required for colonisation and elaboration of enzyme activity, and subsequent rates of loss are related to the crystallinity of the preparations as determined by x-ray diffraction and infra-red techniques (Chesson, 1981). Contrary to many suggestions, pretreatment with alkali at twice the level normally applied to crop residues does not influence the crystallinity of the preparation, its lag time or its rate of degradation.

Crystallinity values are a mean of regions of close molecular packing or high crystallinity and those of a more amorphous nature (see Chapter 4.4 and Figure 4.5 for structure of cellulose). Although overall crystallinity values relate directly to the rate of digestion of cellulose preparations, amorphous and crystalline regions within preparations appear to be degraded at a common rate. Loss of cellulose from nylon bags is linear with time and the crystallinity value of the residue does not increase during the course of digestion as would be expected if amorphous regions were preferentially attacked (Beveridge and Richards, 1975).

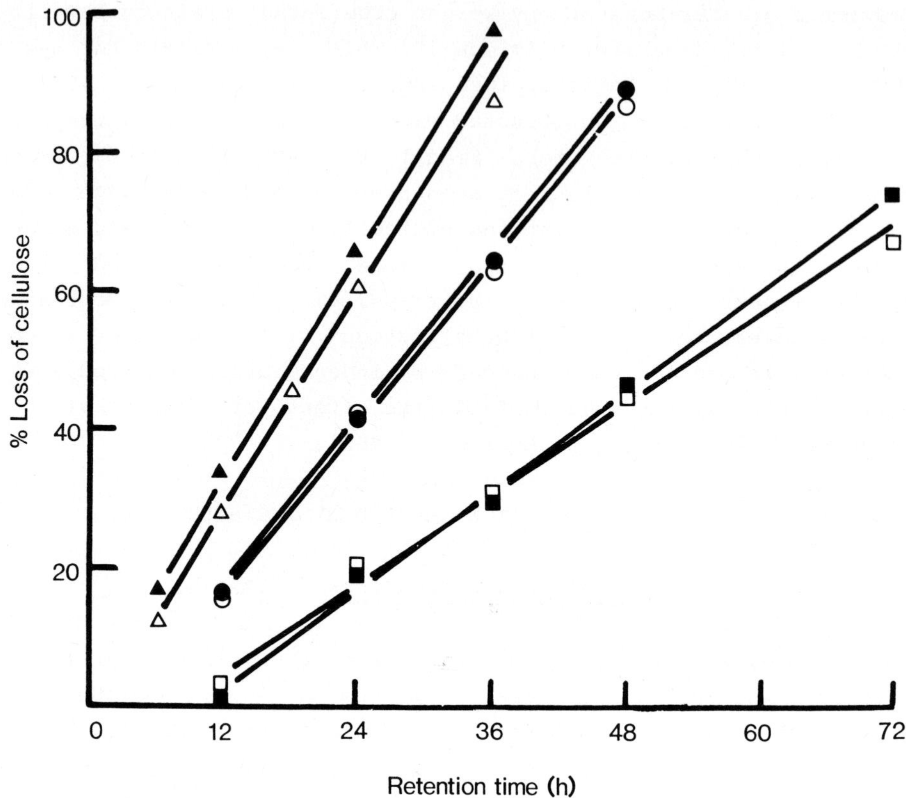

Figure 11.1. Loss of isolated wheat straw cellulose (▲ △), wood pulp cellulose (● ○) and cotton cellulose (■ □) from nylon bags incubated in the rumen of a sheep before (open symbols) and after pretreatment with 20 g NaOH 100g^{-1} cellulose (closed symbols).

Isolated hemicellulose polymers are generally degraded at a faster rate than isolated cellulose. However, results have to be treated with considerable caution since hemicellulose polymers are inevitably modified during extraction and purification. Differences in methods of preparation have led, in the past, to considerable confusion, particularly in relation to the rates of digestion of branched (highly substituted) versus linear hemicellulose fractions. Suggestions by Gaillard et al. (1965) and Bailey and Gaillard (1965) that branched xylans are more resistant to digestion have been questioned by Beveridge and Richards (1973a). They believed that the apparent more rapid rate of breakdown of linear hemicellulose was due to the presence of a contaminating glucan and, as they found, linear and branched hemicellulose were degraded at the same rate. However more recent evidence suggests that the extent to which xylans are substituted may influence rates of digestion (Brice and Morrison, 1982).

11.3.2. <u>Polysaccharides of intact cell walls</u>

The physico-chemical nature of isolated cellulose, indicated by its crystallinity index value, determines its rate of digestion in the rumen. Such a rate represents the maximum rate at which a particular cellulose can be degraded under these conditions. However when the loss of cellulose from intact straw incubated in the rumen is examined, it is evident that this upper rate operates for the initial stages of cell wall breakdown only. Thereafter the rate of loss steadily falls below the maximum value, approaching zero at 72 h in the case of cereal straw. The rate and the ultimate extent of loss is thus determined, not by the chemical nature of cellulose itself, but by its specific environment within the cell wall and by its associations with other wall components (Chesson et al., 1983a).

It cannot be stressed too strongly that information obtained from studies of the behaviour of an isolated polysaccharide cannot be extrapolated to predict its behaviour when it forms part of an intact wall. There are no major intrinsic limitations to the microbial degradation of structural polysaccharides in the rumen which derive from the polysaccharides themselves. Any resistance to microbial attack is a product of the whole cell wall organisation and the behaviour of polysaccharides can only be understood when considered in this context.

As Beveridge and Richards (1973a, b) found for spear grass and we have found for a number of forage species and for cereal straws (Chesson et al., 1983a; Chesson, 1983) there is little difference in the proportion of monosaccharides found in acid hydrolysates of the original undigested material, material lost as a result of microbial action and residues remaining in nylon bags in the rumen. The glycosidic linkage pattern of the undigested straw (see Table 11.2) and its undigested residue are also very similar. Analysis of samples of straw recovered from the rumen at intermediate stages in digestion show that the wall polysaccharides are all degraded at a common rate (Bacon et al., 1981).

The proportion of hemicellulose in digested residues, particularly from forage legumes, is often greater than in the parent material. This has been taken as an indication of a slower rate of hemicellulose degradation and of a possible association with lignin. This may be a misinterpretation. Microbial colonisation and attack favours cell surfaces showing limited lignification. Such cells are relatively cellulose-rich (Gordon et al., 1977; Gordon and Bacon, 1981) while more refractory cell types show a higher degree of secondary thickening and lignification and have a higher hemicellulose content. Differences in the rate at which specific cell types are degraded may lead to an accumulation of relatively hemicellulose-rich cells although the rates at which cellulose and hemicellulose are lost from a single cell-type may be the same.

Phenolic material is also lost from lignified residues such as straws, either as fine particulate material or as soluble polyphenolic-carbohydrate complexes (Gaillard and Richards, 1975; Neilson and Richards, 1982). The rate of loss is only slightly less than that of the wall polysaccharides. The proportion of phenolic material in barley straw increased from the 19% present in undigested straw to 22% of the dry residue recovered after 72 h incubation in a sheep rumen (Chesson et al., 1983a).

11.3.3. Alkali-labile linkages and constraints to digestion

Clearly constraints to the total digestion of plant polysaccharide do not lie directly with the polysaccharides themselves, but with their association with other wall components. It is often asserted that the association of polysaccharide, particularly hemicellulose, with polyphenolic material has a role in the control of digestion. Certainly the difficulty in separating lignin from hemicellulose suggests bonding between these polymers and a number of workers have provided indirect evidence for presence of covalent linkages (Morrison, 1974; Gordon and Gaillard, 1976; Chesson et al., 1983b, Chapter 4.10).

It is evident from the known ability of sodium hydroxide to overcome constraints to digestibility and to enhance the extent of degradation of polysaccharides of Gramineae and, to a lesser extent, legume and hardwood cell walls, that to function as a major determinant of digestibility such linkages must also satisfy the criterion of alkali-lability. The recognition by Hartley and his associates (Hartley and Jones, 1977; Hartley, 1981) that simple phenolic acids (p-coumaric and ferulic acids.), closely resembling the precursors of lignin itself, were, in part at least, ester-linked to hemicellulose, suggested one possible form of lignin-carbohydrate bond which would meet this criterion. Indirect evidence that lignin-carbohydrate linkages resembling those to phenolic acids, might be important was also provided by the direct relationship found between the loss of phenolic acids and the improvement to digestibility which occurred when alkali was applied to cereal straw (Hartley and Jones, 1978; Chesson, 1981).

A method to detect and quantify alkali-labile linkages, including any formed between hemicellulose and lignin, was developed and applied to cereal straw hemicellulose without the prior extraction of the polymer from the wall (Lomax et al., 1983). The ring hydroxyls of glucuronic acid residues were found to carry few alkali-labile linkages and only 0-5 of arabinose residues was substituted in this manner (Chesson et al., 1983b). Xylose residues were, however, extensively substituted on carbons 2 and 3, with approximately 40% of all residues carrying one or more side chains linked by alkali-labile (non-glycosidic) bonds (Table 11.3). Approximately 45% of alkali-labile linkages to straw xylan and 20% to grass xylan could not be accounted for after allowance was made for the known contribution of ester-linked acetyl and phenolic acid groups. These linkages may represent bonding to polyphenolic material (Chesson et al., 1983a). However, as Table 11.3 shows, no accumulation of glycosidic or alkali-labile linkages occurred during the course of hemicellulose digestion in the rumen. It thus seems unlikely that the extent of hemicellulose bonding, whether to minor groups (acetyl, phenolic acids) or to polyphenolic material, represents a major determinant of digestibility.

Table 11.3. The effect of microbial degradation on the extent to which xylose residues from barley straw are substituted with alkali-stable (glycosidic) and alkali-labile linked side chains. Two sites (C2 and C3) on each residue are potentially available for substitution. Undigested fraction was recovered from the sheep rumen after 72 h (data from Chesson et al., 1983a)

	Original straw		Undigested fraction	
	Glyco-sidic	Alkali-labile	Glyco-sidic	Alkali-labile
	% of available sites substituted			
Disubstituted (C2 + C3)	0.9	4.0	0.2	2.9
Monosubstituted (C2 or C3)	8.6	22.1	8.4	21.6
Total substitutions	35.6		33.1	
Unsubstituted (free-OH)	64.4		66.9	

11.3.4. Lignin internal linkages, digestibility and the effect of alkali

Alkali produces multiple changes to cell wall structure including the loss of acetyl and phenolic acid groups, the solubilisation of silica and some hemicellulose (Jackson, 1977; Evans, 1979; Chesson, 1981) and the probable hydrolysis of hemicellulose-lignin likages. All of these modifications to structure appear to be associated phenomena not directly related to improvements in the digestibility of wall polysaccharides. It is the effect of alkali and other hydrolytic procedures on lignin itself which now seems to offer the most likely explanation of the enhancement of digestibility.

Approximately 40% of acid detergent lignin or 55% of total phenolics initially present in barley straw can be washed from straw treated with 10 g NaOH 100 g^{-1} straw. No additional release of phenolic material occurs when higher levels of alkali are applied. At application levels below this figure, the extent to which phenolics are released closely parallels the improvement to cellulose digestibility. The lignin remaining associated with the cell wall after sodium hydroxide treatment and washing is not released until over 50% of the cell wall polysaccharides are degraded by rumen microorganisms. Thereafter release is rapid and does not appear to be related to the further degradation of polysaccharide (Chesson, 1981).

Although lignin is thought of as a polymer highly resistant to degradation, a number of internal linkages, notably α-aryl and α-alkyl ether bonds, are susceptible to relatively mild hydrolysis (see Chapter 4.5). Johansson and Miksche (1982) found that phenolic benzyl aryl ethers are cleaved by cold dilute alkali and that both phenolic and non-phenolic benzyl aryl ethers are selectively hydrolysed by a mild acid treatment. Estimates of the effects of mild hydrolysis indicate that, in wood at least, the average molecular weight of lignin fragments would be reduced to approximately 3000 mol. wt. (Sarkanen, 1982), around the critical molecular weight for the penetration of ethylene glycols through the pores of the cell wall matrix (Tarkow and Feist, 1968). Thus lignin fragments with molecular weights of this magnitude would be readily released from cell surfaces whilst those located deeper in the wall may remain encased by the polysaccharide network and unable to migrate through the wall to the surface.

11.4. THE PLANT WALL-MICROORGANISM INTERFACE

Although the nature of the protection offered by lignin remains unclear, it must operate at the surface of the plant cell wall. Enzymes secreted by the cellulolytic bacteria of the gastro-intestinal tract are either vesicle bound or, if soluble and freely diffusable, of a molecular weight far greater than the low molecular weight solutes able to penetrate the pores of the wall. Adhesion of microbes and their enzymes is probably to the substrate itself, which also must be located at the plant cell surface. During the course of digestion, as available substrates at the surface are degraded, material unable to act as a substrate for microbial enzymes or as a binding site for the organisms themselves, becomes exposed and would be expected to accumulate. The primary candidate for such a material is the polyphenolic polymer, lignin (or condensed tannins). Certainly this is the only fraction of the cell wall to accumulate during digestion.

In highly lignified cell walls, with a high ratio of lignin to polysaccharide, accumulation of polyphenolics of the surface would be expected to occur faster and the rate of cell wall degradation to reduce earlier than in less lignified walls. In such material

much of the wall would remain in an unmodified from protected by
the surface layer. Gross analysis of plant material before and
after digestion might not be expected to reveal large changes in
composition, since the contribution of the modified surface layer
would be small. Thus a limited accumulation of phenolic material,
from 19% to 22% in the case of barley straw, might not seem great
when seen in terms of the whole wall, but may be highly significant if largely restricted to the surface layers.

If this picture of wall degradation is correct then the effect
of alkali and other hydrolytic procedures which enhance digestibility would seem to be the cleavage of the lignin macromolecule,
enabling phenolic material at the surface or that exposed by microbial action to be solubilised. Cleavage of ester linkages to carbohydrate would aid this process. The effectiveness of various
hydrolytic pretreatments would be a measure of their ability to
hydrolyse lignin internal linkages. Physical treatments, such as
fine grinding, would also be expected to expose more surface and
hence a greater number of potential binding sites for microorganisms, but not necessarily to modify wall structure. Rates of degradation would be expected to increase but not the ultimate extent
of degradation. This is found to be the case (Table 11.4).

Table 11.4. Loss of dry matter from barley straw dry ground to
             pass various mesh sizes incubated in nylon bags in
             the sheep rumen for 72 h

Mesh size (mm)		
Passed	Retained	% loss of dry matter
1.00	0.50	31.6
0.50	0.25	34.1
0.25	0.10	33.4
0.10	0.04	31.6

Other wall components resistant to degradation might also be
expected to influence adhesion at surfaces (see Chapter 4.7).
Clearly the cuticle layer operates in this matter and its disruption by alkali or other treatments would be expected to promote
microbial activity. Silica also occurs in cell walls in variable
amounts, generally low in temperate straws but high in some tropical by-products such as rice-straw. Recent evidence would suggest, that in temperate by-products at least, the levels of silica

normally encountered do not influence digestibility (Hartley, 1981; 1982).

11.5. MICROBIAL DEGRADATION AND HOST ANIMAL INTERACTIONS

As far as the host ruminant animal is concerned, the most important characteristic of straw diets is the amount of metabolizable energy (ME) consumed and microbial protein (MP) synthesized per unit of time, usually expressed per day. The uptake of ME and synthesis of MP will largely determine the animal's ability to grow, to produce power, to sustain foetal development, to produce milk or wool, or indeed the extent to which it will loose body energy. While much has been learned about factors affecting uptake of straw or other highly fibrous by-products or roughages, the picture is still far from complete. However a number of important factors determining intake have now been recognized, and these are discussed below in terms of their possible manipulation to maximise digestion of high-fibre diets. Finally some general comments about the interaction of the various factors will be made.

11.5.1. <u>Rate of degradation of straw</u>

The mechanism of microbial breakdown of fibre particles has already been discussed in some detail in earlier sections of this chapter. The importance of rate of degradation to the host animal is due to the fact that it determines the speed at which the digestible components are removed from feedstuffs and thus the time which the feed occupies space in the gut. Intake of straw is almost invariably limited by the physical size of the gut, at least as far as the ruminant is concerned. For straw diets the content of soluble carbohydrate is usually very low, generally about 1-2%, thus only a small fraction is immediately available to invading microorganisms. Since the attachment of bacteria to the surfaces of the substrate and the subsequent formation of colonies does not occur spontaneously, there is a lag phase where apparently little or no digestion takes place, or at least the rate at which it occurs is hardly measurable. An illustration of disappearance of straw from nylon bags incubated in the rumen is given in Figure 11.2.

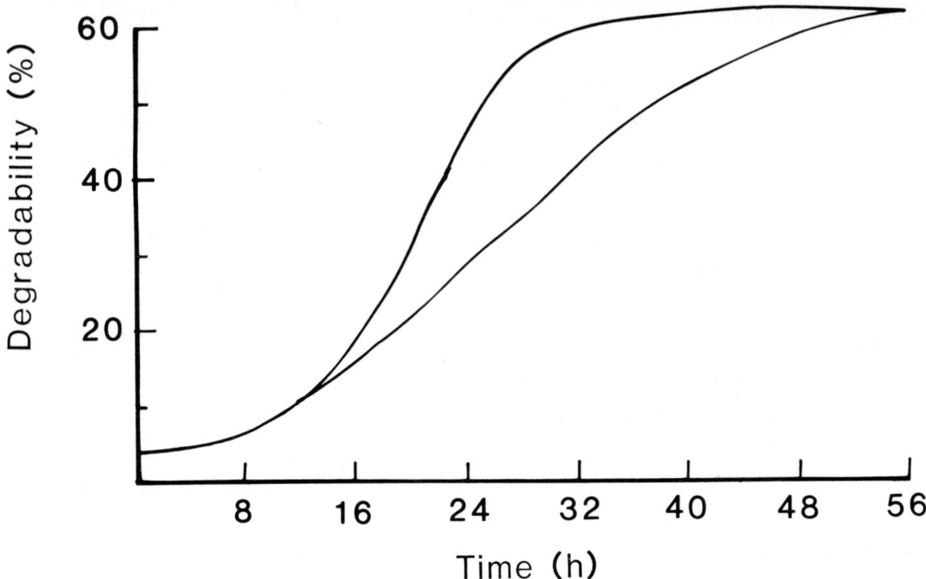

Figure 11.2. Illustration of two feeds with similar extents of degradation but following a different course of digestion. Both curves are represented by the equation $a+b(1-e^{ct})$ with similar values for (a+b) but different values for the degradation rate c.

After 8-10 hours the straw in a normally ingested feed begins to break up into small particles, a process due more to rumination and mastication than directly to microbial action (although microbial activity contributes to the increased fragility of the particles). The illustration in Figure 11.2 must therefore be considered only in a relative sense since no physical comminution of the straw takes place when it is incubated in nylon bags.

It is important that the maximal rate of degradation of straw is achieved to enable the animal to maintain a high turnover rate. A great deal is known about how to affect the rate of degradation and a few of the most important factors will be mentioned below.

The effect of chemical treatment will also be briefly referred to, since chemical treatment normally has the effect of increasing both the rate and the extent of digestion of straw.

11.5.2. Nitrogen limitation

Fibrous diets and straws in particular are often low in N, with the result that fermentation or extent of microbial proliferation is limited by the rumen concentration of ammonia. This is a well known fact and the responses to N supplementation have often been demonstrated (Campling et al., 1962). It should be pointed out however that the amount of N required by the rumen microflora is related to the amount of fermentable energy potentially available. Although the N content may be low in straw diets it may well be adequate due to the low fermentability. This is well illustrated in Table 11.8, taken from Ørskov and Grubb (1978), and shows that neither digestibility nor intake of untreated barley straw containing 6% crude protein responded to urea supplementation.

11.5.3. Effect of soluble carbohydrate

The rate of degradation of cellulose can be significantly reduced if it is fermented in the rumen together with substantial amount of other substrates, particularly soluble carbohydrates or lipids.

The effect of starch in the diet on the disappearance of straw is illustrated in Table 11.5, from Mould and Ørskov (1983). Straw from which the soluble carbohydrate was removed was incubated in the rumen of sheep receiving straw alone or different cereal supplements. It can be seen that the amount of straw digested after 24 h was almost halved when the animals were given cereals. The results also show that the reduction in rate of straw digestion could to a large extent be avoided if the rumen pH was prevented from dropping below about 6.2. In fact, Mould et al. (1983) showed that the depression in digestibility of roughages often observed when cereals and roughages are given together could be largely avoided if bicarbonate was added in sufficient quantities to maintain rumen pH above about 6.2.

Table 11.5. The effect of rumen environment on 24h degradation of straw from which the soluble carbohydrate had been removed by washing (Mould and Ørskov, 1983)

Rumen environment	Rumen pH without bicarbonate (range)	Rumen pH with bicarbonate (range)	Degradability at 24 h	
			Without bicarbo-nate	With bicarbo-nate
			----- % of DM ----	
Straw alone	6.5-6.9		39.8	
Straw + 65% whole barley	6.0-6.2	6.5-6.9	24.0	38.8
Straw + 65% pelle-ted barley	5.6-6.5	6.3-7.0	22.0	30.6
Straw + 65% pelle-ted maize	5.6-6.5	6.6-7.0	19.1	24.9

11.5.4. Effect of lipids

Lipids also inhibit roughage digestion. Kowalczyk et al. (1977) showed that digestion of cellulose from grass in 24 h was 72, 63, 54 and 40% when 0, 5, 10 or 15% lipid was added to the feed. In both types of work the roughage intake was reduced as a result of a reduction in rate of degradations.

11.5.5. Effect of increasing cellulolysis

Recent observations by A. Silva and E.R. Ørskov, illustrate some interesting possibilities. Rumen cannulated sheep were given either untreated straw, ammonia-treated straw or a poor quality hay. Urea and sulphur were added so that no N or S limitation was anticipated. The results are given in Table 11.6. A significantly greater rate of digestion of straw was achieved when, for instance, untreated straw was incubated in the rumens of sheep fed on ammonia-treated rather than untreated straw. These results have enormous practical implications and are at present being examined in greater detail. The differences are probably due to an increased concentration of cellulolytic organisms giving rise to a greater bacterial attachment to the straw particles. Similar observations were made when Ørskov and Hovell (1979) incubated hay in the rumen of Zebu cattle receiving either pangola grass or whole-crop sugar cane. The rate of digestion of hay was much lower in the Zebu cattle receiving whole crop sugar cane even though the rumen pH was about 7, indicating also that plants can contain quantities of soluble carbohydrates which can inhibit cellulose digestion.

Table 11.6. The effect of rumen environment on the degradation of straw and hay from which the soluble materials had been removed and subsequently incubated for 48 h in the rumen of sheep (Silva and Ørskov, unpublished results , 1983)

Rumen environment			48 hrs degradation of:		
Substrate	NH_3 mg/100 ml	pH	Untreated straw	NH_3 treated straw	Hay
			---------- % of DM -----------		
Untreated straw	26.8	6.9	31.3	41.1	37.1
NH_3 treated straw	24.4	6.9	46.6	50.6	48.8
Hay	21.2	6.5	32.4	37.1	35.7

11.6. DIGESTIBILITY

11.6.1. Extent of digestion

The extent of digestion is the characteristic of feed which is most often known, as it is normally taken to be equal to the value obtained from a digestibility determination. However the extent to which feeds are digested in vivo rarely reach that predicted by digestibility determinations, whether by the nylon bag technique or by in vitro assay. This is due to in vivo passage of small digstible particles out of the rumen and inadequate compensation by microbial activity in the large intestine because of the low retention time in that organ.

Extent of digestion is usually, but not always, correlated with rate of digestion. Since both extent and rate of digestion will influence the space taken up by the material during digestion it is little wonder that the voluntary food intake can only be pre-dicted inadequately from information on the extent of digestion or digestibility alone. It is possible to have the digestion of two different fibrous roughages proceeding in the manner described in

Figure 11.2 which illustrates different rates of digestion but a similar potential digestibility.

A method of describing both the rate and potential extent of digestion was suggested by Ørskov and McDonald (1979). They found that the disappearance of substrate from nylon bags incubated in the rumen could be adequately described by the equation:

$$p = a + b (1 - e^{-ct}).$$

This equation may be usefully applied to cereal straw and other by-products. The intercept a, represents soluble material which is immediately digestible, b, the fraction which given time will be digested and c, the rate constant for digestion of b. p is the amount degraded at time (t) and (a + b) is the asymptote or the potential extent of digestion. In fact a simple incubation trial of this kind may give a much more useful description of the potential value of a feed than a digestibility trial.

When the rate of digestion is depressed by the factors referred to earlier, the extent of digestion, i.e. digestibility, is usually depressed as well. In a recent trial for instance, Fahmy and Ørskov (unpublished) determined the digestibility of ammoniated straw fed alone and the digestibility when it was included at the rate of 30% in a sugar-beet-pulp diet and a rolled-barley diet. The digestibility of both the sugar beet pulp and rolled barley was 83%. The results are given in Table 11.7. When the straw was included in diets at 30%, the digestibility of straws was reduced from 53.8 when fed alone to 40.4 and 22.1 when it was included in the sugar-beet-pulp diet and rolled-barley diet respectively. The The results illustrate that, due mainly to differences in rumen pH, both the intake of the total diet and the digestibilities favoured the sugar beet pulp.

When the rate of cellulose digestion is inhibited, the digestibility is normally reduced to the greatest extent if the diets consist of small particles while the voluntary food intake is reduced to the greatest extent with diets consisting of long or large particles.

Table 11.7.   The effect on degradability of NH_3 treated straw of feeding it to sheep in combination with 70% sugar beet pulp and 70% rolled barley. The digestibility of sugar beet pulp and rolled barley was estimated to be 83.4 and 83.9 respectively (Fahmy and Ørskov, unpublished results, 1982)

Treatment	DMD %	Intake of DDM g/d	Estimated DMD of straw, %
NH_3 straw	53.8 ± 3.8	402	
NH_3 straw + 70% sugar beet pulp	70.5 ± 1.6	876	40.4
NH_3 straw + 70% rolled barley	65.4 ± 2.6	653	22.1

11.6.2.   Chemical treatments to increase digestibility of straw

The effect on microbial attachment of differing chemical treatments has been mentioned earlier and will not be further discussed. Usually chemical treatment increases both the rate and extent of fermentation. It should be stressed here that as more substate is made available for digestion due to chemical treatment so the microbial requirement for nutrients increases proportionally. This is illustrated below for N. In Table 11.8, from Ørskov and Grubb (1978), the addition of urea and S to a diet of untreated barley straw had little or no effect on digestion and intake by sheep. On the other hand, addition of urea to NaOH treated straw greatly enhanced digestion and intake. In fact if the N deficiency was not corrected there was apparently no effect of the chemical treatment either on digestibility or on intake. These observations greatly favour the use of N-containing alkalis such as ammonia or urea when straw is to form the major part of the diet or indeed the whole diet.

11.7.   RATE OF BREAKDOWN OF LONG OR LARGE PARTICLES TO SMALL PARTICLES

In ruminants the arrangement of the stomach, in particular the reticulo-omasal orifice, prevents large particles from passing freely out of the rumen. This of course has the advantage to the animal that it achieves as near as possible the maximum extent of degradation, but also the disadvantage that the time taken to eliminate large, indigestible particles may limit food intake. Although the chemical nature of feeds has received considerable attention, their physical properties, likely to determine fragility and the ease of particle size reduction, have not. As a result, no suitable laboratory method exists to measure this important determinant of digestibility and we are unable to predict rates of particle size reduction.

Table 11.8. The effect of urea supplementation of untreated or NaOH treated barley straw on digestibility and intake by lambs (Ørskov and Grubb, 1978)

Urea supplementation	Untreated straw		NaOH-treated straw	
	Intake of	OMD	Intake of	OMD
g/kg	DM, g/d	%	DM, g/d	%
0	423	45.8	355	42.3
6	451	46.7	402	48.0
12	441	49.0	531	58.9
18	463	48.3	567	62.8
SE of treatment differences	35	2.8	47	4.1

The dynamics of the process were recently described by Mertens and Ely (1982). The long or large particles can accumulate in the rumen and thus restrict food intake. The rate at which the breakdown occurs is often positively correlated with the rate and extent of degradation, i.e. the less digestible the feed the tougher is the indigestible residue, but it is not possible to generalise. For instance, the overall digestibility of sisal pulp is quite high, yet intake is limited by a slow rate of breakdown of the fibre residue which has been left in the pulp after the longest fibre has been extracted for rope making. Many tropical forages contain very tough fibre, together with soluble carbohydrate. Another extreme example is whole-crop sugar cane.

The particle size which permits outflow varies with the species of ruminant and within a species it varies, in particular, with size of the animal. This was demonstrated by Andrews et al.(1969). They fed oats to young lambs, of 20-30 kilos liveweight, and compared this with feeding oats to lambs of 40-50 kilos weight. Oat husk accumulated in the rumen and restricted the intake of digestible energy in the young lambs, but not in the older lambs. There are also large differences between individuals within a group of similar animals. This can often be observed by examining the particle size of the faeces. The animals which allow passage of the largest particles usually show the lowest digestibility and the highest voluntary intake of straw (Van Soest, 1982).

As mentioned before, due to the difficulty of measuring degradation of long to small particles there is little known about the factors which influence it. It is known that some indigestible fibre of sisal pulp is broken down extremely slowly and that, on the whole, indigestible fibres from legumes are less tough than those from grass. Physical processing of straw may eliminate the need for particle size reduction and so increase intake, but then digestibility will almost invariably be reduced because of the passage out of the rumen of potentially digestible material. A great deal of work should be directed towards measuring the rate of breakdown and factors influencing it, so that it may be possible to further increase the precision with which the voluntary intake of straw by ruminants can be predicted.

11.8. ANIMAL FACTORS

11.8.1. Rumen volume

Straw intake cannot normally supply sufficient nutrients to meet the growth potential in ruminants because of the long rumen retention time caused by the factors previously discussed, and the restrictions imposed by rumen volume. If the rumen volume could be increased to adapt to the nature of the feed then there would be fewer problems with utilization of straw and less financial return from improving the nutritive value of untreated straw, since the greatest benefit obtained from chemical treatment of straw is the greater food intake (see, for instance, Table 11.8).

The question of adaptation to consume straw diets is not entirely resolved, (see Hoffman, 1975) but considerable evidence suggests that ruminants, over a long period, could be selected for greater capacity to utilize straw occasioned by a greater rumen volume. It was interesting to note that cattle in Bangladesh, given untreated straw, had a gut content of about 33% of liveweight (Mould et al., 1982). It is quite possible that these cattle had been selected for utilization of straw, since rice straw has been the main source of food for centuries in that region. By the same token, it is possible that European and North American cattle are selected against maximal roughage utilization.

It must be pointed out of course that rumen bacteria and factors affecting rumen environment are unlikely to differ between ruminants consuming the same feeds. However, as discussed earlier, the rumen volume is the factor restricting roughage intake. It is of interest to document clearly whether the ability to consume large amounts of straw is genetic in origin. Selection for beef cattle with a high killing out percentage and selection for dairy cattle with high yields, regardless of the feeds used, are likely to select against high roughage utilization. It is a factor which cannot, and should not, be ignored when exotic breeds of ruminants are brought in to improve indigenous breeds in developing countries, unless the feed resources are changed as well. Without change of feed this exercise can be of little or no benefit and may even have a negative value.

11.8.2. Improvements in value of straw

The possibility of selecting for varieties of cereals, or for that matter other cash crops, which give the maximal yield of digestible straw without sacrificing yields of the cereal grains or other harvested products, has yet to be explored. In order to undertake work in this direction the plant breeder must be provided with a simple routine measure of nutritive value. Cannulated animals can be used for this purpose, and do provide estimates of the extent and rate of degradation; but simple standardised laboratory procedures need to be developed. The in vitro "cellulase" assay, has yet to meet this criterion. We need also to develop methods to assess the physical rate of breakdown of large to small particles in the rumen. Knowledge of this aspect of fibre utilization is almost non-existent and until we have easy methods of measuring the nutritive value, of which digstibility is only one factor, we are unlikely to make much progress in this direction.

The simple chemical treatments with ammonia and urea have opened up great possibilities for increasing the use of straw and other fibrous by-products. A greater understanding of the optimal rumen environment for digestion of straw, and possible improvements in value without sacrificing grain yield, will no doubt enhance the productivity that can be achieved from straw. While, therefore, it seems possible to improve the fibrous materials to the benefit of the animal it may also be possible to select for ruminants more suitable for the straw.

11.9. SUMMARY

This chapter summarises what is known of the complex and inter-related factors affecting the microbial degradation of fibrous plant material in ruminants and non-ruminants. Only in ruminants (and the rumen) is sufficient known about the microorganisms responsible for fibre degradation and the influence of the host animal on intake and rates of digestion to attempt to unravel these interactions. Thus, at present, the rumen must serve as a model for all major sites of plant fermentation. Rumen organisms responsible for plant cell wall breakdown are described and the effects of substrate composition on rates of degradation discussed. Emphasis is placed on the events which occur at the interface between plant fragments and associated microorganisms during the course of digestion. It is suggested that important interactions between fibre degradation and intake by the host animal are characterised by:

1. the rate of fibre degradation, determining the speed at which digestible components leave the gut;
2. the extent of degradation (digestibility), determining the amount of undigestible residue;
3. the rate of reduction of long particles to particles small enough to leave the rumen;
4. the total gut volume in the animal

Factors affecting and the means of manipulating these characteristics are discussed. Attention is also drawn to the fact that ruminants appear to differ in their gut capacity and thus their ability to produce power, meat, milk, etc. from poor quality roughage.

ACKNOWLEDGEMENTS

The authors would like to thank Drs. T.P. King and C.S. Stewart of the Institute for the preparation of transmission electron micrographs.

11.10. REFERENCES

Akin, D.E., 1979. Microscopic evaluation of forage digestion by rumen microorganisms - a review. J. Anim. Sci. 48: 701-710.

Akin, D.E., 1980. Evaluation by electron microscopy and anaerobic culture of types of rumen bacteria associated with digestion of forage cell walls. Appl. Environ. Microbiol. 39: 242-252.

Akin, D.E., Burdick, D. and Michaels, G.E., 1974. Rumen bacterial interrelationships with plant tissue during degradation revealed by transmission electron microscopy. Appl. Microbiol. 27: 1149-1156.

Andrews, R.P., Kay, M. and Ørskov, E.R., 1969. The effect of dietary energy concentration on the voluntary intake and growth of intensively fed lambs. Anim. Prod. 11: 173-187.

Bacon, J.S.D., Chesson, A. and Gordon, A.H., 1981. Deacetylation and enhancement of digestibility. Agric. Environm. 6: 115-126.

Bailey, R.W. and Clarke, R.T.J., 1963. Carbohydrates of the rumen Oligotrich *Eremoplastron bovis*. Nature (Lond.) 199: 1291-1292.

Bailey, R.W. and Gaillard, B.D.E., 1965. Carbohydrases of the rumen ciliate *Epidinium ecaudatum* (Crawley). Hydrolysis of plant hemicellulose fractions and β-linked glucose polymers. Biochem. J. 95: 758-766.

Bailey, R.W. and Gaillard, B.D.E., 1969. Hydrolysis of plant mannans by rumen protozoal enzymes. Aust. J. biol. Sci. 22: 267-269.

Bailey, R.W., Monro, J.A., Pickmere, S.E. and Chesson, A., 1976. Herbage hemicellulose and its digestion by the ruminant. In Carbohydrate Research in Plants and Animals. Misc. Pap. 12 Landbouwhogeschool, Wageningen, The Netherlands. pp. 1-15.

Bailey, R.W., Carke, R.T.J. and Wright, D.E., 1962. Carbohydrases of the rumen ciliate *Epidinium ecaudatum* (Crawley). Action on plant hemicellulose. Biochem. J. 83: 517-523.

Bauchop, T., 1977. Foregut fermentation. In: R.T.J. Clarke and T. Bauchop (eds.): Microbial Ecology of the Gut. Academic Press, London, pp. 223-250.

Bauchop, T., 1979a. The rumen anaerobic fungi: colonizers of plant fibre. Ann. Rech. Vet. 10: 246-248.

Bauchop, T., 1979b. Rumen anaerobic fungi of cattle and sheep. Appl. Environ. Microbiol. 38: 148-158.

Bauchop, T., 1980. Scanning electron microscopy in the study of microbial digestion of plant fragments in the gut. In: D.C. Ellwood, J.N. Hedger, M.J. Latham, J.M. Lynch and J.H. Slater (eds.): Contemporary Microbial Ecology. Academic Press, London, pp. 305-326.

Bauchop, T., 1981. The anaerobic fungi in rumen fibre digestion. Agric. Environm. 6: 339-348.

Bauchop, T. and Mountfort, D.O., 1981. Cellulose fermentation by a rumen anaerobic fungus in both the absence and presence of rumen methanogens. Appl. Environ. Microbiol. 42: 1103-1110.

Beveridge, R.J. and Richards, G.N., 1973a. Digestion of polysaccharide consituents of tropical pasture herbage in the bovine rumen. Part IV. The hydrolysis of hemicelluloses from spear grass by cell-free enzyme systems from rumen fluid. Carbohydr. Res. 29: 79-87.

Beveridge, R.J. and Richards, G.N., 1973b. Digestion of polysaccharide constituents of tropical herbage in the bovine rumen. Part III. Examination of the celluloses and hemicelluloses of spear grass (*Heteropogon contortus*) with resist digestion. Carbohydr.. Res. 28: 39-43.

Beveridge , R.J. and Richards, G.N., 1975. Digestion of polysaccharide constituents of tropical pasture herbage in the bovine rumen. VI. Investigation of the digestion of cell wall polysaccharides of spear grass and of cotton cellulose by viscometry and by x-ray diffraction. Carbohydr. Res. 43: 163-172.

Brice, R.E. and Morrison, I.M., 1982. The degradation of isolated hemicellulose and lignin hemicellulose complexes by cell-free, rumen hemicellulases. Carbohydr. Res. 101: 93-100.

Brown, D.W. and Moore, W.E.C., 1960. Distribution of *Butyrivibrio fibrisolvens* in nature. J. Dairy Sci. 43: 1570-1574.

Bryant, M.P., Small, N., Bouma, C. and Chu, H., 1958. *Bacteroides ruminicola* n.sp. and *Succinimonas amylolytica* the new genus and species. J. Bacteriol. 76: 15-23.

Campling, R.C., Freer, M. and Balch, C.C., 1962. Factors affecting the voluntary intake of food by cows. 3. The effect of urea on the voluntary intake of oat straw. Br. J. Nutr. 16: 115-124.

Cheng, K.-J., Akin, D.E. and Costerton, J.W., 1977. Rumen bacteria: interaction with dietary components and response to dietary variation. Fed. Proc. 36: 193-197.

Cheng, K.-J., Fay, J.P., Coleman, R.N., Milligan, L.P. and Costerton, J.W., 1981. Formation of bacterial microcolonies on feed particles in the rumen. Appl. Environ. Microbiol. 41: 298-305.

Chesson, A., 1981. Effects of sodium hydroxide on cereal straws in relation to the enhanced degradation of structural polysaccharides by rumen microorganisms. J. Sci. Food Agric. 32: 745-758.

Chesson, A., 1983. A holistic approach to plant cell wall structure and degradation. In: Fibre in Human and Animal Nutrition. Royal Society of New Zealand, Wellington (In press).

Chesson, A., Gordon, A.H. and Lomax, J.A., 1983a. Cell wall organisation and the biodegradation of cereal straws. In: Biodeterioration 5. John Wiley & Sons Ltd., Chichester, pp. 652-660.

Chesson, A., Gordon, A.H. and Lomax, J.A., 1983b. Substituent groups linked by alkali-labile bonds to arabinose and xylose residues of legume, grass and cereal straw cell walls and their fate during digestion by rumen microorganisms. J. Sci. Food Agric. (in press).

Clarke, R.T.J., Bailey, R.W. and Gaillard, B.D.E., 1969. Growth of rumen bacteria on plant cell wall polysaccharides. J. gen. Microbiol. 56: 79-86.

Davies, M.E., 1965. Cellulolytic bacteria in some ruminants and herbivores as shown by fluorescent antibody. J. gen. Microbiol. 39: 139-141.

Dehority, B.A., 1967. Rate of isolated hemicellulose degradation and utilization by pure cultures of rumen bacteria. Appl. Microbiol. 15: 987-993.

Dehority, B.A., 1973. Hemicellulose digestion by rumen bacteria. Fed. Proc. Fed. Amer. Soc. Exp. Biol. 32: 1819-1825.

Dehority, B.A. and Grubb, J.A., 1977. Characterization of the predominant bacteria occurring in the rumen of goats (*Capra hircus*) Appl. Environ. Microbiol. 33: 1030-1036.

Dekker, R.F.H., 1976. Hemicellulose degradation in the ruminant. In: Carbohydrate Research in Plants and Animals. Misc. Pap. 12. Landbouwhogeschool Wageningen, The Netherlands, pp. 43-54.

Eriksson, K.-E., 1979. Biosynthesis of polysaccharidases. In: R.C.W. Berkeley, G.W. Gooday and D.C. Ellwood (eds.): Microbial Polysaccharides and Polysaccharidases. Academic Press, London, pp. 288-296.

336

Evans, P.J., 1979. Chemical and physical aspects of the interaction of sodium hydroxide with the cell wall components of straw. In: Grossbard, E. (ed.): Straw Decay and its Effects on Disposal and Utilisation. John Wiley & Sons Ltd., Chichester, pp. 187-197.

Forsberg, C.W., Beveridge, T.J. and Hellstrom, A., 1981. Cellulase and xylanase release from *Bacteroides succinogenes* and its importance in the rumen environment. Appl. Environ. Microbiol. 42: 886-896.

Gaillard, B.D.E. and Richards, G.N., 1975. Presence of soluble lignin-carbohydrate complexes in the bovine rumen. Carbohydr. Res. 42: 135-145.

Gaillard, B.D.E., Bailey, R.W. and Clarke, R.T.J., 1965. The action of rumen bacterial hemicellulases on pasture plant hemicellulose fractions. J. agric. Sci. (Camb.) 64: 449-454.

Gordon, A.H. and Bacon, J.S.D., 1981. Fractionation of cell wall preparations from grass leaves by centrifuging in non-aqueous density gradients. Biochem. J. 193: 765-771.

Gordon, A.H., Hay, A.J., Dinsdale, E. and Bacon, J.S.D., 1977. Polysaccharides and associated components of mesophyll cell-walls prepared from grasses. Carbohydr. Res. 57: 235-248.

Gordon, A.J. and Gaillard, B.D.E., 1976. The relationship between lignin and carbohydrate in the hemicellulose A, B and C fractions extracted from lucerne and wheatstraw with alkali. In: Carbohydrate Research in Plants and Animals. Misc. Pap. 12. Landbouwhogeschool, Wageningen, The Netherlands, pp. 55-65.

Groleau, D. and Forsberg, C.W., 1981. Cellulolytic activity of the rumen bacterium *Bacteroides succinogenes*. Can. J. Microbiol. 27: 517-530.

Hartley, R.D. and Jones, E.C., 1977. Phenolic components and degradability of cell walls of grass and legume species. Phytochemistry 16: 1531-1534.

Hartley, R.D. and Jones, E.C., 1978. Effect of aqueous ammonia and other alkalis on the in vitro digestibility of barley straw. J. Sci. Food Agric. 29: 92-98.

Hartley, R.D., 1981. Chemical constituents and processing of lignocellulosic wastes in relation to nutritional quality for animals. Agric. Environm. 6: 91-113.

Hartley, R.D., 1982. Non-carbohydrate constituents and properties of the plant cell wall in relation to its digestion. In: Fibre in Human and Animal Nutrition. Royal Society of New Zealand, Wellington (in press).

Hoffman, R.R., 1975. The Ruminant Shamock. Earl African Whesakeen Busca. Volume 2. Nairobi.

Hungate, R.E., 1966. The Rumen and its Microbes. Academic Press, London. pp. 1-533.

Hungate, R.E., Phillips, G.D., MacGregor, A., Hungate, D.P. and Buecher, H.K., 1959. Microbial fermentation in certain mammals. Science 130: 1192-11194.

Jackson, M.G., 1977. The alkali treatment of straws. Anim. Feed Sci. Technol. 2: 105-130.

Johansson, B. and Miksche, G.E., 1972. Uber die benzyl-aryläther-bindung im lignin. II. Versuche an modellen. Acta Chem. Scand. 26: 289-380.

Kowalczyk, J., Ørskov, E.R., Robinson, J.J. and Stewart, C.S., 1976. Effect of fat supplementation on voluntary intake and rumen metabolism in sheep. Br. J. Nutr. 37: 251-257.

Latham, M.J., 1978. Quantitative aspects of the adhesion of *Ruminococcus flavefaciens* to plant cell walls. Proc. Soc. Gen. Microbiol. 5: 108.

Latham, M.J., Hobbs, D.G. and Harris, P.J., 1979. Adhesion of rumen bacteria to alkali-treated plant stems. Ann. Rech. Vet. 10: 244-245.

Latham, M.J., Brooker, B.E., Pettipher, G.L. and Harris, P.J., 1978a. *Ruminococcus flavefaciens* cell coat and adhesion to cotton cellulose and cell walls in leaves of perennial ryegrass (*Lolium perenne*). Appl. Environ. Microbiol. 35: 156-165.

Latham, M.J., Brooker, B.E., Pettipher, G.L. and Harris, P.J., 1978b. Adhesion of *Bacteroides succinogenes* in pure culture and in the presence of *Ruminococcus flavefaciens* to cell walls in leaves of ryegrass (*Lolium perenne*). Appl. Environ. Microbiol. 35: 1166-1173.

Leatherwood, J.M., 1965. Cellulase complex of *Ruminococcus* and a new mechanism for cellulose degradation. Advan. Chem. Ser. 95: 53-59.

Lomax, J.A., Gordon, A.H. and Chesson, A., 1983. Methylation of unfractionated primary and secondary plant cell walls and the location of alkali-labile substituents. Carbohydr. Res. 122: 11-22.

Mackie, R.I., Gilchrist, F.M., Robberts, A.M., Hannah, P.E. and Schwartz, H.M., 1978. Microbiological and chemical changes in the rumen during the stepwise adaption of sheep to high concentrate diets. J. agric. Sci. (Camb.) 90: 241-254.

Mould, F. and Ørskov, E.R., 1983. Associative effects of mixed feeds. 1. Effects of type and level of supplementation and the influence of rumen pH and cellulolysis and dry-matter degradation of various roughages. Anim. Feed Sci. Technol. (in press).

Mould, F., Ørskov, E.R. and Gauld, S.A., 1983. Associative effects of mixed feeds. 2. The effect of dietary additions of bicarbonate salts on the voluntary intake and digestibility of diets containing various proportions of hay and barley. Anim. Feed Sci. Technol. (in press).

Mould, F., Saadullah, M., Haque, M., Davis, C., Dolberg, F. and Ørskov, E.R., 1982. Investigation of some of the physiological factors influencing intake and digestion of rice straw by native cattle in Bangladesh. Trop. Anim. Prod. 7: 174-181.

Mertens, G.R. and Ely, L.O., 1982. The relationship of role and extent of digestion to forage utilization. A dynamic model valuation. J. Anim. Sci. 54: 895-905.

Minato, H. and Suto, T., 1978. Technique for fractionation of bacteria in the rumen microbial ecosystem. II. Attachment of bacteria isolated from bovine rumen to cellulose in vitro and elution of bacteria attached therefrom. J. gen. Appl. Microbiol. 24: 1-16.

Morrison, I.M., 1974. Structural investigations on the lignin-carbohydrate complexes of *Lolium perenne*. Biochem. J. 139: 197-204.

Morrison, I.M., 1979. The degradation and utilization of straw in the rumen. In: E. Grossbard (ed.): Straw Decay and Its Effect on Disposal and Utilization. John Wiley & Sons Ltd., Chichester. pp. 237-245.

Neilson, M.J. and Richards, G.N., 1982. Chemical structures in a lignin-carbohydrate complex isolated from the bovine rumen. Carbohydr. Res. 104: 121-138.

Orpin, C.G., 1975. Studies on the rumen flagellate *Neocallimastrix frontalis*. J. gen. Microbiol. 91: 249-262.

Orpin, C.G., 1981. Isolation of cellulolytic phycomycete fungi from the caecum of the horse. J. gen. Microbiol. 123: 287-296.

Orpin, C.G. and Letcher, A.J., 1979. Utilization of cellulose, starch, xylan, and other hemicelluloses for growth by the rumen phycomycete *Neocallimastrix frontalis*. Curr. Microbiol. 3: 121-124.

Ørskov, E.R. and Grubb, D.A., 1978. Validation of new systems for protein evaluation in ruminants using untreated and NaOH treated barley straw. J. agric. Sci. (Camb.) 91: 483-486.

Ørskov, E.R. and Hovell, F.D., 1978. Rumen digestion of hay by cattle given sugar-cane or pangola hay. Trop. Anim. Prod. 3: 9-11.

Ørskov, E.R. and McDonald, I., 1979. The estimation of protein degradability in the rumen from incubation measurements weighed according to rate of passage. J. agric. Sci. (Camb.) 92: 499-503.

Parra, R., 1973. Comparison of foregut and hindgut fermentation in herbivores. In: Montgomery, C.G. (ed.): The Ecology of Arboreal Folivores. Smithsonian Institution Press, Washington, D.C. pp. 205-229.

Patterson, H., Irvin, R., Costerton, J.W. and Cheng, K.J., 1975. Ultrastructure and adhesion properties of *Ruminococcus albus*. J. Bacteriol. 122: 278-287.

Pettipher, G.L. and Latham, M.J., 1979. Characteristics of enzymes produced by *Ruminococcus flavefaciens* which degrade plant cell walls. J. gen. Microbiol. 110: 21-27.

Prins, R.A., 1976. Microbiological aspects of plant cell wall digestion in the rumen and caecum: some recent developments. In: Carbohydrate Research in Plants and Animals. Misc. Pap. 12. Landbouwhogeschool Wageningen, The Netherlands. pp. 115-128.

Prins, R.A., Van Vugt, F., Hungate, R.E. and Van Vorstenbosch, C.J.A.H.V., 1972. A comparison of strains of *Eubacterium cellulosolvens* from the rumen. Antonie van Leeuwenhoek 38: 153-161.

Robinson, I.M., Allison, M.J. and Bucklin, J.A., 1981. Characterisation of the cecal bacteria of normal pigs. Appl. Environ. Microbiol. 41: 950-955.

Russell, E.G., 1979. Types and distribution of anaerobic bacteria in the large intestine of pigs. Appl. Environ. Microbiol. 37: 187-193.

Sarkanen, K.V., 1982. Principles and practical approaches to chemical and hydrothermal delignification. In: Improved Utilization of Lignocellulosic Materials for Animal Feed. OECD/COST Workshop. pp. 19-35.

Sijpesteijin, A.K., 1951. On *Ruminococcus flavefaciens*, a cellulose-decomposing bacterium from the rumen of sheep and cattle. J. gen. Microbiol. 5: 869-879.

Smith, W.R., Yu, I. and Hungate, R.E., 1973. Factors affecting cellulolysis by *Ruminococcus albus*. J. Bacteriol. 114: 729-737.

Stewart, C.S., Dinsdale, D., Cheng, K.-J. and Paniagua, C., 1979. The digestion of straw in the rumen. In: E. Grossbard (ed.): Straw Decay and Its Effects on Disposal and Utilization. John Wiley & Sons Ltd., Chichester, pp. 123-130.

Stewart, C.S., Paniagua, C., Dinsdale, D., Cheng, K.-J. and Garrow, S.H., 1981. Selective isolation and characteristics of *Bacteroides succinogenes* from the rumen of a cow. Appl. Environ. Microbiol. 41: 504-510.

Tarkow, H. and Feist, W.C., 1968. The superswollen state of wood. TAPPI 51: 80-83.

Van Gylswyk, N.O. and Hoffman, J.P.L., 1970. Characteristics of cellulolytic cillobacteria from rumens of sheep fed Teff (*Eragrostis tef*) hay diets. J. Microbiol. 60: 381-386.

Van Soest, P.J., 1982. Nutritional Ecology of the Ruminant. O & B Books Inc., Corvallis, Oregon, USA. pp. 1-373.

Waghmare, P.S. and Nambudripad, V.K.N., 1978. Growth and activity of bacterial strains isolated from the buffalo rumen. Indian J. Dairy Sci. 31: 65-72.

Williams, A.G. and Withers, S.E., 1982a. The production of plant cell wall polysaccharide-degrading enzymes by hemicellulolytic rumen bacterial isolates grown on a range of carbohydrate substrates. J. appl. Bact. 52: 377-387.

Williams, A.G. and Withers, S.E., 1982b. The effect of the carbohydrate growth substrate on the glycosidase activity of hemicellulose-degrading rumen bacterial isolates. J. appl. Bact. 52: 389-401.

Wood, T.M., 1981. Aspects of the degradation of plant cell wall carbohydrate in the rumen. In: Degradation of Plant Cell-Wall Material. Agric. Res. Council, London, pp. 24-34.

Wood, T.M. and McCrae, S.I., 1979. Synergism between enzymes involved in the solubilization of native cellulose. Advan. Chem. Ser. 181: 181-209.

Wood, T.M., Wilson, C.A. and Stewart, C.S., 1982. Preparation of the cellulase from the cellulolytic anaerobic rumen bacterium *Ruminococcus albus* and its release from the bacterial cell wall. Biochem. J. 205: 129-137.

Yu, F. and Hungate, R.E., 1979. The extracellular cellulases of *Ruminococcus albus*. Ann. Rech. Vet. 10: 251-257.

Ziolecki, A., Tomerska, H. and Wojciechowicz, M., 1972. Pectinolytic activity of rumen streptococci. Acta Microbiologica Polonica Ser-A, 4: 183-188.

Chapter 12

DIGESTIBILITY, NUTRITIVE VALUE AND FEED INTAKE

by

Jim W. G. Nicholson

Agriculture Canada
Fredericton, N.B., Canada

12.1. INTRODUCTION

Straw and other fibrous by-products which can be used as feed are highly variable in nutritive content. For example, Eriksson (1981) found the in vitro organic matter digestibility (IVOMD) of 250 samples of cereal straw grown in Sweden ranged from 28 to 50%. While it is tempting to discuss wheat straw as if it was a consistent product, this is far from true. Much of the literature, however, does not fully describe the straw being used other than identifying its species, i.e., wheat, oat, barley, etc. For this reason and for convenience these feeds will often be discussed in this chapter in general terms and it is hoped that the reader will keep in mind that variation in chemical composition and nutritional parameters does occur from batch to batch of straw.

12.2. SPECIES OF STRAW

Among the common cereal grains it is generally accepted that oat straw is more digestible than barley straw, and barley straw is more digestible than wheat straw. For example, White et al.

(1981) found the in vitro dry matter digestibility (IVDMD) of straw from winter wheat, spring wheat, barley and oats grown in Montana, USA, in 1976 averaged 36, 36, 40 and 45%. They noted that differences in straw digestibility among crop cultivars within a species were greater than those between species of cereals. Also, there was no significant correlation between straw digestibility and other traits of agronomic importance, such as grain yield and lodging. Therefore, a selection program to improve straw digestibility would not necessarily lead to less desirable grain production capacity.

12.2.1. Wheat straw

Wheat straw is available in North America in greater quantities than any other straw but it is less desirable than the others as a feed. Straw from spring wheat usually has somewhat higher crude protein and dry matter digestibilities than winter wheat (Table 12.1). The utility spring wheats grown in North America (cultivars such as Pitic 62 and Glenlea) are higher in crude protein and organic matter digestibility than hard red spring wheats such as Neepawa (Knipfel and Townley-Smith, 1982).

In the study conducted in Montana, White et al. (1981) found on average that semidwarf and solid stem cultivars had higher IVDMD (40.8 and 40.3%) than did durum varieties (37.0%) or standard height varieties (33.1%) of spring wheat. This suggests that quality of straw for feed is inversely related to height but this was significant in only one of the two years studied. However, they were not able to demonstrate any relationship between straw quality and lodging resistance.

12.2.2. Oat straw

Oat straw is preferred by most cattle producers for feeding and the results of most comparisons between cereal species bears out this belief (Kernan et al., 1979; White et al., 1981; Jackson, 1978a). The fact that oats do not have awns, as do some cultivars of wheat and barley, is another reason why feeders prefer to use oat straw.

12.2.3. Barley straw

Barley straw is usually ranked second to oat straw in preference by feeders although many cultivars of barley will give higher quality straw than some cultivars of oats (see Pearce et al., 1979 and Horton and Stacey, 1979). Barley straw, particularly the two-row cultivars, tend to be higher in crude protein than the other cereal straws (Table 12.1). Erickson et al. (1982) found straw from two-rowed genotypes in most environments in North Dakota was higher in feeding value (5.2% crude protein, 42.9% IVDMD) than sixrowed genotypes (4.8% crude protein, 41.3% IVDMD) while straw from feed and malting barleys was similar in feed value. They also concluded that genotypic selections can be made for other agronomic characteristics without affecting feed quality components.

12.2.4. Rye straw

Rye straw is ranked about equal to wheat and less desirable than barley or oat straw by feeders. The area devoted to growing rye is not as extensive as for the other cereals and consequently there are fewer data on its feeding value. Eriksson (1981) determined the in vitro organic matter digestibility (IVOMD) of 11 samples of one cultivar of rye and found an average value of 42% with a range of 31-50%. This was a lower average than found for oats, barley or wheat straws in the same study.

12.2.5. Rice straw

Rice or paddy straw is produced in large quantities in many countries. It is unique in several ways. The stems are more digestible than the leaves (Jackson, 1978b) which is the opposite of what is found in other straws (Coxworth et al., 1981). Rice straw contains much more silica (12-16%) and less lignin (6-7%) than other straws which contain 3-5% silica and 10-12% lignin (Jackson, 1978b). For these reasons, paddy straw should be cut as close to the ground as possible if it is to be fed to livestock. Jackson (1978b) notes that 30% of the rice straw silica is absorbed and excreted in the urine, yet urinary calculi are not, in general, a serious problem. Rice straw can be upgraded in feeding value to be

Table 12.1. Range of values reported for straw in vitro digestibility, crude protein and acid detergent fibre contents

Species	n	IVOMD %*		CP %		ADF %		Reference
		$\bar{x}$	Range	$\bar{x}$	Range	$\bar{x}$	Range	
Winter wheat 1977	14	36	30-40	2.5	2.1-3.0			White et al.,
Winter wheat 1978	25	40	37-46	2.5	1.9-4.3			1981
Spring wheat 1976	20	36	28-42	2.7	1.9-3.7			
Spring wheat 1977	20	42	36-47	4.0	3.0-6.1			
Oats 1976	20	45	40-48	3.3	2.8-4.1			
Oats 1977	20	45	43-47	5.8	5.2-6.8			
Barley 1976	8	40	33-44	4.4	3.7-5.0			
Barley 1977	19	47	41-42	5.9	5.1-7.0			
Winter wheat	1			2.7				Theander and Åman,
Spring wheat	1			4.0				1978
Oats	1			3.6				
Barley	5			3.1	1.9-5.0			
Rye	1			1.8				
Spring wheat	16	36	30-45	2.6	1.5-4.9	57	52-61	Pearce et al.,
Spring wheat rain damaged	2	31	28-34	2.6	2.2-3.0	60	60-61	1979
Barley	5	45	40-48	2.9	2.5-4.0	54	50-56	
Oats	3	40	34-46	3.9	0.9-4.7	55	54-56	
Oats rain damaged	2	25	25-26	2.7	2.5-2.8	60	60-61	
Wheat	24	37	34-39	3.6	3.3-3.9			Kernan et al.,
Oats	12	40	37-43	3.8	3.7-4.1			1979
Barley	20	38	37-40	4.9	4.2-5.4			
Oats	67	54	40-68					Eriksson, 1981
Barley	82	48	32-61					
Spring wheat	12	46	28-51					
Winter wheat	22	48	35-58					
Rye	11	42	31-50					
Wheat	144	41	34-48	4.1	2.4-6.2			Acock et al., 1978
		IVDMD						
Barley 1977 (6 row)	27	41	34-50	4.8	3.7-8.6	53	47-55	Erickson et al.,
Barley 1977 (2 row)	10	43	38-48	5.2	4.7-6.2	51	50-53	1982
Barley 1978 (6 row)	19	42	38-46	6.2	5.2-7.9	50	46-52	
Barley 1978 (2 row)	12	43	40-46	6.1	5.3-7.4	49	46-51	

*IVOMD = in vitro organic matter digestibility . Values cited from White et al. (1981) and Erickson et al. (1982) are for in vitro dry matter digestibility.

the equivalent to a medium quality hay by appropriate processing
(Jayasuriya, 1979).

12.2.6. Forage straw

The production of forage seed results in large quantities of
grass and legume straw in some areas. As the production of forage
seed is often from perennial stands, it is usually desirable to
remove the straw to help control diseases and insects and to avoid
smothering the next crop. This was often done by burning but, more
and more, this practice is being prohibited by ordinances against
air pollution.

Generally, forage straws are higher in crude protein and organic matter digestibility than are cereal straws but this depends
upon the species of forage, degree of maturation, degree of weathering and method of handling. Anderson and Anderson (1980) have
reviewed possible uses of forge straw including use as a feed.
They state that the digestibility of all constiuents of forage
straws are low and their lack of palatability and low nutrient
density further reduce their value as livestock feeds. Chemical
and/or physical processing are necessary before forage straw can
be used in growing rations for beef cattle above about 25% of the
diet. Early and Anderson (1978) found Kentucky bluegrass straw fed
ad libitum with a commercial liquid feed supplement was an adequate diet for wintering pregnant beef cows. Five varieties of bluegrass straw tested ranged from 3.6 to 7.6% crude protein, 47.9 to
56.6% acid detergent fiber and 24 to 44% IVOMD.

12.2.7. Maize stover

The nutrient content and digestibility of maize stover is higher than that of many other straws. In North America stover is
frequently used as a major part of the diet for dry-pregnant beef
cows. The cows may be turned into the maize fields after harvest
of the grain to glean the crop residue, or the stover may be chopped, ensiled and fed similar to maize silage. When the weather
turns cold, the stover can be stacked or harvested as large round
bales but it will mould in these forms when the weather is warm.
It is desirable to harvest the stover soon after grain harvest as
it loses nutrients while standing in the field. Berger et al.

(1979b) used beef steers to compare ensiled maize stover taken after high-moisture grain was harvested or about one month later after dry grain was harvested. The stovers were compared with normal maize silage. Daily gains of the steers averaged over two trials in different years were 0.65 kg for early harvested stover, 0.48 for late stover and 0.88 for normal maize silage. The feeding value of maize stover has been reviewed by Mowat (1981).

12.2.8. Other straws and by-products

The list of other straws and by-products is almost endless. Kernan et al. (1981) looked at a number of crop residues in Western Canada and grouped them according to usefulness. The IVOMD of sunflower crop residue (68%) and mature Jerusalem-artichoke forage (66%) were higher than for alfalfa and were not improved appreciably by ammonia treatment. In vivo studies with sunflower residue (cited by Coxworth et al., 1981) showed that the intake of digestible energy by sheep fed a sunflower residue diet was only 73% of that from a hay diet in spite of similar energy digestibility of the sunflower residue (57%) to the hay (56%). They attribute the lack of palatability to contamination of the sunflower residue with soil. Faba bean straw was about equal to alfalfa in IVOMD at 61% and field pea straw averaged 50%.

In trials carried out by Thorlacius et al. (1979) the intake of dry matter from faba bean crop residue was equal to that of a medium quality alfalfa-brome hay and organic matter digestibilities were similar, 55.5% (faba bean) and 54.6% (hay). Intake and digestibility of the faba bean residue by the sheep were significantly higher than for a wheat straw diet supplemented with protein.

The digestibility of cotton stalk organic matter as determined with cattle was only 32% compared with 47% for wheat straw (Levy et al., 1980). Ammonia treatment of the cotton stalks increased organic matter digestibility by only 4.4 percentage units while the wheat straw was increased 11.1 units. Treatment of cotton straw with ozone was much more effective in improving IVOMD than treatment with NaOH (Miron and Ben-Ghedalia, 1981). However, when the cotton stalks were compared to wheat straw at a level of 35% of a complete diet there was no difference in the average daily gain of fattening cattle (Levy et al., 1980). Treatment of cotton seed hulls with 60 g NaOH/kg increased the organic matter digestibility by cattle from 33% to 69% (Holzer et al., 1978).

Devendra (1981) has reviewed non-conventional feed resources in the S.E. Asian region. Among the feeds identified as having a high potential are cassava leaves which contain over 20% crude protein. Two problems limit their use. One is the high level of cyanogenic glycosides in some strains; the other is that the crude protein digestibility is often low even when the HCN level is low. Reed et al. (1982) concluded that this is due to the presence of tannins which bind the protein or affect enzyme activity. The availability of protein from cassava leaf meal is probably higher than that from spent tea leaves which contain about 35% crude protein. Jayasuriya et al. (1982) found the solubility of protein suspended in the rumen in nylon bags followed by acid-pepsin digestion was less than 50% for spent tea leaves but reached 77% for cassava leaf meal. These workers suggested spent tea leaves would be a useful source of by-pass protein and cite trials showing it can be included in the diet of growing calves at 7-10% of the dry matter.

The feeding values of by-products of the sugar industry have been reviewed recently by Van Niekerk (1981). The cane tops which are discarded at harvest have a TDN content of about 50% and a crude protein content of about 5% and can be used fresh or after ensiling. Bagasse, the fibrous residue which remains after sugar juice extraction, is somewhat variable in composition. It has very little protein and its TDN content as listed in different tables ranges from 26 to 46%. Many different processing methods have been proposed for improving the nutritive value of bagasse but none has received widespread application.

The dry matter digestibility of soybean straw by sheep was 40%, with nitrogen digestibility ranging from 30 to 46% (Gupta et al., 1978). Both dry matter intake and dry matter digestibility were increased by pelleting or by collecting only the high-pod fraction coming off the lower sieves of the combine. The dry matter intake ranged from 14.7 to 30.6 g/kg live weight. The pelleted straw or chopped high-pod fractions were voluntary consumed at levels which maintained energy balance by sheep (Gupta et al., 1978).

The use of wood wastes as feed has been reviewed by Baker et al. (1975) and Nicholson (1981). The leaves and twigs represent the most digestible parts of these wastes but they are also the most difficult to collect for use as feed as they are usually discarded in the forest. Many species of wild ruminants and domestic goats often use browse as a major part of their diets. The availability of nutrients from wood processing wastes such as bark and sawdust is too low to be useful as feed without processing. The fines from chemical pulping mills are nearly pure cellulose and are well digested by ruminants but chemical pulping to produce feed as a primary product has proven to be too expensive. The process developed by Stake Technology Ltd., Ottawa, for producing feed from the deciduous wood with steam and pressure has received limited commercial application. There is also limited production of "Muka" from tree foliage and wood "molasses" and sulfite liquor as by-products of wood processing but these are quite small in amount. While technology exists to produce feed from many forest industry wastes, its application is rarely economically feasible.

12.3. FACTORS AFFECTING QUALITY OF STRAWS

Coxworth et al. (1981) list 12 factors which influence crop residue feed value. The most important of these are discussed below.

12.3.1. <u>Maturity</u>

The stalks and leaves of growing plants gradually decrease in digestibility and in their content of desirable constituents such as protein as the plant matures. Because most straws are produced as the by-product of grain or seed production there is little opportunity to harvest the straw at a less mature stage. Grains intended for livestock feed can be harvested early and stored in silos or with chemical preservatives as high-moisture grain. The straw from such crops is more digestible and consumed in greater amounts per day than that from crops where the grain is allowed to dry down before harvest. Berger et al. (1979b) found the IVDMD of maize stover declined linearly with time. The regression equation was $Y = 1.93X + 63.22$, where Y is IVDMD % and X is weeks after maize silage harvest stage.

Anderson and Anderson (1980) point out that during maturation (ripening) the nutrients are concentrated in the seed. When ripening is interrupted for any reason, such as drought, hail, etc., the nutrients remain in the straw since the filling of the seed is not completed. In such cases, the protein content of the straw may be double normal values. Kernan et al. (1979) noted that the IVOMD of cereal straw changed very little from the mid dough stage until 10 days after the crop was mature enough to swath. Acock et al. (1978), on the other hand, found IVOMD and crude protein of wheat straw in Nebraska, USA declined from 44.5 and 6.7% one week before combine harvest, to 38.3 and 5.7 at harvest and to 35.0 and 5.8% one week after combining.

Harvesting only the leaf, chaff and top of the cereal plants, leaving behind a high stubble, should result in a higher quality feed. Several studies have shown that leaves and chaff are more digestible than the stems of the common cereal crops (except rice) (Coxworth et al., 1981; Thorlacius et al., 1979). Coxworth et al. (1981) found sheep would consume 845 kJ digestible energy/kg body weight$^{0.75}$ from an untreated Neepawa wheat chaff diet compared with only 636 kJ from a straw diet. The diets contained 25% ground oat grain.

Chaff composed 26-37% of the total weight of the crop residue (Coxworth et al., 1981). The feed value of chaff can be considerably augmented by weed seeds and broken grain kernels. It should be noted that some weed seeds are undesirable because of taste or toxins. Wagons have been developed to tow behind grain combines and collect chaff for animal feed.

12.3.2. Irrigation, climate

A comparison of two cultivars each of wheat, barley and oats grown under irrigation at Saskatoon, Saskatchewan in 1976 compared to the same cultivars at the same location the previous year without irrigation or at another location the same year without irrigation shows substantially higher crude protein content in the straw under irrigation (5.6 vs 3.5%). There was little difference in in vitro digestible organic matter (IVDOM) between the locations for barley and oats but the irrigated wheat straw was higher (41%, irrigated, vs 34%, non-irrigated) (Kernan et al., 1979). This is contrary to the generally held belief that dry land straw has higher feeding value than irrigated straw (Effertz, 1982) and that straw from a crop in which yield is reduced by drought stress is better than from a normal crop. The effect of irrigation and drought stress has not been adequately investigated. The IVOMD of wheat straw grown in Nebraska (Acock et al., 1978) varied much more among 12 locations within the State (34-48%) than among 12 varieties grown at each location (30-42%) but the growing conditions at each location were not described.

12.3.3. Fertilization

Nitrogen fertilization can significantly increase the nitrogen content of straw but has little effect on digestibility. Applying 0, 56 or 224 kg N/ha increased crude protein content of wheat straw from 3.5 to 4.0 to 5.5% but IVOMD only increased from 36 to 38% (Coxworth et al., 1981).

12.3.4. Weathering

Coxworth et al. (1981) found that late harvesting of fababean residue resulted in most of the leaf being lost and a lower crude protein (6.5%) and IVOMD (40%) than found in material harvested at the normal stage. As previously noted, Berger et al. (1979b) found the IVDMD of maize stover declined linearly from the date when maize would normally be harvested as silage.

12.4. METHODS OF FEEDING

12.4.1. Mechanical processing

This topic has been covered in depth by a number of review articles including: Campling (1969), Campling and Milne (1972), Greenhalgh and Wainman (1972), Owen (1978), Nicholson (1981) - see Chapter 5.

One of the major factors limiting intake by ruminants of low-quality forages is the rate of passage of the undigested residue through the digestive tract. It is thought that the main control is exercised at the reticulo-omasal orifice which functions to restrict passage of particles above a certain size (Balch and Campling, 1965). What the critical size is has not been well defined and it likely varies with size of animal and possibly with species of plant and animal. For example, legumes are considered to have a faster rate of passage than grasses of similar digestibility partly because they tend to break down into rectangular shaped particles while grasses give long narrow particles (Troelsen and Campbell, 1968). Campling (1969) suggests that voluntary intake of cows fed ground diets was limited by the large amount of digesta in the abomasum and intestines inhibiting flow of digesta from the reticulo-rumen.

Greenhalgh and Reid (1973) in an elegant experiment involving sheep and cattle at each of three ages, compared the dry matter intake of a high-quality and a low-quality grass as either the long hay or ground and pelleted. Pelleting increased the dry matter intake by 55% in sheep but only by 13% in cattle (Table 12.2) and the intake of the low quality grass diet increased by 44% but only by 19% for the high-quality grass diet. Pelleting increased intake by the youngest animals (6 months of age) by 41%, the middle-aged animals (18 months of age) by 21%, and the oldest (36 months of age) by 28%. These observations support the idea that the critical size of feed particles for passge from the rumen is smaller in sheep than in cattle, and smaller in young (small) animals than in older ones. A similar explanation can be applied to the difference in intake of the low and high-quality grass diets. The low-quality diet would be more resistant to mechanical and microbial comminution in the rumen and passage would be restricted more than for the high-quality forage when both are fed long. Mechanical grinding would remove this restriction on rate of passage (Greenhalgh and Reid, 1973).

Table 12.2. Effect of grinding and pelleting on dry matter intake ($g/kg^{0.75}$ per day) of dried grass by sheep and cattle of three ages (Greenhalgh and Reid, 1973)

Age months	Mean body weight, kg	High quality grass		Low quality grass	
		Long	Pelleted	Long	Pelleted
		Sheep			
6	49	58	92	47	83
18	72	53	77	44	77
36	83	72	92	54	87
		Cattle			
6	272	83	93	80	104
18	464	92	76	74	89
36	614	78	82	68	88

The hays used by Greenhalgh and Reid (1973) were ground through a 1.44 cm screen and pelleted through a 16 mm die. The majority of the particles were less than 0.6 mm long. Troelsen and Campbell (1968) found about 95% of the dry matter in omasal and abomasal digesta of sheep fed various long roughages passed through sieves of 0.5 mm. However, the exact degree of mechanical comminution required to show a response in intake with different sized animals is not known. Innes and Kay (1979) did not find any benefit from shredding barley straw into 2.5 cm lengths for steers while Pickard et al. (1969) did not find any advantage of grinding through screens finer than 0.8 cm. Swan and Clarke (1974) concluded that a screen size of 0.6 cm gave optimum performance of beef cattle fed a diet of 30% ground straw. The screen sizes compared ranged from 0.3 to 1.3 cm.

Fine grinding alone without pelleting has given variable responses. This may be related to dustiness of the unpelleted ground feed and may be influenced by feeding arrangements and dryness of the roughage. There does not seem to be any conclusive evidence to partition the benefit from grinding and pelleting between the two processes. However, Owen (1978) in his review concludes that pelleting or cobbing per se are not important.

Mechanical processing of low-quality forages results in a reduction in digestibility. In the experiment of Greenhalgh and Reid (1973) the reduction in dry matter digestibility was 15 percentage units with high-quality hay and 12 percentage units with the poor-quality hay. The effect was particularly marked for the acid detergent fiber fraction (about 22 percentage units) and was less for nitrogen (about 12 percentage units). The greater loss of feed energy in the feces due to grinding is offset by reduced energy losses as methane and heat. Trials summarized by Wainman et al. (1972) suggest that the increased efficiency of energy metabolism more than offsets the loss of energy through lower digestibility This results in no change in metabolizable energy value of the diet when it is fed at controlled levels of intake but the metabolizable energy was used more efficiently. Owen (1978) concluded it was probably safest to assume no improvement in the metabolizable energy value due to mechanical processing. However, Greenhalgh and Wainman (1972) summarized data from several experiments which showed grinding roughage increased the net energy value of the feed for growth and fattening, the increase being large when its roughage quality was low.

12.4.2. Absorbed protein

The entire effect on intake of grinding and pelleting low-quality forages cannot be explained by the size of particles. Pelleting forages which are unusually low in protein seems to have a relatively small effect (Greenhalgh and Wainman, 1972). A possible explanation is offered in the work of Egan (1965) which showed infusing casein into the duodenum of sheep fed low-nitrogen forages increased the voluntary intake. In later work Egan (1977) showed that when protein digested in the small intestine contributed less than about 10% of the digested energy (5.5 g digested protein /MJ digested energy), intake was likely to increase when protein was infused. No responses in intake were observed when casein was infused with diets which supplied about 13% of the digested energy from protein digested in the intestines.

Kempton and Leng (1979) and Kempton et al. (1979) further developed this concept and concluded that the first limitation to appetite in ruminants given many low-protein feeds is the quantity of absorbed amino acids and not primarily rumen distention. In their work, supplementing a low-protein-cellulosic diet of oat hulls and solka floc with urea increased intake by lambs, but a greater response was obtained with a supplement of both urea and formaldehyde-treated casein. Ferreiro et al. (1979) suggested that available glucose may be the limiting nutrient when cattle are given low-protein roughage diets. Possibly the greater amount of protein absorbed from the small intestine during duodenal infusion (Egan, 1965) or from feeding formaldehyde-treated casein (Kempton and Leng, 1979) served as glucose precursors. Sriskandarajah et al. (1982) fed 200 kg heifers wheat straw ad libitum and one of five pelleted supplements, each of which supplied 40 g of nitrogen per day from urea. The supplements contained either no additional protein or equal protein from varying ratios of casein and formaldehyde-treated casein. Supplying additional protein increased intake of straw but only in the first 5 week period did the formaldehyde-treated casein give higher intakes than the untreated casein. In all cases the extra protein resulted in increased daily weight gains. They concluded that protein supplements do not consistently stimulate intake of low-quality roughages when the requirements for rumen degradable nitrogen have been met. However, changes in live weight gain of the heifers were not significantly correlated with dry matter intake and Sriskandarajah et al. (1982) suggested that this may be attributed to effects of absorbed amino acids on efficiency of tissue protein synthesis either directly or through gluconeogenesis.

Grinding and pelleting of roughages will likely increase the amount of protein escaping digestion in the rumen. Some of the fine particles will rapidly pass out of the rumen thereby decreasing the opportunity for bacterial degradation of the protein (Thomson et al., 1969). The pelleting process also renders some of the protein less soluble and hence more resistant to degradation (Beever et al., 1971; Goering and Waldo, 1978). McAllen et al. (1982) found that bacteria recovered from the rumens of steers fed diets likely to give low rumen ammonia concentrations were higher in starch content but lower in protein than rumen bacteria from steers fed diets with readily degradable protein. All these effects could combine to give a significant increase in the supply of available glucose reaching the intestines following grinding and pelleting. The fact that this would not occur with low-protein roughages could explain the observation of Greenhalgh and Wainman (1972) that intake of such roughages is not improved greatly by grinding and pelleting.

If these hypotheses are correct, grinding a low-protein roughage should increase intake and feeding a high-bypass protein should stimulate intake even more. Unpublished results from our laboratory (Nicholson, 1982) support this hypothesis. A low-protein (6.9% DM basis), mature timothy hay was fed to dairy heifers either chopped (5-10 cm) or ground (0.8 cm screen) each with iso-nitrogenous supplements based on urea or soybean meal. The animals fed the ground hay consumed significantly more dry matter (P < 0.01) than those fed the chopped hay (Table 12.3) and those fed the soybean-based supplement consumed significantly more (P < 0.01) than those fed the urea-based supplement. The dry matter digestibility as determined with sheep was higher for the chopped hay (52.6%) than for the ground hay (48.2%) but the intakes of digestible dry matter were still significantly higher (P < 0.01) for the ground hay and soybean meal supplemented heifers.

Table 12.3. Voluntary intake by dairy heifers of low-protein, mature hay fed with two nitrogen supplements (Nicholson, 1982, unpublished results)

Hay form	Supplement	Dry matter consumption kg/100 kg BW	Digstible dry matter consumption, kg/100 kg BW
Chopped	Urea	1.38	0.69
	Soybean	1.64	0.82
	Average	1.51	0.76
Ground	Urea	1.72	0.86
	Soybean	1.82	0.93
	Average	1.79	0.90
Average	Urea	1.55	0.78
Average	Soybean	1.75	0.88

12.4.3. <u>Supplementation</u>

The extent of digestion in the rumen depends upon adequate nitrogen being available for the rumen microorganisms while the dry matter intake depends, among other things, upon the amount of amino acids absorbed as discussed above. Donefer (1968), for example, demonstrated that adequate nitrogen is necessary for maximum digstion in vivo. The nitrogen requirement has been expressed in various ways. Roy et al. (1977) estimated a rumen degradable protein requirement of 7.8 g/MJ metabolizable energy. Satter and Roffler (1975) suggest rumen ammonia nitrogen levels should be 5 mg/100 ml. Pigden and Bender (1972) suggest that 1% nitrogen in the diet appears adequate for cellulose digestion in the reticulo-rumen for feeds up to 50% digestibility of the energy; for more digestible feeds the requirements may be increased to 1.5% nitrogen. Starch or sugar supplements may raise the requirement to 2%. Smith et al. (1980b) concluded that little change in digestibility of fiber-rich diets in young cattle can be expected when crude protein concentrations are above 85 g/kg dry matter.

The form in which the nitrogen is supplied is important. For example, Hasimoglu et al. (1969) found that lambs fed a diet of 70% wheat straw gained faster and more efficiently when supplemented with soybean meal than with urea, both when the straw fed was treated with NaOH or left untreated.

Smith et al. (1980a) found that animals fed high-fiber, lowprotein diets did not gain much more rapidly when fed additional digestible organic matter per day. However, increasing the protein level, particularly with fishmeal (Smith et al., 1980b), gave increased rates of weight gain.

Anderson (1978) concluded that maximum intakes of cereal residues occurred when small amounts of starch-type carbohydrates are fed in addition to protein supplementation. This agrees with the observation of Barry and Johnstone (1976) that sheep fed chopped barley straw consumed more digestible organic matter from straw when it was supplemented with a ground wheat grain plus urea mixture than when it was unsupplemented or supplemented with urea alone. Higher levels of starchy supplements reduce crude fiber digestibility (Smith et al., 1980b).

Both adequate nitrogen and limited amounts of readily available carbohydrate are needed for optimum fiber digesting activity by rumen microorganisms. It is likely that adequate mineral supplementation is also necessary but there does not seem to be any reports demonstrating responses in intake or digestibility of straws due to mineral supplementation. Klopfenstein et al. (1979) suggested that a chloride anion deficiency and mineral imbalance may be involved in causing lower in vivo digestibility compared with in vitro results when straws are treated with high levels of NaOH. However, the addition of a mineral supplement formulated to supply several other minerals in specified ratios to sodium failed to improve digestibility by lambs of a wheat straw treated with 4% NaOH (Acock et al., 1979).

No information is available concerning ruminant animal response to vitamin supplementation when straws are fed. Because of the very low levels of fat soluble vitamins in most straws, it would only be prudent to supplement animals with vitamins A and D.

Significant improvement in the organic matter digestibility of a beef cow diet of 2 kg of barley fortified with minerals and vitamins and oat straw fed ad libitum was found when urea was replaced with a liquid supplement containing urea, phosphorous, calcium, sodium, molasses and water (Fishwick et al., 1978). The improved digestibility was not associated with a significant increase in voluntary straw intake or in concentrations of ammonia in the rumen liquor or blood of the cows. They concluded that the use of a soluble liquid supplement would seem to be an attractive method of including supplements in straw at baling. However, in a subsequent paper (Bass et al., 1981) they compared three methods of providing the liquid supplement other than when mixed in the straw. The liquid supplement increased straw intake compared with that of cows fed the same straw and barley diet without nitrogen supplementation. The three methods of feeding the supplement were: mixed with the barley, added to the drinking water, or mixed with molasses and provided as a lick. Digestibility data were not presented for the second experiment and the possibility exists that some of the urea in the liquid supplement added to the straw in the first experiment was converted to ammonia with a subsequent improvement in digestibility of the straw. The comparisons reported do not permit separating the effect of the nitrogen from the effect of the other nutrients in the liquid supplement but it is likely that the major benefit was from the supplemental nitrogen.

12.5. CHEMICAL PROCESSING

A great many papers have been published in recent years on the effects of various chemical processes on straws and other lignocellulosic wastes. A number of reviews are available: Jackson (1977, 1978a), Owen (1978), Klopfenstein (1978) and Nicholson (1981). The two most commonly used chemicals are NaOH and ammonia provided as anhydrous NH_3 or as aqueous NH_4OH and both are being used in commercial applications - see Chapters 6 and 7.

12.5.1. Sodium hydroxide treatment

The original work with NaOH treatments was done by soaking the straw in solutions of NaOH followed by washing to remove excess alkali. This process increased digestibility of the straw but a considerable proportion of the soluble nutrients was lost in the washing. Most of the recent work with NaOH has been done with variations of the "dry" process proposed by Wilson and Pigden (1964). With this treatment the excess alkali is not neutralized nor are the excess sodium or other cations removed before feeding.

The use of 10-12% NaOH (unless otherwise specified, treatment levels are expressed as weight on dry matter basis) has resulted in nearly doubling the IVDMD of many straws from the 30% range to 60%. Generally, the poorer the forage quality initially, the greater is its response to chemical processing. The responses obtained with in vivo digestibility correspond with the IVDMD quite well at treatment levels up to 4-6% NaOH. Above this level in vivo digestibility, voluntary daily intake and increased daily liveweight gains all tend to level off or even decrease while the IVDMD continues to increase.

Most of the trials involving in vivo evaluations have used levels of 3-6% NaOH which was considered optimum by Jackson (1977). The most effective level may vary with the material being treated, the type of animal to which it is fed, the level of feeding, degree of dilution with other feeds, and the conditions under which the processing occurs.

12.5.2. Neutralization

Neutralization of the excess alkali with organic or inorganic acids has yielded variable responses in dry matter intake and it is not clear if the high pH has a detrimental effect (Owen, 1978). Cattle seem to be able to tolerate the high levels of alkali for quite long periods of time. Singh and Jackson (1971) fed 9 to 18 month old calves diets based on wheat straw treated with 0, 3.3, 6.7 or 10% NaOH for 23 weeks without any ill effects. The 3.3% level of NaOH resulted in the highest dry matter intakes and weight gains by the calves. Digestibility coefficients for organic matter were significantly increased by the 3.3% level but they levelled off with the higher levels of treatment (55, 70, 71, 71% for the levels of 0, 3.3, 6.7, 10% NaOH, respectively). Palmer (1976) in his review concluded that unneutralized straw fed in large quantities does not appear to cause symptoms of alkalosis in cattle but a change in rumen pH could affect metabolic pathways. The pH of alkali treated straw is about 12.

Work with mature cows by Rexen et al. (cited by Palmer, 1976) did not show any marked change in daily feed intake or rumen pH when NaOH treated straw was fed neutralized or unneutralized. Cows fed the unneutralized straw with a high-urea supplement consumed this feed more slowly. Neutralizing with inorganic acid before feeding did not cause any significant difference in dry matter digestibility in the work of Mowat and Ololade (1970). Levy et al. (1980) treated wheat straw and cotton stalks with 3% NaOH and fed them to cattle with or without neutralization with sulphuric acid.

358

Neutralization reduced the pH of both treated roughages by approximately 3 units but it had no effect on either IVDMD or on the
performance of the animals in the feeding trial. They concluded
that this lack of response to neutralization is consistent with
other reports in the literature and with previous work by their
group (Holzer et al., 1978) in which neutralization of wheat straw
and cotton hulls treated with either 3 or 6% NaOH had no effect.

Jayasuriya and Owen (1975) neutralized barley straw treated
with 4.5 or 9.0% NaOH before feeding to sheep. Although the dry
matter digestibility was higher for the 9% treated straw it was
consumed in lesser amounts per day than the 4.5% treated straw.
Thus, neutralizing the excess hydroxide with HCl did not prevent
the depressed intakes from high levels of NaOH residues observed
by others. In later work with rice straw, Jayasuriya (1979) found
the organic matter digestibility of the straw was increased by
17 to 20 percentage units and organic matter intake by 44% when
the straw was treated with 4% NaOH. Doubling the NaOH to 8% did
not bring about a proportionate increase in digestibility and
neutralization did not significantly affect digestibility or
voluntary intakes of the treated material by sheep.

12.5.3. <u>The influence of NaOH on mineral metabolism and the health
of the animals</u>*

Straw treated with 5% NaOH, contains approximately 25 g sodium
per kg straw. This quantity of sodium is effectively absorbed by
animals and excreted in the urine (Voigt and Piatkowski, 1974;
Stigsen, 1975; Boisen, 1982). Stigsen (1975) suggested that base
excretion at high intake of sodium is mainly regulated by changes
in diuresis. Data from sheep and rabbits, however, showed that the
animals strove to maintain homeostosis by excreting more urine and
concentrating the sodium in the urine (Ali et al., 1977; Boisen,
1982). This process affects the metabolism of the other minerals
and is energy consuming. In an experiment where 25% alfalfa meal
was replaced by NaOH treated straw in rabbit diets, an increase in
heat production and a consequent lower energy utilization was
seen. From this experiment it was concluded that the lower utilization was caused by an increase in diuresis following feeding
with alkaline treated straw (Eggum et al., 1982; Boisen, 1982).

*Section 12.5.3 was written by F.P. Rexen and K.E. Bach Knudsen.

It is still a matter of dispute whether the intake of sodium much above the maximum requirement levels will have any adverse effect on the health of the animals. Ali et al. (1977) stated that the upper limits for the intake of sodium in sheep fed industrially produced NaOH treated straw was in the range of 2.2-2.4 g sodium per kg body weight$^{0.75}$ and that an increase above this level reduced energy intake. Even though the sheep reached this hypothetical threshold value, no health problems were reported during the trials (12 weeks). Similar conclusions were drawn in an experiment with non-lactating dairy cows, at maintenance levels, consuming up to 1.7 g sodium per kg$^{0.75}$ daily (Stigsen, 1975) and in a growth experiment with young rabbits with a daily sodium intake of 0.7 g per kg$^{0.75}$ (Boisen, 1982). Mineral metabolism data supported these conclusions. The mineral retention in the cows and rabbits were positive and the blood plasma values normal (Stigsen, 1975; Boisen, 1982). However, Voigt and Piatkowski (1974) in non-lactating dairy cows fed approximately 1.1 g sodium per kg$^{0.75}$ daily, reported negative retention values for Ca and Mg and that the Na excretory capacity of the kidneys seemed to be exhausted as the protein content of the urine increased. This was the case even though serum plasma concentrations were normal and no health problems reported.

In contrast to these findings, Friis Kristensen et al. (1978) in an investigation with lactating dairy cows consuming 1.2-1.7 g sodium per kg$^{0.75}$ daily observed symptoms similar to grass tetany. Blood samples showed strongly reduced serum Mg levels in a few cows. Similar observations were seen by Arnason (1980) in young bulls fed chopped NaOH-treated straw produced with farm-scale equipment and fed at a level of 2.7 g sodium per kg$^{0.75}$. In the first 12 weeks they grew quickly, then health problems arose. The animals developed diarrhoea and did not thrive although their feed intake remained at a high level. The concentration of P and Mg in blood were significantly reduced.

Thus, even though these results are variable they indicate that health problems can arise when the animals consume high amounts of sodium.

It should be noted that these problems were reported for growing and lactating animals who need a delicately balanced diet in large amounts. This indicates that the upper limits for daily sodium intake is lower than expected by Ali et al. (1977) in animals at a high productional level.

12.5.4. Dilution with other feed

Maeng et al. (1971) determined the energy digestibility by sheep of NaOH (6%) treated barley straw fed as the only forage or when mixed with alfalfa silage at 50:50 or 25:75 straw:silage ratios, dry matter basis. The calculated energy digestibility coefficients for the straw were 66.5, 69.1 and 77.9 for the all straw, 50:50, or 25:50 straw:silage diets. They suggested that cellulose digestibility may be reduced when high levels of treated straw are fed due to the large increase in water intake which possibly results in a faster rate of passage from the rumen. Jackson (1977) also suggested that the lower digestibility often observed in vivo compared to in vitro with high levels of NaOH may be related to the rate of feed passage.

These hypotheses were confirmed by several experiments at the University of Nebraska summarized by Klopfenstein et al. (1979) and Berger et al. (1979a). The results are reproduced in Table 12.4. Rumen fistulated lambs were fed ground maize cob rations consisting of 80% cobs and 20% supplement. The cobs were treated with NaOH to give a complete, mixed diet containing 0, 2, 4, 6 or 8% NaOH. As the level of NaOH increased so did the in vivo dry matter digestibility (24.3% units), IVDMD (38.0% units) and rate of passage of material from the rumen. There was a corresponding decline in mean rumen retention time from 47 hours at 0% NaOH to 30 h for 8% NaOH and rate of digestion of cotton suspended in nylon bags in the rumen declined from 5.4%/h to 2.2%/h.

Soofi et al. (1982) tested with the nylon bag technique a range of concentrations of alkali (NaOH + $Ca(OH)_2$) for treating soybean straw. A concentration of 6% NaOH plus 2% $Ca(OH)_2$ was considered optimum and increased the disappearance of dry matter from the nylon bags by 16.3 percentage units compared with untreated straw. When the treated straw was fed to sheep as the only forage, there was no increse in intake or digestibility compared to other experiments using untreated soybean straw. However, dramatic improvements both in intakes and digestibilities occurred when the alkali treated soybean straw was blended with alfalfa.

Table 12.4. Effect of level of NaOH used to treat maize cobs on rumen parameters of lambs (Klopfenstein et al., 1979; Berger et al., 1979a)

Level of NaOH	In vivo DMD	In vitro DMD	Rate of passage	Rentention time	Rate of digestion
%	%	%	%/h	h	%/h
0	53.6	45.1	2.2	47	5.4
2	55.7	51.5	2.4	43	4.7
4	59.8	64.9	3.2	31	4.2
6	66.2	79.6	2.8	35	2.3
8	77.9	83.1	3.4	30	2.2

Klopfenstein et al. (1979) concluded: (1) the maximum amount of NaOH residue that can be fed is about 2% of the ration dry matter; and (2) excess NaOH residue reduces digestibility due to increased rate of passage and decreased rate of fiber digestion.

12.5.5. _Ammonia treatment_

The use of NH_3 as the anhydrous gas or as aqueous NH_4OH has improved digestibility and voluntary intake similar to that achieved with NaOH when conditions for treatment were appropriate. Not only does the use of NH_3 avoid the problems arising from sodium and alkali residues found with NaOH, but the NH_3 also supplies non-protein nitrogen which can be nutritionally useful under many feeding conditions. The excess NH_3 quickly volatilizes when the treated material is exposed to air. However, this also is a disadvantage as the straw has to be enclosed in an air-tight storage during treatment. The duration of treatment time ranges from several weeks at low ambient temperatures to several hours at high temperatures.

Sundstøl et al. (1978b) concluded that in vitro digestibility determinations and enzyme solubility of organic matter could be used for evaluating NH_3 treated straw. In vivo testing of the product is necessary for mixtures containing less than 50% straw. The energy value of low-quality roughages can be increased to that of medium-quality hay by the use of NH_3 treatment (Sundstøl et al., 1978a). Considerable variation is found in the degree of response obtained from ammonia and this is probably due to less than optimum conditions during treatment (Waagepetersen and Vestergaard

Thomsen, 1977; Sundstøl et al., 1979) and to variation in the initial composition of the material (Kernan et al., 1979; 1981). If the straw is too dry, treatment with anhydrous NH_3 has no effect (Sundstøl et al., 1979).

The combination of treatments involving grinding and ammoniation seems to be especially effective in increasing dry matter intake. For example, Morris and Mowat (1980) showed that grinding (1.3 cm screen) maize stover increased its dry matter intake by steers by 47%, ammoniation (3%) increased it by 12% and the combination by 63%. Grinding decreased the digestibility of energy an average of 6% while ammoniation increased it an average of 14%. As a result, intake of digestible energy was increased an average of 39% by grinding, 31% by ammoniation and 85% by the combination. Such complementary effects of pelleting and ammoniation were not apparent in the trials of Horton et al. (1982) when wheat straw composed only 40% of the diet.

Another advantage of using ammonia treatment is its effectiveness in preventing mould growth in damp forages (those containing more than 16% moisture) (Ørskov and Greenhalgh, 1977). Fermentation was prevented in maize stacks stored at 30% moisture (Mowat, 1981). Sundstøl et al. (1978) noted that ammonia has a definite fungicidal effect in closed stacks of straw containing up to 50% moisture. Thus, the ammoniation not only improves digestibility but also proves a conservation and storage method. They warn that if a high-moisture material is stacked, ammonia should be added immediately to avoid mould and other damage.

12.5.6. Alkali treated silages

The use of NH_3 as a silage additive primarily for maize silage is the basis of a patented process developed in the USA. The development of a simplified, on-farm ensiling process using NaOH, mixtures of NaOH with $Ca(OH)_2$ or NH_4OH was reviewed by Klopfenstein (1978). It would seem to be an economically feasible process, especially for feeds such as maize stover which are commonly ensiled on farms. It is not essential to add grain or other soluble carbohydrate to ensiled, alkali-treated straw nor to heat alkali-treated roughage during the main alkali reaction phase to initiate a useful fermentation and the formation of silage (McManus et al., 1979).

A number of trials with ensiled alkali-treated maize stover have been conducted by Mowat and co-workers at the University of Guelph. In one trial (Mowat, 1971) the dry matter digestibility of maize stover treated with 6% NaOH when ensiled was comparable to that of conventional maize silages (75 vs 76%) and could replace 25% of the maize silage in the diet of beef cattle. When used to replace 50% of the maize silage feed intake was reduced. In another trial (Oji et al., 1977) energy digestibility was increased 12-14% by pre-ensiling treatments of maize stover with 2%NaOH plus 2% $Ca(OH)_2$ at 30% moisture level; or 3% NH_3 at 30% moisture; or 5% NH_3 at 30% moisture with no significant differences between treatments. The organic matter intake of wethers fed the treated silages was 45-50% greater than when fed the untreated control. The pH of the alkali-treated silages ranged from 6.8-8.8. Although lactic acid content was negligible for the NH_3 treated silages compared with the control silage, the content of volatile fatty acids was increased. This was attributed to the liberation of acetyl groups as acetic acid rather than due to fermentation. No moulds were observed even with the high pH silage.

The effects of adding alkali during ensiling are similar to that obtained when it is added shortly before feeding (Wilkinson, 1977). The ease of administration of the alkali during ensiling makes this application method attractive (Kellaway et al., 1978; Klopfenstein, 1978).

A variation of this procedure is ensiling straws after mixing them with urine. Apparently there is sufficient production of NH_3 to improve the digestibility of the forage as occurs with the direct application of NH_3. It is particularly interesting for third world farmers because the ingredients are readily available without cash outlay and the technology can be operated on a small scale.

12.5.7. Microbial processing

Many microbial fermentation techniques have been shown at the laboratory demonstration scale to improve the digestibility of lignocellulosic wastes. Difficulties are encountered when it is attempted to scale up the process for commercial application. The white rot fungi are considered to be especially useful as they attack both lignin and cellulose.

Fermentation with lignolytic fungi under lab-scale conditions improved the IVDMD of poplar shavings by 30 to 40 percentage units and decreased lignin content about 2 percentage units (McQueen and Reade, 1980). However, when the process was scaled up the improvement in IVDMD was less and when the fermented wood was fed to sheep at 30% of the diet the energy digestibility was in the order of 40%. When the fermented wood was included at a level of 45% the sheep failed to thrive and the digestibility of the fermented wood by difference was essentially 0 (McQueen et al., 1981). The process used in this work was semisolid fermentation which Han (1978) concluded from an extensive review as being more promising than submerged liquid fermentation for likely practical application. There is always the danger of undesirable, toxin-producing fungi becoming established in the process, especially when it is carried out at ambient temperatures.

There is increasing interest in North America in the production of edible, lignolytic mushrooms such as *Lentinus edodes*, *Pleurotus florida* and *Pleurotus ostreatus* (Han, 1978). Kaneshiro (1977) showed that *Pleurotus ostreatus* fermentation of feedlot waste fiber reduced the lignin content 10 to 40% and increased cellulose content 15 to 40%. There is need to evaluate the nutritive value of the spent substrate from lignolytic mushroom production by in vitro and in vivo means.

12.6. RUMINANTS AND NON-RUMINANTS

A justification for encouraging ruminant animal production, in some parts of the world at least, is their ability to transform products that humans cannot or will not consume into desirable forms of human food. Ruminants, however, are not the only herbivorous animals and some others, e.g., the horse, may actually be better able to survive on low-quality, lignocellulosic material. This is not due to superior capacity to digest cellulose, because ruminants, with the fermentation compartment at the anterior end of the tract, digest cellulose and related complex carbohydrates more completely than animals with a hind-gut fermentation. The control exercised by the reticulo-omasal orifice on the size of particles allowed to pass down the digestive tract of ruminants ensures that fibrous feeds stay in the fermentation compartments long enough for a reasonable degree of digestion. However, it also limits the amount of feed that can be consumed per day. Animals without this constriction may consume more feed per day and, even though they digest each unit of feed less completely, they end up consuming more digestible nutrients per unit of metabolic body size per day.

Von Engelhardt (1981) discussed the relative efficiencies of large and small ruminants in digesting poor-quality, fibrous diets. He concluded that large herbivores have more spacious fermentation chambers relative to their body weights than smaller animals. Thus, the bigger herbivores achieve a longer retention time of the feed in the rumen and that facilitates more extensive fiber digestion. However, he notes some of the medium sized ruminants seem to be able to adapt to fibrous, low-quality roughages by increasing their forestomach capacity. Lactating dairy cows have 30-40% greater volume of reticulorumen than when not lactating (Campling, 1969). It is interesting to speculate on the success of a genetic selection program for increased rumen size. Von Engelhardt points out that there is a remarkable difference in the capacity of the forestomach within ruminants of similar body size and also within one species. Grazing ruminants, for instance, have in general a more voluminous reticulo-rumen compared to browsers and concentrate selectors.

In addition to researching better methods of processing and feeding low-quality, lignocellulosic feeds, perhaps we should look more carefully at the physiological capability of the animal we are feeding.

12.7. SUMMARY

Straw and other lignocellulosic wastes vary greatly in nutritive value depending upon species of plant, cultivar, maturity, fertilizer, weathering after cutting, and other factors. Among the common cereal crops the straws are generally ranked (best to poorest) oats>barley>wheat = rye>rice but numerous studies have shown greater variation among batches of the same kind of straw than among species.

Physical processing by grinding to reduce the size of particle speeds up rate of passage and increases voluntary dry matter intake per day by ruminants. The exact degree of comminution required depends upon the size of the animal but generally improvement is attained by grinding through a 0.6 cm screen. Optimum digestion requires adequate degradable protein to meet the needs of the rumen microorganisms and adequate absorbed amino acids are needed to maintain high daily intake. Supplemental minerals and vitamins are considered to be required but the benefits are not well quantified.

Chemical treatment of lignocellulosic wastes with NaOH or NH_3 has been developed into commercial application. Levels of 3-6% NaOH are considered optimum for improving in vivo utilization of straw. Higher levels of NaOH show improved IVDMD but not in vivo unless the treated product is diluted with other feed. It is thought that the high level of sodium increases water intake and rate of passage of the forage from the rumen. The high pH per se of treated forage does not appear to be detrimental to animal health and neutralization with inorganic acids is not worthwhile.

The addition of alkalies to lignocellulosic wastes before ensiling has considerable promise for on-farm application. It seems to give much the same benefit as treating with alkali shortly before feeding. The use of NH_3 seems to have some advantage over NaOH as it does not leave an unreacted residue, does not increase water intake, can be used as a source of non-protein nitrogen in appropriate feeding situations, and acts as a preservative. Microbial fermentation of low-quality forages to improve digestibility shows promise but difficulties in scaling up successful laboratory scale demonstrations have so far prevented widespread commercial application.

12.8. REFERENCES

Acock, C.W., Ward, J.K., Rush, I.G. and Klopfenstein, T.J., 1978. Effect of location, variety and maturity on characteristics of wheat straw. J. Anim. Sci. 47, Supplement 1: 327 (Abstract).

Acock, C.W., Ward, J.K., Rush, I.G. and Klopfenstein, T.J., 1979. Wheat straw and sodium hydroxide treatment in beef cow rations. J. Anim. Sci. 49: 354-360.

Ali, C.S., Mason, V.C. and Waagepetersen, J., 1977. The voluntary intake of pelleted diets containing sodium hydroxide treated wheat straw by sheep. 1. The effect of the alkali concentrations in the straw. Z. Tierphysiol., Tierernahr. Futtermittelk. 39: 173-182.

Anderson, D.C., 1978. Use of cereal residues in beef cattle production systems. J. Anim. Sci. 46: 849-861.

Anderson, A.W. and Anderson, J.F., 1980. On finding a use for straw. In: M.L. Shuler (ed.): Utilization and Recycle of Agricultural Wastes and Residues. CRC Press, Boca Raton, Florida, 237-272.

Arnason, J., 1980. Treated straw as feed for young cattle. Thesis, Department of Animal Science, Agricultural University of Norway. 83 pp.

Baker, A.J., Millett, M.A. and Satter, L.D., 1975. Wood and wood-based residues in animal feeds. In A.F. Turbak (ed.): Cellulose Technology Research. Am. Chem. Soc., Washington, D.C., 75-105.

Balch, C.C. and Campling, R.C., 1965. Rate of passage of digesta through the ruminant digestive tract. In R.W. Dougherty (ed.): Physiology of Digestion in the Ruminant. Butterworths, Washington, D.C., 108-123.

Barry, T.N. and Johnstone, P.D., 1976. A comparison of supplementary sources of nitrogen and energy for increasing the voluntary intake and utilization of barley straw by sheep. J. Agric. Sci. (Camb) 86: 163-169.

Bass, J.M., Fishwick, G. and Parkins, J.J., 1981. The effect of the method of presentation of a concentrated solution containing urea, minerals, trace elements and vitamins on the voluntary intake of oat straw by beef cattle. Anim. Prod. 33: 15-17.

Beever, D.E., Thomson, D.J. and Harrison, D.J., 1971. The effects of drying and the comminution of red clover on its subsequent digestion by sheep. Proc. Nutr. Soc. 30: 86A-87A.

Berger, L., Klopfenstein, T. and Britton, R., 1979a. Effect of sodium hydroxide on efficiency of rumen digestion. J. Anim. Sci. 49: 1317-1323.

Berger, L.L., Paterson, J.A., Klopfenstein, T.J. and Britton, R.A., 1979b. Effect of harvest date and chemical treatment on the feeding value of corn stalkage. J. Anim. Sci. 49: 1312-1316.

Boisen, S., 1982. Mineralomsætning og syre-base balance hos voksende kaniner fodret med natriumhydroxyd-behandlet halm. Statens Husdyrbrugsforsøg, Copenhagen. Beretn. 522.

Campling, R.C., 1969. Physical regulation of voluntary intake. In Physiology of Dig. and Met. in the Rumen. Proc. 3rd Internat. Symp., Cambridge, 226-234.

Campling, R.C. and Milne, J.A., 1972. The nutritive value of processed roughages for milking cattle. Proc. Br. Soc. Anim. Prod. 53-60.

Coxworth, E., Kernan, J., Knipfel, J., Thorlacius, O. and Crowle, L., 1981. Review: crop residues and forages in western Canada; potential for feed use either with or without chemical or physical processing. Agric. Environm. 6: 245-256.

Devendra, C., 1981. Non-conventional feed resources in the S.E. Asian Region. Wld Rev. Anim. Prod. 17: 65-80.

Donefer, E., 1968. Effect of sodium hydroxide treatment on the digestibility and voluntary intake of straw. Proc. 2nd World Conf. Anim. Prod., Univ. Maryland, 446-447.

Early, R.J. and Anderson, D.C., 1978. Kentucky bluegrass straw composition, digestibility and utilization in wintering cow rations. J. Anim. Sci. 46: 787-796.

Effertz, N., 1982. Straw varieties that picky cows prefer. Farm Journal, Sept., Beef 12-14.

Egan, A.R., 1965. Nutritional status and intake regulation in sheep. III. The relationship between improvement of nitrogen status and increase in voluntary intake of low-protein roughages by sheep. Aust. J. Agric. Res. 16: 463-472.

Egan, A.R., 1977. Nutritional status and intake regulation in sheep. VIII. Relationships between the voluntary intake of herbage by sheep and the protein/energy ratio in the digestion products. Aust. J. Agric. Res. 28: 907-915.

Eggum, B.O., Chwalibog, A., Jensen, N.E. and Boisen, S., 1982. Protein and energy metabolism in growing rabbits fed sodium hydroxide treated straw. Arch. Tierernähr. 32: 539-549.

Engelhardt, W. von, 1981. Some physiolgocial aspects on the digestion of poor quality, fibrous diets in ruminants. Agric. Environm. 6: 145-152.

Erickson, D.O., Meyer, D.W. and Foster, A.E., 1982. The effect of genotypes on the feed value of barley straws. J. Anim. Sci. 55: 1015-1026.

Eriksson, S., 1981. Nutritive value of cereal straw. Agric. Environm. 6: 257-260.

Ferreiro, H.M., Priego, A., Lopez, J., Preston, T.R. and Leng, R.A., 1979. Glucose metabolism in cattle given sugar cane based diet with varying quantities of rice polishings. Br. J. Nutr. 42: 341-347.

Fishwick, G., Parkins, J.J., Hemingway, R.G. and Ritchie, N.S., 1978. A comparison of the voluntary intake and digestibility by beef cows of diets based on oat straw and supplemented with different forms of non-protein nitrogen. Anim. Prod. 26: 135-141.

Friis Kristensen, V., Andersen, P.E., Stigsen, P., Vestergaard Thomsen, K., Refsgaard Andersen, H., Sørensen, M., Ali, C.S., Mason, V.C., Rexen, F., Israelsen, M. and Wolstrup, J., 1978. Natriumhydroxyd-behandlet halm som foder til kvæg og får. Statens Husdyrbrugsforsøk, Copenhagen. Beretn. 464.

Goering, H.K. and Waldo, D.R., 1978. The effects of dehydration on protein utilization in ruminants. Proc. 2nd Int. Green Crop Drying Congr., Saskatoon. Univ. Sask. Print. Serv. 277-291.

Greenhalgh, J.F.D. and Reid, G.W., 1973. The effects of pelleting various diets on intake and digestibility in sheep and cattle. Anim. Prod. 16: 223-233.

Greenhalgh, J.F.D. and Wainman, F.W., 1972. The nutritive value of processed roughages for fattening cattle and sheep. Proc. Br. Soc. Anim. Prod. 61-72.

Gupta, B.S., Johnson, D.E. and Hinds, F.C., 1978. Soybean straw intake and nutrient digestibility by sheep. J. Anim. Sci. 46: 1086-1090.

Han, Y.W., 1978. Microbial utilization of straw (a review). In D. Perlman (ed.): Advances in Applied Microbiology. Academic Press, New York, 119-153.

Hasimoglu, S., Klopfenstein, T.J. and Doane, T.H., 1969. Nitrogen source with sodium hydroxide treated wheat straw. J. Anim. Sci. 29: 160 (Abstract).

Holzer, Z., Levy, D. and Folman, Y., 1978. Chemical processing of wheat straw and cotton by-products for fattening cattle. 2. Performance of animals receiving material after drying and pelleting. Anim. Prod. 27: 147-159.

Horton, G.M.J., Nicholson, H.H. and Christensen, D.A., 1982. Ammonia and sodium hydroxide treatment of wheat straw in diets for fattening steers. Anim. Feed Sci. Technol. 7: 1-10.

Horton, G.M.J. and Steacy, G.M., 1979. Effect of anhydrous ammonia treatment on the intake and digestibility of cereal straws by steers. J. Anim. Sci. 48: 1239-1249.

Innes, G.M. and Kay, M., 1979. The effect of mechanical processing and urea on straw intake using steers. Anim. Prod. 28: 433 (Abstract).

Jackson, M.G., 1977. Review article: The alkali treatment of straws. Anim. Feed Sci. Technol. 2: 105-130.

Jackson, M.G., 1978a. Treating straw for animal feeding. An assessment of its technical and economic feasibility. Anim. Prod. and Health Paper 10, FAO, Rome, 81 pp.

Jackson, M.G., 1978b. Rice straw as livestock feed. The Green Revolution, pp. 95-98.

Jayasuriya, M.C.N., 1979. Sodium hydroxide treatment of rice straw to improve its nutritive value for ruminants. Trop. Agric. (Trinidad) 56: 75-80.

Jayasuriya, M.C.N. and Owen, E., 1975. Sodium hydroxide treatment of barley straw; effect of volume and concentration of solution on digestibility and intake by sheep. Anim. Prod. 21: 313-322.

Jayasuriya, M.C.N., Wijeyatunge, C. and Perera, H.G.D., 1982. Rumen and post-rumen fermentation of spent tea leaf protein and other protein sources studied by the nylon bag method. Anim. Feed Sci. Technol. 7: 221-224.

Kaneshiro, T., 1977. Lignocellulosic agricultural wastes degraded by *Pleurotus ostreatus*. Dev. Ind. Microbiol. 18: 591-597.

Kellaway, R.C., Crofts, F.C., Thiago, L.R.L., Redman, R.G. and Leibholz, J.M.L., 1978. A new technique for upgrading the nutritive value of roughages under field conditions. Anim. Feed Sci. Technol. 3: 201-210.

Kempton, T.J. and Leng, R.A., 1979. Protein nutrition of growing lambs. 1. Responses in growth and rumen function to supplementation of low-protein-cellulosic diet with either urea, casein or formaldehyde-treated casein. Br. J. Nutr. 42: 289-302.

Kempton, T.J., Nolan, J.V. and Leng, R.A., 1979. Protein nutrition of growing lambs. 11. Effect on nitrogen digestion of supplementing a low-protein cellulosic diet with either urea, casein or formaldehyde-treated casein. Br. J. Nutr. 42: 303-315.

Kernan, J.A., Coxworth, E.C. and Spurr, D.T., 1981. New crop residues and forages for western Canada: assessment of feeding value in vitro and response to ammonia treatment. Anim. Feed Sci. Technol. 6: 257-271.

Kernan, J.A., Crowle, W.L., Spurr, D.T. and Coxworth, E.C., 1979. Straw quality of cereal cultivars before and after treatment with anhydrous ammonia. Can. J. Anim. Sci. 59: 511-517.

Klopfenstein, T., 1978. Chemical treatment of crop residues. J. Anim. Sci. 46: 841-848.

Klopfenstein, T., Berger, L. and Paterson, J., 1979. Performance of animals fed crop residues. Fed. Proc. 38: 1939-1943.

Knipfel, J.G. and Townley-Smith, T.F., 1982. Nutritive values of straw and chaff from different species and classes of wheat. Can J. Plant Sci. 62: 256-257 (Abstract).

Levy, D., Holzer, Z. and Folman, Y., 1980. Chemical processing of wheat straw and cotton by-products for fattening cattle. 3. Performance of animals receiving material in complete feeds. Anim. Prod. 31: 27-33.

Maeng, W.J., Mowat, D.N. and Bilanski, W.K., 1971. Digestibility of sodium hydroxide-treated straw fed alone or in combination with alfalfa silage. Can. J. Anim. Sci. 51: 743-747.

McAllan, A.B., Williams, A.P., Merry, R.J. and Smith, R.H., 1982. Effect of different levels of casein, with or without formaldehyde treatment, on carbohydrate metabolism between mouth and duodenum of steers. J. Sci. Food Agric. 33: 722-728.

McManus, W.R., Grout, L.L., Robinson, V.N.E., Southwell-Keely, P. and Woodhart, P.N., 1979. Ensilage from alkali-treated roughages. Aust. J. Exp. Agric. Anim. Husb. 19: 354-361.

McQueen, R.E. and Reade, A.E., 1980. Changes in composition and digestibility of poplar by fungal fermentation. Can. J. Anim. Sci. 60: 571-572 (Abstract).

McQueen, R.E., Reade, A.E. and Nicholson, J.W.G., 1981. In vivo digestibility of wood fermented by white rot fungus *Ganoderma*. Can. J. Anim. Sci. 61: 1088-1089 (Abstract).

Miron, J. and Ben-Ghedalia, D., 1981. Effect of chemical treatments on the degradability of cotton straw by rumen microorganisms and by fungal cellulase. Biotechnol. Bioeng. 23: 2863-2873.

Morris, P.J. and Mowat, D.N., 1980. Nutritive value of ground and/or ammoniated corn stover. Can. J. Anim. Sci. 60: 327-336.

Mowat, D.N., 1971. NaOH-stover or straw silage in growing rations. J. Anim. Sci. 33: 1155 (Abstract).

Mowat, D.N., 1981. Alternative processes for improving nutritive value of maize stover. Agric. Environm. 6: 153-160.

Mowat, D.N. and Ololade, B.G., 1970. Effect of level of sodium hydroxide treatment on digestibility and voluntary intake of straw. Proc. Can. Soc. Anim. Prod. 35 (Abstract).

Nicholson, J.W.G., 1981. Nutrition and feeding aspects of the utilization of processed lignocellulosic waste materials by animals. Agric. Environm. 6: 205-228.

Niekerk, B.D.H., van, 1981. Byproducts of the sugar industry as animal feeds. S. Afr. J. Anim. Sci. 11: 119-137.

Oji, U.I., Mowat, D.N. and Winch, J.E., 1977. Alkali treatments of corn stover to increase nutritive value. J. Anim. Sci. 44: 798-802.

Ørskov, E.R. and Greenhalgh, J.F.D., 1977. Alkali treatment as a method of processing whole grain for cattle. J. agric. Sci. (Camb.) 89: 253-255.

Owen, E., 1978. Processing of roughages. In: W. Haresign and D. Lewis (eds.): Recent Advances in Animal Nutrition. Butterworths, London, 127-148.

Palmer, F.G., 1976. The feeding value and worthwhileness of chemically processed straw for ruminants. ADAS Q. Rev. 22: 247-266.

Pearce, G.R., Beard, J. and Hilliard, E.P., 1979. Variability in the chemical composition of cereal straws and in vitro digestibility with and without sodium hydroxide treatment. Aust. J. Exp. Agric. Anim. Husb. 19: 350-353.

Pickard, D.W., Swan, H. and Lamming, G.E., 1969. Studies on the nutrition of ruminants. 4. The use of ground straw of different particle sizes for cattle from twelve weeks of age. Anim. Prod. 11: 543-550.

Pigden, W.J. and Bender, F., 1972. Utilization of lignocellulose by ruminants. Wld. Anim. Rev. 4: 7-10.
Reed, J.D., McDowell, R.E., Van Soest, P.J. and Horvath, P.J., 1982. Condensed tannins: a factor limiting the use of cassava forage. J. Sci. Food Agric. 33: 213-220.
Roy, J.H.B., Balch, C.C., Miller, E.L., Ørskov, E.R. and Smith, R.H., 1977. Calculation of the N-requirement for ruminants from nitrogen metabolism studies. In: Protein Metabolism and Nutrition, Pudoc, Wageningen, 126-129.
Satter, L.D. and Roffler, R.E., 1975. Nitrogen requirement and utilization in dairy cattle. J. Dairy Sci. 58: 1219-1237.
Singh, M. and Jackson, M.G., 1971. The effect of different levels of sodium hydroxide spray treatment of wheat straw on consumption and digestibility by cattle. J. Agric. Sci. (Camb.) 77: 5-10.
Smith, T., Broster, W.H. and Siviter, J.W., 1980a. An assessment of barley straw and oat hulls as energy sources for yearling cattle. J. agric. Sci. (Camb.) 95: 677-686.
Smith, T., Broster, V.J. and Hill, R.E., 1980b. A comparison of sources of supplementary nitrogen for young cattle receiving fibre-rich diets. J. agric. Sci. (Camb.) 95: 687-695.
Soofi, R., Fahey, G.C., Jr. and Berger, L.L., 1982. In situ and in vivo digestibilities and nutritive intakes by sheep of alkali-treated soybean stover. J. Anim. Sci. 55: 1206-1213.
Sriskandarajah, N., Kellaway, R.C. and Leibholz, J., 1982. Utilization of low-quality roughages; effects of supplementing with casein treated or untreated with formaldehyde on digesta flows, intake and growth rate of cattle eating wheat straw. Br. J. Nutr. 47: 553-563.
Stigsen, P., 1975. Kulhydratkildens og neutralisationens betydning for udnyttelse af NaOH behandlet halm hos malkekøer. Thesis. Den Kgl. Veterinær- og Landbohøjskole, Copenhagen.
Sundstøl, F., Said, A.N. and Arnason, J., 1979. Factors influencing the effect of chemical treatment on the nutritive value of straw. Acta Agric. Scand. 29: 179-190.
Sundstøl, F., Coxworth, E. and Mowat, D.N., 1978a. Improving the nutritive value of straw and other low-quality roughages by treatment with ammonia. Wld Anim. Rev. 26: 13-21.
Sundstøl, F., Kossila, V., Theander, O. and Vestergaard Thomsen, K., 1978b. Evaluation of the feeding value of straw. Acta Agric. Scand. 28: 10-16.
Swan, H. and Clarke, V.J., 1974. The use of processed straw in rations for ruminants. In: H. Swan and D. Lewis (eds.): Univ. Nottingham Nutr. Conf. Feed Manuf. -8, Butterworths, London, 205-233.
Theander, O. and Åman, P., 1978. Chemical composition of some Swedish cereal straws. Swed. J. Agric. Res. 8: 189-194.
Thomson, D.J., Beever, D.E., Coehlo Da Silva, J.F. and Armstrong, D.J., 1969. Sites of digestion in sheep of a dried lucerne fed in three different physical forms. Proc. Nutr. Soc. 28: 24A (Abstract).
Thorlacius, S.O., Coxworth, E. and Thompson, D., 1979. Intake and digestibility of fababean crop residue by sheep. Can. J. Anim. Sci. 59: 459-462.
Troelson, J.E. and Campbell, J.B., 1968. Voluntary consumption of forage by sheep and its relation to the size and shape of particles in the digestive tract. Anim. Prod. 10: 289-296.

Voigt, J. and Piatkowski, B., 1974. Untersuchungen zum Aufschluss von Getreidestroh. 5. Die Wirkung des Natriums im NaOH-behandelten Stroh auf die Zusammensetzung von Blut und Harn sowie auf die Exkretion verschiedener Verbindungen.Arch. Tierernähr. 24: 589-600.
Waagepetersen, J. and Vestergaard Thomsen, K., 1977. Effect on digestibility and nitrogen content of barley straw of different ammonia treatments. Anim. Feed Sci. Technol. 2: 131-142.
Wainman, F.W., Blaxter, K.L. and Smith, J.S., 1972. The utilization of the energy of artificially dried grass prepared in different ways. J. agric. Sci. (Camb.) 78: 441-447.
White, L.M., Hartman, G.P. and Bergman, J.W., 1981. In vitro digestibility, crude protein and phosphorus content of straw of winter wheat, spring wheat, barley and oat cultivars in eastern Montana. Agron. J. 73: 117-121.
Wilkinson, J.M., 1977. Ensiling alkali treated straw. Rep. Straw Utiliz. Conf., Oxford, MAFF, 32-35.
Wilson, R.K. and Pigden, W.J., 1964. Effect of a sodium hydroxide treatment on the utilization of wheat straw and poplar wood by rumen microroganisms. Can. J. Anim. Sci. 44: 122-123.

Chapter 13

SUPPLEMENTATION OF DIETS BASED ON FIBROUS RESIDUES AND BY-PRODUCTS

by

T. Reg. Preston[1] and Ron. A. Leng[2]

[1]Department of Tropical Veterinary Science
James Cook University, Townsville, Q 4811
Australia

[2]Department of Biochemistry and Nutrition
University of New England, Armidale, NSW 2351
Australia

13.1. INTRODUCTION

Straws from cereal crops are the basal feed for a large proportion of the ruminants in developing countries. Straws are considered to be of such low value that in developed countries the majority is burnt. Nevertheless, the mature draught animal appears to be able to maintain a low body condition and work satisfactorily on straw alone. For the majority of productive purposes, straw must be supplemented and even then it is unlikely that production would reach economic levels in developed countries. In most of the studies of straw utilisation by ruminants in developed countries the straw only represents a small proportion of the diet, usually a filler or a source of roughage in grain-based diets. It is not the intention here to discuss this aspect of straw utilization. Discussion is restricted to situations where straw is the basis of the diet and supplements are used to maximize productivity.

Because of the large variety of crop residues that are available and their variable quality, it is the intention to emphasise the application of principles for the maximum utilisation of such diets. In order to establish these principles, information is drawn from research which has been carried out on a range of other crop residues and byproducts.

13.2. GENERAL DESCRIPTION OF FIBROUS RESIDUES AND BYPRODUCTS USED IN ANIMAL FEEDING

Whilst there are considerable differences between different types of cereal straws and also between the same cereal straw as affected by time of harvesting, length of storage, etc. all cereal straws have in common two factors: 1) they are extremely low in nitrogen; and 2) they are composed of cell wall components of plants with little soluble cell contents and therefore have to be digested by microbial fermentation.

The necessity for fermentation in the utilization of straw-based diets gives advantages to large ruminant animals. The reason for this is that both the size of the fermentation compartments and the rate of rumination increase linearly with liveweight (W); this gives an advantage to the large ruminants because the requirements for essential nutrients increase with metabolic liveweight ($W^{0.75}$) (see Van Soest, 1982). This review therefore is concerned largely with cattle and buffaloes.

There appear to be no toxic substances in straws, except when there is growth of moulds. Rice straw, which forms a large proportion of the available forage in developing countries has a high level of silica, which is reported to be concentrated in the leaf rather than the stem and which may place limitations on the utilisation of this material.

When straws are fed to ruminants the primiary limitations are: 1) the slow rate of, and total digestibility; 2) the rate at which straw particles break down to sizes which can leave the rumen; and 3) the factors that determine or influence the rate of breakdown or rate of passage of straw residues from the rumen. These are discussed in detail in Chapter 11.

Mineral content of straws is generally low and inbalanced but this may be quite adequate for the maintenance and for work by ruminants. For production of meat and milk, requirements for minerals by the animal are increased many-fold and the mineral content of straw is likely to be insufficient. The mineral composition of plants depends largely on the availability of minerals in the soil and closeness to the oceans (for Na). Calcium and phosphorus contents of straw are usually below recommended levels and cobalt, copper, sulphur and sodium may also be limiting. The high concentrations of oxalates, in some tropical forages and the presence of silicates in rice straw also suggest that there is the opportunity for considerable losses of salts as silicates and oxalates in faeces.

13.3. SALIENT FEATURES OF DIGESTION AND METABOLISM IN RUMINANT ANIMALS GIVEN BASAL DIETS OF FIBROUS FEEDS

13.3.1. <u>Rumen fermentation</u>

In order to discuss appropriate strategies for supplementation of straw-based diets given to cattle and buffaloes, it is necessary to present our view of ruminant digestion and the associated constraints.

13.3.1.1. <u>Energy transactions in the rumen</u>

The major energy-providing feed materials in straws are the cell wall constituents comprising mainly cellulose and hemicellulose compounds. These comprise α-glucose molecules combined in chains with a 1-6 linkage. Animals do not have enzymes for the hydrolysis of these bonds and depend on micro-organisms in the rumen to carry out this process. In general these carbohydrates are insoluble and the initial step in their breakdown is for the cellulolytic organisms to split off cellobiose (a 2 glucose compound) which is then degraded to give glucose which is further fermented by a complex series of reactions to give rise to short chain volatile fatty acids (VFA), methane and carbon dioxide (see Leng, 1974). The intermediates of this breakdown together with a nitrogenous source are used for microbial cell production (Figure 13.1). Many species of organisms may be involved in the complete breakdown of these carbohydrates. The overall energy losses (as heat) are low and the majority of the energy present in the fermented carbohydrate is retained in the products of fermentation i.e. VFA and microbial cells.

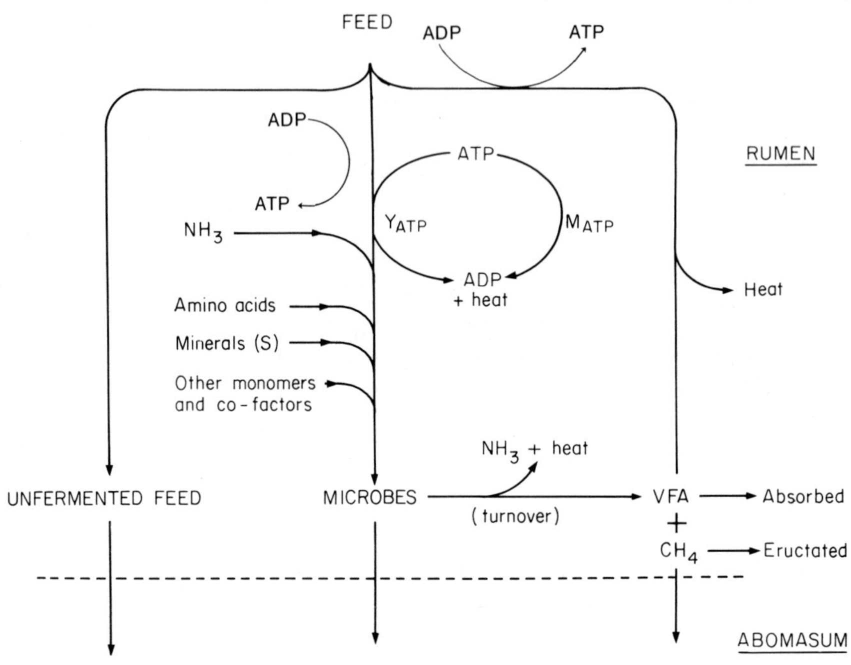

Figure 13.1. Representation of the partitioning of feed materials in the rumen into volatile fatty acids and methane and microbial cells. The feed bypassing rumen fermentation is shown as unfermented feed. The generation of ATP and its utilisation in cell growth (Y_{ATP} as g dry cells/mole ATP generated) is indicated together with the wasteful processes involved in maintenance of microbes in the rumen (M_{ATP}) and the lysis of microbes (from Leng, 1982)

13.3.1.2. Microbial cell synthesis

The synthesis of cellular compounds of microorganisms in the rumen has not been studied extensively. Protein constitutes 30-60% of the dry matter of bacteria and the ash content is often very high (13%) (see Czerkawski, 1976) perhaps indicating a large requirement for minerals by rumen organisms.

Most of the amino acids in bacteria are synthesized from the rumen ammonia pool, but at a maximum some 30% may be synthesized from amino acids of dietary or endogenous origin (see Nolan and

Leng, 1972). The major energy requiring reactions are those associated with amino acid synthesis and polymerisation into proteins.
It can be calculated that roughly 70% of the available ATP is used
in microbial protein synthesis (see Leng, 1982).

13.3.1.3. <u>Relationship between VFA production and synthesis of microbial cells</u>

The substrates for microbial cell synthesis are the intermediates of fermentation, thus there is an inverse relationship between
VFA production and cell synthesis in the rumen. The higher the VFA
production, the lower the microbial cell yield. The factors that
influence this relationship are discussed as this information may
be important in order to manipulate the efficiency with which large ruminants utilise straw based diets (see Figure 13.2).

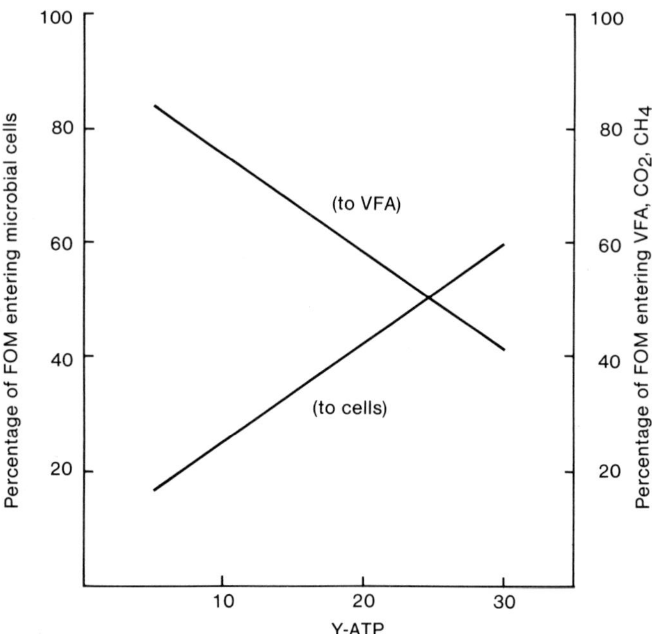

Figure 13.2. Relationship between proportions of organic matter
fermented to microbial cells or to VFA (Leng, 1982)

When the objective of the feeding strategy is the production of meat, milk and wool then any manipulation of the diet, or the animal, must be aimed at maximizing microbial protein output from the rumen relative to the energy in volatile fatty acids. A much lower microbial protein to energy ratio may be required by working animals and therefore in these animals it may be advantageous to have a fermentation which maximizes VFA production and therefore minimizes microbial protein output. For example this might be achieved by having less than optimal fermentable nitrogen in the diet.

13.3.2. Factors that influence the yield and activity of rumen microorganisms

The major factors that affect microbial protein synthesis in the rumen of animals on straw-based diets are:

(1) Intake of potentially fermentable food.
(2) Availability and/or concentration in rumen fluid of precursors of microbes (such as glucose, nucleic acids, amino acids, ammonia and minerals and perhaps other co-factors).
(3) The mainenance ATP requirements of the microbes.
(4) The turnover and lysis of microbial cells.
(5) The activity of protozoa and possibly fungi.

13.3.2.1. Availability of substrates

A limitation to the initial breakdown of straw is probably the pool size of cellulolytic organisms and perhaps of the phycomycetous fungi (see Bauchop, 1981) which appear to control the first steps in the breakdown of plant materials. Although little is known of the role of the fungi, the bacteria are believed to depend on ammonia for most of their cellular N requirements (see Bryant and Robinson, 1961, 1963). The colonisation of ingested plant material requires transfer of organisms between food residues and newly ingested food via the fluid medium. A population (even quite small) of "grazing" protozoa may reduce this "free in solution" population and therefore reduce the rate of colonisation of new feed particles in the rumen.

The cellulolytic organisms have been reported to have specific requirements for branch-chained amino or fatty acids. In one study, Helmsley and Moir (1963) demonstrated a marked increase in digestion by adding these substances to a straw diet. Hume (1970) also showed that addition of a soluble protein increased microbial outflow from the rumen on purified diets. Whilst not disputing a need for these amino acids or their precursors, the significant amount of lysis of microbial cells in the rumen on low quality diets, should provide these compounds (Nolan and Stachiw, 1979); and in a number of studies where the dietary content of fermentable N was adequate no requirements for soluble proteins have been demonstrated (see Kempton and Leng, 1979).

Following ingestion of feed the available substrate present in the rumen will be high for a few hours and will decrease with time. Thus there is probably a period when the organisms have exhausted their substrate supply and begin to die (see Hespell, 1979) and in their turn are fermented by other organisms (Nolan and Leng, 1972). Any manipulation which maintains a large pool of organisms in the rumen will increase the rate of digestion of ingested plant material. Addition of a readily and totally fermentable cellulose source to the diet may possibly increase the microbial biomass after feeding and thus potentially increase the rate of colonisation and degradation of the straw particles (Nielsen, 1981). The continuity of the intake of feed should also be beneficial.

13.3.2.2. <u>Rumen ammonia and amino acids</u>

The rate of fermentation must be synchronized with the rate of uptake of ammonia and/or peptides and amino acids. Maximum microbial synthesis rates apparently occur at ammonia concentrations between 5-8 mg/100 ml (Satter and Slyter, 1974; Pisulewski et al., 1981) although different optima have been found by other researchers (see Stern and Hoover, 1979). For example, rate of dry matter loss from nylon bags in the rumen increased linearly as rumen NH_3 concentration rose from 3 to 12 mg/100 ml when the substrate was alkali-treated maize cobs, while for hay the maximum rate of DM loss was reached at 7 mg/100 ml (Dixon, 1982, unpublished data). In practice, the level of rumen ammonia should be maintained above this minimum since a slight excess is not likely to be a problem as compared to a small deficiency.

A number of studies have emphasized the need for a continuous supply of ammonia in the rumen in order to maintain high intakes and digestibility of a fibrous diet (see Romero et al., 1976; Campling et al., 1962).

Kennedy and Milligan (1980) ascribed a large recycling of urea in sheep and cattle on forage diets to the presence of sugar which in some way apparently stimulated urea entry from blood via the rumen wall. If this is true then continous supply of small amounts of molasses (which contains sucrose, glucose and fructose) may be beneficial as a means of increasing recycling of urea N to the rumen thus ensuring a continuous supply of ammonia for the rumen organisms.

The maintenance of relatively high rumen ammonia levels on straw-based diets may be exceedingly important in maintaining a high rate of fermentation and maximizing intake and digestibility.

13.3.2.3. Availability of other factors and food materials

It is generally considered that the rumen organisms are largely independent of dietary sources of B-vitamins or other co-factors. However, recent work indicates that supplementation of a diet with nicotinic acid increased the rate of microbial cell production (Riddell et al., 1980). The recognition of the stimulatory effect of a small amount of fresh green forage on the rate of cellulose digestion on a sisal pulp basal diet implied that a number of factors may at times be required to maximize rumen digestion (Guttierez et al., 1983). Thus on straw-based diets there is a possibility that a small supplement of green forage may have an influence on microbial growth by providing essential co-factors.

13.3.2.4. The factors that influence the maintenance ATP requirements of rumen organisms

The definition of M-ATP of rumen organisms is that ATP which is directed from utilization in growth of microbes (synthesis) to utilization in non-productive processes (see Leng, 1982).

A major factor which influences the M-ATP is the length of time an organism spends in the rumen. A large pool of microorganisms growing at a slow rate will be less efficient than a smaller pool of microbes turning over rapidly. This has been clearly shown in continuous culture where increasing the turnover of the fluid contents increases markedly microbial cell synthesis (Isaacson et al., 1975). Where rumen digesta flow rates have been increased by artificial means, such as feeding high levels of salts the micro-

bial protein output has increased (see Harrison et al., 1976;
Helmsley, 1975).

The point to be made here is that an increase in rumen volume
and a high retention time of feed in the rumen will tend to decrease the ratio of microbial cells to VFA produced in the rumen
(see Figure 13.2). Under these circumstances more energy substrates (VFA) and less microbial protein would be available from rumen
fermentation which may be an advantage to the draught animal. Any
manipulation that increases rumen outflow rate is therefore likely
to increase the efficiency of conversion of feed to products where
the animal is producing protein-containing end products (milk,
wool or meat).

13.3.2.5. <u>Microbial cell turnover in the rumen</u>

Nolan and Stachiw (1979) have indicated that in sheep on wheat
straw-based diets, some 50% of the microbial-N is recycled within
the rumen. This may result from death from starvation of the bacteria when the rumen supply of fermentable substrates is exhausted
(see Hespell, 1979). On straw diets this could be the major explanation for the apparent low output of protein from the rumen, however, bacteriophages (Adams et al., 1966; Hoogenraad et al., 1967)
or mycoplasma may also destroy bacteria and the ingestion of bacteria by protozoa (Coleman, 1975) is likely to account for a further proportion of the microbial cell turnover depending on the
population density of protozoa in the rumen.

The suggestion made here is that patterns of feed intake on
straw diets may cause microbial populations in the rumen to fluc-
tuate with an intermittent death rate and lysis of microbes, largely through lack of substrate. Immediately prior to feeding the
microbial population in the rumen of straw-fed ruminants may be
low, particularly those free in solution, which may result in a
slow colonization of feed materials (or lag phase) which will decrease the rate of fermentation. This could also lead to low appetite. This situation could be aggravated by the predation of protozoa on bacteria, the presence of protozoa reducing the population of colonizing bacteria in the fluid in the rumen.

13.3.3. <u>Post-ruminal digestion</u>

Digestion in the abomasum and small intestine of ruminants,
proceeds in the same way as in monogastric animals. Ørskov (1980)

has suggested that there may be an upper limit to the amounts of starch and protein that can be digested in the small intestine of ruminants. One constraint to the efficient digestion of starch in the intestine of ruminants was considered to be the low intestinal pH, associated with the feeding of high-concentrate diets (Wheeler et al., 1981). Addition of alkaline buffers to such feeds led to increases in starch digestibility.

The above problems are less likely to occur when concentrates are used to supplement straw-based diets. First, the pH throughout the digestive tract is higher on straw diets due to the buffering effect of: (1) greater salivary secretion, caused by extended eating and ruminating times; and (2) the relatively high levels of urea/ammonia required in such diets. Secondly, because of the long retention time of digesta in the rumen the quantities of dietary starch and protein reaching the intestine will almost never exceed 20% of the diet DM.

The digestion of starch (and/or sugar), proteins and even fats in the intestine avoids the heat and methane losses which occur when these nutrients are fermented in the rumen. Moreover, it is the most effective way of providing the required amounts of amino acids and glucogenic compounds needed by the high producing animal. Therefore, feeding strategies should always be designed so that dietary nutrients which can be digested post rumen do, in fact, reach that part of the digestive tract in a way that avoids or minimizes rumen fermentative losses.

Treatment of protein meals with heat, aldehydes or protective coatings with oils or other materials, is one approach. The other is to take maximum advantage of the oesophageal groove closure, through suckling, since this permits highly digestible nutrients to truly by-pass the rumen. The advantages of this technique can be seen in the relatively high growth rates of calves raised by restricted suckling (e.g. Alvarez et al., 1980; Fernandez et al., 1977) even though the amounts of milk consumed were small (<2 litres/day) and the remainder of the basal diet was of relatively low nutritive value (e.g. molasses/urea and forage).

13.4. BALANCE OF NUTRIENTS IN THE END-PRODUCTS OF DIGESTION. AMINO ACIDS, GLUCOSE AND GLUCOGENIC COMPOUNDS IN RELATION TO ENERGY EXPENDITURE

13.4.1. Nutrient requirements for production

13.4.1.1. The overall balance of nutrients

The principal functions of large ruminants in Asia and many parts of Africa, where straws and similar crop residues are the main energy resource, are workpower, growth, pregnancy and milk production, with beef very much as a by-product (residual or salvage value of the unproductive animals). In order to derive feeding systems for these different productive functions it must be appreciated that the nutrient requirements cannot simply be satisfied by maximizing rumen microbial activity (i.e. total rumen digestion).

First, even when rumen function is apparently maximized by providing the ruminant with feeds of high fermentability such as sugar/urea, often only moderate rates of growth and milk production are achieved. For high productivity it is necessary to provide additional supplements. These are usually concentrates that could also be used by simple-stomach animals including people. Therefore, to justify the inclusion of concentrates in ruminant diets requires that they should be utilized with extremely high efficiency. This means that such a supplement should maintain, or preferably increase, voluntary intake of the basal low cost diet and also provide essential nutrients which complement and balance the absorbed products of rumen fermentation. Secondly, account must be taken of the need to have an appropriate balance of the end-products of digestion. To understand this latter aspect, it is necessary to identify the specific nutrients that are needed for metabolism when the purposes of the feeding systems are to support such diverse functions as work, growth, pregnancy and milk production (see Figure 13.3.).

The greater part of the digested energy is undoubtedly needed for the animal to work both in the physical and metabolic sense. This can be defined as oxidation energy provided by nutrients which give rise to two-carbon fragments (C_2) catabolised in the Krebs or tricarboxylic acid cycle. However, small amounts of energy in the form of glucose, or as compounds which can substitute for glucose and are glucogenic, are needed to catalyze the tricarboxylic acid cycle (the mechanism by which oxidation energy is made available to the animals). There is an even larger requirement for these compounds to supply reduced (NADPH) cofactors needed for the elongation of the fatty acid chains in the synthesis of body fat and also milk fat.

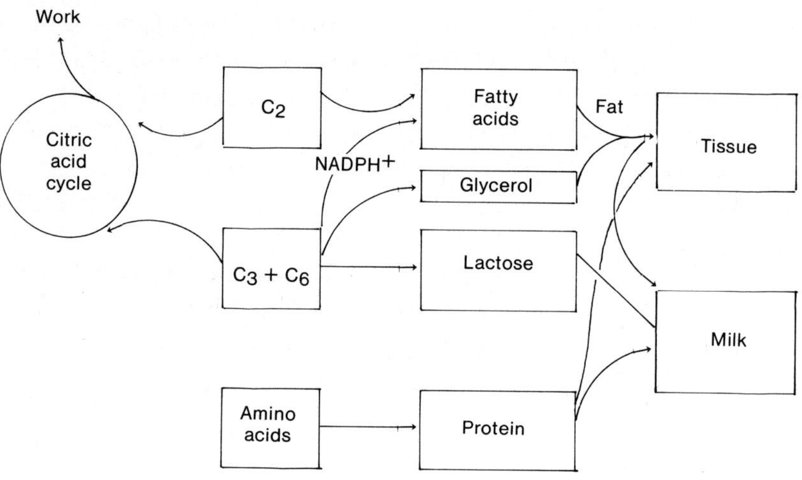

Figure 13.3. Ruminant requirements for major metabolites according to productive state

Glucose is the major energy nutrient for utilization in the brain and the central nervous system. It is also needed for macromolecule synthesis in tissue growth and to form glycerol in fats and lactose in milk. Glucogenic compounds may also spare the catabolism of some of the essential amino acids.

For simplicity, glucose and glucogenic compounds can be referred to collectively as energy for synthesis or C_6 energy (C_6-E).

The remaining major nutrients are the amino acids which are also needed for synthetic purposes, namely of body tissues and the proteins in milk.

All digested nutrients can give rise to oxidation energy but not all of them can supply glucogenic compounds and glucose or amino acids. The main glucogenic precursor in ruminants is propionic acid produced in rumen fermentation. Glucose can also arise from post-ruminal digestion of starch (by-pass or escape starch); certain amino acids are also glucogenic and can provide precursors for glucose synthesis following deamination (see for review Leng, 1970). Amino acids are mainly derived from microbes grown in the rumen and the subsequent digestion of these organisms in the intestine; the second important source is from the post-ruminal digestion of dietary protein (by-pass or escape proteins).

These three groups of nutrients - oxidation energy, glucogenic compounds and essential amino acids - are required in different proportions according to the productive state of the animal. The total requirement for energy is perhaps known with most precision and, for the purposes of ration formulation, can be described in terms of metabolizable energy, or simply digestible energy intake, since ME is usually calculated from DE. The capacity of feed ingredients to give rise to glucogenic compounds, glucose and to the amino acids will depend on a number of factors many of which interact with each other (see Figure 13.4). This makes it difficult to establish a simple additive system for ration formulation.

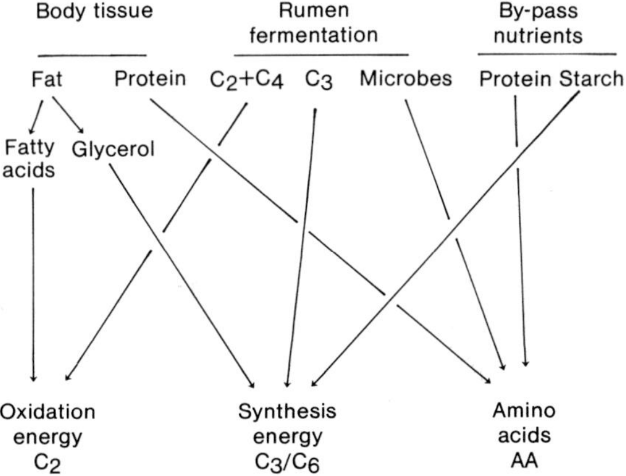

Figure 13.4. Origin of principal metabolites in ruminants

13.4.1.2. <u>Amino acids</u>

The supply of amino acids of microbial origin to the animal will depend on the intake of fermentable feed and the efficiency of net microbial growth in fermentation. The latter is controlled by a number of variables including rumen dilution rate, the presence of protozoa, the supply of precursors including fermentable N and minerals. For all practical purposes the amount of microbial protein derived from fermentable energy can be predicted on the basis of 3 g nitrogen per 100 g of digestible organic matter consumed. The other source of amino acids to the animal which becomes increasingly important as the productive rate increases (especially milk production) is the protein which escapes fermentation and contributes amino acids directly following digestion in the small intestine. The capacity of a protein to by-pass the rumen fermentation is negatively related to its solubility or degradation rate in rumen fluid in vitro. The capacity of a protein meal to escape the fermentation is not only a function of the potential rate of breakdown of the protein by micro-organisms but it is also influenced by the dilution rate of rumen digestion as obviously the greater the flow of fluid and particles from the rumen, the greater the likelihood that they will escape rumen fermentation.

13.4.1.3. <u>Glucose and glucogenic compounds</u>

The dietary nutrients which give rise to glucose and glucogenic compounds are those that lead to high rates of production of propionic acid in rumen fermentation. In diets containing starch glucose may become available directly if the starch has a capacity to escape to the lower tract. The availability of glucogenic compounds is also enhanced by the amount of protein reaching the intestine as a number of amino acids provide three carbon units capable of being converted to glucose. Propionic acid production in rumen fermentation can be altered by chemical additives, such as monensin, which is widely used for this purpose in feedlot rations in Europe and North America.

Reviews of glucose metabolism in ruminants are available (Leng, 1970) and this topic will be discussed here only briefly. It is assumed here that requirements and synthesis rates of glucose are closely correlated, since any unneeded synthesis of glucose would be energetically wasteful since it is expensive in terms of requirements for ATP. Synthesis of glucose in ruminants is related to digestible energy intake (Judson and Leng, 1968; see also Herbein et al., 1978) stage of growth, stage of pregnancy and lactation (Figure 13.5).

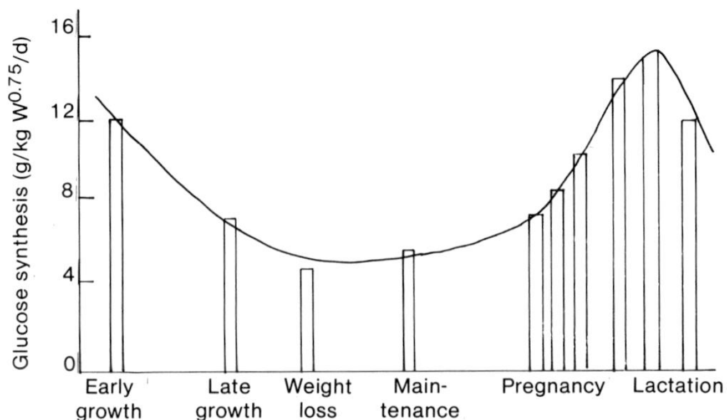

Figure 13.5. Rate of glucose synthesis according to productive state in sheep (from Leng et al., 1977)

When amino acid requirements are high, glucose synthesis rates (and therefore apparent requirements) are high (Figures 13.5 and 13.6). The pattern of apparent requirements for glucose follows closely that for amino acids suggesting that part of the requirement for amino acids may be a requirement for glucogenic compounds.

During growth and lactation there may be competing needs for amino acids for glucose synthesis and for protein deposition. The important point here is that in growing, pregnant or lactating ruminants there is a high demand for amino acids for protein deposition, and for both amino acids and propionate for glucose synthesis. The central importance of glucose is indicated by the fact that about 20% of energy of the digestible nutrients available to mature sheep is synthesized into and pass through the glucose pool in blood (Judson and Leng, 1968). Similarly in growing steers and dry or lactating cows with highly variable feed intakes around 23% of the digestible energy is synthesized into glucose (see Figure 13.7) (Herbein et al., 1978).

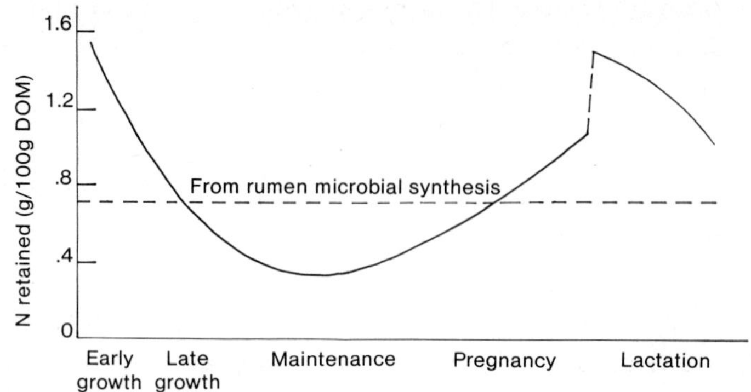

Figure 13.6. Requirements for amino acids in sheep according to productive state (from Ørskov, 1970)

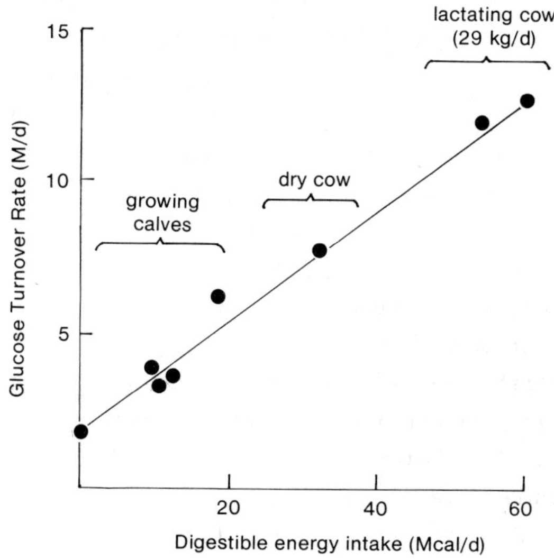

Figure 13.7. Glucose turnover rate (Y) in relation to the digestible energy intake (X) where Y = .228X + 1.12 (after Herbein et al., 1978)

13.4.2. Is productive rate and feed conversion efficiency likely to be constrained by the availability of glucose and glucogenic compounds?

The expression of nutrient requirements for ruminants in terms of the balance of ME and absorbed amino acids is widely accepted (e.g. ARC, 1980). The need to manipulate/supplement diets in order to ensure adequate amounts of glucose and/or glucogenic compounds in the digestion end-products was first proposed by Leng and Preston (1976) as an essential step in order to obtain high ruminant productivity from low-protein tropical feed resources composed of mixtures of soluble sugars and insoluble cell-walls. As this thesis is widely disputed at the present time (see Ørskov, 1980) it is necessary at this stage to summarize the evidence which we consider justifies the need to take account of the glucogenic potential of a diet according to the productive state of the animal:

Thomson (1978) found that the efficiency of utilization of ME for tissue synthesis was higher for concentrate/forage combinations of maize and clover than for barley and rye grass even though the metabolizability of the DM (ME/DM) was the same on all diets. One explanation is the proportionately greater post-ruminal digestion on maize/clover compared with barley/grass.

Work in Australia showed that pangola grass was used more efficienty (28%) than setaria grass (17%) for tissue synthesis in sheep even although both had the same digestibility and were fed at the same rates (Tudor and Minson, 1982); the authors mention the superior glucogenic potential of pangola as one possible explanation for the difference.

Rice polishings (with a large proportion of broken rice grains) was better than cassava root meal for supplementing sugar cane. The starch in rice grains in rice polishings escapes rumen fermentation almost quantitatively (Elliot et al., 1978) whereas the starch of cassava root meal is fermented rapidly in the rumen (Santana and Hovell, 1979). Glucose entry rates were higher and production higher when rice polishings rather than cassava root meal was the supplement in sugar cane diets (Ravelo et al., 1978).

Supplementary energy as maize grain (with good rumen escape characteristics) improved feed conversion efficiency in cattle fed sugar cane whereas the same amount of molasses energy (completely fermented in the rumen) depressed feed conversion efficiency (Donefer E., cited by Pigden, 1972).

Cattle nourished by rumen infusion of VFA and abomasal infusion of casein increased their nitrogen retention as the proportion of propionic acid in the infused VFA was increased (see Figure 13.8) (Ørskov et al., 1979).

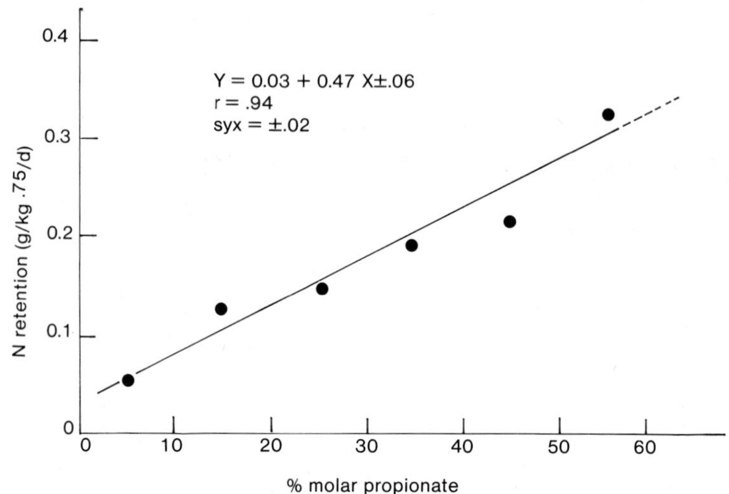

Figure 13.8. Relationship between molar proportion of propionic acid in the mixture of VFA infused into the rumen of sheep and the N retention (diet comprised only VFA infused into the rumen at maintenance level of feeding and casein infused into the abomasum) (Adapted from Ørskov et al., 1979)

The most convincing evidence concerning the need for glucogenic compounds in the end-products of digestion was provided by Tyrrell et al. (1979) (Figure 13.9). Acetic acid infused into the rumen of animals receiving a basal diet of low glucogenic potential (alfalfa hay) was less efficiently utilized for tissue synthesis than when the infusion was given with a basal diet of high potential for providing glucose (60% maize grain and only 40% hay).

The superior nutritive value of propionic acid compared with acetic acid, observed in the original work of Armstrong and Blaxter (1957a, b) and Armstrong et al. (1958) and the absence of differences between these two fatty acids in the experiments of Ørskov and Allen (1962) can also be explained in terms of the glucogenic potential of the basal diet. The diet used by Armstrong et

al. (1958) was dried grass whereas Ørskov and Allen (1962) gave
the different VFA mixtures to animals fed mainly on barley grain.

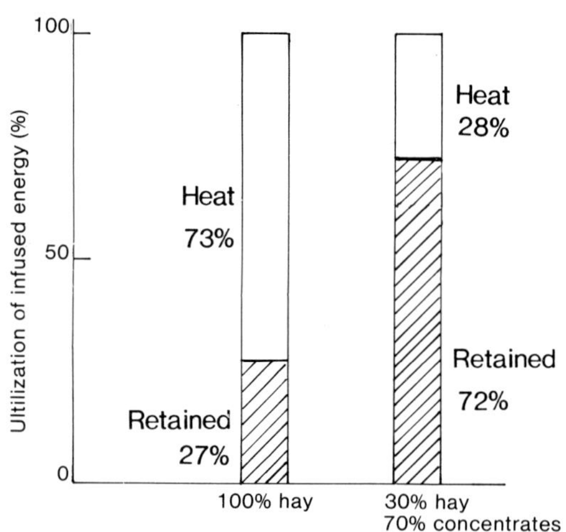

Figure 13.9. Effect of the basal diet on the efficiency of utilization of infused acetic acid in cattle (adapted from
Tyrrell et al., 1979)

13.4.3. Tropical versus temperate feed resources

The reason for the greater awareness in the tropics of the role
of glucogenic compounds in the end-products of digestion of ruminant diets is easily explained. In almost all temperate countries,
the diets for high producing ruminant animals contain at least 50%
of cereal grains. Such rations give rise to relatively large quantities of propionic acid in the rumen fermentation and there can
be considerable escape of the starch for post ruminal digestion to
glucose. Furthermore, most of the dietary nitrogen is in the form
of true protein, and absorbed amino acids can spare glucose. There
is thus rarely a deficiency of glucogenic compounds in the end products of digestion. In contrast, tropical feed resources of high
digestibility usually are of low glucogenic potential since they
frequently contain carbohydrates which are totally fermented, and
give rise to VFA mixtures high in butyric and acetic rather than
propionic acid (Marty and Preston, 1970).

Thus with feeds of temperate origin, increases in nutritive value are brought about by an increase in the proportion of cereal grain and oilseed meal in the diet and this gives rise to concommitant increases in the concentration of glucogenic compounds in the end-products of digestion. In most tropical countries, however, cereal grains and protein-rich meals are in short supply and competed for strongly for human and monogastric nutrition. They are available for ruminant feeding in only small amounts which should be used to catalyze feed intake. Improvements in digestibility of feeds in such situations are brought about either by the inclusion of sugar-rich by-products (e.g. molasses) or by treatment with alkali/acid to hydrolyse resistant lignocellulose. Improvements in digestibility brought about by raising the amount of sugar in the diet, or by hydrolysing the cell walls in crop residues do not lead to an increase in the glucogenic compounds in the end-products of digestion, since neither of these strategies increases the likelihood of rumen escape of digestible carbohydrate. In addition rumen fermentation of such modified feeds is characterized by relatively low proportions of propionate at the expense of butyrate (for sugar-containing feeds) or acetate (for alkali or acid-treated crop residues (see Table 13.1).

Because of the above phenomena it can be extremely misleading to estimate the nutritive value of crop residues and agro-industrial feed resources on the basis of the concentration of digestible energy in the final mixed feed. The poor predictability of livestock performance on tropical feed resources, using standards derived in temperate countries, is mainly due to the situation described above.

Table 13.1. Effect of alkali (NaOH) treatment of rice straw on VFA proportion in rumen fluid of cattle (Kayouli, 1980)

Straw	VFA molar %		
	C_2	C_3	C_4
Untreated	70	20	10
NaOH-treated	76	14	10
NaOH-treated + molasses/urea	65	20	15

13.5. PREDICTION OF NUTRITIVE VALUE AND NUTRIENT REQUIREMENTS

13.5.1. <u>Nutritive value</u>

In order to predict more accurately the feeding value of a particular combination of feed ingredients, it is necessary to introduce constraints (or factors) in order to take account of the likely proportions in the end-products of digestion of the two essential groups of nutrients, namely amino acids and glucose and its precursors, since this is what determines the real productive capacity of the diet.

There is no adequate analytical means of assessing the potential value of feed ingredients as a source of amino acids, glucose and glucogenic precursors other than from knowledge of (1) their composition, in terms of starch and protein; (2) the rate with which they are fermented by micro-organisms; and (3) the rumen dilution rate. The latter two criteria will determine the amount of the protein and/or starch which escapes the rumen fermentation.

The cheapest feed ingredients for ruminants in the majority of developing countries are usually in the form of crop residues and to a lesser extent as agroindustrial by-products which constitute the energy sources, and non-protein nitrogen, in the form of urea and ammonia. The expensive feeds are the protein meals, derived from oilseed residues and the by-products from processing of animals and fish, and the starch-containing cereal grains which are also the staples of human and monogastric animal nutrition. In general terms, therefore, oxidative energy and fermentable nitrogen are relatively inexpensive ingredients, while the sources of amino acids and glucogenic compounds (the protein meals, cereal grains and cereal by-products) are very expensive.

Since it is the combination of fermentable energy and fermentable nitrogen which gives rise to amino acids in the form of microbial protein, and as feeding level is associated positively with rumen dilution rate, it is generally desirable to supply fermentable energy on an ad libitum basis. It is, therefore, counter-indicated to restrict the basal diet. The fermentable nitrogen requirement can be predicted on the basis that 100 g of fermentable carbohydrate will give rise to microbial protein containing 3 g of nitrogen. Of course, it is not always necessary to provide this total amount as NPN, since a proportion of the feed protein will always be fermented to ammonia and in addition considerable blood urea-N may be secreted into the rumen. These processes reduce the amount of non-protein, fermentable nitrogen needed.

The potential of the final mixed diet to satisfy the requirements of the animal for amino acids and glucogenic precursors depends principally on its content of protein- and starch-containing ingredients capable of escaping rumen fermentation and which are able to be digested in and the products absorbed from the intestine.

The extent to which the protein in a particular supplement escapes the rumen fermentation is partly a function of its rate of degradation (solubility) in the rumen, but is likely to be influenced greatly by the rate of flow of fluid and small particles out of the rumen. This latter characteristic will be influenced by processing of the feed (by physical or chemical means), the presence of some good quality forage, the amount of protein reaching the duodenum and external factors such as temperature and excercise/ work.

The same factors will influence the supply of glucose and glucogenic precursors in terms of the likely by-pass of starch and other glucogenic precursors in the duodenum. However, the nature of rumen fermentation will have a major influence in terms of supply of propionic acid.

13.5.2. Requirements for amino acids and glucogenic compounds according to the productive state of the animal

As stated earlier, there is insufficient information available to permit the precise quantification of the proportions of the different nutrients required at the metabolic level for different productive states. Nevertheless, an approximation of the needs can be attempted, based on a scheme attaching relative priorities to the groups of nutrients together with knowledge of the physiological and biochemical processes underlying the expression of the particular productive state. Such an indicative scheme is set out in Table 13.2.

Work primarily involves a need for oxidation energy with minimum requirements for both glucogenic compounds (almost exclusively for the TCA cycle) and for amino acids (to repair the wear and tear of tissues and replace secretions). Maintenance alone obviously requires less energy expenditure. Late growth and gestation imply a relatively small increase in essential amino acid requirements. The requirements for growth in the young animal approaches that in animals with high milk production, in terms of needs for amino acids and C_3-C_6 relative to C_2 energy. Lower levels of milk production are obviously less demanding in terms of essential nutrients.

Table 13.2. Relative priorities attached to the principal metabolite groups according to the productive state of the animal

| | Relative priorities | | |
| | Oxidation energy | Synthesis energy | Amino acids |
Function	(C_2)	(C_3:C_6)	
Work	xxxxx	x	x
Maintenance	xxxx	x	x
Late growth and gestation	xxxx	x	xx
Early growth	xxxxx	xx	xxx
High milk	xxxxx	xxx	xxxx
Medium milk	xxxxx	xx	xxx
Low milk	xxxx	x	xx

These priority ratings can be further simplified by assigning a ranking for each metabolite group according to the productive state being considered. Synthesis energy (C_6-E) means the requirements for glucose and glucogenic compounds while the needs for amino acids are described in terms of by-pass protein (B-P) (see Table 13.3).

Table 13.3. Arbitrary rankings for synthesis energy (C_6-E) and by-pass protein (B-P) according to productive needs

| | Rating (on scale 0 to 5) | |
Function	C_6-E	B-P
Work	0.1	0.2
Maintenance/gestation	0.1	0.5
Beef fattening	0.5	1.5
Low milk production	1.0	1.0
Medium milk production	2.0	2.0

13.6. SUPPLEMENTATION OF STRAW-BASED DIETS

13.6.1. Formulation of complete rations

The capacity of straw-based diets to provide oxidative energy can be estimated relatively precisely from the digestion coefficients determined by a number of means; (i.e. by total collection of faeces from animals fed ad libitum; from the 24-hour DM loss from nylon bags in the rumen; or from some appropriate in vitro fermentation test). By contrast, it is almost impossible at present to predict the capacity of straw-based diets to provide amino acids and/or glucogenic compounds at the level required for efficient metabolism in the animals.

Two schemes are proposed in order to devise suitable supplementation programmes for achieving particular rates of productivity. The first scheme is for the computerised formulation of least-cost complete rations, with straws or similar fibrous residues as the principle energy source. In this programme the available basic feed resources and supplements are rated according to their potential to give rise to amino acids and glucogenic compounds in the end - products of digestion (Table 13.4). Similar constraints are then applied to the nutrient requirements according to the productive state expected of the animal (Table 13.3). These two sets of data are then incorporated in a computer model for least-cost ration formulation together with the traditional standards for the anticipated level of production for digestible energy, crude protein, calcium and phosphorus etc.

13.6.2. Principles of supplementation

The second more empirical scheme, which is probably more appropriate for the conditions of most developing countries, is to select the basal energy resource according to availability, potential fermentability and price and then to provide supplementary nutrients in accordance with their relative priorities (Table 13.5) and costs. To optimise economic returns some information is required relating input/response to such supplements.

Table 13.4. Ratings of some feed ingredients according to their potential to supply by-pass protein and glucogenic compounds

	By-pass protein (B-P)	Glucogenic compounds (C_6-E)
Sugar cane bagasse	0	0
Straws (rice, wheat)	0	0
Elephant grass	0	0
Molasses (sugar cane)	0	0
Sugar cane juice[1]	1	2
Sorghum grain	1	4
Maize grain	1	5
Alfalfa/Berseem hay	2	1
Leucaena leaf	2	1
Wheat bran	3	3
Maize gluten meal	4	4
Soybean meal	4	4
Meat meal	4	1
Fish meal	5	2
Cottonseed meal	5	4
Rice polishings	4	5

[1]Contains no by-pass protein but it appears to support highly efficient rumen microbial protein production

Table 13.5. Priorities for nutritional supplements on straw-based diets

1. High concentration of fermentable carbohydrates
2. Fermentable nitrogen (3 g N for every 100 g fermentable CHO)
3. Adequate rumen ecosystem
    i)   Roughage characteristics
    ii)   Micro-nutrients
    iii) Control of protozoal activity
4. By-pass nutrients
    i)   Protein
    ii)   Energy
5. Balance of end-products of digestion. Supply of:
    i)   Amino acids
    ii)   Glucose and glucogenic compounds
          in relation to total oxidative energy, and animal needs

In this system, the first step is to provide a source of non-protein nitrogen (usually urea or ammonia) to raise the level of fermentable nitrogen to a minimum of 3% of the digestible organic matter. It is desirable that this is done in a way which will ensure an almost continous supply of ammonia-nitrogen to the rumen micro-organisms.

The second priority is to give green forage, preferably legume, up to a maximum of about 0.7% (DM basis) of liveweight (about 25% of the diet.

Finally, a by-product of oilseed crushing or cereal processing, or an animal by-product meal should be given in amounts not to exceed 20% of the total diet DM. The 20% limit is to prevent depression/substitution of the digestible energy of the basal diet (see Chapter 11). Lesser amounts may be more economical, and it is imperative that response trials be carried out so that the amount of supplement can be related to the rate of animal productivity. This optimum level (in economic rather than biological terms) and the degree of response to the supplement, will depend upon the relative nutritive value of the basal diet. For example (Table 13.6), when the basal diet was sugar cane juice then provision of a by-pass protein (fishmeal) at 8% of the diet DM increased growth rates by only 20% and feed conversion efficiency not at all (Duarte et al., 1983). By contrast, on an NH_3-treated straw diet 150 g/d of fishmeal (5% of diet DM) increased growth rate by 300% and feed conversion efficiency by 500% (Saadullah et al., 1982a). Responses to fishmeal on a basal diet of untreated straw (70% barley straw: 30% barley grain) were similar (Smith et al., 1980).

Table 13.6. Responses of growing cattle to by-pass protein from fish meal according to the nature of the basal diet

Basal diet	Initial live- weight, kg	Fish meal in diet DM, %	Liveweight gain of controls, g/d	Improvement due to fish meal, %	
				Growth	Feed conversion
Sugarcane[1] juice/urea	180	5	800	20	0
Straw (70)[2]	282	4.5	212	260	250
barley (30)	282	8.5	175	370	350
Urea-treated straw[3]	85	5	100	300	500

[1]Duarte et al. (1983)   [2]Smith et al. (1980)
[3]Saadullah et al. (1982a)

13.7. CASE STUDIES OF FEEDING SYSTEMS BASED ON STRAW AND FIBROUS CROP RESIDUES

13.7.1. Draught power

Draught animals seem to perform well on a wide range of basal diets and to be little affected by either the digestibility of the diet or its nitrogen content. Thus, working bullocks in Bangladesh ate greater quantities of alkali-treated rice straw (ensiled with urea), than their pair mates given untreated straw but there were no apparent differences in work output or bodyweight change (Dolberg et al., 1981). Further evidence for considering that wor- king animals require minimum supplementation of an energy-rich diet is the finding that young horses which were exercised grew at the same rate on a low-protein diet as un-exercised horses fed a high-protein diet (Orton, Hume and Leng, unpublished data). In Pakistan, bullocks driving a press, crushing 400 kg sugar cane per hour, were apparently able to work continously for a 6 month peri- iod in successive shifts of 3 hours work followed by 3 hours for feeding and rest, on an exclusive diet of sugar cane tops (Preston, 1982, personal observation). In contrast, a diet of derinded sugar cane stalk, of higher digestibility and supplemen- ted with urea and minerals, barely supported maintenance in "gro- wing" steers in Mexico (Preston et al., 1976). Obviously the inf- luence of excercise on nutrient requirements is a subject requi- ring much more research.

13.7.2. Growth

13.7.2.1. Good quality forage

It has already been pointed out (13.3.2.3) that small quanti- ties of high quality forage appear to have a beneficial effect on rumen function on diets based on crop residues. The results in Table 13.7 indicate that such improvements in the rumen ecosystem carry through to improved animal performance. *Azolla pinnata*, a water plant which grows symbiotically with the N-fixing alga *Anaboena azolla* appeared to be even more effective than a mixed con- centrate meal in promoting liveweight gain on a diet of wheat straw and sugar cane tops.

400

Alfalfa hay, added at 13% of a basal diet of maize cobs increased gain by 100% on the untreated cobs and by 50% on cobs treated with ammonia (Cook et al., 1982).

Table 13.7. Effect of a green forage (*Azolla pinnata*) in a straw-based diet given to growing cattle (Singh, 1980)

	Supplement	
	Concentrate[1]	Azolla[2]
Liveweight gain, g/d	140	330
Feed intake, kg DM/d:		
Wheat straw + sugar cane tops	2.1	2.1
Supplement	1.5	0.9
Total	3.6	3.0
Feed conversion	26	9
OMD, %	45	59

[1]Contained: 65% maize meal, 15% rice bran, 16% groundnut cake, 4% minerals

[2]2.8% N in DM; 81% digestibility of DM

13.7.2.2. By-pass nutrients and glucogenic compounds

Results from Bangladesh and Sri Lanka, of recent experiments designed to assess growth responses in cattle to by-pass supplements, using both untreated and alkali (mainly wet urea-process) - treated rice straw are summarized in Table 13.8. The data in Table 13.9 are from an experiment in tropical Australia using a native grass hay of very low digestibility ($<45\%$) and N content ($<0.8\%$ in DM). Figure 13.10 relates to one of the few "response" trials where different levels of the by-pass supplement were given to sheep on alkali- treated straw diet; similar response relationships were obtained in sheep on sugar-chaff diets by Bird and Leng (1978).

There were consistent responses to by-pass protein supplements (mainly oilseed meals) but the effects were more pronounced when the basal straw diet was alkali-treated and/or supplemented with urea (Abidin and Kempton, 1981). Where different levels of supplements were used, the response was distinctly curvilinear with the major effect corresponding to a supplement concentration in the diet of about 25% and no apparent economic benefit from greater quantities. Table 13.10 contains data from an experiment in which cattle were fed on sugar cane bagasse which was steam-treated

Table 13.8. Responses of young growing cattle fed rice straw with or without alkali (NaOH) or urea treatment and by-pass supplement

Authors	Straw treatment	Straw DM intake (kg/d)	By-pass supplements (g/d)	Initial live-weight (kg)	Live-weight gain (g/d)	Feed conversion
1	Untreated	1.7	None	54	35	50
	Untreated + urea	1.7	None	58	75	24
	Urea-treated	1.9	None	58	110	18
	NaOH-treated	1.6	None	56	120	14
2	Untreated + urea	2.2	200 oilcake 300 rice bran	64	207	13.5
	Urea-treated	2.4	"	63	297	10.3
	NaOH-treated	2.1	"	62	279	9.7
3	Untreated	2.9	180 oilcake 180 rice bran	121	124	28
	Urea-treated	3.0		120	305	12
	Urea-treated	3.7	"	122	340	14
4	Urea-treated	3.8	None	158	84	47
	Urea-treated	3.8	250 oilcake	158	371	11
	Urea-treated	3.8	400 oilcake	158	373	11
	Urea-treated	3.8	600 oilcake	159	508	9
	Urea-treated	2.8	None	90	143	21
	Urea-treated	2.9	150 fish meal	90	357	9
5	Untreated	3.4	260 oilcake	105	210	18
	Urea-treated	3.6	260 oilcake	105	300	13
6	Untreated		500 concentrate		73	
	Urea-treated		500 concentrate		346	
	Urea-treated		None		333	
	Urea-treated		180 fish meal		428	
	Urea-treated		1100 rice bran		512	
	Urea-treated		540 coconut cake		494	
7	Untreated	2.1	500 concentrate	165	73	53
	Urea-treated	2.8	500 concentrate	167	346	13

1   Saadullah et al. (1981)

2-4   Davis, C.H. (1982), unpublished data

5   Saadullah et al. (1982b)

6   Perdok, H.B. (1982), unpublished data

7   Perdok et al. (1982)

(acid hydrolysis) (Wong et al., 1974) and supplemented with ingredients rich in by-pass protein (fish meal), or with a high potential to supply glucogenic compounds (maize meal). Bodyweight gain of cattle were better with maize than with fish meal. However, the best performance was with the two supplements combined, when a loss of 170 g/d (unsupplemented diet) was converted into a gain of 300 g/d.

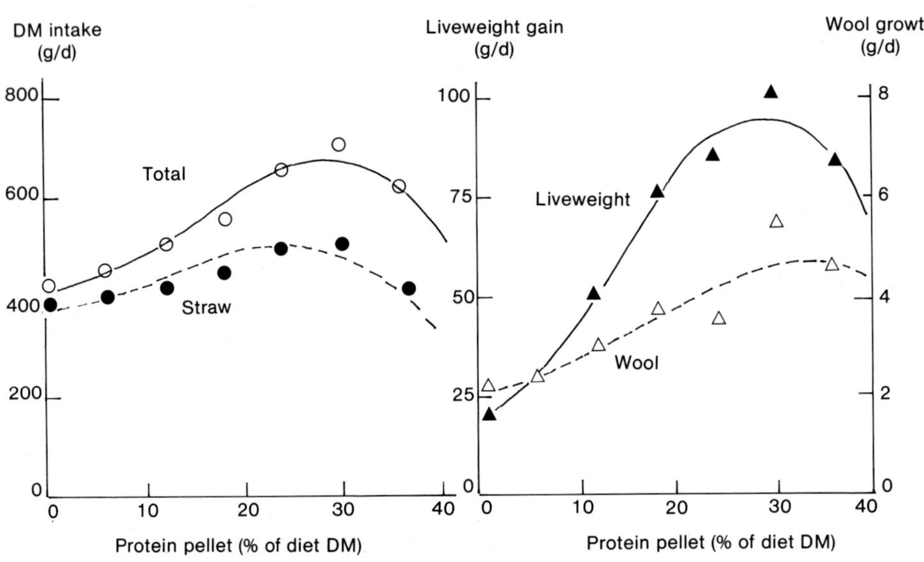

Figure 13.10. Feed intake, wool growth and liveweight gain of lambs given barley straw/urea ad libitum and a by-pass protein pellet (cottonseed meal 80, meat meal 8, soybean meal 10, minerals 2) (Adapted from Abidin and Kempton, 1981)

Table 13.9. Dry matter intake and live weight change of cattle (170 kg W) fed poor quality spear grass hay (*Heteropogon contortus*) and different sources of nitrogen (Lindsay and Loxton, 1981)

	Dry matter, kg/d	Live weight change, g/d
Hay	2.26	-0.41
Hay + urea	3.01	-0.32
Hay + protected casein	3.34	-0.21
Hay + protected cottonseed meal	3.72	+0.11
Hay + urea + cottonseed meal	4.43	+0.22

Table 13.10. Intake, liveweight change and feed conversion of cattle fed steam-treated sugar cane bagasse (200°C for 10 minutes) and minerals alone (control) or with fish meal and maize meal, singly or in combination (Naidoo et al., 1977)

	Control	Fish meal, 250 g/d	Maize meal, 1000 g/d	Fish meal +maize (250+1000)
Bagasse DM, kg/d	4.0	4.4	4.5	4.5
Live-weight change, g/d	-170	+80	+160	+330
Feed (DM) conversion		34	23	12

13.7.3. Milk production

13.7.3.1. Effect of alkali treatment of straw (by ensiling with urea)

In view of the nutritional limitations of most crop residues set by low digestibiity, and the high demands for amino acids and glucogenic compounds for milk production, it is not surprising that there is little published information on this subject. Table 13.11 summarizes some recent findings from Sri Lanka and Bangladesh where rice straw, without or with alkali treatment (by urea ensiling) was the basis of the diet for zebu and buffalo cows. There were significant responses to the urea treatment of the straw, both in milk yield and in liveweight change. Urea treatment of the straw obviously spared the need for by-pass nutrients since the amount of milk produced per unit concentrate fed was increased

by an average of 53%, while the proportion of concentrates in the
diet fell from an average of 14% to only 9% of the total diet DM
(Table 13.12).

Table 13.11. Effects on cattle and buffaloes of feeding untreated
or urea-treated rice straw

Authors	Animals	Milk yield, kg/d		Body-weight change, g/d	
		Untreated	Treated	Untreated	Treated
1	Cows[4]	1.1	2.3	-149	+109
2	Cows[5]	2.4(4.6)[7]	3.4(4.9)	-266	+93
3	Buffaloes[6]	2.4(6.8)	3.2(7.6)	-17	+93

1   Nurazzamal Khan and Davis (1981)
2-3 Perdok et al. (1983)
[4]All cows had 500 g/d rice bran and 200 g oil cake/kg of milk
[5]All cows had 1.5 kg/d of concentrates
[6]All buffaloes had 1.0 kg/d of concentrates
[7]Figures in brackets are milk fat percentages

Table 13.12. Effect of urea treatment of straw on concentrate
proportion of diet and on milk produced per unit
concentrate fed

	Untreated	Urea-treated	Author[1]
Concentrate as % diet DM:			
Zebu x Friesian crosses	11	8	1
Zebu	21	14	2
Buffalo	11	5	3
Milk produced, kg/kg concentrate:			
Zebu x Friesian crosses	1.71	2.7	1
Zebu	2.8	4.2	2
Buffalo	4.0	6.0	3

[1]See Table 13.11 for explanation

13.7.3.2. Effect of high quality forage

The effect of supplementary green forage, in the form of the
leaves of the tree forages gliricidia and leucaena, was investigated in trials carried out by Perdok et al. (1982a, 1982b, unpub-

lished data). On untreated straw, the feeding of gliricidia forage at approximately 15% of the dietary DM, increased milk yield by 22%; on urea-treated straw the level was 10% of the DM and the increase in milk yield was 14%. The average responses to gliricidia (1 kg DM/d) supplementation in the second experiment were 10% and 3% in the absence or presence of supplementary coconut cake (1 kg/d). Comparable data for supplementation with the same amount of leucaena were 9% and 9% respectively.

13.7.3.3. Effect of by-pass nutrients

Only one trial was identified where the design permitted the estimation of response to a protein supplement (Table 13.13). Milk yield was increased by 23%, fat percentage by 8% and liveweight gain by 110%, when 1000 g/d of coconut cake were fed to lactating buffaloes in Sri Lanka. The diet was urea-ensiled rice straw and minerals.

Table 13.13. Effects on milking buffaloes of supplementing urea-treated rice straw with coconut cake (Perdok et al., 1982, unpublished data)

	Coconut cake, g/day	
	0	1000
Milk yield, kg/day	2.6	3.2
Fat, %	9.1	9.8
Body-weight change, g/day	+99	+211

13.8. CONCLUSIONS

In this chapter we have attempted to test the hypothesis that it is "knowledge of the principles of digestive physiology rather than of feeding standards" that is the best guide for the development of efficient systems of supplementation of ruminant diets based on crop residues.

13.8.1. Growth and milk production

The information that has been reviewed indicates that in order to maximize rate of animal production and efficiency of straw/-residue utilization, the following factors must be considered in

406

order of importance:

  i) Maximization of intake of low cost, available fermentable
      carbohydrate

 ii) Addition of a source of non-protein nitrogen to raise the
      nitrogen content of the basal diet to 30 g per 1000 g of
      digestible dry matter

iii) A source of good quality green forage (preferably legumi-
      nous) at between 10 and 20% of the diet dry matter

 iv) A source of highly digestible by-pass nutrients (to provide
      both amino acids and glucogenic compounds at the sites of
      metabolism) at levels in the range of 10% to 20% of the
      total diet dry matter; the actual level being determined
      according to the cost/response relationships.

Two experiments carried out with agro-industrial residues composed almost entirely of soluble sugars and with extremely low
contents of nitrogen and protein, corroborate the relevance of the
above principles (see Figure 13.11 and 13.12), and indicate that
the best results from these basic feeds are achieved when all the
supplements are combined into one feeding system.

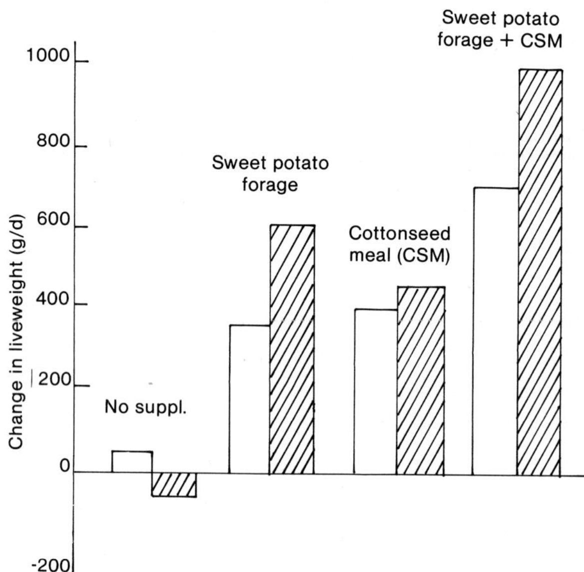

Figure 13.11. Liveweight gain of cattle fed a basal diet of derinded
       sugar cane and minerals with low or high urea levels
       and supplements of sweet potato forage (25% of diet
       DM) and/or cottonseed meal (12% of diet DM) (from
       Meyreles et al., 1979)

Thus, in the case of the derined sugarcane, animal response was increased from 0 to 500 g/day liveweight gain by a supplement of urea and either high quality forage or cottonseed meal. Nevertheless, performance was further increased by another 100% to reach 1 kg/day when all the supplements were combined in the same ration. It is appropriate to note that urea only gave a response when both the good quality forage and/or by-pass nutrients were also provided.

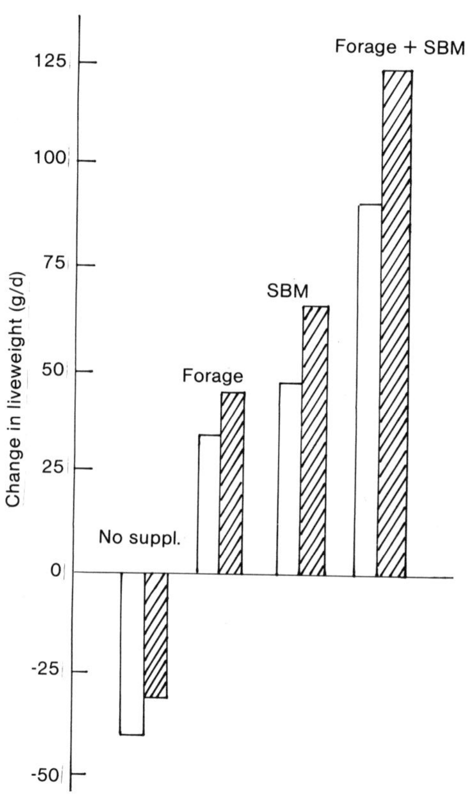

Figure 13.12. Liveweight gain of lambs fed a basal diet of sisal (henequen) pulp/urea, with and without a complete mineral mixture, and supplements of good quality forage (*Brosimum alicastrum*) and/or soybean meal (SBM) (from Rodriguez, 1982)

The second trial (Figure 13.12) was with ensiled henequen (sisal pulp).Although this feed resource is (1) deficient in phosphorus and trace elements and highly imbalanced with respect to calcium; and 2) provokes a metabolic acidosis, there was no response in animal performance either from neutralizing (Harrison, 1982, personal communication) or from providing an appropriate mineral supplement (see Figure 13.12). However, limited amounts of either a good quality green forage (*Brosimum alicastrum*) or a by-pass supplement (soybean meal) changed the liveweight loss in lambs of -40 g/d to a gain of 55 g/d. When all supplements were combined, the growth rate was increased to 125 g/d.

As in the case of the urea in the previous trial, there was no response (this time to minerals known to be deficient in the basal diet) until both rumen function and by-pass nutrient status were provided for by the combination of green forage and soybean meal.

13.8.2. <u>Draught animal power</u>

This area has received little attention during the past decade particularly with regard to the development of appropriate feeding systems.

From work carried out with horses on low protein diets, and from observations in the field in many tropical countries, the hypothesis is advanced: (1) that it may be advantageous to deliberately provide less than the optimal levels of fermentable nitrogen in the diets of working animals in order to maximize total energy release in the rumen in the form of VFA; and (2) that work/exercise accelerates the flow of nutrients along the digestive tract, in a similar way to the effects brought about by low ambient temperatures (Kennedy et al., 1982). This compensates for the normally slow rate of rumen degradability of the low quality crop residues which traditionally form the sole diet of working animals and permits a higher voluntary feed intake.

This field of research merits much more detailed research than it has received both with respect to the digestive physiology of working animals, and the study of the effects of increasing nutritive value either by treatment of the basic feed resource or through appropriate supplementation.

13.9. REFERENCES

Abidin, Z., Kempton, T.J., 1981. Effects of treatment of barley straw with anhydrous ammonia and supplementation with heat treated protein meals on feed intake and liveweight performance of growing lambs. Anim. Feed Sci. Technol. 6: 145-155.

Adams, J.C., Gazaway, J.A., Brailsford, M.D., Hartman, P.A. and Jacobson, J.L., 1966. Isolation of bacteriophages from the bovine rumen. Experientia 22: 717-718.

Alvarez, F.J., Sauecdo, G., Arriaga, A and Preston, T.R., 1980. Effect on milk production and calf performance of milking cross bred European/Zebu cattle in the absence or presence of the calf, and of rearing their calves artificially. Trop. Anim. Prod. 5: 25-37.

ARC, 1980. Nutrient Requirements of Ruminant Livestock. Agricultural Research Council: Commonwealth Agricultural Bureaux, 351 pp.

Armstrong, D.G. and Blaxter, K.L., 1957a. The heat increment of steam volatile fatty acids in fasting sheep. Br. J. Nutr. 11: 247-272.

Armstrong, D.G. and Blaxter, K.L., 1957b. The heat increment of mixtures of steam volatile fatty acids in fasting sheep. Br. J. Nutr. 11: 392-408.

Armstrong, D.G., Blaxter, K.L., Graham, N.McC. and Wainman, T.W., 1958. The utilisation of the energy of two mixtures of steam volatile fatty acids by fattening sheep. Br. J. Nutr. 12: 177-188.

Bauchop, T., 1981. The anaerobic fungi in rumen fibre digestion. Agr. Environm. 6: 339-348.

Bird, S.H. and Leng, R.A., 1978. The effects of defaunation of the rumen on the growth of cattle on low-protein high-energy diets. Br. J. Nutr. 40: 163-167.

Bryant, M.P. and Robinson, I.M., 1961. Some nutritional requirements of the genus *Ruminococcus*. Appl. Microbiol. 9: 91.

Bryant, M.P. and Robinson, I.M., 1963. Apparent incorporation of ammonia and amino acid carbon during growth of selected species of ruminal bacteria. J. Dairy Sci. 46: 150.

Campling, R.C., Freer, M. and Balch, C.C., 1962. Factors affecting the voluntary intake of food by cows. 3. The effect of urea on the voluntary intake of straw. Br. J. Nutr. 16: 115-124.

Coleman, G.S., 1975. Interrelationship between rumen ciliate protozoa and bacteria. In: McDonald,I.W. and Warner, A.C. (eds.) Digestion and metabolism in the ruminant. pp. 149-164. University of New England Publishing Unit, Armidale, Australia.

Cook, F., Brink, D., Klopfenstein, T., Merrill, J., Stock, R. and McDonnell, M., 1982. Sodium hydroxide treatment of corn cobs. Nebraska Beef Cattle Report.

Czerkawski, J.W., 1976. Chemical composition of microbial matter in the rumen. J. Sci. Food Agric. 27: 621-632.

Dolberg, F., Saadullah, M., Haque, M. and Haque, R., 1981. Straw treatment in a village in Noakhli district, Bangladesh. In: Jackson, M.G., Dolberg, F., Davis, C.H., Haque, M. and Saadullah, M. (eds.): Proceedings Seminar Maximum Livestock Production from Minimum Land. Bangladesh Agricultural University, Mymensingh.

Duarte, F., Elliot, R. and Preston, T.R., 1983. Sugar cane juice as cattle feed: use of leucaena as combined source of protein and roughage. Trop. Anim. Prod. 8: (in press).

Elliott, R., Ferreiro, H.M., Priego, A. and Preston, T.R., 1978. Rice polishings as a supplement in sugar cane diets: the quantities of starch (glucose polymers) entering to proximal duodenum. Trop. Anim. Prod. 3: 30-35.

Fernandez, A., MacLeod, N.A. and Preston, T.R., 1977. Production coefficients in a dual purpose herd managed for milk and weaned calf production. Trop. Anim. Prod. 2: 44-48.

Guttierrez, E., Elliott, R., Harrison, D. and Preston, T.R., 1983. Effect of a supplement of fresh grass added to a basal diet of ensiled henequen pulp on cellulose digestion in nylon bags in the rumen of sheep. Trop. Anim. Prod. 8: (in press).

Harrison, D.G., Beever, D.E., Thomson, D.J. and Osbourn, D.F., 1976. Manipulation of fermentation in the rumen. J. Sci. Food Agric. 27: 617-620.

Helmsley, J.A., 1975. Effect of high intakes of sodium chloride on the utilization of a protein concentrate by sheep. 1. Wool growth. Aust.J. Agric. Res. 26: 709-714.

Helmsley, J.A. and Moir, R.J., 1963. The influence of higher volatile fatty acids on the intake of urea-supplemented low quality cereal hay by sheep. Aust. J. Agric. Res. 14: 509-514.

Herbein, J.A., Van Maanen, R.W., McGilliard, A.D. and Young, J.W., 1978. Rumen propionate and blood glucose kinetics in growing cattle fed isoenergetic diets. J. Nutr. 108: 994-1001.

Hespell, R.B., 1979. Efficiency of growth by ruminal bacteria. Fed. Proc. 38: 2707.

Hoogenraad, N.J., Hird, F.J.R., Holmes, I. and Millis, N.F., 1967. Bacteriophages in rumen contents of sheep. J. Gen. Virol. 1: 575-576.

Hume, I.D., 1970. Synthesis of microbial protein in the rumen. 3. The effect of dietary protein. Aust.J. Agric. Res. 21: 305-314.

Isaacson, H.R., Hinds, F.C., Bryant, M.P. and Owens, F.N., 1975. Efficiency of energy utilizatin by mixed rumen bacteria in continuous culture. J. Dairy Sci. 58: 1645-1659.

Judson, G.J. and Leng, R.A., 1968. Effect of diet on glucose synthesis in sheep. Proc. Aust. Soc. Anim. Prod. 7: 354.

Kayouli, A., 1980. Studies on the nutritive value of alkali-treated rice straw. M.Sc. Thesis, University of Tunis.

Kempton, T.J. and Leng, R.A., 1979. Protein nutrition of growing lambs. 1. Responses in growth and rumen function to supplementation of a low-protein-cellulosic diet with either urea, casein or formaldehyde treated casein. Br. J. Nutr., 42: 289-302.

Kennedy, P.M., Christopherson, B. and Milligan, L.P., 1982. Effects of cold exposure on feed protein degradation, microbial protein synthesis and transfer of plasma urea to the rumen of sheep. Br. J. Nutr. 47: 521-536.

Kennedy, P.M. and Milligan, L.P., 1980. The degradation and utilization of endogenous urea in the gastrointestinal tract of ruminants. Can. J. Anim. Sci. 60: 205-221.

Leng, R.A., 1970. Glucose synthesis in ruminants. Ad. Vet. Sci. Comp. Med. 14: 209-260.

Leng, R.A., 1974. Salient features of the digestion of pastures by ruminants and other herbivores. In: Butler, G.W. and Bailey, R.W. (eds.): Chemistry and Biochemistry of Herbage. London & New York, Academic Press, Vol. 3, pp. 81-129.

Leng, R.A., 1982. Modification of rumen fermentation. In: Hacker, J.B. (ed.): Nutritonal Limits to Animal Production from Pastures. pp. 427-481, Commonwealth Agricultural Bureaux, Farnham Royal, U.K.

Leng, R.A., Kempton, T.J. and Nolan, J.V., 1977. Non-protein nitrogen and by-pass proteins in ruminant diets. Aust. Meat Res. Com. Reviews 33: 1-21.

Leng, R.A. and Preston, T.R., 1976. Sugar cane for cattle production: present constraints, perspectives and research priorities. Trop. Anim. Prod. 1: 1-22.

Lindsay, J.A. and Loxton, I.D., 1981. Supplementation of tropical forage diets with protected proteins. In: Farrell, D.J. (ed.): Recent Advances in Animal Nutrition in Australia, pp. 1A. University of New England Publishing Unit, Armidale, Australia.

Marty, R.J. and Preston, T.R., 1970. Molar proportions of the short chain volatile fatty acids (VFA) produced in the rumen of cattle given high-molasses diet. Rev. Cub. Cienc. Agr. 4: 183-187.

Meyreles, L., Rowe, J.B. and Preston, T.R., 1979. The effect on the performance of fattening bulls of supplementing basal diet of derinded sugar cane stalk with urea, sweet potato forage and cottonseed meal. Trop. Anim. Prod. 4: 255-262.

Naidoo, G., Hulman, B. and Preston, T.R., 1977. Effect of fish meal and maize meal supplements on growth of calves fed steam-treated bagasse and urea. Technical Report UNDP/FAO Project Mar/75/004, FAO, Rome.

Nielsen, J.J., 1981. Nutritional principles and productive capacity of the Danish strawmix system for ruminants. In: Jackson, M.G., Dolberg, F., Davis, C.H., Haque, M. and Saadullah, M. (eds.): Proceedings of Seminar on Maximum Livestock Production from Minimum land. Bangladesh Agricultural University, Mymensingh.

Nolan, J.V. and Leng, R.A., 1972. Dynamic aspects of ammonia and urea metabolism in sheep. Br. J. Nutr. 27: 177-194.

Nolan, J.V. and Leng, R.A., 1974. Isotope techniques for studying the dynamics of nitrogen metabolism in ruminants. Proc. Nutr. Soc. 33: 1-8.

Nolan, J.V. and Stachiw, S., 1979. Fermentation and nitrogen dynamics in Merino sheep given a low-quality-roughage diet. Br. J. Nutr. 42: 63-80.

Nurrazzamal Khan, A.K.M. and Davis, C.H., 1981. Effect of treating paddy straw with ammonia (generated from urea) on the performance of local and crossbred lactating cows. In: Jackson, M.G., Dolberg, F., Davis, C.H., Haque, M. and Saadullah, M. (eds.): Proceedings Seminar Maximum Livestock Production from Minimum land. Bangladesh Agricultural University, Mymensingh.

Ørskov, E.R., 1970. Nitrogen utilization by the young ruminant. In: Swan, H. and Lewis, D. (eds.): Proceedings of the 4th nutrition conference for feed manufacturers. pp. 20-35. J. & A. Churchill, U.K.

Ørskov, E.R., 1980. Possible nutritional constraints in meeting energy and amino acid requirements of the highly productive ruminant. In: Ruckebusch, Y. and Thivend, P. (eds.): Digestive Physiology and Metabolism in Ruminants. pp. 309-324. MTP Press Ltd.

Ørskov, E.R. and Allen, D.M., 1962. Utilisation of salts of volatile fatty acids by growing sheep. 1. Acetate, propionate and butyrate as sources of energy for young growing lambs. Br. J. Nutr. 20: 295-301.

Ørskov, E.R., Grubb, D.A., Smith, J.S., Webster, A.J.F. and Corrigal, W., 1979. Efficiency of utilisation of volatile fatty acids for maintenance and energy retention by sheep. Br. J. Nutr. 41: 541-552.

412

Perdok, A.B., Thamotharam, M., Blom, J.J., Van Den Born, H. and Van Veluw, C., 1983. Practical experiences with urea-ensiled straw in Sri Lanka. 2nd Annual Seminar. Maximum Livestock Production from Minimum Land. Bangladesh Agricultural University and B.A.R.C., Dacca.

Pigden, W.J., 1972. Sugar cane as livestock feed. Report to Carribean Development Bank, Barbados.

Pisulewski, P.M., Okome, A.J., Buttery, P.J., Haresign, W.R. and Lewis, D., 1981. Ammonia concentrations and protein synthesis in the rumen. J. Sci. Food Agric. 32: 759-766.

Preston, T.R., Carcano, C., Alvarez, F.J. and Guttierez, D.G., 1976. Rice polishings as a supplement in sugar cane diet: effect of level of rice polishings and of processing the sugar cane by derinding or chopping. Trop. Anim. Prod. 1: 150-163.

Preston, T.R. and Leng, R.A., 1980. Utilization of tropical feeds by ruminants. In: Ruckebush, Y. and Thivend, P. (eds.): Digestive Physiology and Metabolism in Ruminants. MTP Press Ltd., Lancaster. pp. 621-640.

Ravelo, G., Fernandez, A., Bobadilla, M., MacLeod, N.A., Preston, T.R. and Leng, R.A., 1978. Glucose metabolism in cattle on sugar cane based diets: a comparison of supplements of rice polishings and cassava root meal. Trop. Anim. Prod. 3: 12-18.

Riddell, D.O., Bartley, E.E. and Dayton, A.D., 1980. Effect of nicotinic acid on rumen fermentation in vitro and in vivo. J. Dairy Sci. 63: 1429-1436.

Rodriguez, A., 1982. Studies on the supplementation of ensiled henequen (sisal) pulp for growing lambs. Trop. Anim. Prod. 7 (in press).

Romero, V.A., Siebert,B.D. and Murray, R.M., 1976. A study on the effect of frequency of urea ingestion on the utilization of low quality roughage by steers. Aust. J. Exp. Agr. Anim. Husb. 16: 308-314.

Saadullah, M., Haque, M. and Dolberg, F., 1981. Treated and untreated paddy straw for growing cattle. In: Jackson, M.G., Dolberg, F., Davis, C.H., Haque, M. and Saadullah, M. (eds.): Maximum production from Minimum Land. Bangladesh Agricultural University, Mymensingh.

Saadullah, M., Haque, M. and Dolberg, F., 1982a. Effect of chelated minerals supplemented with untreated and urea-treated rice straw on performance of calves. 2nd Annual Seminar. Maximum Livestock Production from Minimum Land. Bangladesh Agricultural University and B.A.R.C., Dacca.

Saadullah, M., Haque, M. and Dolberg, F., 1982b. Effect of fish meal on growth of Zebu cattle calves fed on a basal diet of urea-treated rice straw. Trop. Anim. Prod. 7 (in press).

Santana, A. and Hovell, F.O.DeB., 1979. Degradation of various sources of starch in the rumen of Zebu bulls fed sugar cane. Trop. Anim. Prod. 4: 107-108 (Abstract).

Santana, A., Peralta, G. and Gill, M., 1980. Rate of degradation in the rumen of different protein forages. Trop. Anim. Prod. 5: 82 (Abstract).

Satter, L.D. and Slyter, L.L., 1974. Effect of ammonia concentration on rumen microbial protein production in vitro. Br. J. Nutr. 32: 199-208.

Silvestere, R., MacLeod, N.A. and Preston, T.K., 1977. Effect of meat meal, dried cassava root and groundnut oil in diets based on sugar cane/urea or molasses/urea. Trop. Anim. Prod. 2: 151-157.

Singh, Y.P., 1980. Feasibility, nutritive value and economics of Azolla-amabaena as an animal feed. M.Sc. Thesis, Pant, G.G., University, Pantnagar UP, India.

Smith, T., Broster, V.J. and Hill, R.E., 1980. A comparison of sources of supplementary nitrogen for young cattle receiving fibre-rich diets. J. agric. Sci (Camb.) 95: 687-695.

Stern, M.D. and Hoover, W.H., 1979. Methods for determining, and factors affecting rumen microbial protein synthesis: a review. J. Anim. Sci. 49: 1590-1603.

Thomson, D.J., 1978. Utilisation of the end products of digestion for growth. In: Osbourne, D.F., Beever, D.E. and Thomson, D.J. (eds.): Ruminant Digestion and Feed Evaluation. Agric. Res. Council, London.

Tudor, G.D. and Minson, D.J., 1982. The utilisation of the dietary energy of pangola and setaria by young growing beef cattle. J. Agric. Sci. (Camb.) 98: 395-404.

Tyrrell, H.F., Reynolds, P.J. and Moe, P.W., 1979. Effect of diet on partial efficiency of acetate use for body tissue synthesis by mature cattle. J. Anim. Sci. 48: 598-605.

Van Soest, P.J., 1982. Nutritional ecology of the ruminant. O. & B. Books, Inc., Oregon.

Wheeler, W.E., Noller, C.H. and White, J.L., 1981. Effect of level of calcium and rate of reactivity of calcitic limestone on utilization of high concentrate diets by beef steers. J. Anim. Sci. 52: 882-894.

Wong You Cheong, Y., d'Espaignet, J.T., Deville, P.J., Sansoucy, R. and Preston, T.R., 1974. The effect of steam treatment on cane bagasse in relation to its digestibility and furfural production. ISSCT Congress, February.

Chapter 14.1

STRAW ETC. IN PRACTICAL RATIONS FOR CATTLE AND BUFFALOES

WITH SPECIAL REFERENCE TO DEVELOPING COUNTRIES

by

Manohar L. Verma and Michael G. Jackson

Department of Animal Science, College of Agriculture
G.B. Pant University of Agriculture & Technology
Pantnagar, District Nainital, U.P.,
India, Pin 263145

14.1.1. INTRODUCTION

Straw has probably been used as a livestock feed ever since the advent of cereal cultivation. In virtually all parts of the world and in all ages it has been an important, often major item in the livestock diet. In terms of tonnage straw is today second only to grass herbage in importance as a feed for livestock. Until very recently scientific interest in straw feeding was not at all commensurate with its importance in practice.

Although straw is an important feedstuff, and indeed the staple feed in large parts of the world, it is not a preferred feed. Because of its low feeding value, straw is only used when better quality feeds are not available. Traditionally it has been used to carry stock over periods when fresh and/or conserved grass and tree herbage was inadequate eg. winter in temperate climates and dry season in the arid tropics and subtropics. Although the use of petroleum in temperate, industrialised nations has perhaps tempo-rarily made straw feeding largely unnecessary, in the arid tropics and subtropics straw is steadily becoming more important in the livestock diet. This is because increasing population density and failure to modify traditional grazing practices have caused serious deterioration of natural vegetative cover. In many parts of the world today straw makes up 60 to 90 per cent of the bovine diet. Nearly half the world's bovines is reared and maintained on diets of 50 per cent or more straw. Whilst these bovines are the world's least productive in terms of annual output per animal, they are crucial to the energy economy of the farming systems where they occur. This chapter therefore, deals primarily with the improved feeding of these animals.

Livestock feeding in India is an example of straw as staple feed. A year-long survey was conducted at representative places throughout the country in which feeds actually fed by farmers were regularly weighed (Amble et al., 1965). The results are presented in Table 14.1.1. Very little concentrate feeds are fed because they are simply not available. Green herbage consists of cultivated fodder, hand-harvested tree leaves, weeds from cultivated fields and sugarcane tops. Grazed grass is not accounted for in Table 14.1.1 but estimated to be about 3 kg (air-dry basis) per animal daily. Most of this is obtained by animals during the four monsoon months. This pattern of feeding prevails, with minor variations, over the entire Indian peninsula. In the Northern plains of the Indus and Ganges, where irrigation is widespread, more green fodder is grown and fed, including leguminous winter fodder (berseem and lucerne). In the Deccan plateau and in the East straw is a larger component of the diet than indicated by Table 14.1.1.

South-east Asia, including Indonesia and the Phillipines, is still forested to a greater extent than the Indian peninsula and relatively less reliance is placed on straw. However deforestation is proceeding at a rapid rate and one result will be more straw in the livestock diet. In West Asia, the Mediterranean basin and East Africa straw is nearly as prominent in cattle diets as in India.

In the Indian peninsula all straw barring that unavoidably lost after harvest due to bad weather, is fed to livestock. In Bangladesh, some straw is even burnt as household fuel because of shortage of fire-wood thus seriously reducing supplies for livestock. In West Asia, it is estimated that 50-70 per cent of the straw produced is fed to livestock (Arnason, 1980; Jackson, 1980).

Table 14.1.1. Availability of bovine livestock feeds in India (Amble et al., 1965)

Class of animal	Type of bovine	Amount of feed consumed[1], kg/animal.day		
		Dry roughage	Green herbage[2]	Concentrates
In milk	Cow	3.5	1.5	0.3
	Buffalo	5.9	2.3	0.8
Dry	Cow	2.8	0.9	0.1
	Buffalo	4.0	1.4	0.1
Adult male	Cow	5.7	1.7	0.3
	Buffalo	5.4	2.2	0.2
Young stock	Cow	1.5	0.5	
	Buffalo	1.7	0.5	

[1]Excludes grazing estimated to provide 3 kg/day (air-dry equivalent basis)

[2]On air-dry basis

The task of improving the productivity of straw-fed livestock in developing countries presents us with a unique challenge. The concept of feeding standards, so useful in temperate, industrialised countries, is less relevant in this situation. Straw-fed animals of the Indian peninsula and elsewhere are maintained at nutritional levels far below those prescribed by temperate feeding standards. Even by indigenously-developed standards feeding levels are low. Furthermore there is no way animals can be fed up to standards because of insufficient high quality feeds. With feeding standards in mind, the approach to date when dealing with straw-fed livestock has been to try and feed all avilable high-quality supplementary feeds to only a portion of the bovine population. The soundness of this is questionable because most bovine livestock are crucial to the energy economy of farms in countries such as India. Ploughing, threshing and carting must be done on every farm, and done promptly as timeliness is crucial, especially with two crops a year. Millions of farmers means millions of bullocks and also cows to produce them, all these need to be fed as well as possible. Even from the point of view of milk production, concentrating the use of high-quality feeds on only a portion of the population will lower, not raise a nation's total milk production (Jackson, 1981). Furthermore such programmes are not in keeping

with the policy of most nations to provide conditions of economic
equality to all their citizens.

An alternative way of looking at the problem, which will be
used in this chapter, is to consider straw a vital energy resource
in agriculture. The task is to maximise the efficiency with which
it is converted to animal products and work. Various ways of doing
this will be considered, eg. use of urea to treat straw and provide a nitrogen supplement, appropriate allocation of scarce concentrate feeds, provision of supplementary minerals and rehabilitation of grassland and forest to enhance green herbage supplies.

14.1.2. IMPROVING THE EFFICIENCY OF STRAW-FED LIVESTOCK

14.1.2.1. <u>Constraints to increased productivity</u>

The nutritive constraints inherent in straw are dealt with in
Chapters 11-13. The main constraints are high cell wall and low
nitrogen contents. These factors limit intake and digestibility
and hence limit animal performance. Low mineral content, commonly
calcium and phosphorus, but also in places sulphur, copper, cobalt
and iodine, also limits performance. Straw has no carotene. Animals on typical straw diets (Table 14.1.1) respond to additional
nitrogen, protein and energy. They will usually respond to supplements of calcium and phosphorus and in specific localities, supplements of trace minerals notably sulphur, copper, cobalt and
iodine. During the dry season in more arid tracts, animals also
respond to carotene/vitamin A supplement.

Under experimental conditions bovine animals lose weight on
diets of straw alone (Rekib et al., 1970; Sharma et al., 1972;
Khurana, 1978). In three experiments average weight loss in calves
was 0.15 kg/day (range 0.13-0.16 kg). In Southern Africa cattle on
the veldt during the dry season typically lose weight, mature
veldt grass being similar to cereal straw. In practice, Indian
bovine livestock usually obtain enough supplementary natural herbage during the dry season to about maintain weight. Figure 14.1.1
shows the typical growth pattern of village young stock throughout
the year.

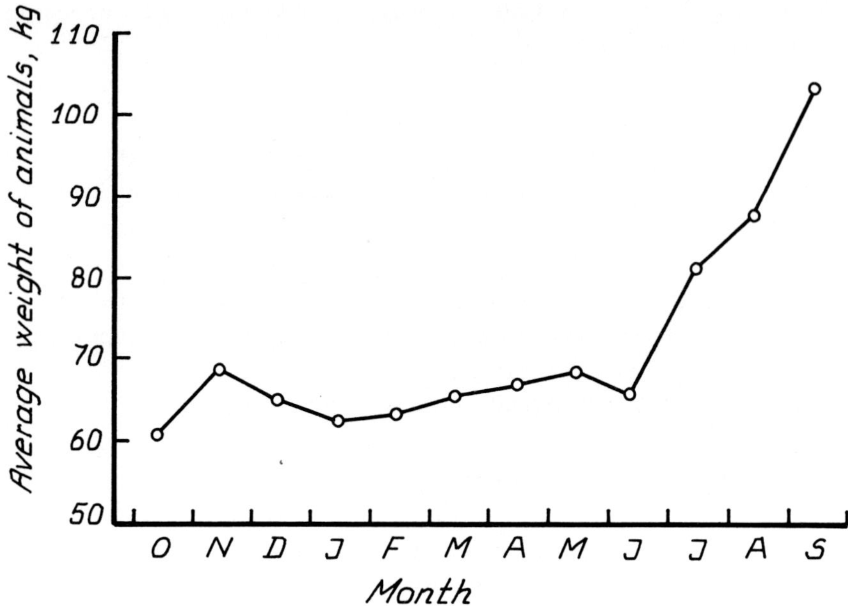

Figure 14.1.1. Average weight of 8 growing calves (cattle males and females and buffalo females) in consecutive months

Under dry-season conditions, it appears that nitrogen, and not energy, is the first factor limiting animal performance. This may be inferred from the experiment of Donefer et al. (1969). They fed sheep oat straw with sucrose and urea, either alone or in combination and observed that the addition of sucrose alone did not affect intake of digestible energy but increased it by about 25% in the case of urea-supplemented untreated straw and 250% with urea-supplemented NaOH-treated straw. Similarly, Khurana (1978) fed buffalo heifers on wheat straw sprayed with urea solution (20 kg urea in 1000 litre water/t field-dry straw) and observed weight gains of 31 g/day. Without urea the heifers lost 145 g/day. Much experimental data from Southern Africa (reviewed by Topps, 1972) on grazing cattle fed supplements of urea, molasses or maize grain also indicate the primacy of nitrogen. Urea supplementation caused an appreciable improvement in the conception rate of cows. The change in weight during winter was also higher (12.3 kg) in supplemented compared to control (7.3 kg) groups. Molasses and maize grain were not found to give any additional improvement.

It should be pointed out that nitrogen also continues to be the first limiting factor during the rainy season. The green fodders during this season are natural grass and cultivated sorghum; the nitrogen content of the former is about 50% higher than that of the latter.

Wherever cropland is irrigated and farm size is large enough, farmers grow some leguminous winter fodder crop like berseem and lucerne. The dry season depression in productivity is thus avoided. A typical growth pattern for buffalo calves fed such leguminous winter fodder and urea supplement the rest of the year is given in Figure 14.1.2. However better-fed livestock such as these account for no more than about 20 percent of all livestock in the Indian peninsula.

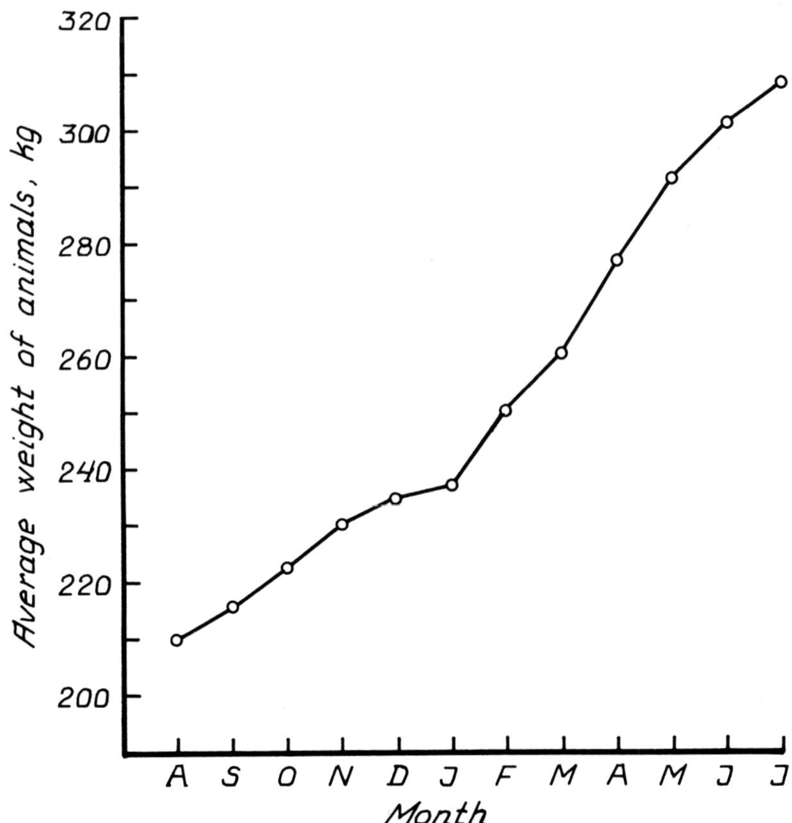

Figure 14.1.2. Average weight of 16 growing buffalo calves on urea supplemented straw diet under village conditions

The quantity of straw available to feed animals, may be a factor limiting productivity. In the densely populated rural areas of Eastern India and Bangladesh, animals are usually not fed as much straw as they could consume (Dolberg, 1981). This, plus the high proportion of the diet constituted by straw, seems to be responsible for the extremely small size and low output of animals in these areas. Straw shortage relative to animal numbers results from each farm family attempting to maintain its bullocks in the face of ever-decreasing farm size and also from shortage of domestic firewood forcing people to burn straw (Scoville, 1976; Ward et al., 1980).

14.1.2.2. Supplementation of the straw diet

The principles of supplementation have been discussed previously (Chapter 13).

Over the Indian peninsula the averge growth rate of cattle and buffalo young stock is 0.1-0.2 kg per day. This means mature body weight is achieved at 3-5 years of age. Milk yields are 300-500 kg per year with animals calving every other year. A pair of bullocks provides the power for farming about 3 ha - about 1.2 kw per ha. For efficient farming 0.373 kw per ha is needed, and could be provided by a pair of well-grown and vigorous bullocks. The corresponding figures for feed efficiency are 20-30 kg dry matter per kg of live-weight gain (not counting the feed consumed by all the calves that die - 50 per cent between birth and 2 years of age), 10-15 kg per kg of milk produced and 1.5-2.0 tonnes of feed dry matter to farm one ha of land. Practical ways by which these levels of efficiency can be improved will be considered.

14.1.2.2.1. Urea supplementation

A supplement of urea at the rate of 1 per cent of diet dry matter can be expected to increase the rate of gain of young stock by about 0.1 kg per day (Gupta et al., 1971; Khurana, 1978; Jaiswal et al., 1982; Saadullah et al., 1982). This amounts to a doubling of growth rate in many instances and thus a halving of time taken to achieve maturity.

There are less precise estimates of the effect of urea supplementation on milk production. Agarwal and Verma (1982) reported increased milk production in village bovines given urea supplemented straw or non-leguminous green forages.

One reason for the paucity of information on responses to supplements like urea is that trials must be done on farmers' animals. The typical diet and management used by farmers cannot be simulated satisfactorily on the experiment station. Trials with farmers' animals, especially where farms are small, are more difficult to conduct than experiment station trials and the results are often less definitive. The general procedure for conducting these trials has been given in a recent FAO publication (Jackson, 1978).

It is usually advocated that urea supplements be accompanied by molasses. However there is little evidence that molasses at the rates usually recommended (50 to 100 g fresh molasses per kg diet dry matter) improves animal performance over that with a simple urea supplement. Furthermore the evidence from trials with cattle on dry-season veldt is that neither molasses nor maize grain supplements with urea have any effect on animal performance (Topps, 1972). The lack of response to these energy supplements is apparently due to the depressing effect sugar and starch have on cellulose digestibility; the energy gained from the sugar or starch no more than offsets the reduction in energy from cellulose. In general, small supplements of sugar or starch appear to be a waste of feed resources. Of course it is assumed that the urea supplements can be mixed with other ingredients of the diet, or at least with one of them. With animals maintained purely on grazing, a carrier must be found for urea and molasses is useful in this respect. A ratio of 1:10 urea:molasses has been found satisfactory.

Urea supplements provide the animal more protein via microbial protein synthesis in the rumen. Thus rate of gain and milk production increase. Urea also increases dry-matter intake, increases that are often in the range 14-35 per cent (Rekib et al., 1970; Sharma et al., 1972; Khurana, 1978; Gadre, 1979). The change in digestible energy intake however may range from (-)15 to (+)50 per cent because increased DM intake is sometimes accompanied by a fall in digestibility. The results of Gadre (1979) illustrate this. Needless to say, the latter is a waste of straw. There is need therefore to determine the optimum intake of urea-supplemented straw diets for various categories of livestock so as to maximize the availability of nutrients without much loss in efficiency of utilization.

It is essential to consider the feasibility of feeding a urea supplement to the large numbers of bovine animals in a country like India. If all animals with the exception of those fed seasonally-cultivated leguminous fodder crops were fed a urea supplement, about 4 million tonnes per year would be needed. Current production of urea is 6.3 million tonnes per year, all of which is used to fertilize crops. Thus widespread use of urea as a feed supplement could be advocated only if effective methods of nitrogen recycling were available eg. processing of dung and urine using a dung gas plant. No such method is available to date suitable for the individual small farmer.

It can be concluded that a supplement of urea, at one percent of diet dry matter, dissolved in water and sprinkled over straw and/ or green fodder, can boost production markedly. At current market prices for urea in India it is economical for the individual farmer to feed it. However on a national scale, it offers no solution to the problem of improving bovine livestock feeding. The only hope appears to be in developing dung gas plants of appropriate size and cost or introducing such changes in village life to make possible community ownership of large current-design plants.

14.1.2.2.2. Concentrate supplements

Concentrate supplements over and above the meagre amounts already fed can boost output of bovine livestock. However available evidence suggests that such supplements are inefficiently used. For 100 kg growing calves on a basal diet of straw and a little green fodder, a supplement of conventional mixed concentrate at 1.2 kg/day gives a weight gain of about 0.35 kg per day (Rekib et al., 1970; Jaiswal et al., 1982). Additional concentrate supplements also appear to be inefficiently used by milking animals (Table 14.1.2).

Both the energy and the protein of the concentrate mixture appear to be inefficiently used. To a level of about 1.5 per cent nitrogen in diet DM, the animal (or more correctly the rumen microflora) needs only a source of ammonia. Protein in excess of the amount needed to supply this level of nitrogen is inefficiently used, being deaminated for the most part. This is brought out by data given later on the remarkable effect on weight gain of calves of very small supplements containing protein which is relatively

insoluble in the rumen. As for the energy of the concentrate supplement, it also has a low apparent efficiency of utilization
because of the depression that starch causes in the digestibility
of straw fibre (Chimwano et al., 1976).

Table 14.1.2. Relationship between level of concentrate feeding to
buffaloes and level of milk yield (Nair and Jackson,
1981, based on Mellor and De Ponteves, 1964)

Concentrates fed/day, kg/animal	Milk yield/day, kg/animal
0	1.2
0.5	1.9
1.0	2.5
1.5	3.0
2.0	3.4
2.5	3.7
3.0	3.9

Although concentrates are inefficiently used for milk production when given in larger amounts than present average levels,
there are some situations in which it pays to do so. One is the
traditional urban milk producer, ie. the operator who keeps cattle
or buffaloes in the city. Another is the modern milk project with
its "milkshed" area in which farmers are encouraged to produce and
market milk by being supplied concentrates conveniently. That some
producers can monopolise limited supplies of concentrate feeds and
make a profit is due to distortions in the market system, the
nature of which cannot be gone into here. An analysis has been
presented by Nair and Jackson (1981). This sort of heavy concentrate feeding to only a segment of the bovine population in a
situation in which the basal feed is straw can hardly be justified
in most countries. In India, for example, only about 20 million
tonnes of concentrate feeds are produced every year. The average
for each animal is thus only about 0.25 kg per day. Using the data
in Table 14.1.2 one can show that feeding the available concentrates in small equal shares to all animals will result in more milk
than if it is fed to only a few animals in larger amounts (and
thus depriving many more animals of even their meagre average sha-
re).

Most farmers feed little or no concentrates to their calves,
preferring to feed what they have to their adult animals in milk

424

and at work (Table. 14.1.1). This is the main reason for the slow
rate of gain by calves, already referred to, and the high calf
mortality rate (about 50 per cent between birth and two years of
age). There are definite economic reasons for this neglect of calves as described by Crotty (1980). Some form of community management of livestock appears necessary if better feeding of calves is
to be possible. There is reason to believe that if a portion of
the concentrate feeds now given to adult bovine stock were given
instead to selected growing calves, particularly pre-ruminant calves, overall efficiency of straw usage would be improved. Apart
from the low output from animals malnourished when young, the calves that die do so only after consuming feed. This is feed from
which there is no return at all. Preston (1977) has shown that the
pre-ruminant calf responds remarkably to dietary glucose and amino
acids, nutrients which can be supplied only by concentrate feeds.
Long-term, on-farm trials will have to be done to determine the
possible benefits to overall efficiency of re-allocating concentrate feeds to calves.

14.1.2.3. <u>Straw treatment</u>

Chemical treatment of straw increases digestibility and hence
boosts animal output. Methods of treatment are described in Chapters 6-8. Urea/ammonia treatment seems to be the only economical-
ly-feasible method in developing countries where straw is a prominent feedstuff. Feeding trials in which the straw portion of a
predominantly-straw diet has been treated with urea (ammonia
treatment) reveals that treatment increases rate of daily gain of
calves by 0.2-0.3 kg (Chauhan, 1982; Jaiswal et al., 1982; Perdok
et al., 1982; Saadullah et al., 1982). This is the combined effect
of increased digestibility and additional nitrogen. This degree of
improvement may be contrasted with the values of 0.03 kg for urea
supplement alone (Khurana, 1978). Daily milk yields can be boosted
by about 1 kg/head (Table 14.1.3.).

Urea treatment of straw also enables animals to consume more.
It is usually considered that additional consumption is an important factor contributing to the increases in output. Recent evidence however, suggests this may not be so. Khan and Davis (1982)
compared the effect of restricting the level of intake on urea-
treated straw to that of untreated straw and found average daily

gain on restricted intake to be similar to those in ad lib. fed groups (Table 14.1.4). In view of this and the fact that straw is so limited in countries such as Bangladesh, treated straw should perhaps be rationed to animals. However restriction of intake is only practicable with individually-stall-fed animals.

Table 14.1.3. Effect of urea treatment of rice straw on intake, milk yield and live-weight change in Gir cows and Surti buffaloes and their calves (Perdok et al., 1982)

Rice straw, ad lib.	Gir cows		Surti buffaloes	
	Treated	Untreated	Treated	Untreated
Concentrate, kg/d	1.5	1.5	1.0	1.0
Number of animals	17	17	10	10
Straw DM intake, kg/d	8.6	5.2	10	10
Milk yield, kg/d	3.41^a	2.42^b	2.97^a	2.17^b
Milk fat %	4.91	4.60	7.54	6.71
Live-weight change:				
Cows, g/d	$(+)93^a$	$(-)266^b$		
Calves, g/d	257^a	181^b		

a-b
Means with different superscripts are significant at P<0.01

Table 14.1.4. Feed intake, live-weight gain and feed conversion with untreated or ammonia treated (through urea) rice straw fed ad lib. or restricted (Khan and Davis,1982)

Rice straw	Untreated	Ammonia-treated	
		Restricted	Ad lib.
Dry matter intake, kg/d	3.45	3.48	4.20
" " ", g/kgW$^{0.75}$·d	91.2^a	87.6^a	104.6^b
Live-weight gain, g/d	124.5^a	303.0^b	310.0^b
Feed conversion, kg feed/kg gain	27.7^a	11.5^b	13.5^b

a-b
Means with different superscripts are significant at P<0.01.

Ammonia treatment raises dietary nitrogen to a level needed for maximum microbial protein synthesis in the rumen. Studies with protein supplements to urea-treated straw diets have shown that if the protein is not readily soluble in the rumen, large responses in weight gain result from surprisingly small supplements. This is seen in Table 14.1.5 which shows that a supplement of only 150 g of fishmeal or 325 g of cotton seed cake per day to a calf gives an increase in rate of gain of 0.2 kg. Fishmeal could not be used in this way on a large scale because of limited supplies. If less effective vegetable protein supplements (Table 14.1.5) could be treated to decrease their protein solubility, large improvements in animal output could result. The treatment should preferably be one that could be done easily on farms to put it within reach of the average peasant farmer.

What was said earlier about the use of urea as a simple dietary supplement must also be said here concerning the widespread treatment of straw with urea. It may not be feasible in those countries where straw is a major feedstuff. Indeed a third to half of the nitrogen applied as urea cannot even be recovered, being lost when the treated straw is aerated before feeding.

14.1.2.4. Green fodder supplements

Supplements of green fodder, either from cultivated fodder crops or from natural herbage, greatly improve animal output. Replacement of 30 or 50% of the straw dry matter by good quality leguminous fodder like berseem improves digestibility and intake (Naik, 1975; Bhargava, 1977) and makes a straw-based diet able to support modest daily gains (eg. 400 g/day, Bhargava, 1977). Improved growth rate due to better availability of green herbage in the monsoon season is also seen in Figure 14.1.1. Such improvements have been attributed to increased rate of microbial production due to provision of essential cofactors (Leng, 1982; Nolan and Stachiw, 1979; Preston and Parra, 1981) or more fermentable cell wall carbohydrates from green herbage. Unfortunately in countries where straw is an important feed, very little cultivable land can be devoted to growing fodder crops. Natural vegetation is also scanty due to the destruction of grass and tree cover through mis-management of grazing stock. Here however, there are opportunities for improvement in many countries. Rehabilitation of barren land through reafforestation or regeneration of grass cover is technically possible almost everywhere. It is the legal and social impediments to rehabilitation which need to be removed. In a monsoon

Table 14.1.5. Effect of various supplements to urea-treated straw in different experiments

Experiment, basal diet and supplement feeding rate (kg/d)	DM intake $g/kgW^{0.75} \cdot d$		DM digestibility (%)		Live-weight gain	Feed conversion kg DM/
	Total	Straw	Total	Straw	kg/d	kg gain
(1) Urea-treated wheat straw ad lib and 1 kg green sorghum, 14 heifers/treatment, 189 days						
Fish meal, 0.2 kg	72	69	55	55	0.36	10.6
Concentrate mixture, 1,0 kg	80	60	56	49	0.39	11.6
(2) Urea-treated rice straw ad lib and 2 kg berseem, 5 crossbred heifers/treatment, 90 days						
Control	135	135[a]	43	43[a]	0.25[cd]	31[ab]
Cotton seed cake 0.32 kg	136	128[ab]	45	44[a]	0.43[ab]	14[bc]
Fish meal, 0.23 kg	125	119[ab]	46	44[a]	0.33[abc]	15[bc]
Leucaena leaf meal, 0.37 kg	126	116[b]	47	46[a]	0.29[bc]	18[bc]
Untreated rice straw ad lib and 2 kg berseem, 5 crossbred heifers/treatment, 90 days						
Urea	96	96[c]	45	45[a]	0.11[d]	42[a]
Concentrate mixture, 1.75 kg	109	66[d]	49	32[b]	0.47[a]	10[c]
(3) Rice straw ad lib supplemented with 5.9 kg silage and 0.5 kg concentrate mixture, 17 Sahiwal heifers/treatment, 70 days						
Urea-treated	94	49	–	–	0.35[a]	13
Untreated	83	46	–	–	0.07[b]	53
(4) Urea and lime-treated rice straw ad lib and 1 kg green grass, 4 indigenous calves/treatment, 105 days						
Control	96[a]	90	–	–	0.14[a]	21
Fishmeal, 0,15 kg	98[a]	89	–	–	0.36[b]	9
Fishmeal, 0.15 kg + rice bran, 0.3 kg	108[ab]	91	–	–	0.35[b]	10
Fishmeal, 0.15 kg + rice bran, 0.6 kg	118[bc]	93	–	–	0.34[b]	11
Til oil cake, 0.3 kg	116[bc]	101	–	–	0.19[ac]	18
" " " + rice bran, 0.3 kg	116[bc]	94	–	–	0.25[c]	14
" " " + rice bran, 0.6 kg	127[c]	99	–	–	0.24[c]	16

Means with different superscripts in the same column differ significantly at P < 0.05.
(1) Chauhan (1982); (2) Jaiswal et al. (1982); (3) Perdok et al. (1982);(4) Saadullah et al. (1982)

climate like the Indian peninsula, uncultivated land constitutes about one half of the land area. This could produce two to ten times more herbage than at present, if it were rehabilitated. The techniques for doing this and the challenges to be overcome cannot be described here, but one important point will be made. Where annual rainfall is sufficient to support tree vegetation (original natural cover was forest), there is distinct advantage to live-stock production and output of farming as a whole if multipurpose plantations (fodder and fuelwood trees) are established rather than grass. Trees yield more herbage than grass and also, because of their deeper root system, yield herbage throughout the dry season. Further, many of the most productive fodder-tree species are legumes.

14.1.3. SUMMARY

The heavy dependence on straw as feed in Asia and Africa is largely due to destruction of natural vegetative cover of unculti-vated land by faulty livestock management. Efforts should be con-centrated on reafforestation where rainfall permits and regenera-tion of grassland where it does not - to produce green fodder. Output can be doubled by including 8-10 kg per day of green legu-minous fodder in the diets of average bovines. Supplies must be adequate and used to improve moderately the diet of all animals and not just a few generously. Another means to boost output over large populations of animals is possibly reallocating a substan-tial portion of concentrate feeds to pre-ruminant calves. In India at least this would require radical change from individual to community management of bovine livestock. Urea supplementation could boost output of significant numbers of animals only if effective recycling of nitrogen were common. Urea (ammonia)-treat-ment of straw in India is currently economic for the individual farmer but is wasteful of nitrogen and may neither be feasible nor justified on a national scale.

14.1.4. REFERENCES

Agarwal, I.S. and Verma, M.L., 1982. Experiences in on-farm research and application of by-products use for animal feeding in Asia. In: B. Kiflewahid, G.R. Potts and R.M. Drysdale (eds.): By-product Utilization for Animal Production, IDRC, Ottawa, Canada. 11: 140-147.

Amble, V.N., Murthy, V.V.R., Sathe, K.V. and Goel, B.B.P.S., 1965. Milk production of bovines in India and their feed availability. Indian J. Vet. Sci. Husb. 35: 221-238.

Arnason, J., 1980. Agricultural extension and seed introduction in the Tiham region. FAO mission report, pp. 3.

Bhargava, M., 1977. The effect of feeding alkali-treated paddy straw and berseem on the growth rate of heifers. M.Sc. Thesis, G.B. Pant University, Pantnagar.

Chauhan, M., 1982. Effect of feeding urea treated straw supplemented with concentrate or fishmeal on growth rate and digestibility of nutrients and its comparison with green sorghum and oats in heifers. M.Sc. Thesis. G.B. Pant University, Pantnagar.

Chimwano, A.M., Ørskov, E.R. and Stewart, C.S., 1976. The effect of dietary proportions of roughage and concentrates on the rate of digestion of dried grass and cellulose in the rumen of sheep. Proc. Nutr. Soc. 35: 101A-102A.

Crotty, R., 1980. Cattle, Economics and Development. London, Commonwealth Agricultural Bureaux.

Dolberg, F., 1981. Rural versus Urban Benefits from Cattle production, Feed Resources Utilization, the role of Cattle and their Farming System. In: M.G. Jackson, F. Dolberg, C.H. Davis, M. Haque and M. Saadullah (eds.): Maximum Livestock Production from Minimum Land. Bangladesh Agric. Univ., pp. 430-454.

Donefer, E., Adelaye, I.O.A. and Jones, T.A.O.C., 1969. Effect of urea supplementation on the nutritive value of NaOH treated oat straw. In: R.F. Gould (ed.): Cellulases and Their Applications. Advances in Chemistry series No 95. New York, Am. Chemical Soc., pp. 328-342.

Gadre, K.R., 1979. A study of urea as a source of ammonia for treating straw to increase its digestibility. M.Sc. Thesis, Pant University, Pantnagar.

Gupta, B.S., Satapathy, N., Chhabara, S.S. and Ranjhan, S.K., 1971. Urea as a sole source of nitrogen in growing buffalo calves fed on a basal roughage of wheat straw - its effect on palatability, growth rate and TVFA production in the rumen. Ind. J. Dairy Sci. 24: 7-15.

Jackson, M.G., 1978. Treating straw for animal feeding. FAO Animal Production and Health paper no. 10, FAO, Rome.

Jackson, M.G., 1980. Treatment of straw for animal feeding. Consultancy report for Cyprus and Syria submitted to FAO Near East Regional Office, Cairo. pp. 8 and 15.

Jackson, M.G., 1981. A new livestock development strategy for India. World Anim. Rev. 37: 2-8.

Jaiswal, R.S., Verma, M.L. and Agarwal, I.S., 1982. Effect of different protein supplements fed with urea treated paddy straw on digestibility, N-balance and growth in heifers. A paper presented at Animal Nutrition Research Workers conference, Tirupati, India Nov 29 - Dec 3, 1982. (Abstr.): 153.

Khan, A.K.M.N. and Davis, C.H., 1982. Effect of the level of feeding ammonia treated paddy straw on the performance of growing cattle. Paper presented at the third annual seminar: Maximum Livestock Production from Minimum Land, Joydebpur, Bangladesh, 15-18 Feb.

430

Khurana, M.P., 1978. The effect of akali treatment and urea supplementation on wheat straw both singly and in combination on growth rate of buffalo heifers. M.Sc. Thesis, G.B. Pant University, Pantnagar.

Leng, R.A., 1982. A theoretical discussion of the factors limiting production in cattle fed basal diets of straw. Paper presented at the third annual seminar: Maximum Livestock Production from Minimum Land, Joydebpur, Bangladesh, 15-18 Feb.

Mellor, J.W. and De Ponteves, B., 1964. Effect of growth in demand for milk on the demand for concentrate feeds: India 1951-1976. Indian J. Agric. Econ. 19: 131.

Naik, D.G., 1975. Digestibility of sodium hydroxide spray-treated and untreated paddy straw fed in combination with berseem to crossbred heifers. In: Investigation on Agricultural by-products and Industrial Waste Materials for Evolving Economic Rations for Livestock. Research Progress Report 1975-76. G.P. Pant University, Pantnagar, pp. 35-45.

Nair, K.N. and Jackson, M.G., 1981. Alternative to Operation Flood II. Strategy. Econ. and Political Weekly XVI (52): 2129-2132.

Nolan, J.V. and Stachiw, S., 1979. Fermentation and nitrogen dynamics in Merino sheep given a low-quality roughage diet. Br. J. Nutr. 42: 63-80.

Perdok, H.B., Thamotharam, M., Blom, J.J., Born, H. Van den and Velew, C. Van, 1982. Practical experiences with urea ensiled straw in Sri Lanka. Paper presented at the third annual seminar: Maximum Livestock Production from Minimum Land, Joydepbur, Bangladesh 15-17 Feb.

Preston, T.R., 1977. A strategy for cattle production in the tropics. World Anim. Rev. 21: 11-17.

Preston, T.R. and Parra, R., 1981. Utilization of tropical crop residues and agro-industrial by-products in animal nutrition: constraints and perspectives. FAO/ILCA Workshop on the utilization of agro-industrial by-products and crop-residues in animal feeding. Dakar, Senegal, 21-25 Sept.

Rekib, A., Mangat, A.S., Kochar, A.S. and Bhatia, I.S., 1970. Improvement for feed value of low grade roughage by incorporating urea and molasses into them. J. Res. 7: 347-358.

Saadullah, M., Haque, M. and Dolberg, F., 1982. Treated and untreated rice straw for growing cattle. Trop. Anim. Prod. 7: 20-25.

Saadullah, M., Haque, M. and Dolberg, F., 1982. Supplementation of alkali treated rice straw. Trop. Anim. Prod. 7: 187-190.

Scoville, O.J., 1976. Improving ruminant livestock production on small holdings. Seminar report. The Agricultural Development Council Inc. No 11, Oct. 1976.

Sharma, V.V., Jhanwar, D.M. and Taparia, A.L., 1972. Utilization of sorghum stover by cattle. Indian J. Anim. Sci. 42: 480-487.

Topps, J.H., 1972. Urea or biuret supplements to low-protein grazing in Afric. World Anim. Rev. 3: 14-18.

Ward, G.M., Sutherland, T.M. and Sutherland, J.M., 1980. Animals as an energy source in third world agriculture. Science (USA) 208: 570-574.

Chapter 14.2

STRAW ETC. IN PRACTICAL RATIONS FOR CATTLE

WITH SPECIAL REFERENCE TO DEVELOPED COUNTRIES

by

Verner Friis Kristensen

National Institute of Animal Science
Department of Research in Cattle and Sheep
Research Center Foulum
DK-8833 Ørum Sønderlyng
Denmark

14.2.1. INTRODUCTION

A few decades ago straw was a very essential part of the winter diet of cattle in virtually all countries. However, as more and more intensive production systems became economic in the so-called industrialized countries, the nutritionally-poor straw was replaced by higher quality roughages, where these could be produced in sufficient quantities. Ruminants became even large consumers of grain and other concentrates. Straw played a role only as a structure-rich component in certain diets. Continuous efforts to utilize straw as an essential energy source for cattle were seen only in places where other roughages were scarce.

In recent years the demand for crops which can be used for direct human consumption, as feed for monogastric animals or for industrial purposes has began to increase. Therefore the cost of producing normal high-quality roughages has risen, thus increasing the need to look for other sources of feedstuffs. A great abundance of straw has been present in many areas. Although use of straw for other purposes, e.g. heating, has increased, straw is still an interesting and essential alternative source of feed for cattle and its use as such has again began to increase in some regions. Different classes and ages of cattle have different nutrient requirements and hence need feeds of differing nutritive value. There is therefore space for some low-to-medium-quality feedstuffs in the diets of some cattle.

Because of the low nutritional value of straw, it will, in most cases need to be processed mechanically and/or chemically. Furthermore, it will need to be suitably supplemented with protein, minerals and vitamins.

14.2.2. MECHANICALLY PROCESSED STRAW

Mechanical treatment means a diminution of the particle size of the straw by chopping (shredding) or grinding and possibly pelleting (Chapter 5). The effects on digestion and animal performance with grinding roughages have been reviewed by Minson (1963), Beardsley (1964), Moore (1964), Greenhalgh and Wainman (1972) and Thomson (1972). It is generally concluded that grinding increases the feed intake, the improvement being larger the lower the digestibility of the roughage. This increase is owing to an increased rate of passage of the feed particles through the alimentary tract. Greenhalgh and Wainman (1972) pointed out that the effect of grinding on intake in beef cattle and sheep was small and might even be negative with high-concentrate, low-roughage diets. At this level of roughage, rumen fill is not limiting intake in these animals and an increased rate of passage is not likely to increase intake. On the other hand, absence of physical structure as a result of grinding the roughage could decrease intake.

It has also generally been concluded that the apparent digestibility of roughages is reduced by grinding. This is due to the reduced time of retention of feed particles in the rumen. This is especially the case at high feeding levels, e.g. when the diet is fed ad libitum (Thomsen, 1980).

A third conclusion was that the utilization of digested energy is increased as a result of grinding roughages. This increase may be due to lower energy costs of eating, ruminating and digesting, lower losses of energy in methane and fermentation heat in the rumen, and a change in the pattern of fermentation in the rumen. Reduction of particle size reduces the rumination time, the number and intensity of rumen contractions and the secretion of saliva (Nørgaard, 1981). It causes reduced pH, increased concentration of volatile fatty acids and a reduced ratio of acetate : propionate concentration in the rumen content (Jorgensen and Schultz, 1963). These changes in pattern of rumen fermentation favour deposition of body tissue and reduce synthesis of milk components in the mammary gland of dairy cows.

The increased utilization of digestible energy compensates for the reduction in digestibility. Some authors conclude that the net energy value of straw and other low quality roughages for young growing animals is improved by grinding (Greenhalgh and Wainman, 1972; Arnason, 1980c). Others report that net energy value does not change on grinding (Andersen and Sørensen, 1975).

Kristensen (1975) reviewed the results of experiments involving various amounts of chopped or ground straw in diets for growing steers and young bulls. The results of experiments with ground straw and pelleted diets were further analysed (Kristensen et al., 1981). Total dry matter intake was in general slightly increased when increasing proportions (up to 70%) of straw replaced concentrate in the diet (Figure 14.2.1). Arnason (1980c) found similar results. Andersen and Sørensen (1975) concluded that the increase in intake occurred at small inclusions of straw (20 to 30%) while further increments in the proportion of straw did not change intake. A recti-linear regression based on the results given in Figure 14.2.1 shows that when 50% concentrates was replaced by ground straw intake of total dry matter increased by only 0.5 kg.

In the same experiments daily live-weight gain was on average reduced by 55 g each time the percentage of straw in the diet increased by 10 (Figure 14.2.2). Gut fill is increased by increasing amounts of straw in the diet (Swan and Lamming, 1967; Jahn et al., 1970; Kay et al., 1970), and dressing percentage is reduced by nearly one unit for each 10% increase in straw content of the diet (Swan and Lamming, 1967 and 1970; Forbes et al., 1969; Raven et al., 1969; Kay et al., 1970; Andersen and Sørensen, 1972 and 1975). Because of this the reduction in carcass gain with increasing content of straw in the diet is relatively greater than the reduction in live-weight gain (Andersen and Sørensen, 1975). When

434

increasing proportions of straw in the diet of steers replaced
corn silage, daily live-weight gain was also greatly reduced (Lesoing et al., 1981).

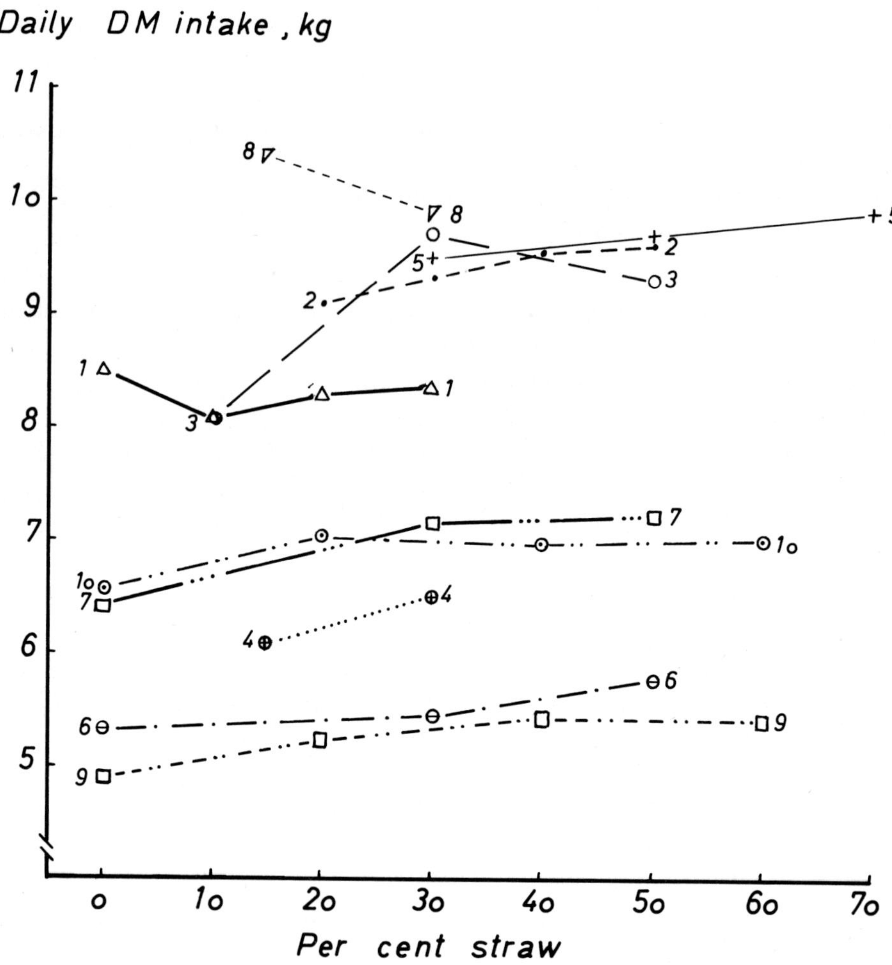

Figure 14.2.1. Effect on dry matter intake of an increasing percentage of ground straw in pelleted diets for growing
steers and bulls.
1) and 2): Lamming et al., 1966; 3) Swan and Lamming,
1967; 4) Pickard et al., 1969; 5) Swan and Lamming,
1970; 6 and 7) Kay et al., 1970; 8) Levy et al.,
1972; 9) Andersen and Sørensen, 1972; 10) Andersen
and Sørensen, 1975.

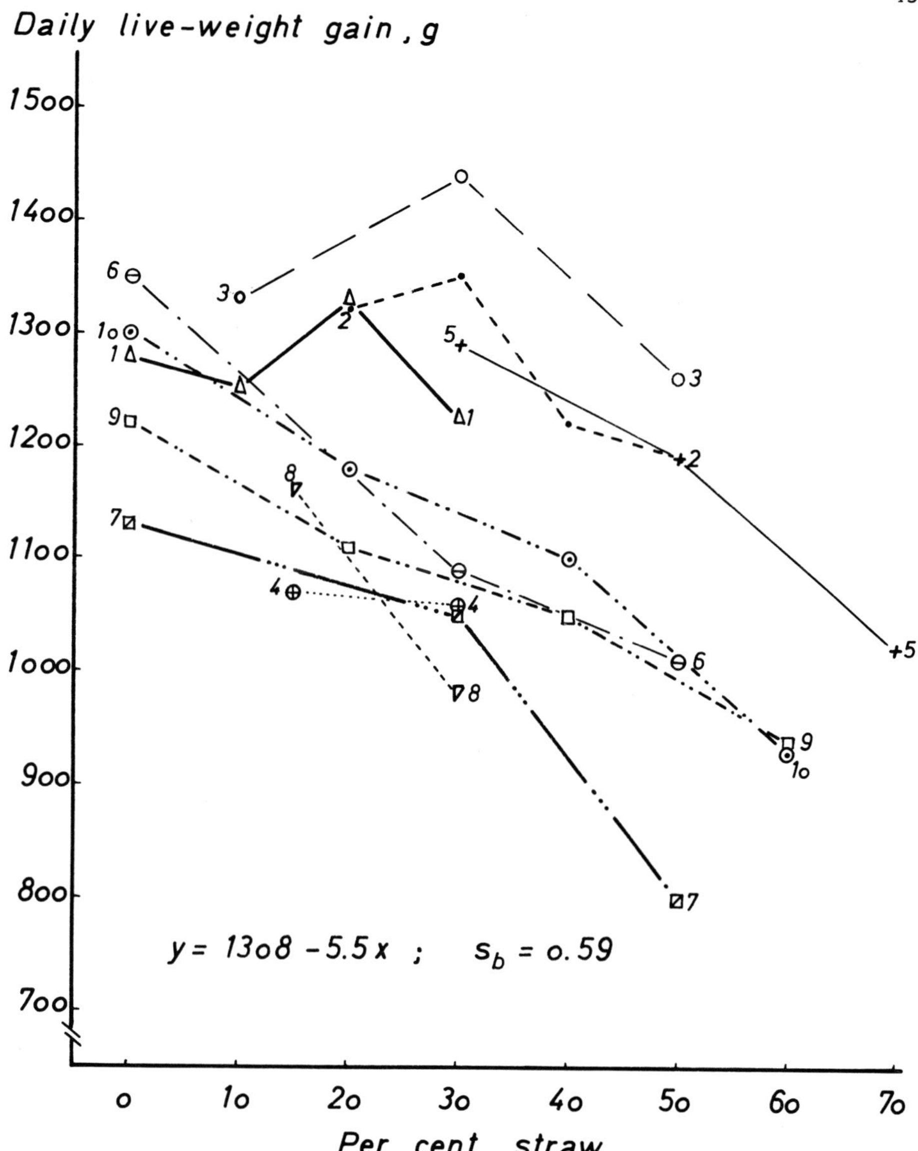

Figure 14.2.2. Effect on live-weight gain of an increasing percentage of ground straw in pelleted diets for growing steers and bulls. References as in Figure 14.2.1.

The utilization of small proportions of straw in a high-concentrate diet is poor because cellulolytic activity in the rumen under these circumstances is reduced (Chimwano et al., 1976) consequently digestibility of cell-wall constituents is lower, especially at higher feeding levels (Dulphy 1977, cited by Jackson. 1978; Thomsen, 1980). At a low level of inclusion, straw should be regarded as a structural component of the diet, rather than as an energy source and therefore should not be finely ground. In order to avoid disturbances of rumen function and animal health, diets with higher contents of ground straw should be supplied with a small amount of long straw or hay (Andersen and Sørensen, 1975; Arnason, 1980c).

The amounts of ground straw which can be used in the diets of growing or fattening animals depends on several factors i.e. the price and availability of alternative feedstuffs, the elongated time of housing, management and feeding and the possible effect on the quality and hence price of slaughtered animals.

Because of its low-energy concentration, unprocessed straw is not suited to form a substantial part of the roughage for high-yielding dairy cows. Thus with straw a high proportion of concentrates are needed in order to reach a sufficient level of nutrition. A high concentrate:roughage ratio results in reduced milk and butterfat synthesis and increased fat deposition in the body (Balch et al., 1955; Flatt et al., 1969). This was also demonstrated in experiments involving replacing hay by straw (Frank, 1982). Even if the straw is milled and/or pelleted feed intake and production is reduced at higher levels of straw in the diet, whereas at a lower straw level, butterfat synthesis is reduced and weight gain increased (Owen et al., 1971; Blair et al., 1974). In contrast to the effect on performance in growing animals, the changes in digestion and pattern of rumen fermentation resulting from grinding roughages have unfavourable consequences for production in dairy cows. Straw in the diet of dairy cows should not be ground, but when given in restricted amounts, it might be advantageous to chop or shred the straw and mix it with other feed components to form a complete diet.

14.2.3. CHEMICALLY TREATED STRAW

14.2.3.1. Diets for growing cattle

The utilization of chemically treated straw also depends on the proportion of straw in the diet, the composition of the diet, and the level of feeding. Furthermore, the nutritional value of chemically treated straw may be affected by the chemicals or their residues.

In some experiments the effect of alkali treatment on growth and energy utilization was found to be small when the diet contained only limited proportions of straw. At 35% rice straw in the diet the effect of alkali treatment was small and non-significant, while treatment when the rice straw constituted 72% of the diet DM significantly improved both growth rate and feed efficiency (Garrett et al., 1979). NaOH treatment of wheat straw or cotton stalks constituting 20 to 30% (Levy et al., 1977) or 35% of the diet of young bulls (Levy et al., 1980) was not reflected in improved live-weight or carcass gain, but the amount of depot fat in the body was significantly increased. Horton (1978) found that treatment of long straw with 3.5% NH_3 increased consumption and digestibility (dry matter and crude fibre) in young growing steers when the straw was fed without supplement. However, when 4 kg/d supplement was given, neither straw intake nor dry matter digestibility were improved.

Lesoing et al. (1981) obtained significant effects of NaOH treatment of wheat straw on the gain and feed efficiency of steers, both at 30% and 60% straw in the diet. In this case the diets contained only 9 to 15% concentrates, while the rest was maize silage.

In a Danish experiment young bulls fed rations containing 20% or 40% untreated or treated (5% NaOH) barley straw to appetite did not respond to treatment at the low level of straw in the diet, but with 40% straw live-weight and carcass gain were significantly improved by treatment (Table 14.2.1) (Kristensen et al., 1977). In another 10-month experiment (Andersen et al., 1980) bull calves were fed rations containing 25% untreated or NaOH-treated (4% NaOH) barley straw ad libitum in loose mixes. Other ingredients were 25% molasses and 50% concentrates. Live-weight gain was significantly improved from 1121 g/day to 1202 g/day due to NaOH-treatment. In a further experiment, a control ration containing 82% barley grain was compared to four rations in which 25% or 50% untreated or NaOH-treated barley straw replaced part of the grain in pelleted complete diets for young bulls (Henriksen, 1978). Feed intake was controlled to achieve equal daily gains. As is shown by the total feed consumption and calculated energy value, the effect of treatment appeared at both levels of straw (Table 14.2.2). Only the diet with 50% untreated straw was fed ad libitum. The decrease in energy utilization occurring when the content of straw was increased from 0 to 25% was greater than the decrease from 25% to 50% straw. Thus utilization of both untreated and treated straw was reduced at the low level of straw in the diet.

In recent experiments (H. Refsgaard Andersen, personal communication, 1983), young bulls were fed complete diets as loose mixes containing 20, 35 and 50% shredded NaOH-treated (4% NaOH) barley straw plus 25% molasses, the remainder being concentrates. These diets were compared to a control ration of pure concentrates supplemented with 1 to 2 kg of long straw and hay. The intake of diets containing treated straw was higher than expected and was considerably higher than that of the control ration. The feed efficiency however was poor. At 20% and 35% straw even negative net energy values for straw were calculated.

Table 14.2.1. Effect of NaOH treatment at different levels of barley straw in pelleted diets for young bulls (Kristensen et al., 1977)

	Untreated straw		NaOH-treated straw	
Straw in the diet, % of DM	20	40	20	40
No of animals	10	10	9	10
Days of experiment	158	176	166	159
Initial live weight, kg	224	234	222	233
Total DM intake, kg/d	7.8	8.5	7.8	8.5
Live-weight gain, g/d	1374	1209	1336	1322
Carcass gain, g/d	736	624	707	685

Table 14.2.2. Effects of NaOH treatment and level of inclusion of straw in pelleted diets for young bulls (Henriksen, 1978)

	Control	Untreated		NaOH-treated	
Straw in the diet, %	0	26	52	25	51
Digestibility of OM, %	73	67	64	68	67
Live-weight gain, g/day	1079	1013	1027	1015	1109
Total feed consumption, kg DM	467	605	765	524	617
Relative net energy value[1]	100	76	60	85	76

[1]Calculated on the basis of carcass gain and feed consumption

Thus the results of alkali treatment at low proportions of straw in the diet show some discrepancies. With small inclusions of straw the change in feeding value of the total diet resulting from alkali treatment of the straw will be small and therefore difficult to detect in feeding experiments. Additionally the discrepancies may partly be explained by differences in level of feeding and composition of the diet. If the diet contains a great deal of other roughages utilization of the straw would not be expected to be depressed as in a high concentrate diet.

Certain limitations on the utilization of alkali-treated straw may be due to reduced time of retention of feed particles in the forestomachs. This seems to be the case on treating with NaOH (Coombe et al., 1979; Berger et al., 1980) or NH_3 (Oji et al., 1979). It results in an increased amount of potentially digestible feed escaping digestion in the rumen (Berger et al., 1979). The higher rate of passage may be due to the increased proportion of organic matter in straw solubilized after alkali treatment. Also in the case of NaOH treatment (dry and semidry methods) water consumption is increased as a result of the high intake of sodium.

It may be concluded that at low proportions of straw and high proportions of easily fermentable carbohydrates, the effect of alkali treatment is small or even absent. Treatment therefore is probably not economic in these situations.

At higher proportions of straw or other roughages the effect of alkali treatment is more consistent. Arnason (1980a) decreased the amount of supplement fed with treated straw, when he compared untreated and NaOH-treated barley straw to young bulls (Table 14.2.3). In experiment K-131 food intake and weight gain were very high during the first 12 weeks of the experimental period. Then, serious health problems arose in the group fed treated straw and the weight gain of this group fell drastically. This problem will be discussed later (14.2.4). In experiment K-145, where straw was fed in restricted amounts, approximately 1 kg DM in concentrate could be saved without reducing the gain, when the straw was treated.

Table 14.2.4 summarizes Norwegian experiments with NH_3-treated straw to growing bulls. In the trial by Arnason (1977) the intake of treated straw was equal to that of grass silage, but the weight gain was lower on straw than on silage. In the trial by Kvåle and Homb (1976) treatment of straw fed in a restricted amount (3.5 kg DM/d) increased daily live-weight gain by 200 g, but in a similar experiment by Kvåle and Homb (1977) no response to treatment appeared. In the experiment by Kvåle (1978) treatment increased the intake of staw by 31% and the weight gain increased from 912 to 1024 g/day.

Table 14.2.3. Comparison of untreated long and NaOH treated[1] chopped straw for young bulls (Arnason, 1980a)

Experiment	K-131		K-145	
NaOH treatment rate, %	0	4.1	0	4.3
Digestibility of OM, %[2]	53	67	55	68
Initial weight, kg	175		275	
Days of experiment	247		176	
Food intake, kg DM/d				
Concentrate	3.8	2.1	3.2	2.3
Hay	–	–	0.9	0.9
Straw	3.0[3]	4.7	2.5[4]	3.4
Live-weight gain, g/d				
0-12 weeks	1225	1225	–	–
Total exp. period	1056	937	906	948

[1]Dry treatment
[2]Data for straw alone. Determined in sheep fed restricted
[3]First 12 weeks ad lib., thereafter maximum 4 and 6 kg untreated and treated straw, respectively
[4]Maximum 3 and 5 kg of untreated and treated straw, respectively

Results of Norwegian experiments with NH_3 treated straw to steers are shown in Table 14.2.5. The straw was fed ad libitum and its intake was increased by 43%, 26% and 67% by treatment in experiments 1, 2 and 3, respectively. In experiment 3 the amount of concentrates fed to animals receiving treated straw was reduced. Daily weight gain was increased by 130 and 315 g in experiments 1 and 2 respectively, and in experiment 3 treatment increased weight gain by nearly 100 g/day even though concentrates were reduced by 1.6 kg/day.

Braman and Abe (1977) fed rations with 50% ground straw, untreated or treated with 4% NaOH, to steers. The intake was increased 21% and the daily gain 270 g by treatment. In steers fed concentrates at 1% of body weight the intake of barley, oat and wheat

Table 14.2.4. Results of experiments with ammonia-treated straw in the diet of growing bulls
in Norway

Experiment	Arnason, 1977		Kvåle and Homb, 1976		Kvåle and Homb, 1977		Kvåle, 1978	
Number of animals	20		28		28		28	
Days of experiment	174		151		172		134	
Initial live weight, kg	284		312		276			
Daily rations, kg:								
Concentrates	2.0	2.0	2.0	2.0	2.0	2.0	2.4	2.4
Hay (Distillery slop)	1.0	1.0	2.0(25)	2.0(25)	(30)	(30)	(30)	(30)
Grass silage (DM)	4.34 (ad lib.)	–			6.5	6.5		
Untreated straw			3.5		3.4		3.5 (ad lib.)	
Ammonia treated straw	4.48 (ad lib.)		3.5		3.4		4.6 (ad lib.)	
Daily gain, g	1025	840	669	855	1161	1075	912	1024

442

straw increased by 10, 17 and 27%, respectively as a result of treatment with ammonia (Horton and Steacy, 1979). Improvements were largest for straws with relatively low intakes. Chopped or pelleted barley straw was fed untreated or treated with 4% NaOH to steers together with 1 kg supplement (Coombe et al., 1979). Treatment increased the intake of chopped straw by 29% while both pelleting and NaOH-treatment increased intake by 72% compared to untreated chopped straw. Horton et al. (1982) fed steers with diets containing 40% wheat straw and measured the intake of total diet DM. Ammonia treatment increased intakes by 11% and 22% when the straw was shredded and ground/pelleted, respectively. The daily gain was increased by 280 to 300 g by treatment. In the same study, treatment with 5% NaOH and pelleting increased total DM intake by 16% and daily live-weight gain by 390 g.

Table 14.2.5. Results of experiments with ammonia-treated straw in the diet of growing steers in Norway

Experiment reference	Pestalozzi and Matre, (1976)		Pestalozzi and Matre, (1977)		Sundstøl and Matre, (1980)	
and number	1		2		3	
Number of animals	8		10		28	
Days of experiment	84		112		168	
Initial live weight, kg	277		282		375	
Daily rations, kg:						
Concentrates	2.0	2.0	2.0	2.0	1.9	0.25
Grass silage	5.0	5.0	3.0	3.0	–	–
Untreated straw	2.8		3.4		6.0	
Ammonia treated straw		4.0		4.3		10.0
Daily gain, g	530	660	215	530	349	434

In an experiment with dairy heifers fed 1.3 kg DM/d of molasses and fodderbeet and 0.6 kg/d soyabean meal, ammonia treatment increased intake of long barley straw by 7% and daily weight gain by 106 g (Foldager, 1978). In dairy heifers the intake of NaOH-treated (wet treatment) barley straw was increased by 33% (Arnason, 1980b) when 1.3 kg/d less concentrate DM was fed with treated straw. Daily weight gain was 707 and 772 g on the diets with untreated and treated straw respectively. The intake of a straw-only diet in heifers (untreated straw supplied with urea) was increased by 51% by ammonia treatment so that the daily live-weight change improved from -447 g to 324 g (Ørskov et al., 1981). Intake of wheat straw in dairy heifers given 0.9 kg of supplement was increased by 30% by ammonia treatment (Saenger et al., 1983).

These results show great potential for using alkali-treated straw for young animals when only limited growth rates are desired. Where straw has constituted more than 40% of the diet alkali treatment has increased intake of straw by about 25%, on average. The increases have varied between 10 and 50%. The daily live-weight gain has increased by 100 to 300 g. The main reasons for variations in the effect of treatment are the digestibility and nutritional value of the untreated straw and the method and efficiencies of alkali-treatment. The effect is greater the lower the feeding value of the untreated straw. Wet treatment with NaOH is more efficient than dry treatment, which in turn is more efficient than ammonia treatment, in increasing the nutritional value of straw (see Chapters 6 and 7).

Dairy heifers required to gain about 600 g/day may have 70 to 80% of the total ration dry matter as alkali-treated straw, provided it is supplemented with easily digestible feeds containing the necessary nutrients. Diets of alkali-treated straw alone supplemented with nitrogen, minerals and vitamins have given daily gains of about 400 g.

At high levels of straw in the diet the supply of amino acids to the animal may be too low to support maximum growth if only non protein nitrogen is given as nitrogen supplement. Smith et al. (1980) found that a supply of rumen-undegradable protein is more critical with high than with low fibre diets. This matter is discussed in Chapter 13. Benefical effects of protein-N over non-protein-N on intake and growth rate have been shown in several experiments with lambs and steers fed diets rich in straw (Hasimoglu et al., 1969; Saxena et al., 1971; Kalinowski, 1974; Rounds et al., 1976; Braman and Abe, 1977).

The utilization of N from ammonia bound in ammonia-treated straw is discussed in Chapter 7. Available evidence indicates that this source of nitrogen acts in the same way as other sources of NPN in ruminants.

444

14.2.3.2. Diets for dairy and beef cows

Untreated chopped barley straw was compared to NaOH-treated (4% NaOH) chopped straw in rations for dairy cows (Table 14.2.6). The intake of straw was increased by 30% by NaOH-treatment. The milk yield and the fat content of the milk was significantly increased, when cows were fed treated straw.

Greenhalgh et al. (1976) fed complete rations containing 50% concentrates and 50% either untreated or NaOH-treated (8% NaOH) barley straw. The DM intake was increased from 10.8 to 13.4 kg by treatment. The milk yield increased from 17.6 to 19.0 kg/d and the fat content of the milk from 3.54 to 3.74%.

The intake of long oat straw in cows was increased by 15% after treatment with NH_3 (Rissanen and Kossila, 1977) (Table 14.2.7). In this experiment, the straw was supplemented with increasing amounts of grass silage. The intake of treated straw was decreased by approximately 0.6 kg DM for each additional kg DM given as grass silage. However, the amount of concentrate was reduced with increasing amounts of silage. Kristensen and Andersen (1980b) fed NH_3-treated, long barley straw ad lib. together with constant amounts of concentrate and increasing amounts of grass silage. The intake of straw DM was decreased by 0.8 kg/kg increase in the amount of DM in grass silage (Table 14.2.8). The in vitro digestibility of the grass silage was 70% and that of straw 58%.

Table 14.2.6. Daily feed intake, milk yield and live-weight change in dairy cows fed untreated and NaOH-treated, chopped barley straw (Kristensen and Andersen, 1980a)

	Untreated	Treated
Concentrates, kg DM	6.9	6.9
Molasses, kg DM	3.8	3.8
Straw ad lib., kg DM	5.7	7.4
Milk, kg/d	24.0	24.6
Milk fat, %	3.34	3.64
4% FCM, kg	21.6	23.3
Weight change, g/d	136	414

Mo (1978) replaced some high quality grass silage by NH_3treated barley straw to dairy cows. The cows consumed 2.7 and 3.2 kg DM NH_3-treated long straw together with 5.2 kg silage DM plus

7.5 kg concentrate (Table 14.2.9). The straw replaced approximately 3.5 kg silage DM and was supplied with approximately 1 kg concentrate more than the pure silage. There was no significant effect on milk yield.

Table 14.2.7. Daily feed intake, milk yield and total live-weight change in dairy cows fed untreated and NH_3-treated long oat straw in rations with varying amounts of grass silage (Rissanen and Kossila, 1977)

Straw	Un-treated	Trea-ted	Un-treated	Trea-ted	Un-treated	Trea-ted
Straw ad lib., kg DM	6.1	6.9	4.4	5.1	2.6	3.0
Silage, kg DM	-	-	3.2	3.2	6.5	6.3
Concentrates, kg	9.3	8.8	7.5	7.3	5.3	5.2
Dry molassed beet pulp, kg	2.0	1.8	2.0	1.9	1.8	1.9
4% FCM, kg	19.0	19.4	18.9	18.5	18.7	19.5
Milk fat, %	4.38	4.41	4.43	4.49	4.30	4.65
Weight change, kg	19	23	16	22	25	31

Table 14.2.8. Daily feed intake, milk yield and live-weight change in dairy cows fed NH_3-treated, long barley straw in rations with varying amounts of grass silage (Kristensen and Andersen, 1980b)

Silage, kg DM	-	3.0	5.9
Straw ad lib., kg DM	6.0	3.5	1.2
Concentrates, kg	7.0	7.0	7.0
Fodder beets, kg DM	2.2	2.2	2.2
Milk, kg	20.9	20.6	21.4
Milk fat, %	3.87	3.79	3.86
Weight change, g	235	314	308

Garmo and Arnason (1980) compared barley straw treated with alkali by different procedures. The straw was either treated with NaOH by soaking and washing of long straw (wet) or by mixing chopped straw with a concentrated alkali solution followed by storing (dry method). The third method was treatment with ammonia by adding liquid ammonia to a sealed stack of long straw. The three types of straw were given to dairy cows in different amounts giving equal calculated amounts of net energy (Table 14.2.10). There were no significant differences in milk yield.

Table 14.2.9. Replacement of grass by NH3-treated barley straw to dairy cows (Mo, 1978)

	Experiment 1		Experiment 2	
Group	1	2	1	2
Silage, kg DM/d	5.2	8.5	5.2	8.8
Straw, kg DM/d	3.2	–	2.7	–
Concentrates, kg/d	7.6	6.8	7.4	6.2
Milk, kg/d	21.4	21.8	20.3	20.4
Milk fat, %	3.64	3.61	3.69	3.77
4% FCM, kg/d	20.2	20.6	19.4	19.6

Hermansen (1983) compared diets containing 28% untreated and 41% ammonia-treated barley straw for dairy cows. The diets were planned to have the same feeding value. The ad libitum feed intakes were equal, but the diet with ammonia-treated straw promoted a significantly higher milk yield (21.0 v. 19.7 kg/d) and fat content in the milk (4.12 v. 3.89%).

Table 14.2.10. Comparison of different methods of alkali treatment of barley straw for dairy cows (Garmo and Arnason, 1980)

	Experiment 1			Experiment 2		
Group	1	2	3	1	2	3
Silage, kg DM/d	4.3	4.3	4.3	4.1	4.1	4.1
Straw, kg DM/d	3.3[1]	3.9[2]	3.8[3]	2.9[1]	3.7[2]	3.5[3]
Concentrates, kg/d	6.4	6.8	6.6	7.9	7.9	8.1
4% FCM, kg/d	18.2	19.4	18.4	19.8	19.3	19.0

[1] NaOH (wet method). DOM in sheep: 74%
[2] NaOH (dry method). DOM in sheep: 70%
[3] NH3 (dry method). DOM in sheep: 61%

Most of these experiments have been carried out with cows in mid-lactation and with a moderate level of milk yield. In such cases it is possible to exchange part of a high quality roughage with alkali-treated straw without negative effects on performance.

In late lactation and the dry period even greater amounts of straw may be used without deleterious effects.

The experiment of Kristensen and Andersen (1980a) (Table 14.2.6) was carried out with cows in early lactation. The results demonstrate that treated straw stimulates milk production and milk-fat synthesis better than untreated straw. However, even with treated straw the pattern of rumen fermentation may not have been optimal for the desired partitioning of nutrients to maximise milk production in preference to deposition of body fat. The fat content of the milk in the group fed treated straw was below the normal and desired level for these cows, and the weight gain was rather high. High-yielding cows require higher quality roughages in order to obtain optimal composition of the ration.

Beef cows for a great part of the year do not require high quality roughages. Acock et al. (1979) found that alkali-treated straw could replace a great part of the conventional hay diet for beef cows.

14.2.4. EFFECTS OF SODIUM HYDROXIDE ON MINERAL METABOLISM AND ANIMAL HEALTH

In the experiment by Arnason (1980a) young bulls fed chopped, NaOH-treated straw consumed large amounts of treated straw and grew fast during the first 12 weeks of the period (Table 14.2.3). Then, health problems arose. The animals developed diarrhoea and did not thrive. Their weight gain was reduced drastically although the feed intake remained high. Similar problems have not seemingly been reported in other places.

In the same experiment the blood serum concentrations of P and Mg were significantly reduced. In another experiment also, the serum Mg concentration was significantly reduced in animals fed NaOH-treated straw (Arnason, 1980d).

The high amount of sodium in NaOH-treated straw is excreted in the urine. This may change the metabolism of other minerals. Arndt et al. (1980) found that increasing ratios of NaOH-treated cotton plant by-products in the diet increased Na-balance and decreased K, Cl and Mg balances. There was no effect on Ca or P balances.

In the case of abrupt changes from a normal diet to a diet high in NaOH-treated straw for high-yielding dairy cows in early lactation, severe cases may arise, with symptoms indicating lack of Mg

(Kristensen et al., 1977). Blood samples taken while acute symptoms occurred showed strongly reduced serum Mg levels. However, in animals accustomed to consuming large amounts of NaOH-treated straw no visible problems normally seem to occur. Recently an experiment has been carried out in which dairy cattle were fed large amounts of NaOH-treated straw as the only roughage from the beginning of the first pregnancy until the third lactation and no clinical problems were observed (P.E. Andersen, personal communication, 1983). However, the serum Mg level of these animals was significantly reduced compared to animals fed normal diets (S. Boisen, personal communication, 1983). Increased supplementation of certain minerals (K, Cl, Mg) to animals fed large amounts of NaOH-treated straw may therefore be helpful. Lesoing et al. (1981) concluded that balancing with minerals increased weight gain in steers.

14.2.5. SUMMARY

Straw may play a role in the diet of cattle either as a structure-rich component or as a source of energy. If straw is to serve as a substantial energy source its intake and/or nutritional value must be increased by processing. The use of mechanically treated or chemically treated straw in cattle diets is discussed.

Because of its low nutritional value straw is not suited as an energy source for intensive beef production nor for high yielding dairy cows. It may form a substantial part of the diet for cattle, which do not require energy-rich diets, i.e. replacement heifers, winter-fed steers (stores), low yielding and dry dairy cows and beef cows. The utilization of straw as an energy source is poor in diets low in roughages and high in concentrates, when the feeding regime is ad libitum. The utilization of NaOH-treated straw is especially poor under these circumstances, probably because of decreased retention time in the rumen. Straw given in high-concentrate diets should be regarded only as a structural component and should therefore be fed long or only roughly comminuted.

Grinding of straw increases intake, reduces digestibility, and causes only minor changes in the energy value when fed ad libitum. Replacing increasing proportions of concentrates in the diet of young growing cattle by ground straw results in small increases in total dry matter intake but reduces daily live-weight gain by about 55 g for each 10 percentage units of straw increase.

When straw consitutes more than 40% of diet dry matter alkali treatment increases straw intake by 10 to 50%, depending on digestibility of the untreated straw and the efficiency of the treatment. This results in daily live-weight gain of growing animals increasing by 100 to 300 g. Replacement heifers may grow sufficiently when alkali-treated straw comprises 70 to 80% of total diet dry matter. Gains of 400 g/d have been observed on alkali-treated straw alone but supplemented with nitrogen, minerals and vitamins. A supplement high in rumen-undegradable protein may be necessary in high-straw diets.

Alkali-treated straw may replace most of the higher-quality roughages in the diets of low-yielding dairy cows or beef cows. Ground straw is unsuitable in the diet of dairy cows.

450

14.2.6. REFERENCES

Acock, C.W., Ward, J.K., Rush, I.G. and Klopfenstein, T.J., 1979. Wheat straw and sodium hydroxide treatment in beef cow rations. J. Anim. Sci. 49: 354-360.
Andersen, H.R., Andersen, B.B., Møller, E., Klastrup, S., Philipsen, H. and Jensen, A.M., 1980. Fuldfoder med ludbehandlet (NaOH) kontra ubehandlet byghalm til ungtyre. Statens Husdyrbrugsforsøg, Copenhagen, Medd. No. 326, 4 pp.
Andersen, H.R. and Sørensen, M., 1972. Formalet halm som bestanddel af fuldfoder til fedekalve. Landøkonomisk Forsøgslaboratoriums Årbog, 356-370.
Andersen, H.R. and Sørensen, M., 1975. Statens Husdyrbrugsforsøg, Copenhagen, Medd. No. 19, 4 pp.
Arnason, J., 1977. Halm i storfekjøttproduksjonen. Husdyrforsøksmøtet, Agric. Univ. Norway.
Arnason, J., 1980a. Tørrlutet (NaOH) halm som fôr til slakte okser. Sci. Rep., Agric. Univ. Norway, 59 (26), 18 pp.
Arnason, J., 1980b. Halm våtlutet og nøytralisert etter Bolidenmetoden som fôr til kviger. Sci. Rep., Agric. Univ. Norwa 59 (27), 17 pp.
Arnason, J., 1980c. Halmfôr (pelletert halm og kraftfôr) som fôr til unge okser. Sci. Rep., Agric. Univ. Norway 59 (28) 20 pp.
Arnason, J., 1980d. Treated straw as feed for young cattle. Thesis. Dept. Anim. Nutr., Agric. Univ. Norway. 83 pp.
Arndt, D.L., Richardson, C.R., Albin, R.C. and Sherrod, L.B., 1980. Digestibility of chemically treated cotton plant byproduct and effect on mineral balance, urine volume and pH. J. Anim. Sci. 51: 215-223.
Balch, C.C., Balch, D.A., Bartlett, S., Bartum, M.P., Johnson, V.W., Rowland, S.J. and Turner, J., 1955. Studies of the secretion of milk of low fat content by cows on diets low in hay and high in concentrate. VI. The effect on the physical and biochemical processes of the reticulorumen. J. Dairy Res. 22: 270-289.
Beardsley, D.W., 1964. Symposium on forage utilization: Nutritive value of forage as affected by physical form. II. Beef cattle and sheep studies. J. Anim. Sci. 23: 239-245.
Berger, L.L., Klopfenstein, T.J. and Britton, R.A., 1979. Effect of sodium hydroxide on efficiency of rumen digestion. J. Anim. Sci. 49: 1317-1323.
Berger, L.L., Klopfenstein, T.J. and Britton, R.A., 1980. Effect of sodium hydroxide treatment on rate of passage and rate of ruminal fiber digestion. J. Anim. Sci. 50: 745-749.
Blair, T., Christensen, D.A. and Manns, J.G., 1974. Performance of lactating dairy cows fed complete pelleted diets based on wheat straw, barley and wheat. Can. J. Anim. Sci. 54: 347-354.
Braman, W.L. and Abe, R.K., 1977. Laboratory and in vivo evaluation of the nutritive value of NaOH-treated wheat straw. J. Anim. Sci. 46: 496-505.
Chimwano, A.M., Ørskov, E.R. and Stewart, C.S., 1976. Effect of dietary proportions of roughage and concentrate on rate of digestion of dried grass and cellulose in the rumen of sheep. Proc. Nutr. Soc. 35: 101A-102A.
Coombe, J.B., Dinus, D.A. and Wheeler, W.E., 1979. Effect of alkali treatment on intake and digestion of barley straw by beef steers. J. Anim. Sci. 49: 169-176.

Flatt, W.P., Moe, P.W., Moore, L.A., Hooven, N.W., Lehmann, R.P. and Ørskov, E.R., 1969. Energy utilization by high producing dairy cows. 1. Experimental design, ration composition, digestibility data, and animal performance during energy balance trials. In: K.L. Blaxter, J. Kielanowski and G. Thorbek (Eds.): Energy Metabolism of Farm Animals. Proc. 4th Symp. Sept. 1967, Warsaw, Poland. EAAP Publication No 12, 221-234.

Foldager, J., 1978. Halm til opdræt. Paper presented at NJF-seminar: Halm; håndtering, behandling og udnyttelse. 28-31 March, Middelfart, Denmark.

Forbes, T.J., Irwin, J.H.D. and Raven, A.M., 1969. The use of coarsely chopped barley straw in high concentrate diets for beef cattle. J. agric. Sci. (Camb.) 73: 347-354.

Frank, B., 1982. Untreated barley straw in dairy cow rations. Substitution of straw for hay. Swed. J. agric. Res. 12: 137-147.

Garmo, T. and Arnason, J., 1980. Beckmann-luta, ammoniakkbehandla og tørrluta (NaOH) halm som fôr til mjølkekyr. Husdyrforsøksmøtet 1980, Aktuelt fra Landbruksdepartementets opplysningstjeneste nr 1, 405.

Garrett, W.N., Walker, H.G., Kohler, G.O. and Hart, M.R., 1979. Response of ruminants to diets containing sodium hydroxide or ammonia treated rice straw. J. Anim. Sci. 48: 92-103.

Greenhalgh, J.F.D., Pirie, R. and Reid, G.W., 1976. Alkali-treated barley straw in complete diets for lambs and dairy cows. Anim. Prod. 22: 159 (Abstr.).

Greenhalgh, J.F.D. and Wainman, F.W., 1972. The nutritive value of processed roughages for fattening cattle and sheep. Proc. Br. Soc. Anim. Prod. 1: 61-72.

Hasimoglu, S., Klopfenstein, T.J. and Doane, T.H., 1969. Nitrogen source with sodium hydroxide treated wheat straw. J. Anim. Sci. 29: 160 (Abstr.).

Henriksen, J., 1978. In vitro teknik til bestemmelse af kvægfoders energiværdi. Thesis. Institute of Animal Science. Royal Veterinary and Agricultural University, Copenhagen. 119 pp.

Hermansen, J.E., 1983. NH_3-behandlet contra ubehandlet halm i blandinger til malkekøer. Statens Husdyrbrugsforsøg, Copenhagen. Medd. No. 451, 4 pp.

Horton, G.M.J., 1978. The intake and digestibility of ammoniated cereal straws by cattle. Can. J. Anim. Sci. 58: 471-478.

Horton, G.M.J., Nicholson, H.H. and Christensen, D.A., 1982. Ammonia and sodium hydroxide treatment of wheat straw in diets for fattening steers. Anim. Feed. Sci. Technol. 7: 1-10.

Horton, G.M.J. and Steacy, G.M., 1979. Effect of anhydrous ammonia treatment on the intake and digestibility of cereal straws by steers. J. Anim. Sci. 48: 1239-1249.

Jackson, M.G., 1978. Treating straw for animal feeding. FAO Animal Production and Health Paper No. 10, FAO, Rome, 81 pp.

Jahn, E., Chandler, P.T. and Polan, C.E., 1970. Effect of fibre and ratio of starch to sugar on performance of ruminating calves. J. Dairy Sci. 53: 460-474.

Jorgensen, N.A. and Schultz, L.H., 1963. Ratio effects on rumen acids, ketogenesis and milk composition. I. Unrestricted roughage feeding. J. Dairy Sci. 46: 437-442.

Kalinowski, J., 1974. Effect of sodium hydroxide treatment and nitrogen supplementation of barley straw upon intake, digestibility and live-weight change in growing lambs. M Phil. Thesis, University of Reading, UK.

Kay, M., McDearmid, A. and Massie, R., 1970. Intensive beef production. II. Replacement of cereals with ground straw. Anim. Prod. 12: 419-424.

Kristensen, V.F., 1975. Chemical and physical treatment of straw. Paper presented at EEC-seminar. Improving the Nutritional Efficiency of Beef Production, Theix, France, 14-17 October, 1975.

Kristensen, V.F. and Andersen, P.E., 1980a. Ubehandlet og natronludbehandlet, snittet byghalm til malkekøer. Statens Husdyrbrugsforsøg, Copenhagen. Medd. No. 334, 4 pp.

Kristensen, V.F. and Andersen, P.E., 1980b. Ammoniakbehandlet byghalm til malkekøer. Statens Husdyrbrugsforsøg, Copenhagen. Medd. No 332, 4 pp.

Kristensen, V.F., Andersen, P.E., Stigsen, P., Thomsen, K.V., Andersen, H.R., Sørensen, M., Ali, C.S., Mason, V.C., Rexen, F., Israelsen, M. and Wolstrup, J., 1977. Sodium hydroxide treated straw as feed for cattle and sheep. 464. beretn., Statens Husdyrbrugsforsøg, Copenhagen, 218 pp.

Kristensen, V.F., Israelsen, M. and Neimann-Sørensen, A., 1981. Processed feed from straw for ruminants. In: Y. Pomeranz and L. Munck (Eds.): Cereals. A renewable resource. The American Association of Cereal Chemists, St. Paul, Minnesota, 589-610.

Kvåle, S.E., 1978. Oppdrettsforsøg med avfall fra potetindustrien som fôr til okser. Mimeographed paper. Dept. Anim. Nutr., Agric. Univ. Norway.

Kvåle, S.E. and Homb, T., 1976. Forsøk med halm og drank til okser. Mimeographed paper, Dept. Anim. Nutr., Agric. Univ. Norway.

Kvåle, S.E. and Homb, T., 1977. Produksjonsforsøk med avfall fra potetindustrien. Mimeographed paper. Dept. Anim. Nutr., Agric. Univ. Norway.

Lamming, G.E., Swan, H. and Clarke, R.T., 1966. Studies on the nutrition of ruminants. 1. Substitution of maize by milled barley straw in a beef fattening diet and its effect on performance and carcass quality. Anim. Prod. 8: 303-311.

Lesoing, G., Rush, I., Klopfenstein, T. and Ward, J., 1981. Wheat straw in growing cattle diets. J. Anim. Sci. 51: 257-262.

Levy, D., Amir, S., Holzer, Z. and Neumark, H., 1972. Ground and pelleted straw and hay for fattening Israeli-Friesian male calves. Anim. Prod. 15: 157-165.

Levy, D., Holzer, Z. and Folman, Y., 1980. Chemical processing of wheat straw and cotton by-products for fattening cattle. Anim. Prod. 31: 27-33.

Levy, D., Holzer, Z., Neumark, H. and Folman, Y., 1977. Chemical processing of wheat straw and cotton by-products for fattening cattle. 1. Performance of animals receiving the wet material shortly after treatment. Anim. Prod. 25: 27-37.

Minson, D.J., 1963. The effect of pelleting and wafering on the feeding value of roughage. A review. J. Br. Grassl. Soc. 18: 39-44.

Mo, M., 1978. Ammoniakkbehandlet halm i fôrrasjonen til melkekyr. Paper presented at NJF seminar, 28-31 March, Middelfart, Denmark, 4 pp.

Moore, L.A., 1964. Symposium on forage utilization: Nutritive value of forage as affected by physical form. 1. General principles involved with ruminants and effect of feeding pelleted or wafered forage to dairy cattle. J. Anim. Sci. 23: 230-238.

Nørgaard, P., 1981. Foderstrukturens indflydelse på lakterende køers tyggeaktivitet og vommotorik. Thesis. Institute of Animal Science, Royal Veterinary and Agricultural University, Copenhagen, 161 pp.

Oji, U.I., Nowat, D.N. and Buchanan-Smith, J.G., 1979. Nutritive value of thermoammoniated and steam-treated maize stover. II. Rumen metabolites and rate of passage. Anim. Feed Sci. Technol. 4: 187-197.

Ørskov, E.R., Tait, C.A.G. and Reid, G.W., 1981. Utilization of ammonia- or urea-treated barley straw as the only feed for dairy heifers. Anim. Prod. 32: 388 (Abstr.).

Owen, J.B., Miller, E.L. and Bridge, P.S., 1971. Complete diets given ad libitum to dairy cows; the effect of straw content and of cubing the diet. J. Agric. Sci. 77: 195-202.

Pestalozzi, M. and Matre, T., 1976. Forsøk med ammoniakkbehandlet halm til kastrater. Mimeographed paper. Dept. Anim. Nutr., Agric. Univ. Norway.

Pestalozzi, M. and Matre, T., 1977. Forsøk med ammoniakkbehandlet halm til kastrater. Mimeographed paper. Dept. Anim. Nutr., Agric. Univ. Norway.

Pickard, D.W., Swan, H. and Lamming, G.E., 1969. Studies on the nutrition of ruminants. 4. The use of ground straw of different particle sizes for cattle from twelve weeks of age. Anim. Prod. 11: 543-550.

Raven, A.M., Forbes, T.J. and Irwin, J.H.D., 1969. The utilization by beef cattle of concentrate diets containing different levels of milled barley straw and of protein. J. agric. Sci. (Camb.) 73: 355-363.

Rissanen, H. and Kossila, V., 1977. Untreated and ammonized straw with or without silage to dairy cows. In: Quality of Forage. Proc. seminar NJF, 20-22 April, Uppsala, Sweden. Inst. Husdjurens utfodring och vård, Rapport No 54: 177-180.

Rounds, W., Klopfenstein, T., Waller, J. and Messersmith, T., 1976. Influence of alkali treatments of corn cobs on in vitro dry matter disappearance and lamb performance. J. Anim. Sci. 43: 478-482.

Saenger, P.F., Lemenager, R.P. and Hendrix, K.S., 1983. Effects of anhydrous ammonia treatment of wheat straw upon in vitro digestion, performance and intake by beef cattle. J. Anim. Sci. 56: 15-20.

Saxena, K., Otterby, D.E., Donker, J.D. and Good, A.L., 1971. Effects of feeding alkali treated oat straw supplemented with soybean meal or nonprotein nitrogen on growth of lambs and on certain blood and rumen liquor parameters. J. Anim. Sci. 33: 485-490.

Smith, T., Broster, V.J. and Hill, R.E., 1980. A comparison of sources of supplementary nitrogen for young cattle receiving fibre-rich diets. J. agric. Sci. (Camb.) 95: 687-690.

Sundstøl, F. and Matre, T., 1980. Bruk av ammoniakkbehandla halm i kjøttproduksjonen. Husdyrforsøksmøtet, Agric. Univ. Noray. 399-404.

Swan, H. and Lamming, G.E., 1967. Studies on the nutrition of ruminants. II. The effect of level of crude fiber in maize-based rations on the carcass composition of Friesian steers. Anim. Prod. 9: 203-208.

Swan, H. and Lamming, G.E., 1970. Studies on the nutrition of ruminants. 5. The effect of diets containing up to 70% ground barley straw on the live-weight gain and carcass composition of yearling Friesian cattle. Anim. Prod. 12: 63-70.

Thomsen, K. Vestergaard, 1980. The nutritional improvement of low quality forages. In: C. Thomas (Ed.): Forage Conservation in the 80's. Occasional Symposium No. 11. British Grassland Society, 164-174.

Thomson, D.J., 1972. Physical form of the diet in relation to rumen fermentation. Proc. Nutr. Soc. 31: 127-134.

Chapter 15

STRAW ETC. IN PRACTICAL RATIONS FOR SHEEP AND GOATS

by

Emyr Owen[1] and Jackson A. Kategile[2]

[1]Department of Agriculture and Horticulture
University of Reading
Earley Gate, Reading, Berks, RG6 2AT
United Kingdom

[2]International Development Research Centre (Canada)
P.O. Box 62084, Nairobi
Kenya

15.1. INTRODUCTION

Sheep and goats have been domesticated for thousands of years and next to cattle are the most important group of ruminants farmed in the temperate and tropical regions. In order to get straw feeding into proper perspective other issues are considered first. These include identifying the attributes and disadvantages of sheep and goats compared with cattle and quantifying their number and products in different global regions. The latter is important as it indicates, in conjunction with Chapter 2, the scope for using straw etc. as feed. Systems of production are also identified for the same reason and the traditional use of straw assessed. The nutrition of sheep and goats is considered, particularly from the aspect of utilizing roughages. This is used as a basis for discussing current and future methods of utilizing straws etc. in practical rations.

15.2. ATTRIBUTES AND DISADVANTAGES

These are outlined only, for a comprehensive discussion the reader is referred to Devendra and Coop (1982) and Oltenacu et al. (1976). Compared to cattle, sheep and goats have different grazing habits. Cattle use the tongue and mouth to tear pasture, whereas sheep and goats especially goats, by greater use of lips, are more selective. Their ability to ingest and utilize roughages is considered later (15.5.1). Goats especially, are considered (unjustly) inferior to cattle, being associated with poverty and desertification due to overgrazing in many tropical areas. Goats especially are also difficult to confine, thus giving problems when grazing near cash crops. The small size of sheep and goats compared to cattle makes them more liable to predation and they are unsuitable for draught purposes. Their small size also requires more labour, especially in dairying. In the developed countries per capita consumption of sheep meat has generally decreased during the past 15 years, giving way to greater poultry meat usage. Consumption of goat meat in these areas is negligible. Compared to cattle internationally, sheep and goats have received less veterinary research and consequently health improvement schemes are less developed. Until recently research and development has tended to ignore the goat completely but this situation is now being rectified (cf. Gall, 1981; Coop, 1982; Ayers and Foote, 1982; Morand-Fehr et al., 1981).

Sheep and goats have many advantages compared to cattle. Their products suffer no taboos and are therefore widely consumed. They tolerate a wider range of environments. Their small size, agility and protective coat allows them to produce in harsh mountainous areas. These characteristics, together with their economy of water and possibly nitrogen usage (especially in goats) make them suited to hot arid regions too. On the other hand goats can thrive in hot humid regions. The production of wool gives sheep a unique attribute. Compared with cattle, sheep and goats have a higher reproductive rate and their production cycle is shorter. Goats have a further advantage over cattle and sheep in maintaining greater lactation length and not necessarily having a dry period. In developing countries the small size of sheep and goats has further advantages. Farms are often small (< 2 ha) and so are feed supplies, making small ruminants more appropriate. Because several

sheep and goats can be kept to each cow, capital invested in sheep
and goats is at less risk from loss due to disease etc. For the
same reason and being smaller currency, sheep and goats are more
appropriate "walking banks". Small ruminants also produce small
carcasses which can be consumed rapidly, without extravagance.
This is important as meat storage is often difficult in less-developed tropical areas. Goat milk has a special role for babies who
are allergic to cow milk.

The distinction between sheep and goats has not been emphasised
so far, but it is important to consider it, as differences between
the two species may alter their potential to use straw and other
low-quality roughages. Mention has already been made of differences in grazing habit, the goat being a browser and the sheep a
grazer. The difficulty of confining goats has also been noted.
Devendra and Coop (1982) note goats to be more sensitive to pneu-
monia than sheep, but more resistant to tse-tse flies.

15.3. POPULATION AND PRODUCTS

Table 15.1 shows world sheep numbers to be 2.5 times those of
goats. More than 80% of goats occur in Africa and Asia with 60% of
sheep in the temperate regions. Product volume in relation to
population is lower in tropical compared to temperate regions.
This is a reflection of smaller, less productive breeds as well as
poorer nutrition and other environmental effects. In the major
sheep-farming regions (Oceania) meat and wool are the main products. Sheep-milk production occurs in the Near East, Southern and
Central Europe as well as Asia. Goats are mainly milk producers in
Europe, USSR and North and Central America, but are also meat producers in Asia and Africa. Not shown in Table 15.1 are within-re-
gion statistics. Differences in these and even within given countries can have large influences on utilization of straw etc. at
local levels. Kossila (Chapter 2) noted the sobering fact that
straw etc. supplies in relation to livestock units and hence
requirement is much greater in developed than developing countries.

Table 15.1. Sheep and goat population and production in different regions
(Based on Devendra and Coop, 1982, from FAO 1978)

Region	Population 10^6		Sheep and goats per human	Meat $tx10^3$		Milk $tx10^3$		Fresh skins $tx10^3$		Greasy wool from sheep $tx10^3$
	Sheep	Goat		Sheep	Goat	Sheep	Goat	Sheep	Goat	
Africa	161	131	0.45	651	425	558	1271	116.2	83.6	201.0
North, Central America	22	12	0.10	189	23	-	233	26.8	5.8	58.2
South America	103	18	0.50	289	64	32	126	82.0	14.2	307.5
Asia	282	231	0.22	1409	1075	3106	3027	270.0	206.6	294.2
Europe	126	11	0.29	1027	92	3472	1491	160.2	12.5	263.7
Oceania	194	0	9.33	1061	1	-	-	207.7	0.2	1005.2
USSR	140	6	0.56	960	40	100	400	120.0	6.6	458.0
World	1028	410	0.35	5586	1720	7268	6548	983.1	329.4	2587.9

15.4. PRODUCTION SYSTEMS AND TRADITIONAL USE OF STRAW ETC.

Coop and Devendra (1982) stress the difficulty of classifying systems of sheep and goat production, but describe a method based on land producing capacity. This is used in Table 15.2 to identify systems which have traditionally used straw and other by-products, either as feed and/or bedding. It has not been possible to quantify, with any confidence, the number of animals associated with each system. However some indications can be given. Over 40% of world sheep are farmed extensively by grazing in temperate areas and use little or no straw etc. Migratory systems involve more than 20% of world sheep and nearly 30% of world goats (Demiruren, 1982). Generally these use little straw although some stubble grazing may be practised (eg. Middle East). Straw and other fibrous by-products play an important role in temperate arable systems and especially so in tropical village/smallholder systems. Purohit (1982) estimated the latter to involve 30% of world sheep and 60% of world goats. Intensive grassland systems in temperate zones, where part of the farm is devoted to grain production (eg. Europe) also use straw, generally as bedding material and some as winter feed.

The role of sheep on intensive arable farms has been discussed by Newton (1982). In Britain their use evolved from the Norfolk four-course rotation in the late 18th century, when sheep were used to add fertility to soil by grazing roots (turnips) in situ. They also grazed stubbles after cereal harvesting, thereby helping to control weeds and promoting tillering in grass which was often undersown to spring barley. In addition, the sheep had a major role as utilizers of the grass ley break-crop. In essence this approach has been practised on intensive arable farms in many temperate countries for the past century. In this system the main contribution of straw to sheep feeding is via stubble grazing. This is particularly important in some regions eg. Middle East. Generally, harvested straw is not a major contributor to winter/-stall feeding of sheep because of its low nutritive value. It is more frequently used as bedding and gut-fill, especially if wheat straw, or as a component of cattle rations. In temperate countries with fodder shortage and long winters, (eg. Norway) straw as a winter feed for sheep and goats has assumed greater importance.

Table 15.2. Production systems (Coop and Devendra, 1982), to illustrate traditional usage of straw etc.

System and land productivity	Location	Flock size	Stocking rate animals/ha	Major feed type	Straw etc. usage	Housed/ stall fed	Animal type
Very extensive Unproductive land	Temperate eg. S.America S.Africa, S.Australia	> 4000	< 0.5	Sparse grazing	None	None	Sheep
Extensive (semi-intensive) Hill land	Temperate, eg. Europe S.America Australia N. Zealand	500 –3000	1–4	Grazing	Some (Europe), for bedding and feeding	Some (Europe), in winter	Sheep
Intensive grassland. Productive easy-contour land	Temperate, eg. Europe S. America Australia N. Zealand	1500–3000 (all-grass farms)	6–20	Grazing	Some (Europe) for bedding and feeding	Sheep – some (Europe) in winter Goats – housed at night and in winter	Sheep mainly, some goats
Intensive arable. Productive grain-growing land	Temperate	100–1000	High on grass area	Grazing and arable by-products	Stubble grazing, winter feeding and bedding	Sheep, mainly in winter. Goats – housed at night and in winter	Sheep mainly, some goats
Very intensive No land	Examples in Europe, USA and oil-rich states	Variable	No land	High concentrate or lucerne	Some, mainly bedding	Completely stall-fed	Sheep mainly, some goats
Government controlled economy	USSR E. Europe China	Variable but often > 20,000	Variable	Grazing	Stubble grazing, winter feeding and bedding	Sheep – in winter Goats – at night and in winter	Sheep mainly, some goats
Transhumance. Unproductive land	Arid tropics mainly, also Middle East, France, Spain, N. America	Variable 300–600		Grazing	Some (Europe) in winter, some stubble grazing	Some (Europe) in winter	Sheep and goats
Nomadic. Unproductive land	Arid tropics- S-W Asia Africa	250–500		Grazing	None	None	Sheep and goats
Village and smallholders. Productive, crop-growing land	Tropics, developing countries	5–20	High and non-arable area	Poor grazing and crop residues	Stubble grazing, stall feeding	At night and often other times, especially milk goats	Sheep and goats

Straw has also gained greater use during shortages caused by war (World Wars I and II) and drought (eg. 1976 in UK).

The tropical equivalence of temperate arable farms is the village and smallholder system. Here crop residues play a vital role in the nutrition (or rather, malnutrition) of sheep and goats as conventional grazing is scarce and generally of low nutritive value (Purohit, 1982; Urio, 1981). In some areas the fodder shortage is accentuated by use of straw (usually wheat) for making clay bricks and for fuel (with manure). The crop residues are either grazed in situ or stall-fed without chopping. Tethering of animals on crop stubbles is common. Purohit (1982) describes how Sri Lankan cultivators allow sheep and goat owners to pen their animals on crop stubbles at no cost in order to reap the benefit of the manure produced.

Some of the advantages of stubble grazing and traditional methods of feeding straw have already been mentioned, cf recycling nutrients, weed control and development of undersown pastures. Further advantages are allowing selective feeding thus reducing the handicap of low nutritive value and incurring no processing costs. The disadvantage of these methods is the inefficient utilization caused by rejection and trampling. A further disadvantage is poor utilization due to nutritional imbalances.

The extent to which sheep and goats are housed in different systems (Table 15.2) merits comment. In some temperate systems sheep are only housed during winter. On the other hand goats are generally housed at night in summer also. Night housing is routinely practised for sheep and goats in tropical villages. This practice is interesting because it provides the opportunity for feeding concentrate supplements. In temperate goat systems concentrates are the major ration constituent because emphasis is on high milk yields.

15.5. ANIMAL SPECIES AND NUTRITIONAL CONSIDERATIONS

15.5.1. <u>Roughage utilization - effect of species size</u>

Although sheep and goats have many attributes (15.2) their ability to utilize low-quality roughages is less than cattle and buffaloes. Van Soest (1982), in an elegant teatise, points to the fact that basal metabolism in all species is proportional to live weight raised to a power function (generally 0.75) and to the increasing evidence showing intake of bulky foods by different ruminants to be proportional to live weight per se. Thus small ruminants are particularly disadvantaged compared to large ones on low-intake promoting, relatively indigestible feeds, such as straw. Evidence of lower roughage utilization by sheep compared to cattle was also produced by Siebert and Kennedy (1972) and Bird (1974). Playne (1978a) further showed that DM digestibility of coarsely milled, unsupplemented low quality tropical grass hay was much higher in cattle (46.9%) than in sheep (34.6%) and the difference was not due to relative intake nor diet selected. Subsequent work (Playne, 1978b) with a range of poor hays of less than 60% DMD, showed that the difference between species was greatest with hays of lowest digestibility. These experiments involved unsupplemented stall-fed animals. Differences might have been less with adequate supplementation (eg. nitrogen and sulphur). Under poor-quality pasture conditions, it is likely that sheep would offset their handicap by more selective grazing (Dudizinski and Arnold, 1973).

The above suggests that care should be taken when extrapolating between species such as cattle and sheep. Greenhalgh and Reid (1973) demonstrated this when intake improvement due to grinding and pelleting high-quality grass was markedly higher in sheep (45%) than cattle (11%). Greenhalgh (1980) also noted the frequency with which sheep have been used to measure intake and digestibility responses to treating straw etc.

15.5.2. <u>Sheep versus goats</u>

Paucity of goat research has led nutritionists to extrapolate from sheep to goats. To what extent this is valid and its implication for utilizing straw etc. in practical rations, needs to be considered.

It has frequently been assumed (Devendra and Burns, 1970) that goats are superior to sheep in their ability to digest roughages, especially fibre. Careful review of the literature (eg. Ndosa, 1980; Louca et al., 1982) indicates no clear advantage to the goat, but comparisons have frequently been suspect due to inadequate replication and inequality eg. intake, age, size, feed types etc. Experiments in the UK (Owen and Ndosa, 1982; Mohammed and

Owen, 1982) comparing British sheep and Saanen goats have consistently shown no difference in digestibility between the species when fed roughages of differing qualities, at restricted and ad libitum intakes. On the other hand Gihad et al. (1980), working with local breeds in Egypt, found goats to have a small advantage over sheep in diet OM digestibility (2% units) but more for ADF digestibility (5% units). This was with untreated rice straw (2/3) and concentrates (1/3) fed ad libitum. Differences virtually disappeared on treating straw with NaOH.

Differences between sheep and goats in their capacity to consume roughages are also unclear from the literature (Ndosa, 1980) although frequent claims are made for the superiority of goats. Many of the experiments showing goats to excel have involved short term trials (14-28 d) with animals of unknown previous nutrition. Long term trials (eg. Mohammed, 1982) with growing animals of comparable growth potential (i.e. equal mature size) show goats to have lower intakes than sheep and consequently to grow less rapidly. The literature generally (eg. Mtenga, 1979; Kyomo, 1977; Gall, 1981) indicates lower growth rates in goats, but few direct comparisons with sheep have been undertaken.

The possibility that breed differences exist in capacity to consume and digest low-quality roughages cannot be ruled out, as has been suggested in cattle (Chapter 11, and Coombe, 1981). Because of differences in nutritional environments tropical and sub-tropical breeds might be considered better roughage utilizers than temperate types. For example, Silanicove et al. (1980) found Black Bedouin goats to have 33% higher energy digestibility than Saanen goats, on a wheat straw-only diet. Bedouin goats maintained weight whereas Saanen lost live weight. The cause of the difference in this experiment was considered to be due to the higher recycling of urea in the Bedouin goats.

Circumstancial evidence (reviewed by Louca et al., 1982) suggests that the superiority of goats over sheep under conditions of poor-quality roughage feeding may be due to more economical use of dietary nitrogen, by greater recycling. There is no disputing their capacity to concentrate urine and use less water than sheep (McFarlane, 1982). Under grazing conditions their browsing habit (Harrington, 1982) is also likely to give them an advantage in providing additonal protein and minerals.

There is a lack of well-established feeding standards for goats

(National Research Council, 1981), unlike sheep (Agricultural Research Council, 1980; National Research Council, 1975). To what extent sheep standards can be applied to goats is rather unclear. Mohammed and Owen (1982), working with winter-housed animals in UK found maintenance energy requirements of Saanen castrates to be up to one third higher than unshorn wethers. With shorn sheep, the species difference was reduced to around 10%. Apart from coat type, other factors contributing to the higher requirement of goats are lower subcutaneous fat thickness and greater activity.

Nutrient requirements for production in sheep and goats have received little comparison and in any case, are difficult to make because of within-species variation. The slower growth of goats compared to sheep has already been noted. Mohammed (1982) found this to be accompanied by poorer feed conversion (feed/live weight gain) presumably because intake relative to maintenance was lower in goats. The leanness of goat meat compared to sheep has been stressed (Owen, Norman, Philbrooks and Jones, 1978; Mtenga, 1979; Throckmorton, 1981), but total body fat (hence energy) might not be too different because of greater visceral and kidney fat in goats (Mohammed, 1982). Milk production by sheep and goats are also difficult to compare because of large breed-differences (Flamant and Morand-Fehr, 1982), but the high intake and yield potential of the goat, particularly in relation to the cow, has been stressed (Devendra, 1975, 1980). The fibre-producing poten-tial of goats is currently under consideration in some countries (eg. Australia, Throckmorton, 1981) especially as young Angoras were shown by Gallagher and Shelton (1972) to be three times more efficient than Rambouillet lambs at converting digestible energy into fibre.

The recently published feeding standards for goats (National Research Council, 1981) acknowledge the dearth of information on the goat but conclude, nevertheless, that requirements for energy, protein and minerals are comparable to sheep for maintenance and production. In view of the earlier discussion it would seem pru-dent to assume higher requirements for maintenance energy in goats, than sheep, particularly in temperate environments.

As far as utilizing straws etc. is concerned, the above review suggests no clear advantage to the goat. Compared with the sheep, the goat may be marginally less able to accomodate high propor-tions of straw etc. in its diet, on account of its greater energy requirement. Furthermore, the goat may be less willing to tolerate high-straw diets on account of its greater capacity to select. Whilst true under grazing conditions, the latter requires validation under stall-feeding conditions.

15.5.3. Level of productivity

Of greater importance than species differences are within-species differences in nutrient requirements due to physiological state. Diet quality, especially its available energy and protein, needs to be good for high-yielding lactating animals and fast-growing animals. Lower nutrient-density diets are possible for near-maintenance needs eg. non-lactating/pregnant animals and fiber-producers. Straws etc. are of low ME content, even after alkali treatment (Chapter 12). The scope for their utilization is going to be greatest in animals at low-to-moderate production levels.

15.6. GRAZING OF CROP RESIDUES - POSSIBILITIES FOR IMPROVEMENT

In view of its wide application this subject has received surprisingly little research, attentions being more often directed to cattle and supplementing low-quality range pastures (eg. Van Niekerk, 1975) and upland grazing in temperate regions (eg. Ducker et al., 1981).

The subject is likely to receive much more attention in the future, as indicated by recent investigations in Syria (E.F.Thomson, personal communication, 1983). The Farming System Programme of the International Centre for Agricultural Research in the Dry Areas (ICARDA) has been conducting surveys of farming systems in Syria since 1977. These surveys have confirmed that cereal and lentil straws are major contributors to the nutrient needs of sheep and goat flocks in Syria. From June to October the only grazing available to flocks is cereal stubbles. In winter, flocks receive supplements containing a high proportion of straw. For example, a study conducted in six villages in Aleppo Province found that cereal and lentil straw together contributed up to 50 per cent of the ME consumed as supplements in winter (ICARDA, 1980). A 1978/81 survey of sheep flocks in the steppe area of Aleppo Provice rceiving about 200 mm annual rainfall showed that cereal straw fed as supplements in winter supplied up to 76% of the estimated ME needs of the ewe (Thomson and Bahhady, 1983). Other studies indicate that along the driest margin of cultivation where the probability of crop failure is high, the economic value of the straw can easily exceed that of the grain.

Clearly, use of low-requirement animals is important. In temperate regions this is normal practice (Table 15.2), with dry ewes and wethers being used for stubble grazing. Where preference can be excercised, choice of crop residue and even cultivar are likely to influence productivity (Chapter 12). The introduction of high-yielding cereal cultivars with shorter stems may be associated with lower digestibility straws, although evidence for this is inconsistent (Coxworth et al., 1981). In some areas eg. Syria, there is concern that new cultivars may give poorer stubble grazing for sheep and goats (B. Capper, personal communication, 1983).

Plate 15.1. Sheep grazing (barley) stubble at Tel Hadya, Syria (Courtesy E.F. Thomson)

Supplementation (cf. Chapter 13) is the other potential method of improving utilization by grazing. Urea and molasses supplementation have received much attention in cattle. In sheep, responses to NPN supplements have sometimes been poor (Mullholand et al., 1976, 1977, cited by Coombe, 1981), presumably due to selective grazing and consequently adequate nitrogen intake. Methods of feeding supplements to grazing sheep and goats are variable eg. pelleted/loose mixtures fed daily, liquid supplements and feed blocks. A disadvantage they have in common is giving rise to large variation in supplement intake between individuals (Foot et al., 1973; Ducker et al., 1981). Devendra (1981) has suggested spraying liquid supplements on to pastures or including them in drinking water. Feed blocks have received considerable commercial application in range/hill pasture supplementation but their usefulness is little-documented, as noted by Ducker et al. (1981). There is need to develop blocks which meet the specifications described by Preston and Leng (Chapter 13). The use of green-forage supplementation (Chapter 13) by grazing, merits investigation with sheep and goats. This approach has been used for supplementing molasses-fed cattle with *Leucaena leucocephala* (Hulman, 1978). Saenz et al. (1982) concluded that a combination of crop residue grazing and stall feeding looked promising for lactating goats in Northern Mexico. Better in situ utilization of crop residues might be obtained by grazing cattle together with sheep (and goats) (Coombe, 1981).

There is clearly need for more research, especially economic evaluation, as noted by Devendra (1982). Coombe (1981) also concludes that the subject merits more attention, especially as harvesting of crop residues may continue to be uneconomic or not feasible in many areas.

15.7. STRAW ETC. IN DIETS FOR STALL-FED ANIMALS

15.7.1. <u>Investigations with treated straws etc.</u>

Despite voluminous research on processing (Chapters 5 to 8) and its frequent evaluation using sheep in digestibility and intake trials, there are relatively few production experiments published on sheep and hardly any reported with goats. This contrasts with the situation in cattle (Chapter 14.2) and may reflect the less frequent stall-feeding of sheep in temperate (more developed) countries, and possibly appreciation of the lower ability of small ruminants to utilize poor-quality roughages (cf. 15.5.1). In assessing the usefulness of straw etc. for production (Tables 15.3 to 15.9), experiments measuring only digestibility and intake have not been included as they are cited elsewhere (eg. Chapters 5 to 8, 12). Trials of the latter type, though essential and useful, need cautionary extrapolation to production situations since they are generally of the short-term, change-over type. Intakes, particularly, may be exaggerated (cf. Greenhalgh and Wainman, 1972). Intake measurements with sheep and goats are also prone to error due to the wasteful feeding behaviour of these animals, particularly when eating long roughages. The confounding effect of gut-fill changes when measuring live weight change must also be borne in mind (cf. Greenhalgh et al., 1976; Hasimoglu and Klopfenstein, 1969; Garrett et al., 1979). The importance of roughage particle size reduction for increasing intake in sheep is well documented (Chapters 5 and 12). Nearly all the experiments reviewed (Tables 15.3 to 15.9) involve some degree of physical processing, even for "untreated" materials. Authors frequently comment that sheep are reluctant to eat straws etc. unless chopped or ground. In view of this it is surprising that the degree of comminution applied is rarely defined adequately (cf. Chapter 5). Another factor contributing to variability in performance is difference between straws, even within cultivars (cf. Chapter 12). Comparison of performances achieved by different investigators is further complicated by the effect that other dietary ingredients might have (cf. Chapter 13). This discussion notwithstanding, Tables 15.3 to 15.9 are presented to demonstrate dietary inclusion rates and responses to treatments. Except for Table 15.3, experiments are grouped according to straw etc. type.

15.7.2. <u>Gestation</u>

The role of straws etc. in diets for low-requirement animals eg. non-lactating pregnant sheep, was noted earlier. Experiments and experiences reviewed in Table 15.3 refer to temperate areas with straw being mainly used as a replacer for hay. Treatments applied in different experiments are not shown, but a common feature is the difficulty of showing treatment effects on live weight changes and lamb birth weights. This difficulty is inherent with this type of animal and is compounded by most experiments having insufficient replicates. Nevertheless data on intake in such

Table 15.3. Examples of straw etc. in diets for pregnant animals

Reference Country Species, weight Feeding regime	Straw type and treatment	Other major diet ingredients % of diet DM	Straw in diet DM, %	Total intake kg DM/d	Period of pregnancy studied (0-147) d	Weight gain kg	No of off- spring
Thonney and Hillers, 1977 USA. Sheep, 80 kg Complete diet, restricted	Wheat straw, milled, untrea- ted	Ground lucerne, 50	50	1.8	75-133	3.9 to 10.6	1.40
Broadbent and Jacklin, 1983. UK. Sheep, 80-85 kg. Straw ad lib, concentrates restricted	Barley straw, long, untreated	Barley, soya- meal	ca 62	ca 1.5	105-147	ca 2.0	2.10
National Research Council 1981, USA, Goats, 30 kg. Theoretical ration	Wheat straw, ? long, untreated	Barley, 32; oat silage, 29.	39	1.3	98-147		
Nedkvitne, J., 1983. Personal communication. Norway, Sheep,ca 75 kg. Straw ad lib, silage and concentrates, restricted	Barley straw, long, NaOH, Beckmann	Grass silage, 30; concentrates, 15	ca 55	ca 1.4	0-147	ca 9	1.70
Owen et al., 1978, UK. Sheep, 68 kg. Complete diet ad lib.	Barley straw, chopped, NaOH, spray method, ensiled	Grass hay, 19; soya meal, 20	58	1.7	102-137	5.8	2.00
Nedkvitne and Maurtvedt, 1980. Norway. Sheep, 75 kg. Straw/hay ad lib.	Barley straw, long, NH$_3$, Stack method. None included	Grass silage, 29; concentrates, 16; grass silage, 33; concentrates, 18; hay, 49	ca 55 0	ca 1.4 ca 1.2	0-147 0-147	10.7 8.7	1.75 1.69
Orr, R. and Treacher, T. 1983, Personal communi- cation. UK. Sheep, 80-85 kg. Straw ad lib., concentrates restricted	Barley straw, untreated, long. Barley straw, NH$_3$, Oven Met- hod, long	Concentrates Concentrates	30-65 50-75	0.8-1.1 1.1-1.6	103-147 103-147	-0.8 to 8.9 0.3 to 10.2	2.70 2.70

trials are useful.

Both Thonney and Hillers (1977) and Broadbent and Jacklin (1983) were optimistic about the economic use of untreated straw for ewes in late pregnancy. Thonney and Hillers (1977) considered it essential to grind the straw and other components, to promote intake and avoid selection, but pelleting was not thought necessary. On the other hand Broadbent and Jacklin (1983) fed unprocessed, long straw, with satisfactory results.

Practical experience over many years in Norway (J. Nedkvitne, personal communication, 1983), first with NaOH-treated straw (Beckmann Method) and recently with NH_3-straw (Stack Method), have shown the need for straw to be treated, if used as a replacer for grass hay when fed with grass silage and concentrates. The complementarity of grass silage and alkali-treated straw should be appreciated (cf. Terry et al., 1975). Beckmann-treated straw has proved to be a successful and safe product for pregnant sheep in Norway, with feeding rates of wet material being 2.5 to 3.5 kg/d. On the other hand there is some concern about the high residual sodium in NaOH-straw from the Spray and Dip Methods. Sundstøl (1981), as a precautionary measure, advocated an upper limit of 0.2-0.3 kg DM/d (1-2 kg wet straw) of Dip Method straw for adult sheep and goats if feeding over long periods (6 months). He noted this was probably being over-cautious.

NH_3-treated straw has now replaced Beckmann-straw for winter feeding on farms in Norway. Use of NH_3-straw for winter feeding of sheep in the UK is at an early stage of application, but it has an advantage over NaOH-straw in that it can be fed, like hay, as a pasture supplement, especially during snow conditions. Its higher nitrogen content also makes it comparable to medium-quality, temperate-grass hay. In the experiment by Orr and Treacher (R. Orr and and T. Treacher, personal communication, 1983) intakes of treated and untreated long straws were low (Table 15.3), but were increased by 67 to 136% by NH_3-treatment depending on quantity of concentrates fed. Increasing the barley and soya bean meal concentrate from 3 to 9 g OM/kg W.d reduced untreated straw consumption from 5.7 to 3.9 g OM/kg W.d. With NH_3-straw, intakes were little affected by the same increase in concentrates (9.5, 9.2 g OM/kg W.d). With untreated straw, concentrates were behaving as a substitute whereas with NH_3-straw they were being supplementary (cf. Chapter 13). With NaOH-straw fed to adult, non-pregnant ewes of 50 kg, Raine and Owen (1978) found a more modest increase in concentrate feeding (4.5 and 7.3 g OM/kg W.d) reduced intakes of both untreated (16.0, 13.7 g OM/kg W.d) and treated (21.2, 18.3 g OM/kg W.d) chopped barley straw.

15.7.3. <u>Lactation</u>

There appear to be no reports of straws etc. being investigated as a major component in diets for lactating sheep and goats. This is understandable in view of their low energy and protein contents, even after treatment with alkali. Untreated chopped barley straw has been included (at 30%) as the roughage component of a complete pelleted diet for lactating ewes in experimental stations in the United Kingdom. However such diets have proved to be expensive, due to pelleting difficulties. Beckmann-straw has been used in Norway for lactating ewes, being fed at 3.5 to 4.0 kg/d (wet straw) (Nedkvitne, 1955). NH_3-straw is currently used for lactating sheep and goats in Norway, but it represents a small component of a diet dominated by grass silage and concentrates.

Plate 15.2. Pregnant sheep being fed aqueous ammonia-treated winter-barley straw in Yorkshire, England (Courtesy Hargreaves Fertilizers Limited)

15.7.4. <u>Growth</u>

Nineteen studies involving growing animals are shown in Tables 15.4 to 15.9. Only two involved goats and only six of the trials were in the tropics. Seven of the experiments were in USA. Only one of the trials involved lambs less than three months of age, most used animals of at least six months of age initially. The animals were therefore likely to be "fattening" rather than "growing" and consequently would have lower capacity for live-weight gain. This is demonstrated by the modest growth rates achieved even after roughage treatment (< 180 g/d). The fact that straw etc. inclusion rates were high (50-80% of diet DM) would also be contributory. A common weakness of the experiments, with notable exceptions (Greenhalgh et al., 1976; Al-Rabbat and Heaney, 1978; Garrett et al., 1979), is the absence of "high" control treatments to enable judgement of capacity to respond to improved nutrition.

It is clear that chopped, untreated barley and oat straws fed with little (Owen, 1981b; and Kalinowski, 1974, Table 15.7) or no supplement (Crabtree and Williams, 1971, Table 15.4) are sub-maintenance feeds. Ground rice hulls are even worse, causing weight loss at only 50% inclusion (Choung and McManus, 1976, Table 15.5). On the other hand chopped, untreated maize stover, with modest supplementation (5.9 g DM concentrates/kg W.d) supports weight gain (Urio, 1981, Table 15.6). Ground, untreated maize cobs appear even better (Table 15.5).

Despite the voluminous literature on grinding roughages (Chapter 5) there is a lack of data for straws showing the effect of grinding on lamb growth. Benefits from optimum steam pressure (Chapter 5) are clearly detectable in better growth rates (Umunna et al., 1972, Table 15.5).

More studies have been undertaken with NaOH-straw (11 Experiments) than NH_3-straw (6 Experiments), reflecting the traditional interest in NaOH (Chapter 6) and the more recent one in NH_3 (Chapter 7). Interest in other chemicals (Chapter 8) is yet to be applied in growth studies except for improving growth by replacing some NaOH with $Ca(OH)_2$ (eg. Lamm, 1976, Table 15.6). Optimum treatment with NaOH or NH_3 produces a large improvement in live-weight gain in nearly all the experiments (Tables 15.4 to 15.9), and these have involved most cereal straws and two major by-products, maize cobs and rice hulls. A range of treatment methods

Table 15.4. Examples of oat straw in diets for growing lambs

Reference Country Animal type Initial age and weight Trial duration Diet type, feeding regime	Straw treatment	Other major diet ingredients	Straw in diet DM %	Total intake kg DM/d	Growth rate g/d
Crabtree and Williams, 1971 UK Lambs ca180 d, 21 kg 98 d Concentrates restricted, straw Ad lib.	Chopped, untreated " " "	Mineral supplement Concentrates " "	100 75 62 32	0.24 0.34 0.45 0.51	-54 -17 8 37
Saxena et al., 1971 USA Lambs ca 180 d, 20-28 kg 56 d Complete diets fed ad lib.	Ground, untreated " Ground, NaOH,Beckmann "	Maize, molasses, urea Maize, soya meal Maize, molasses, urea Maize, soya meal	65 65 65 65	0.82 0.87 1.11 1.29	53 62 125 277

Table 15.5. Examples of ground maize cobs and rice hulls in diets for growing animals

Reference Country Animal type Initial age and weight Trial duration Diet type, feeding regime	Cob treatment	Other major diet ingredients	Cob/ hulls in diet DM %	Total intake kg DM/d	Growth rate g/d
		Maize cobs			
Umunna et al., 1972 USA, Lambs, ca200 d,ca 25 kg, 53 d (Trial 1), 69 d (Trial 2) Complete diet, ad lib	Untreated Steam pressure Untreated Steam pressure	Concentrates " Concentrates "	50 50 70 70	0.94 1.23 0.68 1.19	104 183 56 150
Koers et al., 1969, USA Lambs,ca 200 d,ca 25 kg 56 d. Complete diet, ad lib	Untreated, ensiled 40 g NaOH/kg DM, Spray Method, ensiled	Soya meal "	80 80	0.98 1.23	68 109
Raunds et al., 1976, USA Lambs,ca 200 d, 23 kg 56 d. Complete diet, ad lib	Untreated,ensiled. 40 g NaOH/kg DM, Spray Method, ensiled "	Maize gluten meal, urea " Soya meal	75 75 75	0.63 0.74 0.98	30 82 137
Urio, 1981, Tanzania Goats, ca 365 d, 15 kg 100 d, Cobs ad lib.	Untreated NaOH, 48 h Dip Method	Concentrates (176 g " DM/d)	60	0.41 0.43	33 51
Tubei and Said, 1981, Kenya Lambs,ca 200 d, 29 kg 91 d, Cobs ad lib.	Untreated NH$_3$, Stack Method	Concentrates (352 g " DM/d)	50	0.71 0.99	79 130
		Rice hulls			

Choung and McManus, 1976
Australia, Lambs,
180 d, 23.5 kg
35 d-data used.
Pelleted, complete diets
fed ad lib.
	Untreated, ground
	Ground, 50 g NaOH/
	  kg DM, Spray Method
	Ground, 100 g NaOH/
	  kg DM, Spray Method
		Lucerne, 50
		
		"
		
		"
			50
			
			50
			
			50
				0.87
				
				1.12
				
				1.07
					-71
					
					29
					
					0

Table 15.6. Examples of maize stover in diets for growing animals

Reference Country Animal type Initial age and weight Trial duration Diet type, feeding regime	Stover treatment	Other major diet ingredients	Stover in diet DM %	Total intake kg DM/d	Growth rate g/d
Lamm, 1976. USA Lambs ca 200 d, ca 25 kg 62 d. Complete diet, ad lib.	Ground, untreated, ensiled Ground, 40 g NaOH/kg DM Spray Method, ensiled Ground, 30 g NaOH and 10 g Ca(OH)$_2$/kg DM, Spray Method, ensiled. Ground, 10 g NaOH and 30 g Ca(OH)$_2$/kg DM, Spray Method, ensiled	Brewers grains, urea " " " 	75 75 75 75	0.95 1.46 1.38 1.25	18 90 89 55
Urio, 1981, Tanzania Sheep 550-730 d, ca 30 kg 100 d. Stover ad lib.	Chopped, untreated Chopped, NaOH, 48 h Dip Method	Concentrates (176 g DM/d) "	82 81	0.99 0.93	81 74
Tubei and Said, 1981, Kenya. Lambs, ca 200 d, 29 kg 91 d, Stover ad lib.	Chopped, untreated. NH$_3$, Stack Method	Concentrates (325 g DM/d) "	33 55	0.69 0.79	62 89

Table 15.7. Examples of barley and rye straw in diets for growing lambs

Reference / Country / Animal type / Initial age and weight / Trial duration / Diet type, feeding regime	Straw treatment	Other major diet ingredients	Straw in diet DM (%)	Total intake (kg DM/d)	Growth rate (g/d)
Barley straw					
Greenhalgh et al., 1976, UK	Ground, untreated	Concentrates	50	0.56	77
Lambs, ca 50 d, ca 15 kg	Ground, 80 g NaOH/				
Weaning to slaughter at	kg DM	"			
36 kg.	Spray Method		50	0.85	140
Complete diet, ad lib.	None included	All-concentrate	0	0.91	235
Owen, 1981b, and	Chopped, untreated	Molasses, minerals	70	0.40	−68
Kalinowski, 1974, UK.	"	" " plus urea	62	0.48	−18
Lambs, ca 180 d, 27 kg	Chopped, 70 g NaOH/kg	Molasses, minerals			
72 d.	and 1.8 kg H_2O/kg DM		63	0.37	−64
Straw ad lib., supplements	"	" " plus urea	74	0.70	57
restricted	"	" " plus soya meal	79	0.85	83
Rye straw					
da Silva, 1983, Portugal	Chopped, untreated	Hay (29%), concentrates (45%)	26	0.77	72
Ewe lambs	Baled, NaOH Dip	Hay (27%), concentrates (32%)	41	0.78	71
ca 180 d, 22 kg, 133 d.	Method, then chopped				
Straw ad lib.					
Concentrates added to					
untreated straw to					
achieve same growth rate					
as treated straw					

Table 15.8. Examples of wheat straw in diets for growing lambs

Reference Country Animal type Initial age and weight Trial duration Diet type, feeding regime	Straw treatment	Other major diet ingredients, % of diet DM	Straw in diet DM %	Total intake kg DM/d	Growth rate g/d
Hasimoglu and Klopfenstein, 1969, USA Lambs ca 180 d, 30 kg 56 d Complete diet, ad lib.	Ground, untreated " Ground, 40 g NaOH/kg DM, ensiled "	Wheat, plus urea " , plus soya meal " , plus urea " , plus soya meal	70 70 70 70	0.63 0.92 0.99 1.21	-46 37 80 160
Tejada et al., 1979 Guatemala, Lambs 210-270 d, 18.5 kg 59 d. Forage ad lib, concen- trate at 300 g DM/d	Chopped, untreated " " Chopped, 20 g NH_3/kg, ensiled "	Concentrates, 25; maize silage,34 " 23; " ,32 " 23; " ,31 " 25; " ,34 " 23; " ,32 " 22; " ,30	41 45 46 41 45 49	1.21 1.29 1.31 1.20 1.29 1.38	92 74 56 116 128 130
Al-Rabbat and Heaney, 1978, Canada Lambs ca 180 d, 28 kg 56 d Pelleted complete diets fed ad lib.	Ground, untreated NH_3, Stack Method, ground Ground, untreated None included	Lucerne, 25; cottonseed meal,10 " , 25; " " ,10 " , 25; cereal, 49 " , 100	64 64 25 0	1.41 1.66 1.59 1.66	59 110 183 155

Table 15.9. Examples of rice straw in diets for growing animals

Reference Country Animal type Initial age and weight Trial duration Diet type, feeding regime	Straw/hull treatment	Other major diet ingredients % of diet DM	Straw in diet DM %	Total intake kg DM/d	Growth rate g/d
		Rice straw			
Garrett et al., 1979	Untreated, ground	Lucerne, barley,	72	1.86	89
USA	Ground, 40 g NaOH/kg DM	cottonseed meal			
Lambs	0.5h at 100°C, Dip Method, dried	"	72	1.97	137
180 d, 25 kg	Chopped, 40 g NaOH/kg DM,	"			
63 d,	Spray Method		72	2.09	132
Pelleted, complete	Baled, 70 g NH_3/kg, Stack	"			
diets fed ad lib.	Method, dried, ground		72	2.14	131
	Baled, 47 g NH_3/kg, Stack	"			
	Method, dried, ground		72	1.97	136
	Untreated, ground	"	36	1.87	184
	Ground, 40 g NaOH/kg DM				
	0.5h at 100°C, Dip Method, dried	"	36	1.91	196
	Chopped, 40 g NaOH/kg DM,				
	Spray Method	"	36	2.08	217
	Baled, 70 g NH_3/kg, Stack				
	Method, dried, ground	"	36	1.66	165
	Baled, 47 g NH_3/kg, Stack				
	Method, dried, ground	"	36	1.16	89
Winugroho and Chaniago,	Untreated, ground	Ground cassava leaf,25	75	1,34	45
1983, Indonesia	" "	" ,50	50	1.33	92
Goats	100 g urea and 1.0 kg water	" ,25			
ca 180 d, 22 kg	/kg DM, ensiled 30 d, then				
91 d	dried and ground	25	75	1.51	84
Pelleted, complete diets	"	" ,50	50	1.49	101
fed ad lib.	"		100	0.87	27

have been used, especially with NaOH, but methods for a given chemical have rarely been compared. Only one study (Garrett et al., 1979) has directly compared NaOH and NH_3 in terms of growth effects. As well as showing no difference between the chemicals, the experiment by Garrett et al. is important in that it demonstrates that level of straw in the diet determines whether or not chemical treatment is worthwhile, as discussed by Owen (1981a, 1981b). Only at a high level of inclusion was there a detectable improvement due to treating with either NaOH or NH_3. A further notable point is that all the experiments (Tables 15.4-15.9) have used ad libitum feeding. This is essential, as alkali treatment of straws etc. improves intake as well as digestibilty. The trial by Garrett et al. (1979) and earlier studies (Garrett et al., 1974) are also important because they emphasise the need to measure change in body composition as well as live weight. However, such trials are expensive.

The need for optimum supplementation has been discussed earlier (Chapter 13). Soya bean meal is rated a good source of by-pass protein and glucogenic compounds (Chapter 13, Table 13.4). Its capacity to improve growth is demonstrated in several experiments with both treated and untreated materials (Saxena et al., 1971, Table 15.4; Raunds et al., 1976, Table 15.5; Owen, 1981b and Kalinowski, 1974, Table 15.7; Hasimoglu and Klopfenstein, 1969, Table 15.8). Preston and Leng (Chapter 13) discussed the benefits of supplementing with a green forage and cited the study of Rodriguez (Figure 13.12) showing soyabean meal and good quality forage to be additive in improving live-weight gain in lambs fed a basal diet of sisal pulp/urea. Preston and Leng (Chapter 13) also pointed to the need for more "response" trials of the type reported by Abidin and Kempton (1981) showing the optimum supplementation of barley straw/urea for lambs, using a by-pass protein pellet (Figure 13.10). Apart from Crabtree and Williams (1971) (Table 15.4), experiments shown in Tables 15.4 to 15.9 lack this approach, thus making economic evaluation more difficult. However, the intake:gain data in Tables 15.4 to 15.9 show a large and consistent improvement in feed conversion rate due to straw etc. treatment. Most of the data were recorded with unpelleted diets, which is encouraging as pelleting high-roughage mixtures is expensive. Also promising is the fact that several experiments achieved satisfactory growth with the more practical approach of restricting the concentrates and feeding the untreated/treated straws etc. ad libitum.

15.7.5. Maintenance

Maintenance feeding does not appear to have been studied as such except for the study by Winugroho and Chaniago (1983). They found urea-treated rice straw alone to be satisfactory for maintaining weight in goats (Table 15.9). However, several of the diets used for pregnant animals (Table 15.3) would be satisfactory maintenance feeds. Diets for growing animals (Tables 15.4 to 15.9) could also be satisfactory. Whether or not to treat the straw etc. could depend on its nutritive value, cost of treatment and cost of supplements.

15.8. CONCLUSIONS AND FURTHER RESEARCH

The technical evidence reviewed (Tables 5.3-5.9) indicates that treated and untreated straws etc. could have a greater role in systems already using crop residues i.e. intensive arable farms in temperate regions and crop-producing smallholdings in the tropics (Table 15.2). The importance of goats in the tropics calls for more research to involve this species and to involve straws etc. for lactation as well as maintenance and growth. Nearly all the information on feeding straws etc. relates to sheep. However these data are likely to be applicable to goats provided the somewhat higher maintenance-energy requirements and more pernickerty eating behaviour of goats are appreciated.

Traditional utilization of crop residues by stubble grazing is surprisingly little-researched in view of its wide application. The subject deserves much more research. The efficiency of utilizing residues by this method is ill-defined. Questions relating to the importance of straw quality and supplementation need more answers. Intercropping and growing supplementary crops for grazing (eg. *leucaena* in the tropics) should be investigated to a greater extent.

In stall-feeding situations greater attention should be given to developing more practical systems based on feeding straw etc. ad lib with restricted supplements. Whether or not to treat residues, either physically or chemically, should be given more economic evaluation. In developing countries particularly, the emphasis is likely to be on minimal processing methods. Where chemical treatment is only marginally economic and fodder shortage acute, it might be worth treating the feed refusals only. This would allow selective feeding by sheep and goats and the less digestible refusal, normally discarded, could be refed after chemical upgrading.

Straws etc. may also have a future role in very intensive systems (Table 15.2), particularly where agro-industrial by-products are centrally produced (Chenost and Mayer, 1977; Devendra, 1981; Owen, 1976, 1980), eg. sisal pulp (Rodriguez, 1982), sugar cane bagasse, citrus by-products (Göhl, 1973; Awolumate and Olubajo, 1982). Centrally-planned Economies, with the advantages of large-scale systems are already using processed straw in complete diets for sheep (Bergner, 1981). What is important to remember is that straws etc. even after NaOH or NH_3-treatment, are of only moderate energy-density (< 9.0 MJ ME/kg DM). Their energy contribution in diets for high-producing animals will therefore be limited.

15.9. SUMMARY

The advantages, disadvantages, population, production and systems of sheep and goats in different regions are outlined to provide a perspective for considering traditional and improved methods of straw etc. utilization. The species are compared in relation to roughage utilization capacity and nutrient requirements; differences are small, with goats possibly requiring more maintenance energy. Ways of improving traditional stubble grazing are considered. Stall-feeding studies with untreated and treated straws etc. for pregnant and growing animals (mainly sheep) are reviewed. Including 50 to 75% of diet DM as NaOH or NH_3-treated straws etc. has proved successful with clear improvements in growth produced by alkali treatment. It is concluded that more research is needed on stubble grazing and on stall feeding of straws etc., especially by goats in the tropics. There is also need for more economic evaluation of whether to treat and/or supplement in practical, farm-level situations.

15.10. REFERENCES

Abidin, Z. and Kempton, T.T., 1981. Effects of treatment of barley straw with anhydrous ammonia and supplementation with heat-treated protein meals on feed intake and liveweight performance. Anim. Feed Sci. Technol. 6: 145-155.

Agricultural Research Council, 1980. The Nutrient Requirements of Ruminant Livestock. Commonwealth Agricultural Bureaux, 351 pp.

Al-Rabbat, M.F. and Heaney, D.P., 1978. The effects of anhydrous ammonia and steam cooking of aspen wood on their feeding value and on ruminant microbial activity. 1. Feeding value assessments using sheep. Can. J. Anim. Sci. 58: 443-451.

Awolumate, E.O. and Olubajo, F.O., 1982. The nutritive value of silages made from mixtures of citrus processing wastes and elephant grass as feed for ruminants. World Review Anim. Prod. 18 (4): 15-20.

Ayers, J.L. and Foote, W.C. (Co-chairmen), 1982. Proc. Third International Conference on Goat Production and Disease, Tuscon. Dairy Goat Journal Publishing Co., Scottsdale, Arizona, 604 pp.

Bergner, H., 1981. Chemical treatment of straw. In: Institute for Scientific Co-operation (eds.): Plant Research and Development: A Biannual Collection of Recent German Contributions. Volume 14. Landhausstrasse 18, 7400 Tubingen. pp. 61-81.

Bird, P.R., 1974. Sulphur metabolism and excretion studies in ruminants. XIII. Intake and utilization of wheat straw by sheep and cattle. Aust. J. Agric. Res. 25: 631-642.

Broadbent, J.S. and Jacklin, D., 1983. An evaluation of straw as the sole source of roughage in the diet of pregnant housed ewes. Anim. Prod. 36 (in press).

Chenost, M. and Mayer, L., 1977. Potential contribution and use of agro-industrial by-products in animal feeding. In: New Feed Resources. FAO Production and Health Paper 4.

Choung, C.C. and McManus, W.R., 1976. Studies on forage cell walls. 3. Effects of feeding alkali-treated rice hulls to sheep. J. agric. Sci. (Camb.) 86: 517-530.

Coombe, J.B., 1981. Utilization of low-quality residues. In: A. Neimann-Sørensen and D.E. Tribe (eds. in Chief): World Animal Science, B. Disciplinary Approach. I. F.H.W. Morley (ed.): Grazing Animals. Elsevier Scientific Publishing Company, Amsterdam, pp. 319-334.

Coop, I.E. (ed.): 1982. Sheep and Goat Production. World Animal Science; v: CI. Production-System Approach. In: A. Neimann-Sørensen and D.E. Tribe (eds. in Chief). Elsevier Scientific Publishing Company, Amsterdam, 492 pp.

Coop, I.E. and Devendra, C., 1982. Production systems. In: A. Neimann-Sørensen and D.E. Tribe (eds. in Chief): World Animal Science, C. Production-System Approach. 1. I.E. Coop (ed.): Sheep and Goat Production. Elsevier Scientific Publishing Company, Amsterdam.

Coxworth, E., Kernan,J., Knipfel, J., Thorlacius, O. and Crowle, L., 1981. Review: Crop residues and forages in Western Canada; potential for feed use either with or without chemical or physical processing. In: M.G. Jackson, F. Dolberg, C.H. Davis, M. Haque and M. Saadullah (eds.): Maximum Livestock Production from Minimum Land. Bangladesh Agricultural University, Mymensingh, pp. 181-204.

Crabtree, J.R. and Williams, G.L., 1971. The voluntary intake and utilization of roughage-concentrate diets by sheep. Anim. Prod. 13: 71-82.

482

Demiruren, A., 1982. Migratory (transhumance) systems. In: A. Neimann Sørensen and D.E. Tribe (eds. in Chief): World Animal Science, C. Production-System Approach. 1. I.E. Coop (ed.): Sheep and Goat Production. Elsevier Scientific Publishing Company, Amsterdam. pp. 425-440.

da Silva, A.D., 1983. Dip treated rye straw in growing diets for ewe lambs. In: K. El-Shazly (ed.): Proc. 2nd Workshop on Low Quality Roughage Utilization. University of Alexandria, Egypt, March 16-21.

Devendra, C., 1975. Biological efficiency of milk production in dairy goats. World Review Anim. Prod. 11 (1): 46-53.

Devendra, C., 1980. Milk production in goats compared to buffalo and cattle in the humid tropics. J. Dairy Sci. 63: 1755-1767.

Devendra, C., 1981. Non-conventional feed resources in S.E. Asian Region. World Review Animal Production 17 (3): 65-80.

Devendra, C., 1982. Perspectives in the utilization of untreated rice staw by ruminants in Asia. Paper to 2nd Annual Meeting of the Australian-Asian Fibrous Agricultural Residues Research Network 3-7 May, Selangor, Malaysia.

Devendra, C. and Burns, M., 1970. Goat Production in the Tropics. Commonwealth Agricultural Bureaux, Farnham Royal, England. 184 pp.

Devendra, C. and Coop, I.E., 1982. Ecology and distribution. In: A. Neimann-Sørensen and D.E. Tribe (eds. in Chief): World Animal Science, C. Production-System Approach. 1. I.E. Coop (ed.): Sheep and Goat Production. Elsevier Scientific Publishing Company, Amsterdam. pp. 1-14.

Ducker, M.J., Kendall, P.T., Hemingway, R.G. and McClelland, T.H., 1981. An evaluation of feedblocks as a means of providing supplementary nutrients to ewes grazing upland/hill pastures. Anim. Prod. 33: 51-57.

Dudzinski, M.L. and Arnold, G.W., 1973. Comparisons of diets of sheep and cattle grazing together on sown pastures on the southern tablelands of New South Wales by principle components analysis. Aust. J. Agric. Res. 24: 899.

Flamant, J.C. and Morand-Fehr, P., 1982. Milk production in sheep and goats. In: A. Neimann-Sørensen, and D.E. Tribe (eds. in Chief): World Animal Science, C. Production-System Approach. 1. I.E. Coop (ed.): Sheep and Goat Production. Elsevier Scientific Publishing Company, Amsterdam. pp. 275-295.

Food and Agriculture Organization of the United Nations, 1978. Production Yearbook, FAO, Rome.

Foot, J.Z., Russel, A.J.F., Maxwell, T.J. and Morris, P., 1973. Variation in intake among group-fed pregnant Scottish Blackface ewes given restricted amounts of food. Anim. Prod. 17: 169-177.

Gall, C. (ed.), 1981. Goat Production. Academic Press, London, 619 pp.

Gallagher, J.R. and Shelton, M., 1972. Efficiencies of conversion of feed to fibre of Angora goats and Rambouillet sheep. J. Anim. Sci. 34: 319-321.

Garrett, W.N., Walker, H.G., Köhler, G.O. and Hart, M.R., 1979. Response of ruminants to diets containing sodium hydroxide or ammonia treated rice straw. J. Anim. Sci. 48: 92-103.

Garrett, W.N., Walker, H.G., Köhler, G.O., Waiss, A.C., Graham, R.P., East, N.E. and Hart, M.R., 1974. Nutritive value of NaOH and NH_3 treated rice straw. Proc. W. Section, American Soc. of Animal Science 25: 317-320.

Gihad, E.A., El-Bedawy, T.M. and Mehrez, A.Z., 1980. Fiber digestibility by goats and sheep. J. Dairy Sci. 63: 1701-1706.

Gohl, B.I., 1973. Citrus by-products for animal feed. World Anim. Rev. 6: 24-27.

Greenhalgh, J.F.D., 1980. Use of straw and cellulosic wastes and methods of improving their value. In: E.R. Ørskov (ed.): By-products and Wastes in Animal Feeding. Occ. Publ. B.S.A.P., No 3, pp. 25-31.

Greenhalgh, J.F.D., Pirie, R. and Reid, G.W., 1976. Alkali-treated barley straw in complete diets for lambs and dairy cows. Anim. Prod. 22: 159 (Abstract).

Greenhalgh, J.F.D. and Reid, G.W., 1973. The effects of pelleting various diets on intake and digestibility in sheep and cattle. Anim. Prod. 16: 223-233.

Greenhalgh, J.F.D. and Wainman, F.W., 1972. The nutritive value of processed roughages for fattening cattle and sheep. Proc. Br. Soc. Anim. Prod. 61-72.

Harrington, G.N., 1982. Grazing behaviour of the goat. In: J.L. Ayers and W.C. Foote (Co-chairmen), Proc. Third International Conference on Goat Production and Disease, Tuscon. Dairy Goat Journal Publishing Co., Scottsdale, Arizona. pp 398-403.

Hasimoglu, S., Klopfenstein, T.J. and Doane, T.H., 1969. Nitrogen source with sodium hydroxide treated wheat straw. J. Anim. Sci. 29: 160 (Abstract).

Hulman, B., 1978. *Leucaena leucocephala* as a Source of Protein and Roughage for Growing Cattle. M.Phil. Thesis, University of Reading.

ICARDA, 1980. Studies of Farming Systems in Six Aleppo Villages. Research Report no 2. Farming Systems Program, International Center for Agricultural Research in the Dry Areas (ICARDA), Aleppo, Syria (unpublished report).

Kalinowski, J., 1974. Effect of Sodium Hydroxide Treatment and Nitrogen Supplementation of Barley Straw Upon Intake, Digestibility and Live-weight Change in Growing Lambs. M. Agric. Sci. Thesis, University of Reading.

Kempton, T.J., 1982. Role of feed supplements in the utiization of low protein roughage diets by sheep. World Review Anim. Prod. 18 (2): 7-14.

Koers, W.C., Klopfenstein, T. and Woods, W., 1969. Sodium hydroxide treatment of corn cobs. J. Anim. Sci. 29: 163 (Abstract).

Kyomo, M.L., 1977. Meat from the Goat in Tanzania. PhD. Thesis, University of Dar es Salaam.

Lamm, W.D., 1976. Influence of Nitrogen Supplementation and Hydroxide Treatment Upon the Utilization of Corn Crop Residues by Ruminants. PhD. Thesis, University of Nebraska.

Louca, A., Antoniou, T. and Hatzipanyiotou, M., 1982. Comparative digestibility of feedstuffs by various ruminants, specifically goats. In: J.L. Ayers and W.C. Foote (Co-chairmen): Proc. Third International Conference on Goat Production and Disease, Tuscon. Dairy Goat Journal Publishing Co., Scottsdale, Arizona. pp. 122-132.

McFarlane, W.V., 1982. Concepts in animal adaptation. In: J.L. Ayers and W.C. Foote (Co-chairmen): Proc. Third International Conference on Goat Production and Disease, Tuscon. Dairy Goat Journal Publishing Co., Scottsdale, Arizona. pp. 375-385.

Mohammed, H.H., 1982. Energy Requirements for Maintenance and Growth: Comparison of Goats and Sheep. PhD. Thesis, University of Reading.

Mohammed, H.H. and Owen, E., 1982. Goats versus sheep: effect of coat thickness and body composition on maintenance energy requirement. Anim. Prod. 34: 391 (Abstract).

Morand-Fehr, P., Bourbouze, A. and de Simiane, M. (eds.), 1981. Nutrition and Systems of Goat Feeding. International Symposium, Tours, France.
Mtenga, L.A., 1979. Meat Production from Saanen Goats: Growth and Development. PhD. Thesis, University of Reading.
National Research Council, 1975. Nutrient Requirements of Domestic Animals. No. 5. Nutrient Requirements of Sheep. 5th ed. National Research Council, Washington, D.C., 72 pp.
National Research Council, 1981. Nutrient Requirements of Domestic Animals. No. 15. Nutrient Requirements of Goats: Angora, Dairy and Meat Goats in Temperate and Tropical Countries. National Academy Press, Washington, D.C., 91 pp.
Nedkvitne, J.J., 1955. Luta halm er godt sauefôr. Sau og Geit 5: 93-94.
Nedkvitne, J.J. and Maurtvedt, A., 1980. Ammoniakkbehandla, tørrluta (NaOH) og resirkulasjonsluta halm som fôr til sauer. Husdyrforsøksmøtet, Agric. University Norway, 411-416.
Ndosa, J.E.M., 1980. A Comparative Study of Roughage Utilization by Sheep and Goats. M.Phil. Thesis, University of Reading.
Newton, J.E., 1982. Intensive arable systems. In: A. Neimann-Sørensen and D.E. Tribe (eds. in Chief): World Animal Science, C. Production-System Approach. 1. I.E. Coop (ed.): Sheep and Goat Production. Elsevier Scientific Publishing Company, Amsterdam. pp. 377-399.
Oltenacu, E.A., Martinez, A., Glimp, H.A. and Fitzhugh, H.A. (eds.), 1976. Proc. of a Workshop on the Role of Sheep and Goats in Agricultural Development. Winrock International Centre, Morrilton, Arkansas.
Owen, E., 1976. Farm wastes: straw and other fibrous materials. In: A.N. Duckham, J.G.W. Jones and E.H. Roberts (eds.): Food Production and Consumption. North-Holland Publishing Company, Amsterdam. pp. 299-318.
Owen, E., 1980. Agricultural wastes as foodstuffs. In: M.W.M. Bewick (ed.): Handbook of Organic Wastes. Van Nostrand Reinhold Company, New York. pp. 40-71.
Owen, E., 1981a. Straw: research work. In: B.A. Stark and J.M. Wilkinson (eds.): Upgrading of Crops and By-Products by Chemical or Biological Treatments. Ministry of Agriculture, Fisheries and Food, London. pp. 1-9.
Owen, E., 1981b. Use of alkali-treated low quality roughages to sheep and goats. In: J.A. Kategile, A.N. Said and F. Sundstøl (eds.): Utilization of Low Quality Roughages in Africa. AUN-Agricultural Development Report 1, Aas, Norway. pp. 131-150.
Owen, E., Herrod-Tylor, M., Tetlow, M. and Wilkinson, J.M., 1978. Ensiled alkali-treated straw as a feed for sheep. Anim. Prod. 26: 401 (Abstract).
Owen, E. and Ndosa, J.E.M., 1982. Goats versus sheep: roughage utilization capacity. In: J.L. Ayers and W.C. Foote (Co-chairmen): Proc. Third International Conference on Goat Production and Disease, Tuscon. Dairy Goat Journal Publishing Co., Scottsdale, Arizona. p. 362 (Abstract).
Owen, J.E., Norman, G.A., Philbrooks, C.A. and Jones, N.S.D., 1978. Studies on the meat production characteristics of Botswana goats and sheep. Part III: Carcass tissue composition and distribution. Meat Science 2: 59-74.
Playne, M.J., 1978a. Differences between cattle and sheep in their digestion and relative intake of mature tropical grass hay. Anim. Feed Sci. Technol. 3: 41-49.

Playne, M.J., 1978b. Estimation of the digestibility of low-qua-
     lity hays by cattle from measurements made with sheep. Anim.
     Feed Sci. Technol. 3: 51-55.
Purohit, K., 1982. Village and smallholder systems. In: A. Nei-
     mann-Sørensen and D.E. Tribe (eds. in Chief): World Animal
     Science, C. Production-System Approach. 1. I.E. Coop (ed.):
     Sheep and Goat Production. Elsevier Scientific Publishing Com-
     pany, Amsterdam. pp.. 459-480.
Raine, H.D. and Owen, E., 1978. On-farm treatment of barley
     straws. Effect of sodium hydroxide, post treatment neutraliza-
     tion and concentate level on digestibility and intake by sheep.
     Anim. Prod. 26: 399 (Abstract).
Raunds, W., Klopfenstein, T., Waller, J. and Messersmith, T.,
     1976. Influence of alkali treatments of corn cobs on in vitro
     dry matter disappearance and lamb performance. J. Anim. Sci.
     43: 478-482.
Rodriguez, A., 1982. Studies on the supplementation of ensiled
     henequen (sisal) pulp for growing lambs. Trop. Anim. Prod. 8
     (in press).
Saenz, E.P., Martinez, P.R. and Thomas, N., 1982. Potential of
     crop wastes in goat production in Northern Mexico. In: J.L.
     Ayers and W.C. Foote (Co-chairmen): Proc. Third International
     Conference on Goat Production and Disease, Tuscon. Dairy Goat
     Journal Publishing Co., Scottsdale, Arizona. p. 281 (Abstract).
Saxena, S.K., Otterby, D.E., Donker, J.D. and Good, A.L., 1971.
     Effects of feeding alkali-treated oat straw supplemented with
     soybean meal or non-protein nitrogen on growth of lambs and on
     certain blood and rumen liquor parameters. J. Anim.Sci. 33:
     485-490.
Siebert, B.D. and Kennedy, P.M., 1972. The utilization of spear
     grass (*Heteropogon contortus*). I. Factors limiting intake and
     utilization by cattle and sheep. Aust. J. Agric. Res. 23: 35-
     44.
Silanicove, N., Tagari, H. and Schkolnik, A., 1980. Gross energy
     digestion and urea recycling in the Desert Black Beduin goat.
     Comp. Biochem. Physiol. 67A: 215-218.
Sundstøl, F., 1981. Methods for treatment of low quality rougha-
     ges. In: J.A. Kategile, A.N. Said and F. Sundstøl (eds.): Uti-
     lization of Low Quality Roughages in Africa. AUN-Agricultural
     Development Report 1, Aas, Norway. pp. 61-80.
Tejada, R., Murillo, B. and Cabezas, M.T., 1979. Ammonia treated
     straw as a substitute for maize silage for growing lambs. Trop.
     Anim. Prod. 4: 172-176.
Terry, R.A., Spooner, M.C. and Osbourn, D.F., 1975. The feeding
     value of mixtures of alkali-treated straw and grass silage. J.
     agric. Sci. (Camb.) 84: 373-376.
Thomson, E.F. and Bahhady, F., 1983. Sheep Husbandry Systems in
     the NW Syrian Steppe. Farming Systems Program, International
     Center for Agricultural Research in the Dry Areas (ICARDA),
     Aleppo, Syria (in preparation).
Thonney, S.C. and Hillers, J.K., 1977. Wheat straw feeding as a
     winter maintenance ration for pregnant ewes. Bulletin 855, Was-
     hington State University.
Throckmorton, J.C., 1981. The potential for goat production. In:
     D.T. Farrell (ed.): Recent Advances in Animal Nutrition in Aus-
     tralia. University of New England, Armidale.

Tubei, S.K. and Said, A.N., 1981. The utilization of ammonia-treated maize cobs and maize stover by sheep in Kenya. In: J.A. Kategile, A.N. Said and F. Sundstøl (eds.): Utilization of Low Quality Roughages in Africa. AUN-Agricultural Development Report 1, Aas, Norway.

Umunna, N.N., Klopfenstein, T.J. and Bolsen, K., 1972. Response of lambs fed pressure treated corn cobs. J. Anim. Sci. 35: 277 (Abstract).

Urio, N.A., 1981. Alkali treatment of roughage and energy utilization of treated roughages fed to sheep and goats. PhD. Thesis. University of Dar es Salaam.

Van Niekerk, B.H.D., 1975. Supplementation of grazing cattle. In: Potential to Increase Beef Production in Tropical America. CIAT, Cali, Colombia. pp. 83-97.

Van Soest, P.J., 1982. Nutritional Ecology of the Ruminant. O & B Books Inc., Corvallis, Oregon. 379 pp.

Winugroho, M. and Chaniago, T.D., 1983. Inclusion of cassava leaf in a pelleted rice straw based diet to improve the weight gain of young goats. Australian-Asian Agricultural Residues Research Network. University of Peradeniya, Sri Lanka 18-22 April.

Chapter 16

STRAW ETC. IN THE DIET OF OTHER RUMINANTS AND NON-RUMINANT HERBIVORES

by

David L. Frape

The Priory
Mildenhall, Suffolk
United Kingdom

16.1. INTRODUCTION

For the purpose of classification according to feeding habits land mammals have been divided into carnivores and herbivores. However, this simplistic division ignores the obvious fact that most animals will consume a variety of feeds so that clear cut divisions are impossible. Chivers and Hladik (1980) in recognizing the anomaly organized primates and a number of other animals according to their varied eating habits. A biological continuum resulted which he divided into three groups. This arbitrarily separates them according to the structural carbohydrate content of the diet. It became apparent that the intermediate group could not subsist, or feed effectively, on a balanced mixture of animal and leaf matter, excluding fruit, and these were named frugivores. The other two were named animalivores and folivores.

An allometric differentiation could be successfully applied to distinguish them. He formulated coefficients of gut differentiation in the form of:

a) surface area (stomach plus caecum plus colon) over surface area of small intestine. Also:

b) Combination of: volume (stomach + caecum + colon) over body
    length.

This classification draws a distinction between regions of fermentation and those of digestion and the relative preponderance of
each. The analysis spans a diverse array of species in which no
account of differences in the structure of the absorbing surfaces,
or differences in mechanisms is necessary. Differentiation amongst
ruminants alone, according to their dietary strategies and faculty
for fibre utilization requires an assessment of several apposite
morphological and functional characteristics of the alimentary
canal. Three broad groups differentiated according to feeding
habits and ecological adaptations have been recognized which are:
(1) bulk or roughage eaters or grazers harvesting roughage rapidly
with an agile tongue, (2) concentrate selectors, and a less distinct and more adaptable intermediate group (3) (Hoffmann and
Stuart, 1972). The stomach of the intermediate group resembles
that of the selectors but with a more variable capacity adaptable
to a great variety of dicotyledons and monocotyledons. This group
may be further subdivided into (a) grasseaters, e.g. domestic
goat, impala, Thomson's gazelle and ((b) tree and shrub foliage
eaters, e.g. eland, Grant's gazelle, steinbok. This subdivision is
by no means absolute. Obvious differences exist between the first
two groups, both in terms of behaviour and abilities to utilize
the structural carbohydrates of cell walls. Moreover within group
reproducable differences are apparent amongst individual species,
but with less certainty of prediction. Sheep are, for example,
more selective and less efficient at fibre digestion than are
cattle when compared under the same conditions. Morphological
adaptations to ecological niches which make a stable and satisfactory use of food supply in natural environments and which reflect
the structural carbohydrate content of the forage are many and
have been categorized by Hofmann (1982). These include a wide
variation in the capacity of each region of the gastro-intestinal
tract, enormous variation in the actual surface area of absorption, through papillation (surface enlargement factor), differentiation in the surface relief of the omasal mucosa, delaying or
prehensile mechanisms for extending fermentation time in grazers,
including ruminal pillars and narrow ostia, microbiological diver-

sity and differences in salivary gland development (Kay et al., 1979).

Animals of group (2) above generally possess the least ability to digest structural carbohydrates of plant cell walls, consistent with their adaptation to frequent meals achieving a relatively large daily intake and rapid rate of passage of browse and graze selected as rich sources of cell contents.

A parallel classification of non-ruminant herbivores has not been attempted but between species a vast and disparate range of morphological and physiological characteristics exists. Some possess large complex forestomachs as in hippopotamuses and camels and others small simple stomachs as in the horse, or midgut enlargements as in the hyrax. Despite foregut enlargement colobus monkeys are selective feeders and although the ass has a relatively small stomach it can subsist on poor coarse roughage. A large variation amongst extant species is recognized in their faculty for digesting plant cell wall constituents and in their capacity for feed dry matter per unit of metabolic weight. Nevertheless as expounded by Chivers and Hladik (1980) the total volume for microbial fermentation has more relevance than the capacity of any single gut diverticulum or enlargement. Amongst rabbits and equids, for which much of the non-ruminant arguments will rely, rates of passage of digesta are generally faster than in grazing ruminants (group 1) and capacity less, so that their strategy is to be more selective akin to that of the intermediate ruminant group (Jarvis, 1976). The search for processes enhancing unit digestibility of roughage should also be viewed in the light of their effects on these other factors of economic significance - appetite and voluntary consumption.

Plant fibre is the structural carbohydrate of cell walls. Cell wall constituents are defined as those substances of plant material which are insoluble in neutral detergent (NDF) (Van Soest, 1963). The capacity of herbivores for plant dry matter intake is reasonably closely allied to the NDF content of the feed consumed. By following the procedure for the determination of acid detergent (ADF) insoluble material in cell wall constituents most of the hemicellulose is removed. However, in heavily lignified material some of this is insoluble. The ADF content of forages is reasonably closely correlated with their apparent digestibility in herbivores. Both of these correlations are negative.

As the practical feeding value of any plant fibre source depends not only upon its digestibility, but also upon the appetite of the animal for the material, information about both the NDF and ADF content is germain to the discussion. In drawing conclusions on species differences in roughage utilization allometric evidence will also be weighed owing to a paucity of direct evidence in many cases.

16.2. COMPARATIVE EVIDENCE AMONGST HERVIBORES ON VOLUME OF GI TRACT AND DIGESTIVE EFFICIENCY

Although domesticated ruminants have a considerable faculty for digesting roughage, it has been shown that some wild grazing animals such as scimitar horned oryx compared with the sheep (Miller and King, 1981) and bison and yak compared with domestic cattle (Schaeffer et al., 1978) have a greater ability to digest low quality roughage diets. When both groups received better quality diets the wild species showed no superiority. Thus comparative evdience amongst species should be based on fundamentally similar diets and the practical scope for using fibrous byproducts in certain species depends on the composition of the basal diet and therefore on the purpose for which the stock are kept. The digestibility of poor roughage and the amount consumed by domestic ruminants is decreased when excessive starchy feed is added to the diet. This addition induces a lowering of rumen pH and therefore a partial suppression of cellulolysis with a consequential reduction in voluntary intake. Preliminary evidence suggests that this suppression does not necessarily occur in certain selective feeders possessing a higher rumen buffering capacity.

The larger ruminant species may in general have a greater faculty for digesting cell wall constituents than do non-ruminant herbivores, but the differences are not clear cut. For instance, cellulose and NDF appear to be digested more completely by cattle than by zebras or the horse, but not more so in comparison with onagers (Frape and Chamberlain, 1981). Unfortunately all published comparisons involving wild species describe effects measured in very few individuals and so it is not justified to assume that the values represent the species as a whole or even any definable population within species.

In can be concluded from direct and indirect comparisons that equids and rabbits have a lower capacity for long fibrous byproducts than do domestic ruminants (Joyce et al., 1967; MacLean et al., 1970, and Table 16.1). Capacity and rate of passage are related phenomena and correlated with the extent of unit fibre digestion. Daily intake, as distinct from capacity, and rate of passage are also related. Apart from these mechanical and anatomical differences the digestive powers of the fluid medium differ amongst species. In a comparison of oryx and sheep Miller and King (1981) concluded that the greater efficiency of the oryx in fibre digestion from a high fibre diet was associated with a longer retention time for digesta only in respect of a lucerne diet, not those containing long grass hay.

Table 16.1. Species comparison of stomach and caecum + colon capacity in relation to $W^{0.75}$ (Frape, unpublished results)

	Liveweight kg	Stomach	Caecum + colon
	----------------	ml/kg $W^{0.75}$	----------------
Horse	545	74.5	595[1]
Pig	73	71.3	358
Pig	9	61.6	135
Sheep	36	772	191
Rabbit	2.36	30.5	104
Guinea pig	0.58	71.1	75.4
Rat	0.30	16.9	36.3
Hamster	0.222	5.4	37.1
Gerbil	0.068	16.3	45.8[2]
Mouse	0.030	20.8	34.7

[1]Excluding dorsal colon
[2]Caecum only

Furthermore when both coarsely and finely milled roughage were given to red deer and sheep (Sanchez-Hermosillo and Kay, 1979) both poorer digestive efficiencies and longer retention times were found in the deer. Use of the suspended cotton thread technique or the nylon bag technique demonstrates relative differences between animal species in the digestive capabilities of ruminal or caecal fluid (Table 16.2). However, this technique when used over fixed time periods ignores differences between animals in rate of passage or retention time. Clearly both characteristics play their parts in determining the end result.

Table 16.2. Cell wall digestibility (%) (Koller et al., 1977 and 1978)

	Wheat straw		Timothy hay	
	Nylon bag	In vitro	Nylon bag	In vitro
Cow (ruminal microbes)	23.0 ± 3.5	19.2 ± 2.5	37.3 ± 2.6	35.3 ± 1.2
Pony (caecal microbes)	16.9 ± 1.8	13.1 ± 0.6	30.1 ± 1.5	21.8 ± 2.5

Chemical composition and form of fibre, not only affect its susceptibility to digestion but can also influence feeding habits. Fioramonti et al. (1979) found that a material with a high hemicellulose to cellulose ratio compared with one of low ratio (beet pulp versus lucerne) increased by three fold feeding time in rabbits and reduced their capacity for feed dry matter. Furthermore circadian variations in the caecocolonic motility were enhanced by the beet pulp. Differences in digestibility and capacity for dry matter in both ruminant and non-ruminant herbivores are moreover attributable to the physical characteristic-length of fibre in roughage. Incidence of vices in horses and caecotrophy in rabbits are similarly associated with this physical property and evidence for this will be presented later.

Certainly an important factor in microbial digestion of fibre is gut capacity and its related clearance time. This is complicated in the case of the rabbit and other coprophagous species by the consumption of caecotrophs. Piekarz (1963) showed that the clearance of digesta took between 5.6 and 48.6 hours in normal rabbits, but only 5.0-28.0 hours when coprophagy was prevented. Nevertheless the quantitative impact of this on fibre digestion in the rabbit is questionable (Udén and Van Soest, 1982). Comparative data on the capacity of different species for feed dry matter are scarce because of the necessity of making comparisons amongst animals consuming feed of the same type. Limited evidence suggests that the voluntary intake for rabbits, sheep and cattle are respectively 58-63, 77-93 and 90-100 g/kg $W^{0.75}$ (Joyce et al.,

1967, 1971, MacLean et al., 1970). Table 16.1. gives related data on the capacity of the stomach and large intestine as a function of metabolic size, indicating that the capacity of the organs of microbial activity declines in the order sheep, horse, rabbit. Moreover the table also indicates that the ratio volume to metabolic size tends to decrease with decreasing size of species.

The fasting metabolic rate of mammals has been determined to be a constant per unit $W^{0.75}$ (Kleiber, 1961) when considered over the whole spectrum of interspecific body weights. Subsequent work has however demonstrated significant deviations from the interspecific mean. There is therefore no certainty that smaller species in each family will require more concentrated feed as might be expected if capacity was proportional to $W^{1.0}$. Nevertheless there is some tendency for this to be so within both non-ruminant and ruminant herbivores.

A number of factors disturb the general relationship. Wild ruminants tend to have a higher FMR than domestic ruminants, or the inter-specific mean. Measurements indicate a value some 35-40% higher than this mean (Miller and King, 1981), possibly related to the higher proportion of lean, and hence metabolically active tissue and lower proportion of insulating fatty tissue. Moreover many of these species appear to be temperamentally less placid. The higher energy costs incurred limit their facility for consuming poor quality roughage. However many species within the intermediate group of ruminants vary their energy expenditure and/or consumption with seasonal rhythms of day length and availability of browse. Capacity for this seasonal adjustment relies partly on the size of the animal. Unlike some smaller cervids red deer have a more catholic taste adapting to the available supply both between and within seasons and the stag will consume poorer feed than the hind (Kay and Staines, 1981). The energy costs of work are roughly proportional to $W^{1.0}$ and where domestic herbivores are subjected to hard work the relative advantage of the larger individual for accommodating poor roughage in maintenance is swamped by the demands of the work.

16.3. SIGNIFICANCE OF FIBRE IN THE NORMAL DIET OF SEVERAL NON-
        RUMINANT SPECIES

The products of microbial degradation of dietary fibre which contribute to the energy demands of the host animal, are the volatile fatty acids. An effective absorption of VFA from the large intestine has been demonstrated in all non-ruminant herbivores which have been investigated (Hintz et al., 1978). The extent to which VFA contribute to the maintenance energy demands of a selection of species is given in Table 16.3. VFA may not be essential to survival of adult herbivores, but fibre is apparently necessary for normal health. Much support for this statement could be cited, but two examples will suffice. Franck and Coulmin (1979) showed that by raising the dietary crude fibre level from 9 to 12% in rabbits, mortality was decreased from 19.4 to 5.5% without a depression in performance. Meyer et al. (1981) and may others have demonstrated that in the absence of dietary long fibre sources disturbed behaviour, including crib biting and coprophagy, occurs in horses. In their experiment 0.5 kg straw per 100 kg liveweight was a satisfactory feed source of fibre.

Table 16.3. The contribution of volatile fatty acids to the maintenance needs of some non-ruminant animals

Species	Approximate proportion of maintenance energy expenditure (%)	Authority
Porcupine	16	Johnson and McBee, 1967
Willow ptarmigan	30	McBee and West, 1969
Beaver	19	Hoover and Clarke, 1972
Rabbit	10-20	Hoover and Heitman, 1972
Pony	25	Argenzio et al., 1974

16.4. COMPARISONS OF GASTRO-INTESTINAL FLORA AND THEIR PRODUCTS

Bacterial growth, both in the rumen and in the large intestine of the horse or rabbit, follows a circadian rhythm and therefore realistic comparisons of population size are made difficult. Nevertheless, the data given in Table 16.4 show that some similarities exist both in numbers, in bacterial types and in their products between the pony caecum and the steer rumen. A bacterial population per gram of contents in the rabbit caecum and colon ranging from 10^7 to 10^9 was estimated by Gouet and Fonty (1979).

Table 16.4. The effect of a diet of timothy hay with or without a supplement of oat grain upon the bacterial population of the pony caecum and the steer rumen (Kern et al., 1973)

Oats:	−	−	+	+
	Caecum	Rumen	Caecum	Rumen
Total bacterial counts/ ml x 10^7 of digesta	481	504	702	622
Rods, % Gram −	63.1	31.8	60.0	33.5
Gram +	8.6	2.8	8.5	0.5
Cocci, % Gram −	23.0	49.3	16.5	52.5
Gram +	6.3	13.8	12.6	13.3
Acid, (µM/ml) Acetic	39.2	44.6	34.6	61.9
Propionic	13.4	13.9	10.7	17.9
Butyric	3.5	9.3	3.9	9.9

16.5. COMPARATIVE ASPECTS OF FERMENTATION

Bauchop (1977) reviewed the evidence for fore gut fermentation in a number of mammalian families (Table 16.5). Microbial fermentation of fibre is a feature of variable significance in a wide range of ruminant and non ruminant species and it may occur to some extent throughout the GI tract except in parts of the stomach lined with oxyntic glands and adjacent to the pylorus. It may be a major function of parts of the forestomach, of mid or hind gut, or of all these. Cellulolytic activity is related to pH of the medium and hence its buffering capacity, and the degree of dissolution to the extent of development of delaying mechanisms and associated fermentation volume (Table 16.6).

The substrate fermented varies greatly between species. Amongst smaller cervids of the concentrate selector group, including muntjack, duiker, dik dik, suni and mousedeer, the relatively high ME intake required to sustain their high metabolic demands for maintenance per unit of live weight is achieved by several behavioural and morphological mechanisms. They selectively consume highly digestible roughage, the soluble carbohydrates and protein of which are then subject to rapid fermentation and the residues to a brief retention in the GI tract, allowing a relatively high dry matter intake but little, or no, fibre digestion of economic significance (Hofmann, 1982). The morphological and physiological adaptations required to accommodate the rapid fermentation probably include a copious flow of saliva of high buffering power (in contrast to the low buffering power of equine saliva), extensive ruminal papillation and a highly vascularized propria allowing rapid absorption of VFA.

Table 16.5. Mammalian families with major foregut fermentative digestion (Bauchop, 1977)

Order	Family	
Marsupiala	*Macropodidae*	(kangaroo)
Primates	*Cercopithecidae*	subfamily *Colobinae* (langur, proboscis and colobus monkeys)
Edentata	*Bradypodidae*	(sloth)
Artiodactyla	*Hippopotamidae*	(hippopotamus)
	Giraffidae	(giraffe, okapis)
	Tragulidae	(cherrotains)
	Cervidae	(deer)
	Antelocapridae	(pronghorn antelope)
	Bovidae	(sheep, cattle, gazelle)

Table 16.6. Relative capacity of organs of GI tract (%) (Jacquot et al., 1958, and Lebas, 1974)

	Horse	Cow	Pig	Rabbit
Stomach	9	71	29	34
Small intestine	30	19	33	11
Caecum	16	3	6	49
Colon	45	8	32	6

The intermediate mixed feeding ruminants are roughly classified into two groups, the first of which appears to have on average a greater ability in fibre digestion:

1) grass eaters - for example, domestic goat, impala and Thomson's gazelle and
2) Tree or shrub foliage eaters including eland, Grant's gazelle and steinbok

Table 16.7. Volatile fatty acids in some species of wild or domesticated animals
(Bauchop, 1977)

		Total VFA (mM/l)	Acetic	Propionic	Molar proportions (%)			
					N-butyric	Isobutyric	N-valeric	Iso-valeric
Presbytis cristatus		165	51	24	6	5	7	8
Procolobus verus		181	70.5	16.7	9.9	–	2.5	–
Procolobus		219	63.9	22.8	9.8	–	2.2	–
Setonix brachyurus	mixed	101	85	4.2	8.5	1.0	0.5	0.9
(small macropod marsupial)	grazing	137	71.2	11.0	17.1	trace	1.2	trace
Hippototamus		110–170						
(stomach diverticulum)								
Cattle		137	64	21.9	11.9	–	2.2	–
Sheep		94	58	29	7	2	1	3

498

The stomach resembles that of concentrate selectors but the scope for adjusting capacity and other morphological adaptations is considerable and varieties are legion. In the impala and chamois the surface enlargement factor may increase 3 fold in response to seasonal improvements in roughage quality (Hofmann, 1982). On the other hand when red deer in winter, or antelope in the dry season resort to consuming sparse and poor roughage the ruminal epithelium becomes partly keratinized, cellulolytic bacteria abound and a diverse array of ciliate protozoa can be detected in the rumen (Kay et al., 1979). In contrast, but as would be anticipated, Frape et al. (1982) observed a decline in faecal ciliates of horses with increasing dietary fibre.

This characteristic of an individual animal to adapt to a changing environment is common in varying degrees throughout the animal kingdom. The adaptive mechanism requires the passage of time, hence the change in efficiency of a species to utilize a given fibre source is gradual, and no constant value, or species mean, may be quoted without an inflated error of between species comparisons. For example Engelhardt et al. (1978) determined that cotton threads were digested at similar rates in the stomachs of Tammar wallabies and sheep, but wallabies confined to a straw diet demonstrated a greater rate of thread digestion.

A number of similarities amongst widely different species occur in microbial population density and Table 16.7 shows that correspondence occurs in the fermentation pattern and total acid production amongst species as widely divergent as monkeys, marsupials, hippopotamus, cattle and sheep. The most variable major metabolite appears to be propionate. It should be appreciated however that these characeristics are not only affected by rates of production, but also by rates of metabolism within the lumen of the gut and by rates of absorption into the portal blood.

Despite the apparently poorer utilization of fibre by rabbits than by horses or ruminants, Parker (1976) concluded from ^{14}C studies that VFA absorbed from the caecum in the normal rabbit provided 30% of the maintenance energy requirements, a value similar to that attributed to those products from the caecum and colon of the horse (Glinsky et al., 1976; Meyer, 1980). The average production rate of VFA in the caecum of the pony amounts to approximately 5.7 mM per min (acetate 3.67-3.98, propionate 0.41-1.66, butyrate 0.34-1.12 mM per min, Glinsky et al., 1976).

Production rates of VFA tend to be greater in the horse's caecum and ventral colon than in the dorsal colon, yet the pattern of
fermentation and the levels of total VFA present are rather similar amongst these organs at comparable levels of intake (Table
16.8). Proportionately less propionate in the dorsal colon than in
the caecum or ventral colon undoubtedly arises from less starch
being present in the substrate of the dorsal colon. Between 75 and
85% of the insoluble carbohydrate consumed by horses are fermented
in the large intestine (Meyer, 1980), but the microbial concentration in the caecum and colon and the reaction time in these organs
are generally less than they are in the rumen (Meyer, 1980). Comparisons between individual animals in terms of bacterial numbers
and VFA concentration are, however, influenced by sampling time
relative to time of feeding (Table 16.8). Alexander (1952) demonstrated that between 22 and 58% of a cotton thread was digested in
the ventral colon of fed ponies during 24 hours, but only 0-19%
was digested in those which had been fasted. The situation in the
horse, however, is apparently in contrast to that in the rabbit as
Susmel and Lanari (1977) were unable to demonstrate any clear
relationship between time after feeding, or level of feeding, and
total VFA concentration, or the molar proportions of the acids
(Table 16.9) in that species.

Table 16.8. Proportions and concentrations of VFA in the large
intestines of fed and fasted horses and ponies

	Acetic	Propionic	Butyric	VFA as acetic acid Fed	Fasted
	------- molar % ----------			----- mM/l ------	
Ventral colon (Alexander, 1952; Meyer, 1980)	66.5	25.0	6.0	50-70	8
Dorsal colon (Alexander, 1952)	68.8	16.5	8.2	60-70	35
Caecum (Alexander, 1952; Meyer, 1980)	60-80	15-25	4-10	60-95	16
Colon (Meyer, 1980)	67-80	11-18	5-9	-	-

In a wide range of herbivores dietary fibre not only contributes to the energy demands of the animal but can also play a vital role in health. It has been well established that dietary composition, in particular the ratio of soluble to insoluble carbohydrate, has a notable effect on ruminal pH and VFA proportions and similar phenomena are demonstrable in the horse. A comparison between a diet of hay and one of concentrate based upon maize, oats, soya, wheat bran and molasses clearly showed that caecal pH in the concentrate-fed horses was significantly lower 6-7 hours post-prandially (Willard et al., 1977) (Table 16.10) and that the molar proportion of propionate was greater. These differences were associated with a greater incidence of wood chewing and coprophagy in the concentrate fed horses. A decline in caecal pH would be expected to adeversely affect the activity of celulolytic bacteria.

Table 16.9. Effect of feed composition and system on VFA content in the caecum of the rabbit (Susmel and Lanari, 1977)

Time of day	Low fibre diet (6.1% CF)		High fibre diet (14.7% CF)	
	Fed once per day	Ad lib.	Fed once per day	Ad lib.
	---------------Total VFA, mM/kg DM ---------------			
0900	207	227	125	248
1200	182	207	140	190
1500	236	190	211	203
1800	221	200	170	201
	-------- molar proportions of acids, % --------			
Acetate	61.2	63.5	69.7	68.7
Butyrate	14.5	9.7	11.2	10.7
Propionate	20.2	23.9	15.9	17.2

16.6. RATE OF PASSAGE OF DIGESTA

Few studies have been undertaken of comparative transit times for dietary residues when a number of species have received the same diet; but Table 16.11 gives the mean transit times for 50% and 85% of fluid and particulate markers in 10 animal species receiving a diet of pelleted hay and grain. These data are unexpected to the extent that the values for ponies considerably exceeded those in bullocks, even for the larger particulate marker. Nonetheless clear evidence for a shorter retention time of cell wall material in the gastro-intestinal tracts of equids and rabbits, as compared with domestic ruminants, has been obtained (Udén, 1978) and it is speculated that it is this factor which largely accounts for the species differences in digestibility of fibre.

Table 16.10. Effect of time after feeding and feed type on caecal pH of horses (Willard et al., 1977)

Time after feeding (hours)	Caecal pH	
	Hay	Concentrate
0	7.14	7.22
1	7.16	7.24
2	7.04	7.14
3	7.01	6.77
4	6.92	6.43 ($p < 0.01$)
5	6.89	6.27 ($p < 0.05$)
6	6.87	6.12 ($p < 0.10$)
7	6.75	6.19
8	6.83	6.41
9	6.96	6.44
10	6.92	6.64
11	7.12	6.82

In the rabbit observations are complicated by a disturbance the physical form of feed may have on the normal diurnal variations in activity and the circadian changes in the occurrence of coprophagy, but Lebas and Laplace (1977a) found no significant difference in the passage time for a mixture of soya, maize and straw in the form of meal or of pellets. The rate of gain on pellets exceeded that on meal despite a lower dry matter intake on the former. These investigators attributed the outcome to an abnormal pattern of feeding and excretion in rabbits receiving the meal.

Comparison amongst species which normally consume quite distinct diets are difficult to interpret because of the effect diet type and fibre length have on intestinal movements and rate of passage. Barley straw given to rabbits moves more rapidly than oak sawdust which in turn exceeds the rate for wood cellulose (Lebas and Laplace, 1977b). The grazing ruminant harvests long roughage with an extensive mobile tongue and swallows the material after very minimal mastication. The ruminal pillars and the reticulo-omasal orifice and other ostia have a prehensile effect on the long fibrous material protracting passage, favouring the slow fermentation rate of fibre, until through rumination the particle size becomes sufficiently small for linear passage to proceed. Thus the passage time for ground fibre is less than that for long in cattle and sheep. In the horse and some other non-ruminant herbivores an analogous situation exists in which the grinding and pelleting of chopped or long roughage accelerates its rate of passage, increases the amounts volountarily consumed and influences utization and performance (Haenlein et al., 1966a; Hintz and Loy, 1966). This effect appears to be attributable mainly to the grinding forage receives prior to pelleting. Wolter et al. (1974) found that the mean rate of passage of ground meadow hay was 26 hours in ponies, whereas ground and pelleted meadow hay took 31 hours and the long material 37 hours.

Table 16.11. Gastro-intestinal transit time (h) for a pelleted mixture of hay and grain for 50% and 85% of each digesta marker receovered in faeces using polyethylene glycol (fluid) and radioopaque polyethylene tubing in lengths of 2 and 10 mm (Clemens and Stevens,1980)

| Recovery | Fluid marker | | Particulate marker | | | |
| | | | 2x2 mm | | 2x10 mm | |
	50%	85%	50%	85%	50%	85%
Opossum	10	20	7	14	6	10
Raccoon	11	20	17	22	18	23
Pig	19	43	31	47	34	60
Rat	20	47	22	41	42	62
Dog	32	45	24	59	33	64
Rabbit	39	79	27	59	73	134
Guinea pig	40	78	39	65	48	64
Llama	40	62	72	134	>240	>240
Ox	62	136	93	224	112	>240
Pony	106	168	195	>240	>240	>240

In contrast to sheep and cattle the horse subjects long hay to thorough mastication prior to deglutition, whereas concentrates receive a more summary treatment (3000-3500 chewing movements per kg compared with 800-1200 per kg for concentrates, Meyer et al.,

1975). Particle size of swallowed hay in the horse is normally less than 1.6 mm in length (Meyer et al., 1975), approximating to that of preground material. Thus a different explanation for the shortened transit time of ground fibre in the horse must be advanced. It is proposed that a greater rate of consumption during a discrete meal of ground and pelleted roughage leads to a larger volume of digesta being propelled in a limited time period through the GI tract. A large mass of digesta is likely to stimulate stomach emptying and peristaltic action. If this hypothesis is true then procedures which decrease the consumption rate of ground fibre should increase its digestibility.

It is apparent however that such an explanation does not account for all the evidence of Meyer et al. (1975) on rate of consumption of ground but unpelleted forage; although these workers observed an accelerated rate of passage through the stomach of ingesta resulting from ground hay in comparison with that from long hay. Schurg et al. (1978) demonstrated similar digestibilities for ADF, DM and cell wall constituents of ryegrass straw when given in a long form, or as hard pressed cubes or briquettes requiring greater chewing and consumption time. On the other hand when the ground straw was formed into softer pellets eating time was shortened and the digestibility of the components was significantly depressed.

It has been assumed that the capacity of a horse for bulky feed is in proportion to the rate of passage of digesta. However, with continuous access to feed in the stable meals may occur every two hours. The urge to commence eating and the satiety responses appear to be controlled by the concentration of digestion products, especially of glucose, but also VFA, in the lumen of the intestines (Ralston and Baile, 1982). It is possible that ground and pelleted roughages may delay the satiety response so that more may have been consumed during a meal before the mechanism is triggered. The actual bulk of the digesta appears not to be directly involved in this response (Ralston and Baile, 1982).

The delaying mechanisms for long roughage in the *Ruminantia* differ in degree amongst the species. Those in grazers allow them to be relatively unselective, whereas at the other extreme selection is vital. Amongst these concentrate selectors a simple rumen capable of complete contraction with thin, or weak, pillars, wide ostia, a large reticulum with low crests and a small omasum having few laminae (Hofmann, 1982) all contribute to extensive fermentation of protein and soluble carbohydrates only, affording to incontinent delay to larger fibrous particles. Thus in an environment of abundant lush vegetation the relatively scarce resources of fermentation space is reserved for the more nutritious cellular components. Fibre is largely voided intact. Low cellulolytic activity has, for example, been demonstrated in kudu (Giesecke and van Gylswyck, 1975) and roe deer (Prins and Geelen, 1971). The dik dik and suni, amongst the smallest of ruminant species, select fruit and dicotyledon leaves rich in cell contents (Miller and King, 1981). Digestibility trials (Hoppe, 1977) have shown that they digest well the soluble carbohydrate fraction of the leaves of lucerne hay but digest poorly the crude fibre fraction, a difference associated with very short retention times of 1.58 and 20.5 h for the residues of stained hay leaves and stems respectively in the suni (Hoppe and Gwynne, 1978).

16.7. APPETITE AND ACCEPTABILITY

The acceptability, or palatability, of a fibrous feed is influenced by some imponderable factors, but several more definable and less illusive characteristics include the amount, chemical composition and digestibility of its fibre component. Physical processing can influence this acceptability. Grinding and pelleting of straw may not improve its digestibility but can increase voluntary consumption by horses in contrast to the rabbit. Schurg et al. (1978) compared long stem ryegrass straw with that in pelleted, cubed and briquetted forms and found that the average voluntary intake of each was respectively 5.5, 6.4, 6.0 and 5.9 kg/horse daily.

Several fibrous materials may present potential sources of dietary energy, but their acceptability even when processed may be poor, implying that their practical value is negligible. Wolter et al. (1979) commented that both wheat straw and grape pulp in pelleted forms were not well accepted or digested by ponies under their conditions. Luick (1976, 1977) observed when reindeer were offered hay, lichen, or pelleted feed, that the hay was quite unacceptable and if offered alone could be rejected and lead to inanition and death. Similar observations have been made with muntjac (concentrate selectors) (Norma G. Chapman, personal communication, 1982) which will not accept and eat long hay. Generally

Table 16.12. Apparent digestibility coefficients (%) of cell wall components of timothy hay (Udén and Van Soest, 1982), of straws (Hintz, 1969) and of the crude fibre component of alfalfa (Slade and Hintz, 1969)

| | Cell wall components | | | | Cell wall indigesti-bility | | Straw crude fibre digestibility | | | Alfalfa crude fibre digestibility |
	Cell wall	Cellu-lose	Hemi-cellu-lose	Lignin	Coeff.	SD	Oat	Rye	Wheat	
Heifers	51.4	55.9	56.0	21.1	49.6	2.4	59	–	58	–
Goats, sheep	44.0	46.0	48.9	17.2	56.1	2.1	59	55	59	–
Ponies	37.0	36.6	42.0	20.5	64.0	4.3	–	–	–	38.1
Horses	33.3	33.4	39.5	10.8	64.0	4.3	51	52	59	34.7
Rabbits	9.3	6.9	11.6	9.6	90.4	2.0	25	–	–	16.2
Guinea pig	–	–	–	–	–	–	–	–	–	38.2

poor hay appears to contain too much crude fibre to be acceptable to reindeer.

As proposed earlier there appears to be optimum dietary fibre levels for each herbivorous species. Spreadbury and Davidson (1978) compared oat husk with barley straw and pure cellulose in rabbit diets and determined that daily feed intake increased as the ADF content of the diet was increased from 3.9% to 27%. The rabbits maintained a constant daily ME intake of 1100 kJ by increasing their voluntary intake. However, maximum growth rate occurred at approximately 10% crude fibre or ADF. Cheeke and Patton (1980), in contrast, observed that oat husk, beet pulp or wheat straw given to rabbits in diets containing 7.8% fibre did not increase growth rate in comparison to the control low fibre diet. Nevertheless under practical conditions of extreme competition many herbivores may subsist on diets of which straw, or other poor roughage forms well over half the daily dry matter intake. This applies to both grazing and intermediate, or mixed feeding, ruminants. The camel does not chew the cud and is not classified amongst the *Ruminantia* , but is a member of the *Tylopodia* . It has has a three compartment stomach, with a large rumen but from which the omasum is excluded. Even under working conditions straw can form over half its diet (Yagil and Etzion, 1980).

16.8. COMPARATIVE VALUES FOR FIBRE DIGESTIBILITY

Few studies have been conducted in which several species have received the same diet in the same proportion to metabolic size, but several comparisons have been made concerning the digestibility of dietary fibre in a range of ruminant and non-ruminant herbivores. The complication of adaptation to dietary change has to be accepted as a factor which inflates the error of any comparison between species. The extent to which equids and other non-ruminant herbivores can adapt, it would appear, is less than in members of the intermediate ruminant group.

Despite the existence of inconclusive evidence recent investigation supports the contention that domestic ruminants (cattle, sheep and goats) are on average more efficient than horses and ponies in the digestion of various components of dietary fibre (Udén and Van Soest, 1982). In contrast there is unqualified agreement that rabbits are less able in this respect (Hintz, 1969;

Slade and Hintz, 1969; Udén and Van Soest, 1982). Furthermore the pony may be more efficient than the horse (Slade and Hintz, 1969; Udén and Van Soest, 1982) (Table 16.12), and the donkey may have an even greater faculty for cellulose digestion than either the horse or ruminant (Wolter and Velandia, 1970). Two white rhinoceroses digested poor quality hay fibre with an efficiency approximating to that of the horse (Table 16.13) (Frape et al., 1982). Schurg et al. (1978) reported that the percentage dry matter digestibility for horses of a pelleted diet containing ryegrass straw or fescue hay was respectively 56.0 ± 1.8 and 66.0 ± 1.3.

Table 16.13. Apparent digestion coefficients for modified acid detergent fibre in ponies and white rhinoceroses receiving long hay[1] and pelleted high fibre concentrate[2] in various proportions (Frape et al., 1982)

| | Ponies | | White rino | |
	Pellet proportion, %	Digestibility, %	Pellet proportion, %	Digestibility, %
	29	66	25	64
	43	60	41	60
	71	76	50	47
	86	58		
SE of means	–	3.9	–	4.4[3]

[1]Pony hay contained 332 g MADF/kg
[1]W. Rhino hay contained 386 g MADF/kg
[2]Pellets contained 205 g MADF/kg
[3]Significant linear decrease in digestibility with increasing cube proportion (P<0.05).

In the late 1960s there was a scarcity of evidence on the efficiency with which horses could digest fibre, but a wealth of evidence in this respect on domestic ruminants. Therefore, equations were derived for predicting equine crude fibre (CF) digestibility with feeds containing more than 15% crude fibre from values determined in domestic ruminants and rabbits (Hintz, 1969):

$$\text{Cattle} \quad Y = 5.650 + 0.630X \pm 0.223[1]$$
$$\text{Sheep} \quad Y = 9.423 + 0.5853X \pm 0.167[1]$$
$$\text{Rabbits} \quad Y = 42.831 - 0.439X \pm 0.988[1]$$

508

where   Y = predicted CF digestibility for horses, and
           X = value obtained with other species

[1]SE of regression coefficient.

Similar equations were derived for the percent crude fibre digestibility of forages in horses (Noot and Trout, 1971):
$$Y = 38.09 + 0.04X_1 \pm 5.61^{1}$$
where X_1 = CF digestibility in steers,

$$Y = 57.53 - 0.50X_2 \pm 5.46^{1}$$
where X_2 = per cent CF in diet.

[1]SE of estimate.

Subsequent to that time information in species other than
domestic ruminants and rabbits increased and there was also a move
to use NDF and ADF as measures of dietary fibre which avoided the
criticisms that may be levelled at the use of crude fibre (Van
Soest, 1963). The data in Table 16.14 (Noot and Gilbreath, 1970)
nevertheless show that both crude fibre and cellulose digestibility of three grasses determined in steers exceeded the comparable
values determined in geldings. In Table 16.15 comparisons amongst
a range of wild herbivores in respect of their ability to digest
dietary fibre, described principally as NDF and ADF are given. It
is clear that species differences exist when fibrous material is
presented either in pelleted or in long form in comparable quantities under similar conditions.

In an endeavour to understand differences in efficiency of fibre digestion by five species Udén and Van Soest (1982) made some
interesting calculations. In deriving the data given in Table
16.12 the metabolic faecal output (MFO), assumed to be entirely of
microbial origin, expressed as a proportion of dietary cell wall
material, was 0.167, 0.425 and 2.13 for ruminants, equids and rabbits respectively. Udén and Van Soest (1982) concluded that the
rabbit's intestinal microflora either utilize substrates other
than fibre or their MFO contains substantial non-microbial endogenous matter. The data also indicate that little digestion of microbes occurs in the hind gut of equids or of rabbits in contrast
to extensive post-ruminal digestion of microbes.

Table 16.14. Apparent fibre digestibility (%) of three dried grasses by steers and geldings (Noot and Gilbreath, 1970)

	Crude fibre	Cellulose (ADF-insoluble lignin)
	Field cured orchard grass	
Steer	53.4	64.3
Gelding	43.1	52.1
	Artificially dried timothy	
Steer	61.5	67.2
Gelding	43.9	48.3
	Artificially dried brome grass	
Steer	59.9	63.7
Gelding	34.5	37.8
SE (between species)	0.61	0.76

16.9. HEART DISEASE

The suggestion that some zoo diets alien to the natural preferences of wild herbivores might be instrumental in the precipitation of cardiac ailments could well mean that this aspect of diet and of dietary fibre in particular, is of economic significance where valuable specimens are concerned. Kritchevsky et al. (1977) and Kritchevsky (1978) have reported that dietary lucerne has a greater capacity to suppress cholesterolaemia in rabbits than either wheat straw or cellulose and that both lucerne and wheat straw have a greater capacity to suppress atherogenesis than does cellulose.

16.10. THE PHYSICAL PROCESSING OF DIETARY FIBRE SOURCES

In a number of studies the effects of grinding, or chopping, roughage and of pelleting the products have been assessed. Rougeot et al. (1980) could find no differences in terms of fur yield between groups of rabbits receiving 120 g/week of ground straw incorporated into pellets, or the same quantity chaffed and placed in racks, or even when the chaffed straw was given to appetite. Unfortunately the appetite consumption by rabbits of chaffed straw was extremely variable ranging from 3.5 to 43 g daily. The voluntary consumption of loose chopped straw was also observed by Payne et al. (M. Payne, E. Owen and B.S. Capper, personal communication, 1982) to be minimal in rabbits from 45 to 82 d of age, amounting to 4.5 g/d (5% of DM intake). Thus with species unaccustomed to the consumption of poor roughage, including some intolerant ruminants, a worthwhile intake will obtain only in the context of a complete mixed feed, such as can be furnished by pellets.

Some interest has attached to the chopping of hay and its incorporation in briquettes measuring 5-6 cm in each dimension. Ahlswede (1977) gave Shetland ponies briquettes containing 46% chopped hay, oats, milo, maize, wheat bran, apple skins, linseed and molasses or long hay plus the other materials separately. The briquettes were well accepted without further supplementation with long roughage and the digestibility of their crude protein, lipid and nitrogen free extract was greater than when the components were given in the traditional fashion. Although the rate of consumption of the briquettes was the more rapid, experiments have shown that this rate depends upon the degree of compaction during manufacture. The consumption of hard pressed cobs or briquettes composed entirely of hay was shown by Meyer et al. (1975) to endure for a time equal to that for long hay whereas the consumption of soft pressed cobs was more rapid. Schurg et al. (1978) compared ryegrass straw in the long form with the same material ground and pelleted, or cubed, or in the form of briquettes. The cube and briquette forms required more chewing and extended eating time amongst mature mares. Nevertheless the daily intake of all the three compressed forms was greater than that of the long straw; but most horses given any of them were prone to chew wood. On the other hand Haenlein (Haenlein et al., 1966a; Haenlein, 1969) noticed no wood chewing amongst horses offered wafered or briquetted (5 x 6 cm), in contrast to pelleted, alfalfa hay. Of the three compressed forms compared by Schurg et al. (1978) only the pelleted straw lowered the digestibility of ADF, cell wall constituents and dry matter. A similar effect was reported by Frape et al. (1982) in that a linear reduction in ADF digestibility occurred in white Rhinos when the proportion of fibre derived from long hay was decreased and that derived from pellets was increased. However, no significant change was measured in ponies under similar circumstances (Table 16.13).

Table 16.15. Apparent fibre digestibility of a number of feeds by herbivores (Bracketed values are number of individuals tested)

	1 ADF	1,2 NDF	2 Cellu- lose	3 Crude fibre
Bison		50.2	63.6	
Llama		47.8	58.9	
Persian gazelle		51.2	56.8	
Onager		47.1	58.7	
Przewalski horse		41.6	46.0	
Grevy's zebra		39.0	49.7	
Buffalo (2)				56.8
Reedbuck				56.8
Goat				55.8
Ugandakob				48.0
Sheep (2)	64.9[3]			
Hippopotamus	61.7[3]			
Adax (2)[4]	42	57		
Giraffe (2)[4]	55	54		
Bongo (2)[4]	42	42		
Gaur (2)[4]	41	38		
Banteng (4)[4]	33	29		
Dorcas gazelle (3)[5]	39	40		
Fringe eared oryx (2)[4]	19	41		
Arabian oryx (2)[4]	39	33		
Caucasian tur (1)[6]	47	42		
Elephant (5)[6]	20	31		
Black rhinocerus (2)[6]	22	33		
White rhinocerus (2)[7]	54	67		
Scimitar horned oryx(4)[8]	49.9	58.6		
Sheep (4)[8]	35.6	47.0		
Scimitar horned oryx(2)[9]	60.3	67.7		
Sheep (2)[9]	57.8	65.7		

[1] Ullrey et al., 1979, Ullrey, 1980
[2] Hintz et al., 1976
[3] Grass hay mixture, ADF of Elephant grass hay (Arman and Field, 1973)
[4] Alfalfa of 36% ADF and 42% NDF, 17% crude protein
[5] Alfalfa of 21% ADF and 27% NDF, 25% crude protein
[6] Sudan grass 44% ADF and 66% NDF
[7] Sudan grass 35% ADF and 62% NDF
[8] Alfalfa of 23.8% ADF and 39.8% NDF (Miller and King, 1981)
[9] Grass hay : concentrate, 3 : 1 (Miller and King, 1981)

The faster rate of passage of pelleted hay in comparison with chopped or long hay (Hintz and Loy, 1966; Haenlein et al., 1966b) is associated with a decrease of 9-15% in fibre digesibility (Haenlein et al., 1966a; Hintz and Loy, 1966) without adversely affecting the digestibility of the other components. This eventuality appears to be the consequence both of grinding and of pelleting (Wolter et al., 1975). Nevertheless, horses have a greater capacity for compressed roughage than for loose hay. Haenlein et al. (1966a) determined that 14% more wafered hay and 21% more pelleted hay could be consumed daily than of loose hay. That such procedures may promote a larger daily dry matter consumption of poor quality roughage may be of considerable economic significance in situations where only maintenance of condition and health are of paramount importance. The salient issue is whether the procedures will increase the daily intake of digestible energy. It has been asserted here that dry matter intake may be increased to the extent of 15-20%, but that crude fibre digestibilty can be depressed by 9-15%. Cereal straws contain approximately 45% fibre and thus physical processing may allow an increase of 10-15% in daily digestible energy intake in the horse.

Level of intake itself does not appear to affect digestibility of all roughage diets in horses (Reid and Tyrrell, 1964) whereas in common with cattle and sheep the addition of grain to the roughage may decrease fibre digestibility with increasing dietary intake. These consequences do not appear to be influenced by the number of feeds per day in the range one to six (Butler and Hintz, 1971).

Notwithstanding the greater voluntary consumption of pelleted roughage the more frequent incidence of wood chewing does not seem to be necessarily associated with the time spent eating. For example, Meyer et al. (1975) observed that horses on average took 40 minutes to consume 1 kg of long hay or 1 kg of hard pressed hay cobs. For soft pressed cobs the time taken was less, but they took longer than 40 minutes to consume ground or chopped hay and long straw. It was further noted that ponies took 2-5 times as long as the horse to consume similar feed.

It is perhaps appurtenant to the occurrence of this vice that Tisserand et al. (1980) measured a lower cellulolytic activity in the caecum and colon of a pony when ground or pelleted hay was given alone or with oats in comparison to the activity folllowing the consumption of hay in a long form.

Quite apart from the physical form in which feed is given, recent evidence suggests that aspects of feed management may play their part in its utilization. Where horses are given a meal of straw and concentrate, the order in which these two are consumed appears to influence the consequences (Muuss et al., 1982). The concentrate mixture followed by the straw led to intense mixing of the digesta from the two ration parts which was not observed when the feeding was in the reverse order. Observations on the fermentation taking place in the caecum and ventral colon showed that the concentrate followed by straw led to a constant VFA level of approximately 55 mM/l, whereas by feeding in the reverse order there was an initial decrease in VFA level followed by a gradual rise from 23.5 to 63.7 mM/l.

16.11. DIETARY NITROGEN

Degradable dietary N sources play two roles in respect of the intestinal microflora of herbivores. They form a substrate for microbial growth from which amino acids may be derived for protein synthesis by the host and they stimulate the breakdown of structural carbohydrates when added to N-deficient diets. Thus ruminants receiving a diet depleted of N express a low capacity for fibre digestion as the growth of cellulolytic bacteria is depressed. However the N content of digesta entering the large intestine of the horse is relatively constant despite variations in the amounts of dietary N. On the other hand a small proportion of dietary protein is not degraded until the digesta enter the large intestine and moreover an increase in dietary N probably leads to a greater rate of N secretion into the small intestine from the blood. Thus it can be appreciated that poor quality roughage diets may deprive intestinal bacteria of the horse of adequate N nutrition; and in fact some evidence is forthcoming for a response in horses to protein and non-protein N when they are given access solely to low quality roughage (Godbee and Slade, 1979; Godbee et al., 1979). Pregnant mares receiving 8 kg of poor quality hay with or without a concentrate block containing 20% crude protein including 6% derived from urea showed differences in the dimensions of the hair fibres in the coat. Those receiving the block had longer hair

514

of greater bulb diameter, and the albumin to globulin ratio of the blood was greater.

In rabbits the evidence is at first sight generally less convincing. Lebas and Colin (1973) were unable to demonstrate any significant difference in weight gain, feed intake or feed conversion efficiency in 5 week old rabbits receiving a 12.5% protein diet with or without 1.5% urea. In this regard Hoover and Heitmann (1975) concluded that N disappeared between the caecum and colon in the form of ammonia and that any effective utilization of urea resulted from tissue synthesis of dispensable amino acids and from coprophagy. This mechanism may have accounted for an increase in N retention by rabbits when urea was added to diets containing up to 7% crude protein (Slade and Robinson, 1970).

Nevertheless, Slagsvold et al. (1979) were unable to demonstrate any increase in the utilization of oat straw by Shetland ponies as a consequence of adding either 2% urea or 15% soya to the basal diet (Table 16.16). The basal diet contained per kg 650 g straw, 50 g soyabean meal, 20 g maize, 10 g molasses with a chemical composition of 1.31% N (8.2% crude protein). The basal, therefore, contained some high quality protein and at a higher level than Slade and Robinson (1970) found to be a maximum for obtaining a response to inorganic N in rabbits. Low quality roughage contains no more than 0.6-0.8% N (4-5% crude protein) and the provisional conclusion may be drawn that in neither the rabbit, the horse nor any herbivore may a positive response to dietary NPN be forthcoming in terms of roughage utilization if the diet as a whole contains appreciably more N than this.

This conclusion must hinge upon the potential of a given species for poor roughage digestion. A greater potential, as between species, implies a greater minimal demand for degradable N if this potential is to be realised. Miller and King (1981) calculated from their evidence that the sheep required only 74 g degradable crude protein per kg DM, whereas the scimitar horned oryx required 92 g for a maximum microbial fermentation of a hay diet of given ME value.

Table 16.16.   The effect of dietary N and ammonia treatment of oat straw on the response
of Shetland ponies (Slagsvold et al., 1979)

	Dry matter	Cellu-lose	Hemi-cellulose	Lignin	Nitrogen	Utilization[1] efficiency	N-retention[2]
			Composition of diets (g/kg)				
Untreated	–	262	169	76	13.1	–	–
Ammonia-treated straw	–	262	169	76	15.2	–	–
Untreated straw + urea	–	262	169	76	21.6	–	–
" " + soya	–	262	169	76	18.0	–	–
			Digestibility and utilization of diets (%)				
Untreated straw	50.4	39.1	40.2	4.9	59	41	-13
Ammonia-treated straw	59.6	51.8	60.3	31.1	62	40	– 6
Untreated straw + urea	52.7	36.3	41.6	13.2	73	33	6
" " + soya	50.1	37.0	35.7	17.3	67	44	6
SE of mean	0.66	1.57	1.60	2.00	2.3	4.5	4.6

[1] Absorbed N- (urinary N-endogenous urinary N)
            Absorbed N

[2] N retained/kg bodyweight daily

16.12. THE SODIUM HYDROXIDE TREATMENT OF STRAW AND OTHER BYPRODUCTS

Niekerk and Couvaras (1979) treated coarsely ground lucerne with sodium hydroxide at the rate of 4% and gave the product to foals at the rate of 50% of their diet. No toxic effects, as measured by rates of gain and blood characteristics, including electrolytes, was detected. Most investigators have moreover concluded that alkali treatment of straw yields positive responses in fibre utilization but Jensen (1977) calculated that the inclusion of 10% treated straw in a lucerne diet brought about a faster rate of growth but at an unacceptable cost. Nevertheless costs of processing may be justifiably ignored in the early stages of investigation as it is essential to demonstrate significant and consistent biological responses before work may be warranted on methods to minimise the costs of production of treated material.

With all livestock, and especially with horses, problems of diet acceptability are continually encountered. Mundt (1978) commented that, although the crude fibre digestibility of straw in Shetland ponies was improved by offering it in the form of pellets treated with sodium hydroxide, the treated straw was not well accepted. No problem with electrolyte balance was detected however, whilst water intake increased.

A number of successful experiments with alkali treated straw have been conducted on rabbits. Yet Jensen and Tuxen (1978) asserted that the results depended in some measure on the initial quality of the straw used. They also found that if more than 15% was included in the diet, having tested a range from 10-30%, weight gain was decreased. It has already been pointed out that there are optimum levels of dietary fibre for herbivorous species and Blas et al. (1979) included from 0 to 15% alkali treated straw in rabbit diets in which the crude fibre content was held constant at 14.6%. The digestibility of N and fibre and the rate of gain were all increased linearly with increasing treated straw content of the diet, despite a decrease in the feed to gain ratio. Much other work has shown alfalfa to be a well accepted and utilized forage by non-ruminant herbivores and as the straw in this study was substituted for alfalfa hay and sunflower hulls with ranges of substitution of 15 and 3.6 percentage units respectively the work is encouraging. Lindeman et al. (1982) concluded that 30% sodium hydroxide-treated straw in the diet of growing rabbits has no detrimental effect on production, but that 8-10% was optimum, whereas 8% of untreated ground straw led to a depression in growth rate.

The alkali treatment of rabbit manure has demonstrated similar responses in rabbits to that of straw treatment. Table 16.17 gives the results of a rabbit digestibility study (Swick et al., 1978) in which the effects of sodium hydroxide and autoclaving were measured. The alkali treatment had a much larger effect than the autoclaving, although the former treatment apparently entailed a higher incidence of mortality from enteritis than occurred when untreated faeces were given. The mortality rates were 30%, 20%, 0% and 20% for the low fibre basal, 10% lucerne, untreated faeces and alkali treated faeces groups respectively. A salutary note should, however, be expressed that mortality rates are notoriously variable.

Table 16.17. The effect of processing on the utilization by rabbits of the fibre in dried rabbit manure included in the diet at the level of 75% (Swick et al., 1978)

Treatment of manure	Digestibility (%)	
	ADF	CWC
Untreated	8.1 ± 5.3^a	16.0 ± 3.7^a
Treated (2% wt/wt NaOH)	39.4 ± 3.4^b	41.7 ± 3.8^b
Autoclaved	16.8 ± 3.8^c	23.9 ± 2.3^c
Autoclaved + NaOH	35.8 ± 2.6^b	38.1 ± 3.2^b

abc Means with different superscripts within column are different (P<0.01)

Some concern has been raised that adverse effects may result from altered electrolyte balance in stock receiving sodium hydroxide treated products. Lebas et al. (1979) concluded that their failure to obtain any response to sodium hydroxide-treated straw in young rabbits resulted from the dietary sodium residues. However, there was no significant depression in growth rate. Lindeman et al. (1982) assessed the effect of incorporating 5 or 10 g sodium per kg diet compared with 2.5 g per kg in the standard diet of rabbits. They in fact measured a positive response to the additional sodium and estimated that the maximum feed to gain ratio occurred at 4.6 g sodium per kg diet. The optimum rate of sodium hydroxide addition to straw is approximately 4% of the dry matter, so that treated straw would contain about 2.3% sodium. A 10% inclusion of treated straw in the diet would therefore provide half the desired amount of sodium per 100 g diet. Boisen (1982) examined in some detail the effect of dietary sodium hydroxide on the acid-base balace of growing rabbits. He included 25% sodium hydroxide-treated barley straw in a diet containing 5% alfalfa. This diet had a pH of 7.2 in comparison to 6.2 for a diet containing 30% alfalfa and no treated straw. The extra base of the former diet was neutralised by acids liberated in splitting cellulose and hemicellulose from lignin. The author points out that green plants contain large amounts of neutral salts of organic acids transformed into bases during metabolism. Therefore alfalfa reduces the difference between the total base content (actual and potential base) in the two diets.

The principle of electro-neutrality applies to excreted urine which means that the equivalence of anions in urine corresponds with that of cations. The effective secretion of the additional sodium is thus associated with an increase principally in the bicarbonate content, raising the pH. In the study referred to the urinary pH for treated straw rabbits was 8.1 in comparison to 7.0 for the alfalfa group. Lower plasma bicarbonate in the former group demonstrated that the higher urinary pH was caused principally by the higher sodium excretion and not by the higher base surplus. This implies that in spite of the inclusion of base in the treated diet the lower production and higher excretion of base in that group tended towards metabolic acidosis.

The increased sodium excretion in the treated group of Boisen (1982) initiated diuresis and some increase in chloride, phosphate and potassium excretion, although the chloride and phosphate concentrations were independent of the diuresis. Neither the lower retention of chloride and potassium, nor the higher retention of sodium was reflected in the plasma. The author, however, suggested that lower potassium retention might lead to cellular changes in soft tissue and the higher sodium retention may bring about higher skeletal sodium. The diuresis would raise heat loss causing some increase in heat production and thus a lower net energy utilization. Boisen (1982) suggests that if high levels of sodium hydroxide-treated straw are used then one should replace sodium chloride in the diet by potassium chloride.

16.13. AMMONIA TREATMENT OF STRAW

Although the sodium hydroxide treatment of byproducts has been successful, the potential problems referred to above have led to a search for other chemical treatments. Ammonia has the putative advantage firstly that it is not a fixed base and secondly, its N may be used by the animal, or the symbiont microflora, in protein synthesis and transamination reactions. However, ammonia itself is highly toxic, although it evaporates readily when the product is exposed to air. The effect ammonia treatment has on plant residues rich in cell wall material probably results from ammonolysis and breaking of ester bonds between lignin and structural carbohydrates with the formation of amides.

Meyer et al. (1981) made a number of comparisons in horses in which hay and untreated, or ammonia-treated straw was fed with concentrates in ratios of 1.0 to 0.5 and 1.0 to 2.0, concentrate to forage. They measured no change in energy digestibility as a consequence of ammonia treatment. Mundt (1978) moreover found that the ammonia treatment of long straw was less effective in terms of the digestibility of crude fibre than was the sodium hydroxide treatment of pelleted ground straw, although there was an improvement over the untreated material. Payne, Owen and Capper (personal communication) also measured no positive response by rabbits to the ammonia treatment of straw and although they found that sodium hydroxide treatment improved feed:gain, killing out % and overall digestibility of dry matter there was some reduction in dry matter intake when the straw formed half the dry matter of the diet. Nevertheless, very encouraging results were reported by Slagsvold et al. (1979) in that the anhydrous ammonia treatment of oats straw incorporated at the rate of 65% in the diet of Shetland ponies, to which reference has been made already, led to an increase in dry matter digestibility of 15% and an increase in cellulose digestibility of 28%. The N content of the straw was increased by 50% and this N was used as efficiently as that from the diet containing untreated straw plus soya (Table 16.16). Ammonia treatment did not affect utilization of calcium, phosphorus or magnesium.

Variation in results amongst experiments in which the efficacy of chemically treated roughage has been examined must be expected. The quality of the original material may, as stated above, affect the outcome, but the method of application and duration of treatment are of particular significance for ammonification and probably also for sodium hydroxide use. Only by continued investigation will optimum procedures be established when one may anticipate more consistent biological responses.

16.14. DIET FORMULATION

This discussion has been restricted to a consideration of the chemical treatment of poor quality roughage providing little in the way of nutrients in the strict sense. Dietary formulation and balance should of course be considered carefully by those embarking on feeding systems employing large amounts of chemically processed roughage. Where alkali treatment of cereals or other feeds containing unsaturated fats and tocopherols is undertaken the destruction of the latter can lead to dystrophic myodegeneration in young stock unless approporiate supplementation with vitamin E is given (McMurray et al., 1980).

Where stock are at the maintenance level of energy intake, or are required to achieve only moderate levels of production there is a wide range of fibrous byproducts suitable for feeding, especially in complete mixed feeds of a compressed form. A number of such materials are listed in Table 16.18 together with several vegetable protein concentrates. Better quality protein sources, such as soyabean meal and fishmeal would also be entirely satisfactory.

The ingredient composition of several complete diets is given in Table 16.19. These are in no way intended to be least cost formulae or nutritionally identical, but are given as a general guide. They should be suitable for large non-ruminant herbivores, grazing ruminants and numbers 2 and 4 should be suitable for most members of the intermediate groups of ruminants. As a general rule under maintenance conditions the larger the animal the greater is the dietary concentration of poor roughage which can be imposed.

Table 16.18. Some less commonly used vegetable byproducts suitable for herbivores used for work, slow rates of growth and moderate lactation yields

Byproduct or feed	Group	Comments
Citrus pulp Chinese leaf meal Shea nuts Spent hops (dried)	I	Low palatability, medium protein level (except for citrus pulp) medium fibre level
Grape pulp Soya husk Broad bean pod meal	II	Slightly greater acceptability than group I, medium to high fibre, medium to low protein
Flax chaff Soya straw Soya bean pods Oatfeed	III	Somewhat greater acceptability than group I, medium to high fibre, medium to low protein
Olive pulp Wheat straw[1] Barley straw[1] Oat straw[1] Oat hulls Rice husks Broad bean hulls Linseed chaff Millet chaff and husks Wheat chaff Broad bean straw incl. pods Maize straw Sodium hydroxide treated straw	IV	Somehwat greater acceptability than group I, high to very high fibre and very low protein
Rice bran extr. Sugar beet pulp	V	High acceptability and digestible fibre. If large amounts given the feed should be soaked before feeding
Sesame cake (English) Niger cake Rape seed meal (low glycosinolate) Sorghum gluten meal Corn gluten meal	VI	High protein
Grassmeal (16% protein) Alfalfa meal (16% protein) Brewer's dried grains Corn gluten feed Sorghum gluten feed	VII	Medium protein content, high acceptability
Hydrolysed feather meal	VIII	Very high crude protein

[1]Spring cereal straw more acceptable and lower fibre content than winter cereal straw

Table 16.19. Complete compressed feed mixtures (%) for herbivores in maintenance, work, moderate rates of growth and lactation and for growing and breeding rabbits

	Min	Max	Suggested mixtures			
			Main-tenance	Hard work and selec-tive rumi-nants	Lacta-tion	Rab-bits
Group[1] I	–	10	–	–	–	–
" II	–	35	20.5	20.75	27	18.25
" III	–	35	20	20	20	20
" IV	–	30	30	15	–	–
" V	–	40	–	–	–	–
" VI	–	10	10	8	5	5
" VII	–	20	–	–	10	10
" VIII	–	5	–	–	5	5
Sodium hydroxide treated straw[2]			–	–	–	–
Barley, wheat, maize, milo, oats (grain)	13	60	13	30	25	35
Molasses (cane or beet)	0	10	5	5	5	5
Dicalcium phosphate[3]	0.5	2	1	0.5	0.75	0.5
Limestone	0	2	–	–	1	0.5
Sodium chloride	0.5	1	0.5	0.75	0.75	0.50
Magnesium oxide	0	0.5	–	–	0.5	–
Vitamins/trace elements[4]	+	+	+	+	+	+
Requirements (Per cent of diet):						
crude protein			8	8	12–14	12–14
crude fibre			20–30	16–22	16–22	10–12
calcium			0.35	0.35	0.7	0.6
phosphorus			0.28	0.28	0.5	0.5

[1]See Table 16.18

[2]For each 10 percentage unit addition replace 0.3% units of sodium chloride by 0.3% units of potassium chloride. Maximum 30% treated straw except in case of rabbits for which 15% is maximum.

[3]May be replaced by steamed bone flour or by deflourinated rock phosphate. When using 10% or more of rice bran replace phosphate by limestone.

[4]To provide per kg diet: Vit. A 3000 iu; vit. D 1000 iu; vit. E 15 mg; thiamin 2 mg; riboflavin 2 mg; pantothenic acid 8 mg; nicotinic acid 8 mg; manganese 20 mg; iron 50 mg; zinc 50 mg; copper 6 mg; cobalt 1 mg; iodine 0.5 mg; selenium 0.1 mg.

Table 16.20. Ruminants classified according to feeding characteristics, with approximate body weight (kg). (Compiled from various sources; we are greatly indebted to Professor R.R. Hofmann, Dr. P.P. Hoppe and Dr. J.M. King for access to unpublished data). (From Kay, Engelhardt and White, 1980)

Grazers or roughage eaters:			Browsers or concentrate selectors:		
American Bison	*Bison bison*	800	Giraffe	*Giraffa camelopardalis*	800
African Buffalo	*Syncerus caffer*	700	Moose	*Alces alces*	400
Ox	*Bos taurus*	600	Greater kudu	*Tragelaphus strepsiceros*	250
Zebu	*Bos indicus*	400	Bongo	*Taurotragus eurycerus*	200
Roan antelope	*Hippotragus equinus*	250	Lesser kudu	*Tragelaphus imberbis*	90
Waterbuck	*Kobus ellipsiprymnus*	220	Bushbuck	*Tragelaphus scriptus*	60
Wildebeest	*Cannochaetes taurinus*	220	Gerenuk	*Litocranius walleri*	40
Sable antelope	*Hippotragus niger*	200	Roe deer	*Capreolus capreolus*	20
Oryx	*Oryx gazella*	180	Muntjacs	*Muntiacus spp.*	20
Hartebeest	*Alcelaphus buselaphus*	150	Red duiker	*Cephalophus harveyi*	16
Topi	*Damalisus lunatus*	120	Grey duiker	*Sylvicapra grimmia*	14
Uganda kob	*Adenota kob*	90	Klipspringer	*Oreotragus oreotragus*	12
Nile lechwe	*Kobus megaceros*	80	Dik -dik	*Madoqua spp.*	5
European sheep	*Ovis aries*	50	Suni	*Nesotragus moschatus*	4
Reedbucks	*Redunca spp.*	40	Larger mousedeer	*Tragulus napu*	4
Mounflon	*Ovis musimon*	30	Lesser mousedeer	*Tragulus javanicus*	1.5
Oribi	*Ourebia ourebi*	16			

Intermediate or adaptable mixed feeders:		
European bison	*Bison bonasus*	800
Eland	*Taurotragus oryx*	700
Musk ox	*Ovibos moschatus*	350
Wapiti	*Cervus canadensis*	300
Red deer	*Cervus elaphus*	150
Mule deer	*Odocoileus hemionus*	120
Caribou	*Rangifer tarandus arcticus*	120
Reindeer	*Rangifer tarandus tarandus*	100
Whitetail deer	*Odocoileus virginianus*	100
White sheep	*Ovis dalli*	80
Fallow deer	*Dama dama*	70
Grant's gazelle	*Gazella granti*	60
Impala	*Aepyceros melampus*	60
Pronghorn	*Antilocapra americana*	50
Goat	*Capra hircus*	40
Springbok	*Antidorcas marsupialis*	35
Chamois	*Rupicapra rupicapra*	30
Maasai sheep	*Ovis aries*	30
Thomson's gazelle	*Gazella thomsoni*	20
Chinese water deer	*Hydropotes inermis*	12
Steinbok	*Raphicerus campestris*	10

16.15. SUMMARY AND CONCLUSIONS

Straw is a poor quality roughage and the scope for its use depends upon the daily energy requirements of the animals under consideration and upon their acceptance and effective utilization of
it. High producing or hard worked stock have large energy demands
limiting their capacity for bulky feeds. Therefore, these feeds
have greatest practical value for the energy demands of maintenance, or for those of up to twice maintenance. Nevertheless even at
these rates of intake consideration must be given not only to unit
digestibility but also to bulkiness and capacity as between roughage byproducts and animal species respectively.

Table 16.20 lists species of the Ruminantia classified as grazers, concentrate selectors and an intermediate group. Their appe-
tites for poor quality roughage are generally greatest amongst
grazers and least amongst selectors. Acceptability of a roughage
and the ability to digest it, between species, are considered to
be correlated characteristics. Several experiments in which species of at least two of these groups have been compared provide
some evidence in support of this assertion.

Some of the observations on digestibility do not agree with the
general argument. However, very few individual animals of exotic
species were used in each of these comparisons and as a number of
variable environmental factors are cited as influencing the issue
it is not justified to predicate biological laws and make dogmatic
assertions based upon fragmentary quantitative evidence. Considerable differences amongst and within the groups exist in the morphology and physiology of the gastro-intestinal tract and it is
suggested that the apposite use of these facts should be of predictive value in so far as the use of straw, for a particular species is concerned.

Grazing ruminants tend to retain the ingesta resulting from the
consumption of unprocessed straw longer than do many of their non-
ruminant counterparts. This results from the possession of more
highly evolved delaying mechanisms to the linear flow of ingesta,
a greater fermentation volume per unit of metabolic liveweight and
thus a more complete digestion of fibre. For the same reasons differences between these groups in the ability to digest roughage
byproducts of smaller particle size is probably less. Variation
amongst species in the fermentation efficiency of the microflora

and -fauna can to some extent be reduced by adaptation. Morphological and physiological adaptations of the host also occur widely
amongst herbivores in response to changes in natural food supply
and quality with seasonal and climatic cycles.

As many non-ruminant species will consume straw, or equally
poor roughage, and as ruminant species of the intermediate group
are particularly adaptable and will consume a great variety of
stems and leaves, it is suggested that grazing ruminants, and particularly the larger species of the intermediate group, together
with a number of non-ruminant herbivores, a few of which are described, are likely to accept after adjustment significant amounts
of straw in their diets.

Species which are biologically equipped for coping efficiently
only with higher quality roughage may consume little or no poor
roughage, even in the absence of any other feed. The incorporation
of such roughage with other more nutritious materials in pellets
may entice individuals of, for example, selective ruminants and
rabbits, to indulge in a signficant consumption of the material.
However, four prerequisites must be met:

1. The roughage must represent a useful source of energy or must
    represent some other asset of economic value.
2. The additional costs incurred in the processing of a roughage
    must not outweigh its additional value.
3. The effects of other highly fermentable components of the diet
    on roughage digestibility should be known for the species
    under consideration. Although soluble carbohydrates depress
    fibre digestion in grazing ruminants early evidence indicates
    that this may not occur to the same extent in more selective
    animals - equids, rabbits and selective ruminants. In equids
    additional starch may depress roughage digestion only if dry
    matter intake is also substantially increased. However the
    energy expenditure and expected products of the animal, and
    thus its purpose, should influence the choice of other ingre-
    dients and composition of the remainder of the diet. The eco-
    nomic value of the waste roughage could therefore be affected
    by the relationship between role of the animal and its diet
    composition.
4. The formulation of the complete diet should take into conside-
    ration other interactive effects, e.g. that of low N-roughage

526

upon cellulolytic activity and certain roughage processing procedures on vitamin requirements and electrolyte balance.

Physical processing, including chopping, grinding and the production of briquettes, pellets and cubes influences the potential value of roughage by accelerating the rate of passage of the residue through the GI tract of grazing ruminants and equids. In contrast to the effects in rabbits grinding and pelleting therefore increase the voluntary consumption of roughage in these animals, but with a decrease in digestibility of the fibre component by up to 9-15%. The reason for this reduction in grazing ruminants is understood, and an hypothesis is adduced for analogous effects in some non-ruminant herbivores. One experiment with horses showed that 14% more wafered hay and 21% more pelleted hay could be consumed. Overall this led to a 10-15% greater daily intake of digestible energy for roughage diets.

Optimal concentrations of dietary fibre (ADF) appear to be 27% and 10% of the diet in rabbits for maximal ME intake and maximum gain respectively. The comparable figure for the horse lies between 15 and 20% ADF. The overall digestive capacity for fibre in horses lies between cattle, sheep and goats on the one hand and rabbits on the other.

The treatment of poor quality roughage with sodium hydroxide has failed to reveal any major problems concerning electrolyte balance. Approximately 8-10% of sodium hydroxide treated straw is optimal in the diet of rabbits and 4.6 g of sodium per kg diet (independent of pH) leads to maximum performance. Where large amounts of sodium hydroxide-treated straw are used any supplementary salt provided should be in the form of the potassium salt.

The treatment of straw with ammonia has the advantage of the presence of no fixed base, undesirable excess chemical evaporates when the product is exposed to air and the additional nitrogen in diets containing less than 7% crude protein may stimulate fibre digestion.

Where straw replaces good quality roughage its obvious nutrient deficiencies should be made good. Straw is deficient particularly in protein, calcium and phosphorus, and, especially following alkali treatment, most vitamins are absent, more particularly β-carotene. α-tocopherol and riboflavin.

16.16. REFERENCES

Ahlswede, L., 1977. Untersuchungen uber Pferdealleinfutter in Form von Briketts. Deutsche Tierartzliche Wochenschrift 84: 132-135.

Alexander, F., 1952. Some functions of the large intestine of the horse. Quart. J. Exp. Physiol. 37: 205-214.

Argenzio, R.A., Lowe, J.E., Pickard, D.W. and Stevens, C.E., 1974. Digesta passage and water exchange in the equine large intestine. Amer. J. Physiol. 226: 1035.

Arman, P. and Field, C.R., 1973. Digestion in the hippopotamus. E. Afr. Wildl. J. 11: 9-17.

Arman, P. and Hopcraft, D., 1975. Nutritional studies on East African herbivores. I. Digestibilities of dry matter, crude fibre and crude protein in antelope, cattle and sheep. Br. J. Nutr. 33: 255-264.

Bauchop, T., 1977. Foregut fermentation. In: R.T.J. Clarke and T. Bauchop (eds.): Microbial ecology of the gut. Chapter 5: 223-250. Academic Press, London.

Bayley, H.S., 1978. Comparative physiology of the hind gut and its nutritional significance. J. Anim. Sci. 46: 1800-1802.

Blas, J.C.O. de, Merino, Y., Fraga, M.J. and Galvez, J.F., 1979. A note on the use of sodium hydroxide treated straw pellets in diets for growing rabbits. Anim. Prod. 29: 427-430.

Boisen, S., 1982. Mineralstofomsætning og syre-base balance hos voksende kaniner fodret med natriumhydroxid-behandlet halm. Beretning 522. Statens Husdyrbrugsforsøg, Copenhagen.

Butler, D. and Hintz, H.F., 1971. Effect of feeding frequency on digestibilty by ponies. Anim. Sci. Mimeo., Cornell Univ.

Carregal, R.D., 1977. Crude fibre in rations for growing rabbits. Cientifica 5: 336-339.

Chandrasena, L.G., Emmanuel, B., Hamar, D.W. and Howard, B.R., 1979. A comparative study of ketone body metabolism between the camel (*Camelius dromedarius*) and the sheep (*Ovis aries*). Comp. Biochem. Physiol. B. 64: 109-112.

Cheeke, P.R. and Patton, N.M., 1980. Alfalfa (lucerne) utilization by rabbits. Commercial Rabbit. 8: 8-9.

Chivers, D.J. and Hladik, C.M., 1980. Morphology of the gastrointestinal tract in primates. Comparisons with other mammals in relation to diet. J. Morph. 166: 337-386.

Clemens, E.T. and Stevens, C.E., 1980. A comparison of gastrointestinal transit time in ten species of mammal. J. agric. Sci. (Camb.) 94: 735-737.

Engelhardt, W. Von, Wolter, S., Lawrenz, H. and Hamsley, J.A., 1978. Production of methane in two non ruminant herbivores. Comp. Biochem. Physiol. A 60: 309-311.

Fioramonti, J., Beuno, L. and Candau, M., 1979. Gastro-intestinal motility as affected by dietary cell wall constituents (carbohydrates) in the rabbit. Ann. Zootech. 28: 127.

Fonnesbeck, P.V., Lydman, R.K., Noot, G.W.V. and Symons, L.D., 1967. Digestibility of the proximate nutrients of forage by horses. J. Anim. Sci. 26: 1039-1045.

Franck, Y. and Coulmin, J.P., 1979. Use of ground straw as a source of crude fibre in fattening rabbit feeding, a comparison of two crude fibre levels. Ann. Zootech. 28: 131.

Frape, D.L. and Chamberlain, A.G., 1981. The determination of the nutrient requirements of wild herbivores in captivity. In: Symposium on Advances in the Veterinary Care of Zoo Animals. The Zoological Society of London, Regents Park, 17-18 June 1981.

528

Frape, D.L., Tuck, M.G., Sutcliffe, N.H. and Jones, D.B., 1982. The use of inert markers in the measurement of the digestibility of cubed concentrates and of hay given in several proportions to the pony, horse and white rhinoceros (*Diceros simus*). Comp. Biochem. Physiol. 72A: 77-83.

Giesecke, D. and Van Gylswyck, H.O., 1975. A study of feeding types and certain rumen functions in six species of South African wild ruminants. J. agric. Sci. (Camb.) 85: 75-83.

Glinsky, M.J., Smith, R.M., Spires, H.R. and Davis, C.L., 1976. Measurement of volatile fatty acid production rates in the cecum of the pony. J. Anim. Sci. 42: 1465-1470.

Godbee, R.G. and Slade, L.M., 1979. Range blocks with urea for brood mares increase the nutritional value of pasture feeding. Feedstuffs 51: 34-35.

Godbee, R.G., Slade, L.M. and Lawrence, L.M., 1979. Use of protein blocks containing urea for minimally managed brood mares. J. Anim. Sci. 48: 459-463.

Gouet, P.L. and Fonty, G., 1979. Changes in the digestive microflora of holoxenic rabbits from birth to adulthood. Ann. Biol. anim. Bioch. Biophys. 19: 553-566.

Haenlein, G.F.W., 1969. Nutritive value of a pelleted horse ration. Feedstuffs 41 No. 26: 19-20.

Haenlein, C.F.W., Holdren, R.D. and Yoon, Y.M., 1966a. Comparative responses of horses and sheep to different physical forms of alfalfa hay. J. Anim. Sci. 25: 740.

Haenlein, G.F.W., Smith, R.C. and Yoon, Y.M., 1966b. Determination of the fecal excretion rate of horses with chromic oxide. J. Anim. Sci. 25: 1091.

Hintz, H.F., 1969. Review article: Comparison of digestion coefficients obtained with cattle, sheep, rabbits and horses. Veterinarian 6: 45-51.

Hintz, H.F., Angenzio, R.A. and Schryver, H.F., 1971. Digestion coefficients, blood glucose levels and molar percentage of volatile acids in intestinal fluid of ponies fed varying forage-grain ratios. J. Anim. Sci. 33: 992.

Hintz, H.F. and Loy, R.G., 1966. The effects of pelleting on the nutritive value of horse rations. J. Anim. Sci. 25: 1059.

Hintz, H.F. and Schryver, H.F., 1978. Digestive physiology of the horse. J. Equine Med. Surg. 2: 147-150.

Hintz, J.F., Schryver, H.F. and Stevens, C.E., 1978. Digestion and absorption in the hindgut of nonruminant herbivores. J. Anim. Sci. 46: 1803-1807.

Hintz, H.F., Sedgewick, C.J. and Schryver, H.F., 1976. Inter. Zoo. Yearbook 16: 54-57.

Hofmann, R.R., 1982. Die funktionelle morphologie des Wiederkäuer-Magens: Schleimhaut u. Versorgungsbahnen 65 Abbildungen. Ferdinand Enke Verlag, Stuttgart.

Hofmann, R.R. and Stewart, D.R.M., 1972. Mammalia (Paris) 36: 226-240.

Hoover, W.H. and Clarke, S.D., 1972. Fiber digestion in the beaver. J. Nutr. 102: 9.

Hoover, W.H. and Heitmann, R.N., 1972. Effects of dietary fiber levels on weight gain, cecal volume, and volatile fatty acid production in rabbits. J. Nutr. 102: 375.

Hoover, W.H. and Heitmann, R.N., 1975. Cecal nitrogen metabolism and amino acid absorption in the rabbit. J. Nutr. 105: 245 and 252.

Hoppe, P.P., 1977. Comparisons of voluntary food and water consumption and digestion in Kirk's dik-dik and suni. E. Afr. Wildl. J. 15: 41-48.

Hoppe, P.P. and Gwynne, M.D., 1978. Food retention time in the digestive tract of the suni antelope (*Nesotragus moschatus*). Saugetierkund. Mitt. 3: 236-237.

Jacquot, R., Le Bars, H. and Simonnet, H., 1958. Nutrition animale. Vol. 1 Donnes générales sur la Nutrition et l'alimentation - Bailliere éd. Paris.

Jarvis, C., 1976. The evolutionary strategy of the Equidae and the origins of rumen and caecal digestion. Evolution 30: 757-774.

Jensen, N.E., 1977. Kaninforsøgsstationen 1977. Afkomsprøver; Fodringsforsøg. Beretning No. 473, Statens Husdyrbrugsforsøg, Copenhagen, 32 pp.

Jensen, N.E. and Tuxen, J., 1978. Kaninforsøgsstationen 1978: Afkomsprøver; Fodringsforsøg, Staldforhold. Beretning No. 484, Statens Husdyrbrugsforsøg, Copenhagen. 35 pp.

Johnson, J.L. and MacBee, R.H., 1967. The porcupine cecal fermentation. J. Nutr. 91: 540.

Joyce, J.P., Rattray, P.V. and Parker, J., 1971. The utilization of pasture and barley by rabbits. 1. Feed intakes and live-weight gains. New Zealand J. Agric. Res. 14: 173-179.

Joyce, J.P. and Newth, R.P., 1967. Proc. of the New Zealand Soc. Anim. Prod. 27: 166-180.

Kay, R.N.B., Engelhardt, W.V. and White, R.G., 1980. The digestive physiology of wild ruminants. In: Y. Ruckebusch and P. Thivend (eds.): Digestive physiology and metabolism in ruminants. Proc. 5th Internat. Symp. Ruminant Physiol., Clermont-Ferrand 3-7 Sept. 1979. MTP Press Ltd. pp 743-761.

Kay, R.N.B. and Staines, B.W., 1981. The nutrition of the Red Deer (*Cervus elaphus*). Commonwealth Bureau of Nutrition. Nutr. Abstr. Rev. B, 51 No 9: 601-622.

Kern, D.L., Slyter, L.L., Weaver, J.M., Leffel, E.C. and Samuelson, G., 1973. Pony cecum vs. steer rumen: the effect of oats and hay on the microbial ecosystem. J. Anim. Sci. 37: 463-469.

Kholmirzaev, D., 1980. Fattening of horses in the foothills of Uzbekistan Konevodstvoi Konnyi Sport No 1: 7 (Russian).

Kleiber, M., 1961. The Fire of Life. John Wiley & Sons, New York and London.

Koller, B.L., Hintz, H.F., Robertson, J.B. and Van Soest, P.J., 1977. Comparative cell wall and dry matter digestion in the cecum of the pony and rumen of the cow. Proc. Fifth Equine Nutr. Symp., St. Louis (Abstr.).

Koller, B.L., Hintz, H.F., Robertson, J.B. and Van Soest, P.J., 1978. Comparative cell wall and dry matter digestion in the cecum of the pony and the rumen of the cow using in vitro and nylon bag techniques. J. Anim. Sci. 47: 209-215.

Kritchevsky, D., 1978. Fiber, lipids and atherosclerosis. Am. J. Clinical Nutr. 31: 565-574.

Kritchevsky, D., Tepper, S.A., Williams, D.E. and Storey, J.A., 1977. Experimental atherosclerosis in rabbits fed cholesterol-free diets. VII. Interaction of animal or vegetable protein with fiber. Atherosclerosis 26: 397-403.

Lebas, F., 1974. AEC. Information Review. Other Animals 120. Nitrogenous nutrition of the rabbit, 19 pp.

Lebas, F. and Colin, M., 1973. Effect de l'addition d'urée a un régime pauvre en proteines chez le lapin en croissance. Ann. Zootech. 22: 111-113.

Lebas, F., Colin, M., Mercier, P. and Tremolieres, E., 1979. Use of straw treated with sodium hydroxide in rabbit feeding. Ann. Zootech. 28: 132.

Lebas, F. and Laplace, J.P., 1977a. Le transit digestif chez le lapin. VI - Influence de la granulation des aliments. Ann. Zootech. 26: 83-91.
Lebas, F. and Laplace, J.P., 1977b. Le transit digestif chez le lapin. VIII - Influence de la source de cellulose. Ann. Zootech. 26: 575-584.
Lindeman, M.A., Brigstocke, T.D.A. and Wilson, P.N., 1982. A note on the response of growing rabbits to varying levels of sodium hydroxide treated straw. Anim. Prod. 34: 107-110.
Luick, J.R., 1976/1977. Acceptability of atypical feeds by penfed reindeer. Technical Progress Report, Univ. of Alaska, RLO-2229-T3: 98-122.
McBee, R.H. and West, G.C., 1969. Caecal fermentation in the willow ptarmigan. Condor 71: 54.
McMurray, C.H., Blanchflower, W.J. and Rice, D.A., 1980. The effect of pretreatment on the stability of alpha-tocopherol in moist barley. Proc. Nutr. Soc. 39: 61A.
MacLean, K.S., Joyce, J.P. and Rattray, P.V., 1970. Proc. of the New Zealand Soc. Anim. Prod. 27: 32-40.
Meyer, H., 1980. Neuere Erkenntnisse zur Dickdarmverdauung des Pferdes. Tierernähr. 8: 123-150.
Meyer, H., Ahlswede, L. and Reinhardt, H.J., 1975. Untersuchungen über Fressdauer, Kaufrequenz und Futterzerkleinerung beim Pferd. Deutsche Tierarztlich Wochenschrift 82: 54-58.
Meyer, H., Schmidt, M. and Güldenhaupt, V., 1981. Untersuchungen über Mischfutter für Pferde. Deutsche Tierartzlich Wochenschrift 88: 2-5.
Miller, E.R. and King., C.H., 1981. Studies on the digestive physiology of wild ruminants. Digestibility and rate of passge of digesta in oryx and sheep. In: Symposium on Advances in the veterinary care of zoo animals, The Zoological Society of London, Regents Park, 17-18 June, 1981.
Mundt, H.C., 1978. Untersuchungen über die Verdaulichkeit von aufgeschlossenem Stroh beim Pferd. Publ. Thesis Tierartzliche Hochschule Hannover. 109 pp.
Muuss, H., Meyer, H. and Shcmidt, M., 1982. Beiträge zur Verdauungsphysiologie des Pferdes. In: H. Meyer (ed.): Fortschritte in der Tierphysiologie und Tierernährung. Paul Parey, Hamburg and Berlin. 13-23.
Noot, G.W.V. and Gilbreath, E.B., 1970. Comparative digestibility of components of forages by geldings and steers. J. Anim. Sci. 31: 351-355.
Noot, G.W.V. and Trout, J.R., 1971. Prediction of digestible components of forages by equines. J. Anim. Sci. 33: 38-41.
Olsson, N. and Ruudvere, A., 1955. Nutrition of the horse. Nutr. Abstr. Rev. 25: 1.
Parker, D.S., 1976. The measurement of production rates of volatile fatty acids in the caecum of the conscious rabbit. Br. J. Nutr. 36: 61.
Pickard, D.W. and Stevens, C.E., 1972. Digesta flow through the rabbit large intestine. Amer. J. Physiol. 222: 1161.
Piekarz, R., 1963. Effect of coprophagy on transit time through the digestive tract in domestic rabbits. Acta Physiol. Polon. 14: 359.
Prins, R.A. and Geolen, M.J.H., 1971. Rumen characteristics of red deer, fallow deer and roe deer. J. Wildl. Manage. 35: 673-680.
Ralston, S.L. and Baile, C.A., 1982. Gastrointestinal stimuli in the control of feed intake in ponies. J. Anim. Sci. 55: 243-253.

Reid, J.T. and Tyrrell, H.F., 1964. Effect of level of intake on energetic efficiency of animals. Proc. Cornell Nutr. Conf. p. 25.

Rougeot, J., Colin, M. and Thebault, R.G., 1980. Definition de conditions experimentales pour l'etude des besoins nutritionnels du lapin angora: nature de la litiere et presentation du lest alimentaire. Ann. Zootech. 29: 1-11.

Sanchez-Hermosillo, M. and Kay, R.N.B., 1979. Retention time and digestibility of milled hay in sheep and red deer (*Cervus elaphus*). Proc. Nutr. Soc. 38: 123A.

Schaeffer, A.L., Young, B.A. and Chimwano, A.M., 1978. Ration digestion and retention times of digesta in domestic cattle (*Bos taurus*), American bison (*Bison bison*) and Tibetan Yak (*Bos granniens*). Can. J. Zool. 56: 2355-2358.

Schurg, W.A., Pulse, R.E., Holtan, D.W. and Oldfield, J.E., 1978. Use of various quantities and forms of ryegrass straw in horse diets. J. Anim. Sci. 47: 1287-1291.

Slade, L.M. and Hintz, H.F., 1969. Comparison of digestion in horses, ponies, rabbits and guinea pigs. J. Anim. Sci. 28: 842-843.

Slade, L.M. and Robinson, D.W., 1970. Nitrogen metabolism in rabbits and guinea pigs. J. Anim. Sci. 30: 1044.

Slagsvold, P., Hintz, H.F. and Schryver, H.F., 1979. Digestibility by ponies of oat straw treated with anhydrous ammonia. Anim. Prod. 28: 347-352.

Spreadbury, D. and Davidson, J., 1978. A study of the need for fibre by the growing New Zealand White rabbit. J. Sci. Food Agric. 29: 640-648.

Susmel, P. and Lanari, D., 1977. Variations in volatile fatty acid levels in the cecum of the rabbit. Riv. Zootec. Vet. 3: 282288. (CA. 1978 A49448g).

Swick, R.A., Cheeke, P.R. and Patton, N.M., 1978. Evaluation of dried rabbit manure as a feed for rabbits. Can. J. Anim. Sci. 58: 753-757.

Templeton, G.S. and Kellogg, C.E., 1950. USDA Farmers Bull. No 1730.

Thompson, K.N. and Baker, J.P., 1981. Digestion of hay and grain fed in varying ratios to mature horses. J. Anim. Sci. 53, Suppl. 1. Abstr. p. 267.

Thurston, J.P., Noirot-Timothée, C. and Arman, P., 1968. Fermentative digestion in the stomach of *Hippopotamus amphibius (Artiodactyla: Suiformes*) and associated ciliate protozoa. Nature 218: 882-883.

Tisserand, J.L., Pecchio, M.O. and Rollin, G., 1980. L'activite cellulolytique dans le gros intestin du poney. Reproduction Nutrition Developpement 20: 1685-1689.

Udén, P., 1978. Comparative studies on rate of passage, particle size and rate of digestion in ruminants, equines, rabbits and man. PhD Thesis, Cornell University, Ithaca, NY, USA.

Udén, P. and Van Soest, P.J., 1982. Comparative digestion of Timothy (*Phleum pratense*) fibre by ruminants, equines and rabbits. Br. J. Nutr. 47: 267.

Ullrey, D., 1980. Establishing the nutrient requirements of exotic animals. In: E.R. Maschgan , M.E. Allen and L.E. Fisher (eds.): Proceedings of the First Annual Dr. Scholl Nutrition Conference. A conference on the nutrition of captive wild animals. Lincoln Park Zoological Gardens, Chigaco.

Ullrey, D.E., Robinson, P.T. and Whetter, P.A., 1979. Comparative digestibility studies with zoo herbivores. Proc. Amer. Assoc. of Zoo. Veterinarians, Denver, CO Fall 1979.

Van Niekerk, H.P. and Couvaras, S., 1979. Die effek van bytosa-dabehandelde ruvoere by pferde. 1. Behandelde lusernhooi as bestanddeel van n volledige rantsoen by vullens. J. South African Vet. Assn. 50: 59-60.

Van Soest, P.J., 1963. Use of detergents in the analysis of feeds. J. Ass. Off. Agr. Chem. 46: 825-831.

Van Soest, P.J. and Marcus, W.C., 1964. A method for the determination of cell wall constituents of forages using detergents and the relationship between this fraction and voluntary intake and digestibility. J. Dairy Sci. 47: 704.

Willard, J.G., Willard, J.C., Wolfram, S.A. and Baker, J.P., 1977. Effect of diet on cecal pH and feeding behaviour of horses. J. Anim. Sci. 45: 87-93.

Wolter, R., Durix, A. and Letourneau, J.C., 1974. Influence du mode de presentation du fourrage sur la vitesse du transit digestif chez le poney. Ann. Zootechn. 23: 293-300.

Wolter, R., Durix, A. and Letourneau, J.C., 1975. Influence du mode de presentation du fourrage sur la digestibilité chez le poney. Ann. Zootech. 24: 237-242.

Wolter, R., Durix, A., Letourneau, J.C. and Carcelen, M., 1979. Evaluation chez le poney de la digestibilité du mais-fourrage deshydrate, des pulpes seches de bettirave de la luzerne des-hydratee du son de ble, de la paille de ble et des pulpes de raisins. Ann. Zootech. 28: 93-100.

Wolter, R. and Velandia, J., 1970. Digestion of forages in the donkey. Rec. Med. Vet. 146: 141-152.

Yagil, R. and Etzion, Z., 1980. Milk yield of camels (*Camelus dromedarius*) in drought areas. Comp. Biochem. Physiol. A.67; 207-209.

Chapter 17

LABORATORY METHODS FOR EVALUATING THE NUTRITIVE VALUE OF UNTREATED AND TREATED FIBROUS BY-PRODUCTS

by

Peter Udén

Department of Animal Husbandry
Swedish University of Agricultural Sciences
S-750 07 Uppsala 7
Sweden

17.1. INTRODUCTION

As early as the 18th century attempts were made to obtain crude estimates of how feeds promote growth or milk production (Fonnesbeck, 1976). These trials were limited to a few feeds that could be fed singly. Real progress in this field required knowledge of the various nutrients necessary for body function. Such progress was made during the 19th and 20th centuries in the disciplines of chemistry, physics and medicine. An understanding was gained of the separate roles of energy, protein, mineral elements and vitamins in the body. Laboratory analysis for the estimation of nitrogen, fibre and fat (later called the proximate analysis system) were developed during the 19th century and also some digestion trials were performed. Wolff (see Van Soest, 1981) took into account only N and fibre when calculating "hay equivalents" in 1856, whereas Henry (1898) and Morrison who in 1915 became coauthor to Henry in his book "Feeds and Feeding" developed the TDN (total digestible nutrient) system. The TDN-values were calculated as the sum of the digestible amounts of fat, protein, crude fibre and nitrogen free extract in 100 lbs of feed. Fat was also multiplied by a factor of 2.25 due to its high energy value.

The amount of fibre in the feed has perhaps been the single most important determinant of nutritive value over the centuries. In ruminant research, fibre was considered practically indigestible until 1854 when Haubner (See Van Soest, 1981) discovered otherwise. It is interesting to notice that this idea of fibre being a non-nutritive entity persisted into the 1970's in human research. The idea of Wolff, Henry and others, that digestibility was the most important component of the nutritive value concept has persisted in the 20th century. Protein, minerals, vitamins, etc, have been regarded as important, but separate entities. Two other components have been included in the nutritive value concept. These are the utilization of digestible energy and voluntary intake. Of these two components, the latter has been shown to have the largest individual variation.

This chapter will for the most part treat laboratory estimations of digestibility in general with most of the references coming from forages. Lack of information on the chemical composition of fibrous by-products is evident and few analyses have been specifically developed for by-product analysis. The wide variety of plant materials and need for more in-depth knowledge of factors controlling digestibility make research in this field very challenging. Some attention will be given to the evaluation of alkali-treated straws in terms of their digestibility and "protein value" in case of ammoniated straws.

17.2. RUMINANT DIGESTION

Digestion in the ruminant forestomach is characterized by an extensive anaerobic fermentation. Digestion and absorption of fat, escaped protein and microorganisms takes place in the acid stomach and small intestines. Some further fermentation occurs in the lower tract of escaped starch, fibre, undigested microbial cell walls, mucus from intestinal secretion etc. In the faeces all indigestible residues are found together with endogenous material of which the vast majority is of microbial origin (Mason, 1979). Faecal endogenous matter is subtracted in the calculations of true digestibility whereas it is included in apparent digestibility.

Ruminants are physiologically adapted to handle large quantities of fibrous material and extract energy from carbohydrates available through fermentation. The processes occurring in the forestomachs are dynamic and consist of hydration, fermentation, particle size reduction, passage of digesta and absorption of volatile fatty acids (VFA). Buffering salts are provided by the saliva and removal of end-products (microorganism and VFA) occurs either through the rumen wall (VFA) or by passage down the tract. All these processes can be affected by a variety of factors. Time available for fermentation in the rumen may decrease due to increased intake or altered physical characteristics of the feed (Blaxter et al., 1956; Colucci, 1979). Too little nitrogen and too much starch will also decrease fibre fermentation.

Whole plants usually become less digestible with increasing maturity unless accumulation of storage carbohydrates (starch, fructans, sucrose) offsets this trend such as in maize. The relative amount of fibre generally increases and its digestibility decreases with age. Plant species differ in rate at which this process proceeds depending on environment and the varied need for aerial support. Carbohydrates, protein and lignin are intimately associated in the cell wall and provide a more or less rigid physical structure depending on their relative amounts. The nature of all fibre fractions vary across plant species. Pure cellulose may be rapidly fermented, as in lettuce or very slowly, as in cotton.

Accurate predictions of digestibility under laboratory conditions are not an easy task. Research on prediction of digestibility has followed two general paths. Microdigestion techniques have been developed to simulate in vivo conditions or chemical analyses have been utilized to investigate relationships between certain plant fractions and the digestibility of the whole plant.

17.3. CHEMICAL FRACTIONATION

A logical system for the chemical fractionation of plant material from a nutritional standpoint should at least differentiate between a completely available and a less available fraction. In the proximate analysis, much hemicellulose and lignin is excluded from crude fibre, resulting in an often lower nutritive value of the nitrogen-free extract compared with the crude fibre (Paloheimo, 1953). The use of neutral detergents in fibre analysis (Van Soest, 1963) meant a break-through in this respect. Neutral detergents dissolve a fraction containing soluble carbohydrates, proteins, organic acids, starch, pectin etc. with a digestibility

approaching 100% and leave behind cellulose, hemicellulose, lignin, cutin and some relatively indigestible protein (Hogan and
Lindsay, 1979). Neutral detergent fibre (NDF) can be fractionated
into acid detergent fibre (ADF) containing mainly cellulose and
lignin but also cutin and some acid-resistant hemicellulose
(Goering and Van Soest, 1970). Lignin in the ADF may either be
oxidized directly with permanganate or isolated with the use of
72% H_2SO_4.

Lucas (1964) developed a regression technique to test for nutritional uniformity of chemical fractions. The digested amount of
a fraction per unit of feed is regressed upon its concentration in
the feed. This technique has shown that crude protein and neutral
detergent solubles behave uniformly across feeds with a true
digestibility of ca 95%. Lignin is also a uniform entity nutritionally, but has zero digestibility. Nitrogen-free extract, crude
fibre, NDF and ADF do not behave uniformly and their true digestibilities vary with plant source. Isolation of NDF in the faeces
has also become a means to determine true dry matter digestibility
(DMD) and to separate effects of endogenous excretion and fibre
fermentation. Unfortunately this possibility has not become a routine procedure in in vivo and in vitro digestion trials, which has
made interpretation of results and comparisons and evaluations of
methods difficult.

17.4. PLANT COMPOSITION AND DIGESTIBILITY

17.4.1. <u>Polysaccharides</u>

A division of the plant cell wall into cellulose and hemicellulose (Richards and Reid, 1953; Gaillard, 1962; Beveridge and Richards, 1975), branched and linear hemicellulose (Beveridge and
Richards, 1973) or into various polysaccharide sugar constituents
(Gaillard, 1962; Dekker et al., 1972; Beveridge and Richards,
1975) has not explained the sources of variation in fibre digestibility. A study of cellulose crystallinity gave a very high negative correlation with digestibility (Baker et al., 1959) in cotton
linters and wood celluloses. In a later study, Beveridge and Richards (1975) could find no difference, however, in the digestion
rate of amorphous and crystalline regions of cellulose. Molecular

weight of different hemicelluloses could not account for differences in their digestibility (Beveridge and Richards, 1975).

17.4.2. Lignin

According to paper chemists, wood lignin is a plastic-like, high molecular phenyl propane product (Adler, 1968), which penetrates the fibres to increase their rigidity. The chemical composition of lignin varies with plant species (Van Soest, 1981) and probably also with maturity and environment. So far the only large-scale efforts to characterize lignin have been done on wood. Grass and wood lignins may have similarities but very little is known about legume lignins. Digestibility of legumes is much less improved with alkali treatment (Hartley and Jones, 1977) possibly as a result of fewer ester linkages (internally and to carbohydrates). It seems as if the saponification of ester linkages by alkali treatment is one of the major factors improving digestibility and actual removal of lignin does not seem necessary. Crude lignin isolated from forages or faeces is often contaminated with cutin, tannin-protein complexes, Maillard-products (protein-carbohydrate complexes) etc. The amount of non-artifact lignin seems to limit the amount of fibre which can be fermented anaerobically, rather than determining the rate of utilization (Van Soest, 1964; Mertens, 1973) and removal of lignin makes the fibres totally available. The weight of lignin is probably not the ideal parameter for predicting its affect on digestibility as association and attachment with carbohydrates and differences in composition are not taken into account.

Lignin is truly indigestible even though quantitative recovery in the faeces can be difficult. Soluble lignin - carbohydrate complexes ($M_W > 10,000$) are also found in the rumen (Gaillard and Richards, 1975) and can be precipitated at pH 3 and will not be recovered in NDF.

17.4.3. Silica

Silica occurs in most plants and can be both a structural part of the plant or the result of soil contamination (Van Soest, 1981). It is found in abundance in rice hulls, rice straw and many tropical grasses (Van Soest, 1970) and has been shown to depress dry matter digestibility 1-3 times its own weight (Van Soest and Jones, 1968; Smith et al., 1971). In some grasses the sum of lignin and silica was better related to digestibility than lignin alone (Van Soest, 1970). Removal of silica with alkali will increase digestibility (Van Soest, 1970). Total silica may be estimated as the ADF ash residue which is insoluble in HBr. Soil silica can be determined if this procedure is preceded by a neutral detergent extraction (Van Soest, 1981).

17.4.4. Cutin

Cutin is often found in the cuticle of the plants. In hulls from rice, cotton seeds and peanuts the content may be 10-20% (Van Soest, 1969). As cutin is indigestible it lowers cell wall digestibility, but if it has any affect on digestion of cell wall carbohydrates in fibrous feeds has received little attention. The hydrophobic nature of cutin will probably delay the onset cell wall fermentation. Cutin may be estimated, using the detergent system of analysis, as the organic ADF residue resistant to 72% H_2SO_4 and permanganate oxidation (Goering and Van Soest, 1970)

17.4.5. Phenolic acids and acetyl groups

Phenolic acids (ferulic, diferulic and p-coumaric) constitute 0.52-5% of the cell walls in grasses (Waite and Gorrod, 1959; and Theander et al., 1981), but can be removed with alkali under relatively mild conditions. Bacon and Gordon (1980) showed that successive removal of acetyl groups with alkali improved digestibility but whether this was due to the action of the bases on the carbohydrates or the actual removal of acetyl groups was uncertain. Both phenolic acids and acetyl groups are assumed to be linked to the hemicellulose. The phenolic acids may also be the building-blocks for lignin (Hartley and Jones, 1977), but the question is far from resolved. Acetyl groups and phenolic acids can be determined quantitatively by gas-liquid chromatography (Bethge and Lindstrom, 1973; Salomonsson et al., 1978).

17.4.6. Tannins

Tannins occur in many plant parts such as seed coats, browses, bark etc. Some tannins have the ability to irreversibly bind protein and to inhibit fibre fermentation. They also interfere in fibre analysis (Van Soest, 1981). Solvent extraction of tannin has been shown to increase fibre digestibility (Robbins, 1973). No quantitative analysis has yet been developed to account for tannins specifically interfering with fibre and/or protein digestion. Ruminant microorganisms adapted to high tannin levels seem to digest fibre better than unadapted (Sidahmed et al., 1981).

17.4.7. Maillard products

Maillard products are the result of condensations between carbohydrates and amino acids or protein. The reactions are favoured by high temperature and require water (Van Soest, 1965). Free amino groups, such as in lysine, react most readily. The products are recovered in the lignin residue and will interfere with lignin estimations. A correction of the lignin value is outlined by Goering and Van Soest (1970). Unavailable nitrogen can be easily determined as the nitrogen insoluble in acid detergents (Goering and Van Soest, 1970).

17.5. PREDICTION OF DIGESTIBILITY

17.5.1. Fibre fractions

Neutral detergent fibre has been proposed by Rohweder et al. (1978) to be used for grading hay even though amounts of cellulose, hemicellulose, ADF, NDF, lignin or any of the other inhibitory components mentioned have been poor predictors of digestibility across plant species grown at different locations or at different times of the year (Joshi, 1972; Van Soest et al., 1978; Lindgren, 1979). Vestergaard Thomsen and Henriksen (1976) also concluded that fibre or cell wall components could not adequately predict the energetic value of feeds containing varying proportions of NaOH-treated or untreated straw. McLeod and Minson (1976) working with five legume species found a better relationship between ADF and in vivo DMD than between in vitro and in vivo. Such results are sometimes found when samples are collected within one plant family from a specific climatic region.

Quantifying fibre quality has so far been the only method achieving some success. The lignification of the cell wall and content of silica has been proposed by Goering and Van Soest (1970) and included in a summative equation. The lignin/ADF was curvelinearly related to DMD so a logarithmic relationship was used for purely empirical reasons. ADF was used instead of NDF as it gave a lower variance. (Lignin and ADF are analysed in sequence on the same sample.) Silica was added to the equation as it was found to improve the prediction. The equation correlates reasonably well with DMD (see Table 17.1) when applied to a wide variety of plant species.

Table 17.1. Predictive errors associated with systems to estimate dry matter digestibility (DMD) and total digestible nutrients (TDN) from composition (Van Soest and Robertson, 1980)

Method of estimation	Value predicted	Bias[a] units	S.D. from regression
Crude fibre	DMD	–	11.0
Acid-detergent fibre	DMD	–	9.0
Equations based on crude fibre and protein			
Legume and grass	TDN	+4.0	7.7
Mixed	TDN	+3.3	8.0
Summative equation			
Unmodified	DMD	–1.0	6.1
With silica correction	DMD	+4.5	3.8
In vitro rumen digestibility	DMD	+2.5	3.7
True digestibility	DMD	+0.7	2.8

[a] Mean difference between predicted and observed values

Less satisfactory results may be obtained when the samples are collected from a narrow botanical, geographical or climatic range. Lindgren (pers. comm.) found higher residual standard deviation (RSD) for the summative equation than for multiple regressions which included hemicellulose, cellulose and lignin. The multiple regressions gave RSD's (as % of mean) of 4.1 for grasses and 6.1

for legumes (Lindgren, 1979) when predicting metabolizable energy
contents of 58 grasses and 28 legumes in Sweden. No other fibre
components have, as yet, been included in the summative equation,
partly due to lack of analytical data but also due to a lack of
understanding of the factors limiting digestion.

17.5.2. Physical methods

Near infrared reflectance (NIR) spectrophotometry (1400-2400 A)
has recently been developed to handle large-scale anaysis (Norris
et al., 1976). The technique applied to measure feed components or
potential digestibility is purely empirical and requires similar
samples which have been analysed by separate methods. From the reflectance of wavelengths the computer picks out those wavelenghts
that correlate best with the parameter of interest, e.g. crude
protein. The advantages of the method are that it is non-destruc-
tive, sample preparation is simple and many samples can be analysed per day. The drawbacks are that it requires similar samples of
known composition (or digestibility), equipment is expensive and
the botanical composition of the sample should be known. The technique is also unsuitable for basic research on plant composition
or digestibility.

17.5.3. Biological methods

Three general types of microdigestion techniques exist for the
estimation of digestibility: Digestion with cell-free fungal cellulase, digestion with rumen microorganisms in vitro, or in situ
incubations of samples in synthetic bags in the rumen.

17.5.3.1. The cellulase method utilizes the enzymes from wood
fungi (a mixture or proteases, pectinases, hemicellulases and cellulases). Incubations are made in a buffer usually with pH of 4.6-
4.8 at a temperature of 40-50°C. The residue after digestion
contains no endogenous debris and should therefore be equal to the
true indigestible residue and be less than found in vivo or in
vitro when using rumen microorganisms. Jones (1976) has reviewed
the characteristics of fungal cellulases and their action and
modes of preparation. He points out the unfortunate use of carbox-
ymethylcellulose as an assay for cellulase activity and the wide

variety of conditions used. Donefer et al. (1963) and Jarrige et al. (1970) were among the first to use this method for predicting forage digestibility. The advantages of the cellulase methods, compared to the in vitro method, have been the simple methodology, ease of automation and possibilities for highly standardized enzyme preparations. In order to obtain similar results from different laboratories the experimental conditions have to be standardized - enzyme preparation, pH, type of buffer, incubation time, temperature and ratio of substrate to buffer volume (McQueen and Van Soest, 1975) and pre and post-treatments of the samples. Type of buffer, temperature and pH have been fairly easy to standardize for optimum digestion. Incubation times of 24-72 hours have been most commonly used. Enzymes of Aspergillus and Trichoderma have been most commonly used, but the greatest source of variation seems to be the enzyme preparation (Jones and Hayward, 1975; McQueen and Van Soest, 1975; Clark and Beard, 1977; Rexen, 1977a, Tinnimit and Thomas, 1976; Bartiaux-Thill et al., 1980). Degradation curves from cellulase incubations vary among laboratories and enzyme preparations. The main cell wall digestion occurred during the first 6 hours for Roughan and Holland (1977). They used a Trichoderma selection at 50°C on neutral detergent preparations. Maximum digestibility of the cell wall preparations was obtained if enzyme solutions were replaced after 5 hours and incubations continued for 16-18 hours. McQueen and Van Soest (1975) obtained maximum digestibility at 72 hours whereas Jones and Hayward (1975), Rexen (1977a) and Classon (1980) favoured 24-hour incubations. The accuracy of the method and the extent of digestion were improved by pre-treatment of the samples with pepsin-HCl (Jones and Hayward, 1975), Roughan and Holland (1980) obtained the lowest digestion residues using a pre-treatment with neutral detergents. A post-treatment with pepsin-HCl also lowered the residues somewhat (McQueen and Van Soest, 1975). The amount of fibre digested by the fungal cellulases is generally poor compared with in vivo or good in vitro techniques. Regressions against in vivo data normally include a positive constant of about 30 units of digestibility (see Kellner and Kirchgessner, 1977; Carlier et al., 1979; Lindgren et al., 1980). An exception is the study of Roughan and Holland (1980). Their regression contains a constant of -10.12 which is similar in size to the in vivo excretion of endogenous matter.

These results seem to have been due to a very careful enzyme preparaton.

One of the crucial questions concerning the cellulase method is whether the digestion of fibre is comparable with rumen fermentation. Digestion may be lower as mentioned above, but it should be comparable across plant species and digestibilities. No adaptation of the enzyme to the substrate is possible as in the rumen (McQueen and Van Soest, 1975). Roughan and Holland (1977) found no species interaction possibly due to too few samples (34). Guggolz et al. (1971) and McQueen and Van Soest (1975) found higher correlations with in vivo data after grouping samples according to species. McQueen and Van Soest (1975) also found a lower in vitro - cellulase correlation for cellulose digestion than for hemicellulose digestion.

Carlier et al. (1979) found that the cellulase method underestimated digestibility of less digestible plants and overestimated highly digestible ones. They also noted a much higher variation among laboratories when analysing low quality samples. Considerably higher values for legumes as compared with grasses were found by Terry et al. (1978), using the pepsin-cellulase technique. Jones and Hayward (1975) claimed no difference when using pepsin pretreatment and correlating it to in vitro digestibility. This seems to be mainly the result of an unusually high RSD for the in vitro technique.

17.5.3.2. <u>The in vitro technique.</u> The ideas of rumen fermentation were formulated by Zuntz and cellulose fermentation was demonstrated in vitro by his student von Tapperheimer (see Hungate, 1966). However, good anaerobic techniques were not developed until the 1940's thanks to the efforts of R.E. Hungate. During the 1960's several laboratories developed their own techniques for forage evaluation. The most commonly employed method today was described by Tilley and Terry (1963) and involves a 48-hour fermentation with rumen fluid followed by a 48-hour pepsin digestion. A 96-hour fermentation and elimination of the pepsin digestion was made by Den Braver and Eriksson (1967), whereas Van Soest et al. (1966) exchanged the pepsin digestion with a neutral detergent extaction.

Most rumen microorganisms require anaerobic conditions and have an optimal temperature requirement of 38-39°C. The end-products of fermentation are mainly CO_2, CH_4, VFA and microbial mass.

Build-up of gasses or VFA's can repress fermentation. Therefore, the sample size must be small (generally not more than 1 g per 100 ml), gas relief must be provided and the buffer must maintain a stable pH of around 6.8. In addition, temperature control and some agitation of the samples is required. Marten and Barnes (1980) reviewed sources of variation involved in the method including fermentation vessels, buffer and medium, inoculum (source, processing and amount), anaerobic conditions, pH, temperature, sample size, sample preparation and length of incubation. They recommended the use of blanks and standards of known digestibility. According to Marten and Barnes (1980) aeration of the fresh inocula has little affect on its activity even though they admit that lag times of up to 12 hours are not uncommon. This points to the fact that bacteria are able to recover from improper handling if given enough time. Long lags caused by small amounts of inocula, poorly reduced medium, poor temperature and oxygen control when handling the inoculum etc. will have the greatest influence on the result at short incubation times.

Some laboratories have therefore increased fermentation time (Den Braver and Eriksson, 1967) to improve in vivo predictions. Long in vitro lag periods occur naturally in prepared celluloses and cotton (Van Soest, 1973). It is doubtful whether attempts should be made to reduce this lag by extensive soaking, boiling or other pretreatments.

The residues from the two-stage method of Tilley and Terry (1963) consist of unfermented fibre and bacterial cell walls. The uncorrected digestion coefficients obtained are nearly identical to in vivo values (see Table 17.2). No plant species interactions were observed in 48 grasses and 25 legumes analysed (Terry et al., 1978).

Table 17.2. Linear regression equations of digestibility in vivo upon in vitro

Intercept(a)	Slope(b)	Plant material	Author
1.01	.99	Grasses and legumes	Tilley and Terry (1963)
-12.9	1	" " "	Goering and Van Soest (1970)
14	.75	Legumes (> 80%)	Lindgren (1977)
-2	.90	Grasses (> 80%)	" "
10	.66	Straw	Eriksson (1981)

The one-stage method of Den Braver and Eriksson (1967) as used by Lindgren (1977) yields a residue composed of undigested fibre and some microbes, which have not escaped through the filter. Three separate equations depending on legume percentage are used to predict in vivo digestibility (see Table 17.2). For the analysis of straws, a fourth method is presently used (Eriksson, 1981). A higher a-value and a lower b-value were found for the legume regression compared to grasses. An explanation was offered by Lindgren (1977) who found that the regression of cellulose digestibility in vivo upon in vitro had a slope of .60 and an intercept of 28. The low quality range was dominated by legumes, which may have caused the observed interactions.

The in vitro technique of Van Soest et al. (1966) involves a neutral detergent extraction of the residue after a 48-hour fermentation. As a result of this treatment the value obtained will be a measure of true digestibility. Apparent in vivo digestibility is obtained by subtracting 12.9 units of digestibility (Goering and Van Soest, 1970). Van Soest (1981) has shown with his system, that the highest correlation with in vivo apparent digestibility is at incubation of 36 hours.

Gas production is related to digestibility even though the fermentation pattern (propionate/acetate ratio) will affect the total amount of gas produced per unit carbohydrate fermented (Wolin, 1960). Menke et al. (1979) measured gas production after 24 hours from a 200 mg sample. The digestibility of organic matter was best related to gas production in a multiple regression also including the protein and fat content of the feed. Errors associated with washing and filtering of the residue were avoided with this technique.

17.5.3.3. <u>The in situ method</u>. Feed samples are placed in porous bags made of a synthetic material. The bags are incubated in the rumen for a specific length of time. After incubation, the bags are thoroughly washed in tap water or the content is removed and treated with neutral detergent. The method is often called the nylon bag technique even though a variety of materials have been used such as silk, dacron, nylon, terrylene, polyester etc. (Quin et al., 1938; Fina et al., 1962; Van Keuren and Heineman, 1962; Hopson et al., 1963; Czerkawski, 1967; Meyer et al., 1971; McManus et al., 1972; Udén, 1978). It has been used to predict apparent dry matter digestibility (eg. Neathery, 1969) even though it should be better correlated with true digestibility because microbial matter is partially removed from the bag residues during washing. Mehrez and Ørskov (1977) used the method to study N-disappearance from barley and Playne et al. (1978) followed the release of minerals in hays. Microbial growth inside the bag can mask the effects of feed protein degradation in low protein feeds such as straws (Lindberg, 1981a). Mathers and Aitchison (1981) have quantified this contamination for lucerne and fish meal.

The in situ technique is a very useful tool when studying specific diet effects on the rumen environment, which may affect normal fermentation (McManus et al., 1972; Ørskov and Hovell, 1978; Lindberg, 1981a). These effects can be difficult to determine when using an in vitro method. Ørskov and Hovell (1972) found in their experiment that bulls fed Pangola grass fermented hay better than when fed sugar cane. Two of the problems involved with the in situ technique to measure DM or fibre digestibility are build-up of fermentation end-products in the bag; and in-flow and out-flow of material through the pores. To avoid build-up of fermentation end-products, the sample has to be small relative to the capacity of the bag. A sample size of 5-7 mg/cm^2 bag surface (Udén et al., 1974; Mehrez and Ørskov, 1977; Lindberg, 1981b) is probably adequate. Fine bag porosity will also limit liquid flow (Udén, 1978) and decrease fermentation rate. In-flow and out-flow of particles will be restricted by a small membrane porosity and a coarse grind. The need for a large pore size for maximum fluid exchange and a small pore size for minimal particle migration makes it necessary to compromise. When digestibility was measured over time comparing 5-20 µm pore sizes (Lindberg and Knutsson, 1981) the curves were different, but the final values were very similar. This seems to indicate minimal differences in particle migration for these membranes, whereas a retarded fermentation rate was indicated for the small pore sizes (< 20 µm). Udén and Van Soest (unpublished) found an overall lower cell wall digestibility and a higher measurement error with 5 as compared with 37 µm pore size bags. They recommended that bag pores have an aperture in the range of 39µm, be attached to a weight and suspended freely in the rumen or put in a loose mesh sack containing a weight.

17.6. ANALYSIS OF ALKALI-TREATED MATERIALS

17.6.1. <u>Digestibility</u>

Specific problems arise when trying to predict the digestibility of alkali-treated feeds such as straws and wood. Lignification no longer serves as an indicator of digestibility unless actually removed. In "dry-treated" straws lignin analysis may be identical to original material and only some solubilization of the hemicellulose can be revealed (Braman and Abe, 1977; Kristensen et al., 1978). Removal of soluble lignin with neutral detergents (without sulfite) and subsequent lignin analysis would improve the analysis. Lau and Van Soest (1981) have proposed two alternative methods to analyse for bound and soluble lignin. They either titrated residual lignin with sodium hydroxide after neutral detergent extraction or measured the optical density (280 nm) of the filtrates after refluxing the straws in a phosphate buffer or neutral detergents. The samples consisted of wheat and oat straws and maize stovers. Both methods were highly correlated with digestibility except for the optical density measurements on maize stover.

A common practice is to regress in vitro or cellulase digestibility upon level of alkali used (eg. Kristensen et al., 1978) under given treatment conditions. This technique can only be successful when treatment conditons are constant and the raw materials have little compositional variation. The N-content of NH_3-treated straw can similarily be an indicator for digestibility under standard conditions. The Biotechnical Institute in Kolding (Denmark) uses a simple empirical method (Schopper-Riegler method) to determine the thoroughness of the NaOH treatment of straw (Rexen et al., 1975). It is based on how quickly water can run through a mat of ground straw while placed on a sieve. The more complete the alkali treatment the slower the water will run through the fibre mat. The method is simple and rapid and seems to be of practical use in a treatment plant.

Comparing cellulase and in vitro digestibilies of alkali-treated and untreated straws, Rexen (1977a), Waagepetersen and Westergaard-Thomsen (1977), Kristensen et al. (1978) and Sundstøl et al. (1978) showed in all cases lower values for the cellulase method. The statistical relationships were similar to those for forages in the range of cellulase digstibilty < 50%. In the range above 50% cellulase digestibility Waagepetersen and Vestergaard Thomsen (1977) and Sundstøl et al. (1978) found no further increase of the in vivo digestibility. The mode of action of live rumen microorganisms seems undoubtedly to be different from fungal enzymes.

The in vitro methods have in several cases been found to overestimate in vivo digestibility of NaOH-treated straws (Jayasuriya and Owen, 1975; Berger et al., 1979). This seemed to occur at levels of 4% NaOH and above. Several factors seem to contribute to these results. Berger et al. (1980) showed that ruminal digesta retention decreased with increasing level of NaOH and that the rate of fibre fermentation in situ was reduced by increasing NaOH levels. This could have been caused by changing physical properties of the treated materials combined with an osmotic effect on passage and microbial efficiency. Filtration losses of saponified lignin and lignin-carbohydrate complexes soluble at neutral pH after the in vitro incubations could also have contributed to the elevated in vitro values. The problems of filtration losses should be possible to overcome by an acidification and centrifugation procedure or by measuring gas production (Menke et al., 1979) instead of fermentation residue. The effect of the ration upon rumen environment (rate of passage, inhibitory effects on the microflora) and consequently on digestibility are difficult to predict by laboratory methods. A combination of in vivo and in vitro methods are instead required.

17.6.2. <u>Nitrogen</u>

A variety of feeds will retain nitrogen after ammonia treatment (Hoch and Dargel, 1962) and some of this nitrogen is water insoluble. Formation of ammonium acetate, uronic acid amides, various amines and binding of ammonia to soluble phenols and lignin (Rexen, 1977b) are possible during ammonia treatment. Most of these products are water soluble and will be rapidly utilized by microorganisms.

Solaiman et al. (1979) found that ca 13% of the N adsorbed from the ammonia was unavailable and 43% was analysed as NH_4. Rexen (1977b) reported that the total water soluble N of ammoniated straw was approximately equal to the nitrogen increase due to treatment. He estimated that approximately 20-30% of this N could have come from the plant itself. It is not unlikely that the increase in unavailable nitrogen observed by Solaiman et al. (1979) is the result of Maillard-type reactions and may vary considerably depending on temperature in the stack. From the above it seems appropriate to analyse for ADF nitrogen (Goering and Van Soest, 1970) as a measure of unavailable N in ammoniated straws. The remaining nitrogen from the ammoniation is probably utilized similar to urea.

17.7. SELECTING A SUITABLE METHOD FOR PREDICTING IN VIVO DIGESTIBILITY

The need for accurate information differs and type of equipment and facilities vary with each laboratory. Detergent analysis (using the summative equation) and possibly the cellulase method (if steady supplies of a standardized enzyme are secured) may be the only methods possible for very large-scale evaluations. All the microdigestion techniques are generally more reliable than the summative equation, but the nylon bag and in vitro methods require fistulated animals. The nylon bag method has the advantage of requiring much less laboratory equipment than the in vitro method. A simple refluxing unit and a filtration stand for neutral detergent treatment after incubations are however recommended when studying fibre digestion in situ. Marten and Barnes (1980) concluded that the future of the cellulase method relies on further research in developing and standardizing the enzymes. Which sample pretreatment to use and if each laboratory should produce its own enzymes has not yet been resolved. The in vitro method involves several steps which have to be well standardized and therefore require highly trained personnel. Some degree of automation is possible for the in vitro method (Alexander and McGowan, 1966) and it is superior to the nylon bag method for large scale evaluations.

A comparison of the accuracy of different in vitro methods is difficult to make. Few attempts have been made where large numbers of well distributed samples have been used. The purpose of the methods may also differ. One technique may be superior for evaluation of widely differing feeds whereas other methods conveniently and rapidly handle large numbers of samples of similar origin with high accuracy. Analyses of interaction causing over- and under-estimations in certain digestibility regions or of certain plant families or types of materials are also scarce. The distribution and large number of samples (152 grasses, 82 grass/legume mixtures and 73 legumes) used by Lindgren (1979) may explain the significant differences among the three separate equations whereas Terry et al. (1978) found no differences between 48 grasses and 25 legumes. Residual standard deviation from the regression of in vivo digestibility on in vitro digestibility will also be influenced by the selection of samples. The drawback of the techniques of Den Braver and Eriksson (1967) and of Tilley and Terry (1963) is the time required, even though the Den Braver-Eriksson method requires few manipulations. Van Soest's method is specifically useful when studying plant fibre digestion and rates of degradation. It is important to realize that no laboratory method for feed evaluation can be used blindly. For the in vitro technique, each laboratory has to develop its own standard procedures and test its reliability to the feeds to be analysed (Marten and Barnes, 1980).

Prediction of in vivo digestibility of fibrous by-products or crop residues has not been studied to any great extent. Such feeds may have long natural lag of fermentation combined with slow rate of fermentation and sometimes long rumen retention times due to low intakes. Pigden (1969) has noted that the primary application of the in vitro method is not to measure absolute, but relative digestibility. If feeds differ widely in the previously mentioned characteristics, digestibility may have to be predicted from the fermentation curve (nylon bag or in vitro) combined with some estimates or rumen passage. Special attention has to be given the alkali-treated feeds. The methods of Lau and Van Soest (1981) have to be verified by other laboratories and additional techniques must be developed for the nutritive evaluation of these materials.

17.8. SUMMARY

The ability of ruminants to digest various plant fractions differ. Using the detergent system of analysis, plant material can be divided into a highly digestible, an indigestible and a fraction of varying digestibility. The proximate analyses system fails in this respect. Lignin is the single most important plant component restricting fibre digestibility. Silica, cutin and tannins may also restrict fibre digestibility in certain plants. Cellulose crystallinity and minor constituents such as acetyl groups and phenolic acids are less well understood. Predictions of digestibility using chemical, physical and biological methods are reviewed. Chemical analysis of specific plant fractions is generally less reliable for predicting digestibility, but can give important information as to what limits digestion. Near infrared reflectance spectrophotometry is useful for large scale analyses, but requires calibrations with sets of reference samples of known digestibilities, some knowledge of botanical composition and large capital investments. Biological methods (in vitro, nylon bag and cellulase methods) are generally the most reliable methods for predicting digestibility. The in vitro method and its modifications are discussed and recommended as still the most accurate and versatile of all. The merits of the methods are discussed relative to accuracy, sample size, laboratory facilities, training of personnel and other factors. Some possibilities for estimating digestibility of alkali-treated straw are outlined as well as how to evaluate adsorbed N from ammoniation of straw.

17.9. REFERENCES

Adler, E., 1968. Ligninets kemiska byggnad. Svensk kemisk tidsskrift 80: 279-290.

Alexander, R.H. and McGowan, M., 1966. The routine determination of in vitro digestibility of organic matter in forages. An investigation of the problems associated with continuous large scale operation. J. Br. Grassl. Soc. 21: 140-147.

Bacon, J.S.D. and Gordon, A.H., 1980. The effect of various deacetylation procedures on the nylon bag digestibility of barley straw and of grass cell walls recovered from sheep faeces. J. agric. Sci. (Camb.) 94: 361-367.

Bartiaux,Thill, N., Fabry, J., Briston, R. and Boukharta, M., 1980. Estimation in vitro de la digestibilite de l'herbe de prairie permanante par la cellulase. Bull. Rech. Agron. Gembloux 15: 297-308.

Baker, T.I., Quicke, G.V., Bentley, O.G., Johnson, R. and Moxon, A.L., 1959. The influence of certain physical properties of purified celluloses and forage celluloses on their digestibility by rumen micro-organisms in vitro. J. Anim. Sci. 18: 655-672.

Berger, L.L., Klopfenstein, T.J. and Britton, R.A., 1979. Effect of sodium hydroxide on efficiency of rumen digestion. J. Anim. Sci. 49: 1317-1323.

Berger, L.L., Klopfenstein, T.J. and Britton, R.A., 1980. Effect of sodium hydroxide treatment on rate of passage and rate of ruminal fiber digestion. J. Anim. Sci. 50: 745-749.

Bethge, P.O. and Lindstrøm, K., 1973. Determination of O-acetyl groups in wood. Svensk Papperstidn. 17: 645-649.

Beveridge, R.J. and Richards, G.N., 1973. Digestion of polysaccharide constituents of tropical pasture herbage in the bovine rumen. Part IV. The hydrolysis of hemicelluloses from Spear Grass by cell-free enzyme systems from rumen fluid. Carbohyd. Res. 29: 79-87.

Beveridge, R.J. and Richards, G.N., 1975. Investigation of the digestion of cell wall polysaccharides of Spear grass and of cotton cellulose by viscometry and by X-ray diffraction. Carbohyd. Res. 43: 163-172.

Blaxter, K.L., Graham Mc. C.N. and Wainman, F.W., 1956. Some observations on the digestibility of food by sheep and on related problems. Br. J. Nutr. 10: 69-80.

Braman, W.I. and Abe, R.K., 1977. Laboratory and in vivo evaluation of the nutritive value of NaOH-treated wheat straw. J. Anim. Sci. 46: 496-505.

Carlier, L.A., Van Hee, L.P. and Andries, A.-P., 1979. Estimation de la degestibilite des fourrages grossiers par le traitement a la pepsine - cellulase ou au jus de rumen - pepsine. Rev. de l'Agr. 32: 147-157.

Classon, C., 1980. The relationship between the content of enzyme soluble organic matter and metabolizable energy of forages. Rep. no 49. Swedish Univ. Agr. Sci., Dept. Anim. Nutr., Uppsala, 40 pp.

Clark, J. and Beard, J., 1977. Prediction of the digestibility of ruminant feeds from their solubility in enzyme solutions. Anim. Feed Sci. Technol. 2: 153-159.

Colucci, P.E., 1979. Rate of passage of digesta through the gastro-intestinal tract of dairy cattle. M.S.Thesis, Cornell Univ., Ithaca, NY., USA.

Czerkawski, J.W., 1967. Incubation inside the bovine rumen. Br. J. Nutr. 21: 865-871.

Dekker, R.F.H., Richards, G.N. and Payne, M.J., 1972. Digestion of polysaccharide constituents of tropical pasture herbage in the bovine rumen. Part 1. Townsville Stylo. Carbohyd. Res. 22: 173-185.

Den Braver, E. and Eriksson, S., 1967. Determination of energy in grass hays by in vitro methods. Lantbr.högsk. Annlr. 33: 751-765.

Donefer, E., Niemann, P.J., Crampton, E.W. and Lloyd, L.E., 1963. Dry matter disappearance by enzyme and aqueous solutions to predict the nutritive value of forages. J. Dairy Sci. 46: 965-970.

Eriksson, S., 1981. Nutritive value of cereal straw. Agric. Environm. 6: 257-260.

Fina, L.K., Keith, C.L., Bartley, E.E., Hartman, P.A. and Jacobsen, N.L., 1962. Modified in vivo artificial rumen (VIVAR) techniques. J. Anim. Sci. 5: 201-210.

Fonnesbeck, P.V., 1976. Estimating nutritive value from chemical analysis. In: Proc. 1st Int. Symp. Feed Comp., Animal Nutrient Requirements and Computerization of Diets. July 11-16 at Utah State Univ., USA.

Gaillard Blanche, D.E., 1962. The relationship between the cell-wall constituents of roughages and the digestibility of the organic matter. J. agric. Sci. (Camb.) 59: 369-373.

Gaillard Blanche, D.E. and Richards, G.N., 1975. Presence of soluble lignin - carbohydrate complexes in the bovine rumen. Carbohyd. Res. 42: 135-145.

Goering, H.K. and Van Soest, P.J., 1970. Forage fiber analysis. Agriculture Handbook. Agric. Res. Service, USDA, Washington D.C., 20 pp.

Guggolz, J., Saunders, R.M., Kohler, G.O. and Klopfenstein, T.J., 1971. Enzymatic evaluation of processes for improving agricultural wastes for ruminant feeds. J. Anim. Sci. 33: 167-170.

Hartley, R.D. and Jones, E.C., 1977. Phenolic components and degradability of cell walls of grass and legume species. Phytochem. 16: 1531-1534.

Henry, W.A., 1898. Feeds and Feeding. A handbook for the student and stockman. Madison, Wisconsin.

Hoch, A. and Dargel, D., 1962. Untersuchung über die Bindung von Ammoniak in Trockenschnitzeln. Arch. Tierernähr. 12: 343-350.

Hogan, J.P. and Lindsay, J.R., 1979. The digestion of nitrogen associated with plant cell wall in the stomach and intestine of the sheep. Aust. J. Agric. Res. 31: 147-153.

Hopson, D.J., Johnson, R.R. and Dehority, B.A., 1963. Evaluation of the dacron bag technique as a method for measuring cellulose digestibility and rate of forage digestion. J. Anim. Sci. 22: 448-452.

Hungate, R.E., 1966. The rumen and its microbes. Academic Press, New York and London, pp. 1-7.

Jarrige, R., Thivend, R. and Demarquilly, C., 1970. Development of a cellulolytic enzyme for predicting the nutritive value of forages. Proc. 11th Int. Grassl. Congr., Surfers Paradise, 1970, pp. 762-766.

Jayasuriya, M.C.N. and Owen, E., 1975. Sodium hydroxide treatment of barley straw: Effect of volume and concentration of solution on digestibility and inake of sheep. Anim. Prod. 21: 313-318.

Jones, D.I.H., 1976. Fungal cellulases and hemicellulases and their application to the analysis of forage. In: Carbohydrate Research in Plants and Animals. Miscellaneous Papers 12 (1976). Landbouwhogeschool Wageningen, The Netherlands.

554

Jones, D.I.H. and Hayward, M.V., 1975. The effect of pepsin pretreatment of herbage on the prediction of dry matter digestibility from solubility in fungal cellulase solutions. J. Sci.
Food Agric. 26: 711-718.
Joshi, D.C., 1972. Different measures in the prediction of the
nutritive value of forages. Acta Agric. Scand. 22: 243-247.
Kellner, R.J. and Kirchgessner, M., 1977. Estimation of forage
digestibility by a cellulase method. Z. Tierphysiol. Tierernahr. Futtermittelk. 39: 9-16.
Kristensen, F.V., Andersen, P.E., Stigsen, P., Vestergaard Thomsen, K., Refsgaard Andersen, H., Sørensen, M., Ali, C.S.,
Mason, V.C., Rexen, F., Israelsen, M. and Wolstrup, J., 1978.
Sodium hydroxide-treated straw as feed for cattle and sheep.
Beretning, Statens Husdyrbrugsforsøg, no 464, pp. 53-68.
Lau Mamie, M. and Van Soest, P.J., 1981. Titratable groups and
soluble phenolic compounds as indicators of the digestibility
of chemically treated roughages. Anim. Feed Sci. Technol. 6:
123-131.
Lindberg, J.E., 1981a. The effect of basal diet on the ruminal degradation of dry matter, nitrogenous compounds and cell walls
in nylon bags. Swedish J. Agric. Res. 11:159-169.
Lindberg, J.E., 1981b. The effect of sample size and sample structure on the degradation of dry matter, nitrogen and cell walls
in nylon bags. Swedish J. Agric. Res. 11: 71-76.
Lindberg, J.E. and Knutsson, P.G., 1981. Effect of bag pore size
on the loss of particulate matter and on the degradation of
cell wall fibre. Agric. Environm. 6: 171-182.
Lindgren, E., 1977. Results from a modified one-stge in vitro
technique. Proc. of a Seminar organized by NJF, Uppsala 1977,
pp. 119-126.
Lindgren, E., 1979. The nutritional value of roughages detemined
in vivo and by laboratory methods. Report 45,Swedish Univ.
Agric. Sci., Dept. Anim. Nutr., Uppsala, 61 pp.
Lindgren, E., Theander, O. and Åman, P., 1980. Chemical composition of timothy at different stages of maturity and of residues
from feeding value determinations. Swedish J. Agric. Res. 10:
3-10.
Lucas,H.L., 1964. Stochastic elements in biological models; their
sources and significance. In: I Gurland (ed.): Stochastic
Models in Medicine Biology. The Univ. of Wisconsin Press, Madison Wi., USA, pp. 335-385.
Marten, G.C. and Barnes, R.F., 1980. Prediction of energy digestibility of forages with in vitro rumen fermentation and fungal
enzyme systems. In: Standardization of analytical methodology
for feeds. Proc. of a workshop held in Ottawa, Canada 12-14
March 1979.
Mason, V.C., 1979. The quantitative importance of bacterial residues in the non-dietary faecal nitrogen of sheep. Z. Tierphysiol. Tierernahr. Futtermittelk. 41: 131-140.
Mathers, J.C. and Aitchison, E.M., 1981. Direct estimation of the
extent of contamination of food residues by microbial matter
after incubation within synthetic fibre bags in the rumen. J.
agric. Sci. (Camb.) 96: 691-693.
McLeod, M.N. and Minson, D.J., 1976. The analytical and biological
accuracy of estimating the dry matter digestibility of different legume species. Anim. Feed Sci. Technol. 1: 61-72.
McManus, W.R., Manta, L. and McFarlane, I.D., 1972. The effect of
diet supplementation and gamma irradiation on dissimilation of
low-quality roughages by ruminants. I. Studies on the terryle-
ne-bag technique and effect of supplementation of base ration.
J. agric. Sci. (Camb.) 79: 27-40.

McQueen, R. and Van Soest, P.J., 1975. Fungal cellulase and hemicellulase prediction of forage digestibility. J. Dairy Sci. 58: 1482-1491.

Mehrez, A.Z. and Ørskov, E.R., 1977. A study of the artificial bag technique for determining the digestibility of feeds in rumen. J. agric. Sci. (Camb.) 88: 645-650.

Menke, K.H., Raab, L., Salewski, A., Steingass, H., Fritz, D. and Schneider, W., 1979. The estimation of the digestibility and metabolizable energy content of ruminant feedingstuffs from the gas production when they are incubated with rumen liquor in vitro. J. agric. Sci. (Camb.) 93: 217-222.

Mertens, D.R., 1973. Application of theoretical mathematical models to cell wall digestion and forage intake in ruminants. PhD Thesis, Cornell Univ., Ithaca, NY, USA.

Meyer, R.M., Bartley, E.E., Julius, F. and Fina, L.R., 1971. Comparison of four in vitro methods for predicting in vivo digestibility of forages. J. Anim. Sci. 32: 1030-1036.

Neathery, M.W., 1969. Dry matter disappearance of roughages in nylon bags suspended in the rumen. J. Dairy Sci. 52: 74-78.

Norris, K.H., Barnes, R.F., Moore, I.E. and Shenk, I.S., 1976. Predicting forage quality by infrared reflectance spectroscopy. J. Anim. Sci. 43: 889-897.

Ørskov, E.R. and Hovell, F.D. DeB., 1978. Rumen digestion of hay (measured with dacron bags) by cattle given sugar cane or pangola hay. Trop. Anim. Prod. 3: 9-11.

Paloheimo, L., 1953. Some persistant misconceptions concerning the crude fibre and the nitrogen free extract. J. Sci. Agric. Soc. Finland 25: 16-22.

Pigden, W.J., 1969. Laboratory analyses of herbage used to predict nutritive value. In: Cambell, J.B. (ed.): Experimental methods for evaluating herbage. Ottawa, Public. 1315, Canada Dept. Agr., Queens Printer, pp. 52-72.

Playne, M.J., Echevarria, M.G. and Megarrity, R.G., 1978. Release of nitrogen, sulphur, phosphorus, calcium, magnesium, potassium and sodium from four tropical hays during their digestion in nylon bags in the rumen. J. Sci. Food Agr. 29: 520-526.

Quin, J.I., Van der Vath, J.C. and Myburgh, S., 1938. Studies on the alimentary tract of Merio Sheep in South Africa. IV. Description of experimental technique. Onderspoort J. Vet. Sci. Anim. Ind. 11: 341-360.

Rexen, B., 1977a. Enzyme solubility - a method for evaluating the digestibility of alkali-treated straw. Anim. Feed Sci. Technol. 2: 205-218.

Rexen, F.P., 1977b. Ammoniakbehandling av strå. Report 78. Biotech. Inst. Kolding, Denmark.

Rexen, F., Israelsen, M., Busk, J. and Waagepetersen, J., 1975. Increased digestibility of straw through chemical treatment. Report 73. Biotech. Inst. Kolding, Denmark.

Richards, C.R. and Reid, J.T., 1953. The digestibility and interrelationships of vaious carbohydrate fractions of pasture herbage and a resolution of the components of crude fiber and nitrogen-free extract. J. Dairy Sci. 36: 1006-1115.

Robbins, C.T., 1973. PhD Thesis. Cornell Univ., Ithaca, NY.

Robbins, C.T., Van Soest, P.J., Mautz, W.W. and Moen, A.N., 1975. Feed analyses and digestion with reference to white tailed deer. J. Wildl. Manage. 39: 67-73.

Rohweder, D.A., Barnes, R.F. and Jorgensen,, N., 1978. Proposed hay grading standards based on laboratory analyses for evaluating quality. J. Anim. Sci. 47: 747-759.

Roughan, P.G. and Holland, R., 1977. Predicting in vivo digestibility of herbages by exhaustive enzyme hydrolysis of cell walls. J. Sci. Food Agric. 28: 1057-1064.

Salomonsson, A.-C., Theander, O. and Åman, P., 1978. Quantitative determination by GLC of phenolic acid, as ethyl derivatives in cereal straws. Agric. Food Chem. 26: 830-835.

Sidahmed, A.E., Morris, J.G., Koong, L.J. and Radosevich, S.R., 1981. Contribution of mixtures of three chaparral shrubs to the protein and energy requirement of Spanish goats. J. Anim. Sci. 53: 1391-1400.

Smith, G.S., Nelson, A.B. and Boggino, E.J.A., 1971. Digestibility of forages in vitro as affected by content of silica. J. Anim. Sci. 33: 466-471.

Solaiman, S.G., Horn, G.W. and Owens, F.N., 1979. Ammonium hydroxide treatment on wheat straw. J. Anim. Sci. 49: 802-808.

Sundstøl, F., Kossila, V., Theander, O. and Vestergaard Thomsen, K., 1978. Evaluation of the feeding value of straw. A comparison of laboratory methods in the Nordic countries. Acta Agric. Scand. 28: 10-16.

Terry, R.A., Mundell, D.C. and Osburn, D.F., 1978. Comparison of two in vitro procedures using rumen liquor-pepsin or pepsin-cellulase for prediction of forage digestibility. J. Br. Grassl. Soc. 33: 13-18.

Theander, O., Udén, P. and Åman, P., 1981. Acetyl and phenolic acid substituents in Timothy of different maturity and after digestion with rumen microorganisms or a commercial cellulase. Agric. Environm. 6: 127-133.

Tilley, J.M.A. and Terry, R.A., 1963. A two-stage technique for the in vitro digestion of forage crops. J. Br. Grassl. Soc. 18: 104-111.

Tinnimit, P. and Thomas, J.W., 1976. Forage evaluation using various laboratory techniques. J. Anim. Sci. 43: 1058-1065.

Udén, P., 1978. Comparative studies on rate of passage, particle size and rate of digestion in ruminants, equines, rabbits and man. PhD Thesis, Cornell Univ., Ithaca, NY, USA.

Udén, P., Parra, R. and Van Soest, P.J., 1974. Factors influencing reliability of the nylon bag technique. J. Anim. Sci. 37: 358 (Abstr.).

Van Keuren, R.W. and Heineman, W.W., 1962. Study of a nylon bag technique for in vivo estimation of forage digestibility. J. Anim. Sci. 21: 340-345.

Van Soest, P.J., 1963. Use of detergents in the analysis of feeds. J. Ass. Official Agr. Chem. 46: 825-831.

Van Soest, P.J., 1964. Symposium on nutrition and forage and pastures: New chemical procedures for evaluating forages. J. Anim. Sci. 23: 838-845.

Van Soest, P.J., 1965. Use of detergents in the analysis of fibrous feeds. III. Study of effects of heating and drying on yield of fiber and lignin in forages. J. Assoc. Offic. Agr. Chem. 48: 785-790.

Van Soest, P.J., 1969. Chemical properties of fiber in concentrate feed stuffs. Proc. of the Cornell Nutr. Conf. p. 17-21.

Van Soest, P.J., 1970. The role of silicon in the nutrition of plants and animals. In: Proc. of the Cornell Nutr. Conf. pp. 103-109.

Van Soest, P.J., 1973. The uniformity and nutritive availability of cellulose. Fed. Proc. 32: 1804-1808.

Van Soest, P.J., 1981. Nutritional ecology of the ruminant. O & B Books Inc., Corvallis, Oregon, USA. 374 pp.

Van Soest, P.J. and Jones, L.H.P., 1968. Effect of silica in forages upon digestibility. J. Dairy Sci. 51: 1644-1648.

Van Soest, P.J., Mertens, D.R. and Deinum, B., 1978. Preharvest factors influencing quality of conserved forage. J. Anim. Sci. 47: 712-718.

Van Soest, P.J. and Robertson, J.B., 1980. Systems of analysis for evaluating fibrous feeds. In: Standardization of analytical methodology for feeds. Proc. of a workshop held in Ottawa, Canada, 12-14 March 1979. pp. 49-60.

Van Soest, P.J., Wine, R.H. and Moore, L.A., 1966. Estimation of the true digestibility of forages by the in vitro digestion of cell walls. Proc. Xth int. Grassl. Congr. p. 438-441.

Vestergaard Thomsen, K. and Henriksen, J., 1976. Laboratory methods for evaluation of the energy value of feed stuffs and feed compounds for ruminants. Beretn. fra Statens Husdyrbrugsforsøg no. 436, 29 pp.

Waagepetersen, J. and Vestergaard Thomsen, K., 1977. Effect on digestibility and nitrogen content of barley straw of different ammonia treatments. Anim. Feed Sci. Technol. 2: 131-142.

Waite, R. and Gorrod, A.R.N., 1959. The structural carbohydrates of grasses. J. Sci. Food Agric. 10: 308-316.

Wolin, M.J., 1960. A theoretical rumen fermentation balance. J. Dairy Si. 43: 1452-1459.

Chapter 18

THE ECONOMICS OF USING STRAW AS FEED

by

Harald Giæver

Department of Agricultural Economics
Agricultural University of Norway
P.O. Box 33, 1432 Ås-NLH
Norway

18.1. INTRODUCTION

This chapter will concern itself with the use of untreated or treated straw as feed, however much of what will be said will also be valid for other fibrous by-products like maize cobs, sugar cane tops etc.

Straw is high in fibre and contains little protein. The digestibility of untreated straw is low, and voluntary feed intake of animals fed straw is low compared with animals fed higher-quality roughages like hay and silage made from early-cut grass. Digestibility as well as feed intake can be improved by treating straw by various methods, however at a cost. Even treated straw is high in fiber content and low in protein.

Does it make economic sense to use straw as feed, or to treat straw by one of the methods available, before feeding it? Evidently, no general answers to such questions exist. The intention of this chapter is to discuss some of the factors and relationships which the author believes are important as a basis for supplying answers to the questions in a specific situation. He will also attempt to arrive at some general statements about the type of region where it is likely to be profitable to use straw as feed, or where it is economic to use any of the existing treatment methods. Finally, he will briefly present a few cases where the use of straw is known to have been profitable.

18.2. ANIMAL PRODUCTION AS RELATED TO FEED QUALITY

It is necessary to start with some basic assumptions about relationships between feed quality and animal production.

"Feed quality" is not a simple and concise concept. Among the many characteristics of a feed which have a bearing on "quality", only the following will be considered, fibre content, digestibility, net energy (or some other measure of energy) per unit of dry matter, and protein content. Untreated straw has a high fibre content and falls far short of good hay or silage in all the other aspects.

The animal's reaction to feeds of differing quality will vary with the genetic characteristics of the animal. A convenient point of departure may be a model suggested by Broster (1976) to describe the relationships between feed intake (in energy terms) and milk production for cows of varying ability (Figure 18.1).

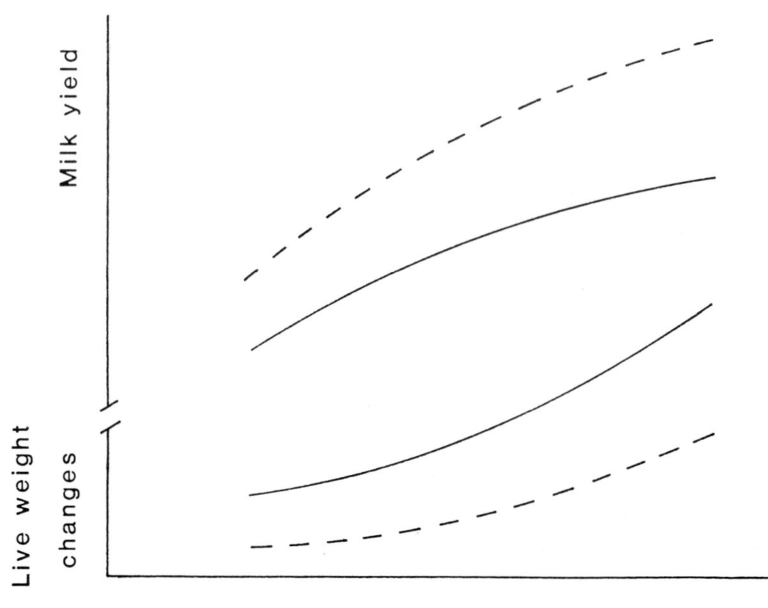

Figure 18.1. Relationships of food to milk and liveweight change in dairy cows (from Broster, 1976)
Cow of high (----) and low (———) dairy merit

At a given level of feed intake, the cow with high genetic endowment for milk production yields more milk. At the same time, increase in milk production per unit increase in feed intake is higher for the better cow. The difference is explained as differences between cows in the way energy supplied is used for milk production and body fat production. At a given level of feed energy intake, the better cow will use more of the energy supplied for milk production and less for body fat deposition. If less energy is supplied by the feed ration, the cow of high dairy merit may use her body fat energy reserves to maintain milk production. If we want to utilize the higher production capacity of the good cow, we have to feed her a larger ration. This can be done in different ways, by feeding roughage to appetite instead of rationing, by feeding roughage of better quality, or by increasing the concentrate ration.

The cow's response to higher concentrate rations depends on the quality of the roughage. If roughage is fed to appetite, the relations will, in principle, look like the curves in Figure 18.2. Here, a cow of given genetic endowment is assumed. If small amounts of concentrate are fed, the response to roughage of high quality will be great. As concentrate rations are increased, the difference in milk production between high and low quality roughage will decrease. However, it is likely that even large rations of concentrates cannot entirely compensate for low-quality roughage.

Whether it is profitable or not to feed much concentrate depends very much, among other things, on the price of concentrate in relation to milk. Under conditions where it is not profitable to feed much concentrate, the economic return on high-quality roughage will be higher. With low concentrate rations, the return on higher-quality roughage will also be higher if the cows are genetically better endowed for milk production.

For growing animals, there may be a similar variation between animals in the genetic potential for growth. If so, the same type of relationship between roughage quality, concentrate rations and animal production can be expected, only that "weight gains" have to be substituted for "milk yields" in Figure 18.2. Young animals of a breed genetically endowed for rapid growth, therefore, will give better returns for high-quality roughage, especially if concentrate rations are very restricted.

One note must be added. While high fibre content is associated with low feed quality, it should not be forgotten that ruminants require certain minimum quantities of fibrous feeds. If concentrate rations are large, this may become a problem, and it is necessary to ensure that such minimum requirements are met.

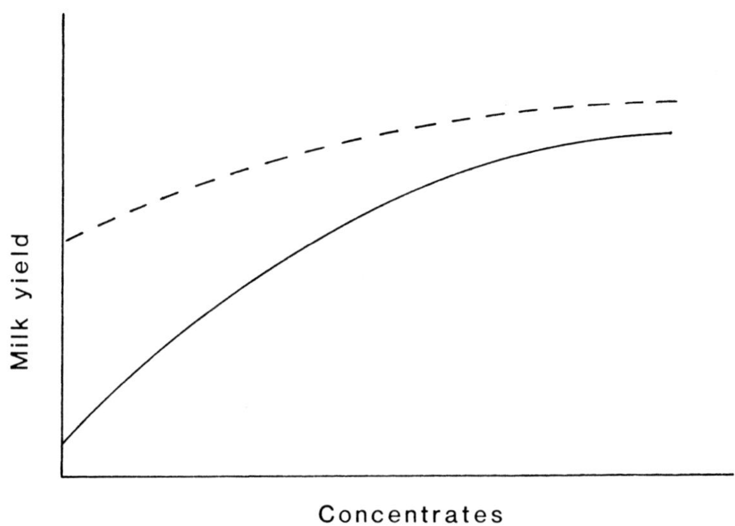

Figure 18.2. Relationship of concentrate to milk yield when roughage is fed to appetite
Roughage of high (----) and low (———) quality

18.3. SOME GENERAL CONSIDERATIONS ABOUT INPUT PRICES AND COSTS

Straw is a by-product from grain production. Under mixed farming systems it may be available on the farms where its use for feed is being considered, or available on neighbouring farms. However, if threshing takes place in the fields, collecting it and transporting it to the farm centre implies extra costs. Storing it until it can be fed may incur additional costs.

Many farming regions are specialized, so that grain-producing regions with straw surpluses are far away from the regions where straw can be fed. Because straw bales are bulky, transportation costs over long distances are high. In some cases, straw has uses other than feed. Large quantities may be needed for bedding. In some countries, straw is used as a fuel. In other cases, it is used as a raw material for paper production. Under some cropping systems ploughing-in of straw may have considerable value or burning may be worthwhile.

The cost of supplying straw for feed, therefore, is its value for the best alternative use plus the extra costs of collecting, transporting and storing it. These costs may be very low in some cases, but considerable in others. Since the feed value is low, the use of straw as feed should hardly be considered if it cannot be made available at low costs.

Feed value can be improved by various treatments. Usually, chemicals like NaOH or NH_3 are involved. Treatments will require varying amounts of capital equipment and labour. The total costs of such treatments are important.

Whether untreated or treated, straw contains very little protein, so its use must be supplemented with protein and/or non-protein nitrogen like urea. Under the highly-intensive farming methods of many industrialized countries, most animals are fed considerable quantities of feed grains or other concentrates. Therefore the total quantity of concentrates fed need not be increased but a higher protein content must be substituted. The cost of protein supplementation is smaller for NH_3-treated straw than for lye-treated straw, since the former contains some NPN while the latter contains none.

The following example of a cost calculation is based on Norwegian conditions:

	$ per 1000 MJ ME
Value of straw in field	2.10
Cost of collecting, baling, transporting	3.10
Cost of NH_3-treatment	6.50
Cost of protein supplementation	3.20
Sum	14.90

The cost of protein supplementation is calculated in such a way that the crude protein content of a complete ration will be the same if treated straw is substituted for some of the grass silage in the original ration. If lye-treated straw is used, the cost of protein supplementation will be $7.60. This demonstrates that cost of protein supplementation may add substantially to the costs of using straw as feed. Under less-intensive animal production systems, urea may be good enough as a source of nitrogen, and may be much cheaper.

Such costs per MJ ME can be compared with the corresponding costs of feed to be replaced, however such comparisons do not tell the whole story. Referring to Figure 18.2, it should be stressed that quality differences between processed straw and the feed to be replaced are important. If processed straw replaces feeds of lower quality, the economic value per MJ ME of processed straw is higher than the value per MJ ME of the feed replaced. Often this may be the case in less-developed regions. If processed straw replaces feeds of higher quality, the effect is the opposite.

Price ratios between different inputs and between inputs and products are of great importance in all economic considerations. Such price ratios may vary widely between regions and over time. The following prices are especially important:

- Land rents (or alternative value of land used for purposes other than feed crops).
- Agricultural wages (or alternative value of the farmer's own time when used outside his farm).
- Price of straw (or costs of supplying straw).
- Price of protein supplements.
- Price (and availability) of chemicals for straw treatment (NH_3, NaOH etc.).
- Prices of alternative feedstuffs (e.g. imported feed grains).
- Prices of products (milk, meat etc.).

Sometimes, there are important differences between what is profitable from the individual farmer's point of view and what is profitable for society. In social cost-benefit calculations, it is often right to include benefits and costs which are outside the reach and responsibility of the individual farmer, or to use prices on inputs and products other than those which are relevant to the farmer. For example, the original Beckmann method gives considerable river pollution the costs of which are carried by people downstream. The farmer may use the going agricultural wage to price labour inputs, while a society which wants to avoid depopulation of rural areas may want to price labour in such areas at a lower wage. The farmer is facing national prices on inputs and products. For society it may well be right to calculate on the basis of other prices, for example if world market prices are different from the national prices or if the country has difficulties with its balance of payments.

18.4. THE PROFITABILITY OF USING STRAW AS FEED IN DIFFERENT PARTS OF THE WORLD

Economic conditions vary so much between countries and regions that it is impossible to generalize about the profitability of some given method or technique. The following is an attempt at least to arrive at some statements of general validity.

Besides the relations and factors which have been discussed in the sections above, the relative merit of using untreated or treated straw as feed may depend on the following factors:
- The ratio between the agricultural land area and the size of the population which is going to be fed from that area.
- The present state of agricultural technology within the area.
- Relative scarcity of agricultural labour and capital as compared with land.

For the purpose of this discussion, I propose to classify agricultural regions of the world into four broad categories:

18.4.1. Less developed regions with high population density

Large parts of the Far East and some Middle East countries fall in this category. A large part of the population is still rural, so there is much agricultural labour available per unit of land to be worked. Capital is scarce and therefore expensive. Under such conditions, land rents can be expected to be high as compared with agricultural wages.

There are strong economic incentives to produce products which yield much human food per unit of land. Many food crops do this directly. Cash crops like cotton or tea may do it indirectly through the money income they bring. Specialized animal production inevitably gives relatively low yields of human food per unit of land. Land suited for food crops or cash crops will be used for such crops. Some land is not suited for such crops and can therefore be used for feed crops or pasture. Apart from this, the justification for animal production is that farm animals can utilize by-products and in this way augment the total food supply.

Of course the state of agricultural technology is important. If land productivity can be improved by modern farming techniques, pressure on land can be relieved, general standards of living can be improved, and there will be greater possibilities of improving human diets by diverting more agricultural resources to animal production. With such development, it will also become possible to use some of the better land for feed crops.

Straw and other low quality by-products are the main sources of feed in these countries. Because animal production has long been restricted to the utilisation of low quality by-products from cash and food crops, there has been no incentive to breed animals for high yields or rapid growth.

The question is whether it pays to treat straw (and other fibrous by-products) by any of the methods available.

The effect of such treatments on animal production will be twofold. Firstly, the available feed supply, measured in digestible energy or net energy terms, will be increased. Secondly, since daily energy intake of each animal will increase, a smaller percentage of intake is necessary for maintenance and a larger percentage is available for body growth or milk production. Growing animals will reach first calving or market weight at a younger age; milking animals will yield more milk per day. This second effect may be the more important. A theoretical example may illustrate this point. If digestibility of untreated straw is 45 per cent and treatment increases this to 60 per cent, digestible energy per kg DM will increase by 33 per cent. If dry matter intake increases by 15 per cent intake of digestible energy will increase by 50 per cent. If intake of untreated straw and other available feeds is enough to meet maintenance and the energy requirements of some small production, the whole increase in digestible energy intake can be used for increased production. With reference to Figure 18.2, since little or no concentrate is fed, an increase in roughage quality, supplemented by protein or some source of non-protein nitrogen, may result in a substantial increase in animal production.

The question of animal breed or genetic improvement of the existing breeds may enter the picture. Animals of higher production capacity will respond more to an increase in energy intake, as suggested by Figure 18.1. Genetic improvement programmes may be complementary to improvements in feed quality and may increase the economic returns of such.

An economic assessment must be done for each region and perhaps each individual case, by economic calculations adjusted to regional conditions and using regional or local prices. Most probably, the straw treatment method used should be simple enough for use at village or farm level and should not require expensive capital equipment. It should be labour-intensive rather than capital-intensive. Of course, the availability and price of necessary inputs like NaOH or NH_3 and of supplements like urea or protein-rich feeds are also very important.

Besides direct imports of feedstuff, treatment of straw and other fibrous by-products under such conditions may be the only way to improve animal feeding It seems likely that treatment of straw in many cases will be profitable therefore. There are few reports, however, dealing with the economic side of the problem. Jackson (1978a) cites two cases, both from India, where the feeding of treated straw to growing animals was found to be very profitable.

18.4.2. <u>Less developed regions with low population density</u>

Large parts of Africa and Latin America fall into this category. A large part of the population is still rural, capital is scarce and expensive, but more agricultural land is available per person to be fed. If this simple model and economic reasoning are correct, land rents can be expected to be lower relative to agricultural wages.

There is not the same economic incentive as in the former category (18.4.1) to produce as much human food per unit of land. Cattle-farming based on natural grazing is much more common. Beef production may be more important than milk production. Harvested feeds are not used to the same extent.

Seasonality of plant growth may be important. Quality of pastures deteriorates during the dry season. It is likely that feeding animals with stored feeds will prove economical during such seasons, though more so for intensive milk production than for meat production. Harvested feeds may be hay, silage or straw. Edelsten and Lijongwa (1981) found that in the Arusha region of Tanzania, alkali-treatment of low-quality roughages was economical for intensive milk production, but not for the fattening of steers. It is impossible to generalize on the basis of one such case. However, since alternative good-quality feeds may exist in these regions, it seems that situations where treatment of straw is profitable will be less frequent than in the first category (18.4.1).

18.4.3. Industrialized countries with ample food supply

In most industrialized countries, modern techniques have increased food production per unit of land many times as compared with the pre-industrial level. The pressure on land therefore has been relieved. International trade has made it possible for countries with little agricultural land per capita to import much of the food they need, in exchange for industrial products. Per capita consumption of grains for food and feed use is many times that in less-developed countries. Most of this grain is used as feedstuff for animals. Feed grain and other concentrates make up a high percentage of the total feed, even for ruminants. Most of the rest is supplied by feed-crops like maize silage, pasture and grassland for hay and silage. Animal products supply a high percentage of human needs for protein as well as for energy. Even with this rich diet only a small part of consumers' expenses (in many countries about 20 per cent) is on food.

Under such conditions, it can be expected that land rents will be modest relative to agricultural wages. Capital is relatively cheap. Agricultural methods commonly used are capital-intensive rather than labour-intensive.

With reference to Figure 18.2, it might seem that roughage quality is of less importance when concentrate rations are large. However, farm animals have been bred for high production and will therefore respond to higher-quality roughages as well as large quantities of concentrates.

Over the last 30 years, biological and technical progress have been greater in grain production than in grassland production. This has influenced the relative costs of producing a feed unit of grain compared with a feed unit of pasture grass, silage or hay. Rising energy prices influence in the same direction, since it takes less energy to produce a grain than a grass crop. In regions of the developed countries with mostly arable land, it may often be cheaper to supply cattle with a feed unit of grain than a feed unit of silage or hay. On the other hand, there has been a tendency for cattle production to move away from such regions and into areas where soil and climatic conditions are more favourable for grassland production.

The digestive system of ruminants requires certain minimum quantities of fibre. In regions with good conditions for grain production, the least-cost feed may often be a mix close to this minimum limit. One reason therefore for using some straw as feed is that it is a cheap way to meet this requirement for fibre. Since untreated straw is a low-quality roughage compared with hay and silage, animal production may be somewhat reduced as compared to what it would have been if hay or silage was fed. It seems likely that the treatment of straw will be profitable in such regions. In regions where natural conditions are good for grass production and less satisfactory for grain production, straw treatment is less likely to be profitable. Straw cannot carry the large costs of transportation from grain-producing regions to grassland regions. If it is profitable to use treated straw, it will usually be economic to choose technical equipment which is labour-saving and capital-intensive for the treatment process.

Apart from Norway, treated straw is not much-used as cattle feed in industrialized countries. Jackson (1978a), in an extensive review paper, cites many studies in industrialized countries as well as a few in less-developed countries. Recent studies not covered by Jackson have been made by Wright and McIlmoyle (1978), Henneberg (1978), Coxworth et al. (1978), Hoffmann et al. (1979), Steffen and Lampe (1979), Berg and Mietz (1980), and Butler (1981). The studies generally show treatment of straw to be profitable in some but not all cases. This is to be expected, since the results of economic calculations depend very much on variables which differ so widely between regions that it is hardly possible to arrive at general conclusions. In addition to these studies Potts (1983) has discussed, in more general terms, the economic aspects to be considered when planning research on by-product utilization.

18.4.4. Industrialized countries with little agricultural land per capita and where priority is given to a high degree of self-sufficiency in foods

Even if industrialized countries with little agricultural land per capita can meet their food demands through imports, some may be reluctant to become too dependent on food imports for reasons of national safety. Some critics argue that concern about self-sufficiency is an excuse, while the underlying reason may be a desire to maintain farm incomes and to avoid a rapid depopulation of rural areas. Concerns of this type have led a country like Norway to adopt a protectionist agricultural policy. Thus prices of imported feedstuffs and agricultural products have been kept at a level considerably above world prices, as a way to increase both farm incomes and the national food supply.

Higher prices of purchased feeds and farm products generally increase land rents and encourage more intensive farming. Utilisation of straw may be one way to increase food production on a limited acreage. When cattle have been bred for high production and concentrates are expensive, roughage quality becomes important. High quality roughage is one way to maintain animal production at high levels without using much concentrates. Untreated straw is not economic as a feed under such conditions. However, the use of treated straw is an interesting alternative. The author believes that this situation is one reason why treated straw has been used as feed in Norway to a much larger extent than in other industrialized countries. Even under such conditions, however, it is not certain that the use of treated straw is profitable. High-yielding dairy cows are sensitive to the quality of roughage. The use of straw treated by the Beckmann method meets serious environmental objections. NH_3-treated straw is clearly not as good as silage made from early cut grass. The new method of dip treatment in NaOH-solution may give a high-quality roughage when the straw is supplemented by adequate quantities of protein, and the method does not give pollution. High prices of protein supplements, however, may be a drawback to the use of treated straw.

The competitive advantage of treated straw is higher when used for growing animals than for high-yielding dairy cows. Growing animals need less protein per unit of ME and do not require such a high concentration of energy per kg of feed dry matter.

18.5. SOME CASE STORIES

In the following, a few cases will be presented where the use
of straw as feed is known to have been profitable. They may serve
as illustrations to the discussion earlier.

18.5.1. Denmark 1947

The author worked as a dairy herdsman in Denmark in 1947. The
following is based on personal observations rather than on scientific studies:
Denmark had an export-orientated agricultural industry. The
agricultural population was still of considerable size and farming
methods were labour-intensive rather than capital-intensive.
The most common farming system was a combination of dairy cows,
pigs and grain production on the same farm. A large part of the
grain crop was usually fed on the farm. Stocking rates were high.
Roots were grown extensively as the main source of cattle forage,
and gave very much higher feed unit yields per hectare than grass
or grains. There was a tendency to restrict the grassland area as
much as possible and feed large quantities of roots, while untreated straw was used as the main source of fibre. While other feeds
were rationed, animals were given straw to appetite. Straw was
used for bedding as well.
This case illustrates one type of situation. The costs of supplying fibre through hay or silage would be high since the alternative value of land used for roots or grain production was high.
Straw, therefore, was the cheapest way of supplying fibre while
energy was supplied mainly through roots and some feed grains.
Such rations are low in protein, and imported protein concentrates
had to be used as supplements.

18.5.2. Norway 1956

Straw treated by the Beckmann method was quite widely used as
feed on dairy farms, for cows as well as young animals. The method
was a source of river pollution. However environmental concerns
were not high at that time and few cared about this problem.
At that time, most dairy farms and herds were small, labour was
relatively abundant and land was scarce. Average yield of recorded

herds was about 3 250 kg per cow annually. Roots were grown for feed to some extent, but root crop yields were not nearly as high as in Denmark, so roots did not have the same competitive advantage. By using sodium hydroxide-treated straw as feed, a farmer could, for example, increase his herd from eight to ten cows without using more of his land for roughage production. He would have to purchase more protein concentrates. Partly due to government price policies, however, the extra cost of using protein-rich concentrate mixes rather than carbohydrates was not high.

Although farmers had to work longer to do the straw-processing and tend the extra cows, many found this was the best way to increase their income. A study of the economics of dairy farming, using linear programming, confirmed this (Langvatn, 1959).

18.5.3. <u>Norway 1982</u>

Environmental considerations have placed severe limitations on the use of straw processed by the Beckmann method. However there is considerable interest in the use of straw treated with anhydrous ammonia.

Labour input in Norwegian agriculture has decreased to about 36 per cent of the 1956 level while total acreage has changed little. However, total volume of agricultural production has increased by more than 50 per cent. On dairy farms, there has been a general shift to the more high-yielding NRF breed of cattle, and average yield of recorded herds has increased to about 5 750 kg per cow annually. Many of the small farms have ceased to exist as production units. Others, while still in existence, have quit milk production and become part-time farms.

The cost of supplementing a roughage ration low in protein with a protein-rich concentrate mix has increased significantly. The price ratio between a protein-rich mix and a carbohydrate concentrate was 1.13 in 1956 and 1.60 in 1982.

Common practice as well as an economic study (Hegrenes, 1981) suggest NH_3-treated straw may be an economic feed for young animals, while it is of limited interest as feed for dairy cows. A number of reasons may contribute to this:

- With the higher-yielding cows of today, roughage quality is of greater importance. Digestibility of ammonia-treated straw is lower than that of straw treated by the Beckmann method. However hopes are pinned on the new method of dip-treating straw in a NaOH-solution, which may give a high-quality roughage.
- Since protein has become more expensive, the cost of supplementing roughage low in protein has increased.
- Agriculture has become more regionally specialized. There are now few dairy cows in the main grain-growing districts, so extensive use of straw as feed would require long-distance transportation.
- The ratio of land to labour has increased. There is no longer the same incentive to increase production per unit of land.

In poor-crop years however, NH_3-treated straw may be economic as feed, even for dairy cows.

18.6. SUMMARY

Untreated straw is a low-quality feed for ruminants, even if its high fibre content can be an advantage in some situations. Various treatments can improve its quality, but it has a very low protein content.

Ruminant production of milk or body growth responds to quality of roughage fed. The response is greater with low-concentrate diets, and when animals are bred for high production.

Profitability of using untreated or treated straw as feed depends on availability and costs of straw, costs of treatment, price of urea or protein supplements, and the general agricultural and technological situation in the country concerned. No general conclusions can be drawn, but some tendencies are likely. In densely populated, less-developed regions, straw and other fibrous by-products are the main feeds for ruminants. In such situations treatment of straw will probably be profitable. In less-developed countries with low population-density, cattle production is often based on natural pastures and land is not limiting. Since more alternatives exist for supplying additional feed, treatment of straw may not always be profitable. It should be considered as one of the alternatives, especially for providing feed during dry seasons and in emergencies. In industrialized countries with ample food supply ruminants are fed considerable quantities of concentrates besides roughages and animals are bred for high production. Untreated straw is sometimes used as a source of fibre instead of hay or silage. In such situations treatment of straw will probably prove profitable. It is more doubtful whether it is profitable to feed treated straw instead of low-cost hay and silage. Some industrialized countries have little agricultural land per capita and a policy of self-sufficiency in foods. In such situations it is more likely that the use of straw will prove profitable since it is a way of increasing food production on a limited area.

Example situations are described. In Denmark 1947, the alternative costs of producing hay and silage were high since most of the land was suited to producing roots and grains. It was profitable to use straw as a source of fibre while roots supplied most of the energy for milk cows and young cattle. In Norway 1956, lye-treated straw was much used to increase stocking rates and milk production on small dairy farms. In Norway 1982, environmental concerns have restricted the traditional lye-treating method. A number of factors have changed and made the use of treated straw less profitable for dairy cows, but ammonia-treated straw is still considered an economic feed for growing animals.

18.7. REFERENCES

Berg, F. and Mietz, M., 1980. Ökonomische Wertung der Futterwert-
    erhöhung von chemisch behandeltem Stroh. Arch. Tierernähr. 30:
    1-3.
Broster, W.H., 1976. Plane of Nutrition for the Dairy Cow. In: H.
    Swan and W.H. Broster (eds.): Principles of Cattle Production.
    Butterworths, London, pp. 271-285.
Butler, B., 1981. Treated straw - a feed for the dairy herd?
    Report 28. Milk Marketing Board, Farm Management Services, Rea-
    ding, England.
Coxworth, E., Kernan, J., Thomson, D., Kullman, P., Darrach, W.
    and Spurr, D., 1978. Processing of straw with ammonia to produ-
    ce new products with increased value as animal feeds. Volume
    II. Saskatchewan Research Council.
Edelsten, P. and Lijongwa, C., 1981. Utilization of low quality
    roughages - Economical considerations in the Tanzanian context.
    In: J.A. Kategile, A.N. Said and F. Sundstøl (eds.): Utiliza-
    tion of Low Quality Roughages in Africa. AUN-Agricultural Deve-
    lopment Report 1, Aas, Norway.
Hegrenes, A., 1981. Halm som fôr. - Ei økonomisk vurdering. Nor-
    ges Landbruksøkonomiske Institutt, Oslo, Norway.
Henneberg, U., 1978. Forsøg med NH_3-behandlet halm til opdræt.
    In: Helårsforsøg med kvæg XVIII. Beretning 474. Statens Husdyr-
    brugsforsøg, Copenhagen, 110-117.
Hoffmann, H., Steinhauser, H. and Heissenhuber, A., 1979. Ökono-
    mische Überlegungen zum Einsatz von aufgeschlossenem Getrei-
    destroh in der Rinderfütterung. Berichte über Landwirtschaft
    57: 272-285.
Jackson, M.G., 1978a. Treating straw for animal feeding. An
    assessment of its technical and economic feasibility. FAO Ani-
    mal Production & Health Paper No 10. Rome.
Jackson, M.G., 1978b. Treating straw for animal feeding - an
    assessment of its technical and economic feasibility. World
    Anim. Rev. (28).
Langvatn, H., 1959. Arbeids- og driftsorganisering på mjølkepro-
    duksjonsbruk. Norges Landbruksøkonomiske Institutt, Oslo, Nor-
    way.
Potts, G.R., 1983. Application of research results on by-product
    utilization: Economic aspects to be considered. In: B. Kiflewa-
    hid, G.R. Potts and R.M. Drysdale (eds.): By-product Utiliza-
    tion for Animal Production. Proc. workshop on applied research,
    Nairobi, Kenya 26-30 Sept. 1982. International Development Cen-
    ter, Ottawa.
Steffen, G. and Lampe, J., 1979. Betriebswirtschaftliche Beurtei-
    lung von Aufgewertetem Stroh in der Tierfütterung. Übersichten
    zur Tierernährung 7,2: 109-132.
Wright, R.C. and McIlmoyle, W.A., 1978. An economic assessment of
    the utilization of alkali-treated straw. Agriculture in Nort-
    hern Ireland 53: 137-138.

Chapter 19

IMPLICATIONS OF A MORE WIDESPREAD USE OF STRAW AND OTHER FIBROUS BY-PRODUCTS AS FEED

by

Tim Smith and C. Clive Balch

National Institute for Research in Dairying
Shinfield, Reading
United Kingdom

The criteria that will determine whether more widespread use is made of straw have already been extensively discussed in the pre-ceeding chapters of this book. This final chapter will consider the feasibility of using straws as feed, both in nutritional and economic terms, the scope for processing straws to enhance nutri-tive value, and the likely consequences of a marked increase in straw feeding. To achieve the optimum benefit from any feed resource demands a combination of nutritional, husbandry and mana-gerial skills. A change in feeding system will only be successful if husbandry and managerial criteria are also met.

19.1. STRAW AS A RAW MATERIAL

The problems associated with feeding an increasing population and matching land use and production systems with energy supplies, technical achievement and consumer demand have been widely discus-sed, as has the conflict between man and his animals for the avai-lable resources (Reid, 1969; Mellanby, 1975; Duckham, Jones and Roberts, 1976; Balch and Cooke, 1982). This conflict can be reduced if the ability of the ruminant to utilize cellulose and fibrous materials is exploited. If the energy derived from these sources is ignored, because they are of no value to other species, then the efficiency of ruminant production from energy-rich feeds increases. Balch (1977) and Owen (1976) estimated from data of FAO (1974) that world production of cereals straws in 1974 was in the order of 1710 Mt of dry matter (see Table 19.1 and also Chapter 2). Most feeds have a gross energy value of about 18 MJ/kg DM (Ministry of Agriculture, Fisheries and Food (MAFF), 1975) and therefore, it follows that the potential gross energy in straw is greater than that in grain. However, the high lignin and low protein content, and costs of collection and storage all reduce the value of straw as feed, especially when alternative materials are available.

Where grain is combine-harvested, so that the straw is left in the field, the farmer is confronted with a number of alternatives: burn; chop and plough in; feed in situ; collect, when it becomes available for a number of uses including feeding. In developed societies there is growing opposition to straw burning because of pollution, and in the U.K. this has led to a code for straw burning (National Farmers Union, 1978), and in the USA to the development of machinery to contain the burn. The advantages of burning were discussed by Hughes (1979) and Ellis (1979). Harper and and Lynch (1979) described the conditions necessary for successful chopping and ploughing of cereal residues. The feeding of straw in situ will depend on the availability of stock, water, adequate fencing or control and the facility to offer supplementary feed if necessary. NLVF (1982) has shown that NH_3-treated barley straw is produced at a lower energy cost than traditional Norwegian conserved forages.

Most of the straw used for feeding will be collected, transported, stored and then fed. It is the cost of these individual operations, together with the availability of other feeds and labour which will determine the primary cost of straw. Other uses should not be ignored but will not be discussed here (Staniforth, 1979).

There will in the foreseeable future be a shortage of cereal grains, but new husbandry techniques, such as direct drilling, will extend the arable land available. Political restrictions apart cereal production and hence straw production should rise, as it has been shown to have done between 1970-1981 (Chapter 2).

Table 19.1. Estimates of World and Regional Production of Straws, 1974 (million tonnes). (Based on: FAO Production Yearbook, 1974. Values for production of individual crops. From Balch, 1977)

Crop	World	Africa	North & Central America	South America	Asia	Europe	Oceania	USSR	Developed	Developing	Centrally Planned Economies
Wheat[1]	360.2(21)	8.5	65.8	10.2	90.0	90.5	11.4	83.8	139.2(39)	72.0(20)	149.1(41)
Paddy Rice[1]	323.2(19)	7.6	6.9	10.2	294.3	1.9	.4	1.9	23.2(7)	174.9(54)	125.1(39)
Barley[1]	170.9(10)	4.8	15.6	1.0	31.7	60.5	3.1	54.2	65.9(39)	16.6(10)	88.3(51)
Maize[1]	586.0(35)	53.5	261.0	58.2	100.3	88.2	.5	24.3	317.3(54)	142.8(24)	125.9(22)
Rye[1]	32.6(2)	< .1	1.0	.3	.6	15.5	<.1	15.2	5.8(18)	.9(3)	25.9(79)
Oats[1]	51.2(3)	.2	13.0	.5	3.3	17.8	1.1	15.3	26.9(52)	1.0(2)	23.3(46)
Millet[2]	92.4(5)	18.1	.4	.5	67.4	.1	.1	5.8	.1(0)	38.6(42)	53.7(58)
Sorghum[2]	93.8(5)	18.5	38.0	14.6	19.3	.9	2.2	.2	36.3(39)	57.1(61)	.4(0)
Total Cereal Straw	1710.3(100)	111.2(6)	401.7(24)	95.5(6)	606.9(35)	275.4(16)	18.8(1)	200.7(12)	614.7(36)	503.9(29)	591.7(35)

(million tonnes dry matter)

Crop	World	Africa	North & Central America	South America	Asia	Europe	Oceania	USSR	Developed	Developing	Centrally Planned Economies
Sugar cane tops[3]	49.9	4.0	12.0	12.2	19.9	–	1.7	–	4.9(10)	41.9(84)	3.1(6)
Bagasse[4]	66.5	5.4	16.0	16.3	26.6	–	2.3	–	6.5(10)	55.9(84)	4.1(6)

[1]Assuming straw:grain ratio of 1:1 (Owen, 1976)

[2]Assuming straw:grain ratio of 2:1 (Owen, 1976)

[3]Assuming tops:cane ratio of 1:4 (Owen, 1976)

[4]Assuming bagasse:cane ratio of 1.5 (Owen, 1976)

19.2. NUTRITIONAL VALUE OF STRAW

Straws are poor-quality feeds, as defined by Balch (1977), in that they have an ME content of less than 7.5 MJ/kg DM; feed tables show the variation between straws of differing origin (McDowell et al., 1974 ; Göhl, 1975; MAFF, 1975, Chapter 12). Variation can also be caused by varietal differences, season of planting, harvesting conditions and height of cut (Wilson and Brigstocke, 1977). The feeding value of straw will depend on intake and digestibility (see Chapter 12). In a mixed diet this will embrace supplementation and substitution, by the animal, of one feed for another. Campling et al. (1961; 1962) demonstrated the beneficial effect on intake of energy of NPN added to an all straw diet, so that amount consumed, digested and rate of digestion were increased. The NPN can be added with equal efficiency, either in a dry concentrate, in the drinking water or in a molasses lick (Bass et al.,1981). To achieve maximum intake of straw a crude protein content of 66-85 g/kg DM is necessary (Elliot and Topps, 1963; Blaxter and Wilson, 1963; Andrews et al., 1972; Smith et al., 1980).

In a mixed roughage:concentrate diet the amount of roughage will depend on its quality and the desired level of production. In diets fed ad libitum total DM intake must increase as roughage proportion increases if level of production is to remain constant. Swan and Clarke (1974) achieved this with up to 30% ground straw in the diet of beef cattle, although other workers have achieved the greatest growth rates on low-roughage diets (Raven et al., 1969; Kay et al., 1970). Broster et al. (1978) and Smith et al. (1980) both found reduced digestibility of the diet as straw proportion increased.

Maximum intake of DM has been observed when roughage forms 16-35% of the diet (Raven et al., 1969; McCullough, 1969; Kay et al, 1970). However, several workers have noted that intake of concentrates falls more when fed together with good forage than when poor quality forage is offered (Blaxter et al., 1961; Montgomery and Baumgardt, 1965; Campling and Murdoch, 1966). Intakes of unprocessed straw of 1.0-1.2% of body weight, when concentrates and adequate protein were fed, were reported by Smith et al. (1981).

Although the addition of NPN to the diet increases roughage intake (Campling et al., 1961) other workers have reported enhanced growth rates from diets containing proteins resistant to rumen breakdown (Smith et al., 1980; Waldo and Tyrrell, 1980; Ekern, 1981). The evidence presented demonstrates the close relationship between the feeding value of roughages and dietary protein concentration and source. Similar responses between protein source and processed straws (both ground and alkali-treated), have also been found (Smith et al., 1981; 1983).

19.2.1. Processing of roughages

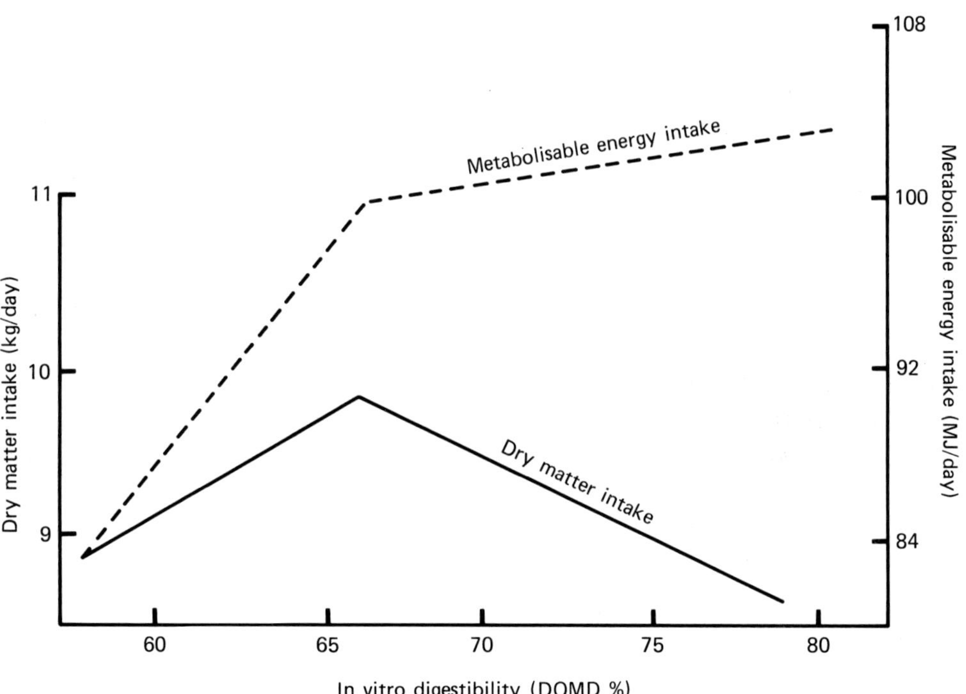

Figure 19.1. Dry matter intake and metabolisable energy intake plotted against in vitro digestibility (Swan and Clarke, 1974)

When considering the advisability of processing roughages there are two fundamental questions which should be asked: firstly, will the digestibility, and, therefore, the ME content, of the straw be improved by the process applied; secondly, will the intake of processed straw, by the animal be increased. It is a combination of these two aspects of processing which will lead to the greatest benefits. In attempts to quantify the likely responses many feeding trials and digestibility and intake studies have been reported in the literature. It is clear that intake, rate and site of digestion, and the nature of the end products of digestion are basic factors determining whether it is worthwhile to process or not. To extend the physical limitation on roughage intake, described by Montgomery and Baumgardt (1965) and confirmed by Swan and Clarke (1974) (Figure 19.1) is one of the aims of processing. Treatments applied are either physical, chemical or a combination of both (Chapters 5 to 8).

19.2.2 Physical treatment

Physical treatment is considered in depth in Chapter 5. Factors relevant to straw feeding will be discussed here. The most commonly used physical treatment has been grinding in order to reduce particle size and increase the surface area of the roughage exposed to rumen microbial action. The energy costs (Shepperson et al., 1972; Osbourn et al.,1976) and machinery costs (Palmer, 1976) are high, both for grinding and the often associated pelleting. Particle size affects the power costs of grinding, Swan and Clarke (1974) finding no advantage in using a screen of less than 7 mm.

Several reviews have summarised the effects of grinding straw (e.g. Greenhalgh and Wainman, 1972; Kay, 1972; Swan and Clarke, 1974) and in general increased DM intake, especially of the poorer materials, and some reduction in digestibility is to be expected. However, several workers (Montgomery and Baumgardt, 1965; Swan and Clarke, 1974; Palmer, 1976) commented on the beneficial effects of pelleting on intake, probably because of the elimination of dust. Younger, smaller animals have shown greater responses to physical treatment of roughage than older larger ones (Nicholson, 1981).

Both Chaturvedi et al. (1973a) and Holzer et al.(1975) reported increased DM intakes, and hence digestible organic matter consumption, after soaking straw in water. Effects on nitrogen retention were variable. In comparisons between soaked and NaOH-treated straws Chaturvedi et al. (1973b) recorded similar DM intakes from both forms of treatment. Where labour is cheap and water readily available this form of treatment may be a practical proposition, especially if NPN could be added to the straw before feeding.

Irradiation and high temperature-high pressure steam treatments are not feasible, either on the farm or commercially at the present time (Sundstøl, 1981; Nicholson, 1981).

19.2.3. Alkali treatment

Before deciding on the chemical to be used and method of application several factors must be considered: cost; availability of chemical and necessary machinery; available technical ability; safety. All the materials at present being used are safe when properly handled and dangerous when not. The dangers are increased where familiarity with written instructions and machinery is low, especially if coupled with scant clothing and the absence of footwear.

At the present time the most widely used chemicals are NaOH and NH_3, with some use of locally produced materials such as lime and urine (Saadullah et al., 1981). It has long been known (e.g. Kellner, 1915) that soaking in lye (alkaline solutions) improves the digestibility of straw. Homb et al. (1977) gave an historical account of the processes used (see also Chapter 6.1), and several recent reviews have summarized the experimental evidence available (Jackson, 1977; Greenhalgh, 1980; El Shazly and Naga, 1981; Kristensen, 1981; Sundstøl, 1981). Chapter 6.2 describes the industrial process and 6.3 the on-farm process for adding NaOH to straw. In general, alkali treatment has resulted in increased digestibility and DM intake leading to an increased output of product. In comparisons of the major treatment methods Sundstøl (1981) ranked the digestion coefficients of DM for treated wheat straw as follows: untreated < stacked (NH_3) < sprayed (NaOH) < ensiled (NaOH) < "dip" treated (NaOH), with a range of 0.45-0.70. Horton, Nicholson and Christensen (1982) found that NaOH treatment resulted in higher growth rates and feed efficiency than NH_3 treatment.

Smith et al. (1982) ranked growth rate from fixed amounts of a diet based on winter barley straw in the order: untreated < ground pelleted (NaOH) < stacked (NH_3) < ensiled (NaOH). In the same experiment ad libitum intake of straw was in the order: untreated < stacked (NH_3) < ensiled (NaOH) ground pelleted (NaOH) (See Table 19.2 for details).

Table 19.2. Live-weight gain in heifers receiving a fixed amount of total diet, and ad libitum intake of straw with a fixed amount of concentrates (Smith et al., 1982)

	Untreated straw + barley	Untreated straw + barley & urea	Untreated straw + barley, urea & fishmeal	Ground-pelleted NaOH straw + barley & urea	Chopped-ensiled NaOH straw + barley & urea	NH_3 straw + barley	NH_3 straw + barley & fishmeal
Daily liveweight gain, g/d	350	478	549	501	602	543	641
Intake of straw, kg DM/100 kg BW		1.04		1.76	1.54	1.27	

From earlier experiments (Smith et al., 1981; 1983), summarized in Table 19.3, it appeared likely that the physical form of straw treated with NaOH was influencing the responses obtained from such straws. Measurements of rate of passage of straw particles through the rumen are now in progress, using the same treatments applied in the feeding trial described above (Grigera-Naón et al., 1982). Smith et al. (1981) were able to predict an extra 250 g/day growth in animals receiving ad libitum spring barley straw, ground, pelleted and treated with NaOH, compared to the same straw fed long and untreated, with a fixed allocation of concentrates. However much of the extra DM intake was due to the grinding and pelleting. No health problems were encountered in these trials (see Chapters 6.2 and 12).

The efficiency with which the treatment process is carried out is fundamental to the success of the processing operation. Economic aspects will be discussed later, and also in Chapter 18. Alternative chemicals to unpgrade roughages are discussed in Chapter 8 and microbial treatments in Chapter 9. The introduction of straws of greater digestibility, possibly with increased length of internode, should be considered as one alternative to processing.

19.3. CONSEQUENCES OF INCREASED STRAW USE

The preceeding sections of this chapter have endeavoured to show that straw has a place in ruminant feeding, although its nutritive value will depend on both supplementation and processing and it will be shown later that the monetary value to be placed on straw either native or processed, can be simply determined in relation to the other feedstuffs available. Estimates of the amount of the straw crop used for feeding range from 50% on a world basis (Jackson, 1977) to about 20% in the U.K. (Hughes, 1979), to the absence of a market for straw-based feeds in parts of the USA (Miles, 1977) (these differences are attributed to the differences in cost and are a practical demonstration of the importance of market forces).

19.3.1. <u>Feed energy supply</u>

An increase in straw feeding brought about by pressure to use alternative feeds, and an awareness of the benefits of processing will increase the amount of feed energy available both on the individual farm and in terms of world supply. Taking the total wheat straw crop (360.2 Mt, Table 19.1) as an example, every 1% of the crop used for feeding contains 20531 x 10^6 MJ ME, and for every 1% improvement in the ME content a similar amount of extra ME becomes available. Where animals are kept almost totally on straw, as is often the case in South-East Asia (Jackson, 1977), the extra feed available will be the sum of the benefits to ME content · and intake, assuming that there is some straw surplus to requirements.

Table 19.3. Daily live-weight gain (g/d) in growing cattle receiving mixed diets containing untreated or NaOH treated straw or hay in fixed amounts of diet. The average ad libitum intake of straw is also shown (Smith et al., 1981; 1983)

Straw form in diet:	Untreated	Ground NaOH	Ensiled NaOH	Hay
Smith et al., (1981)				
a) Spring barley straw				
Expt 1	673	585		
" 2	685	673		
" 3	511	527		
b) Winter wheat straw				
Expt 2	556	570		
Expt 3	432	538		
Smith et al., (1983)				
Spring barley straw				
Expt 1	378		442	456
Expt 2	509	248	475	494
Expt 3	310	379	410	
Ad libitum intake of roughage (spring barley straw or hay) (kg/100 kg liveweight)	1.17	1.84	1.38	1.82

Present processing techniques are making between 33-55% of the gross energy of straw available to the animal, whereas with cereals approximately 75% of the gross energy is utilized as ME. In Table 19.4 some calculations of the percentage of maintenance requirements obtainable from untreated or treated straw are given, using the values for straw intake obtained by Smith et al. (1983)

Table 19.4.   Estimated intake of ME from straw by a 250 kg steer, receiving a
mixed diet, based on the straw intakes (kg/DM/day) from (A),
Table 19.2 and (B), Table 19.3. (Numbers in parenthesis denote %
of maintenance requirements[1])

	Untreated straw	Ground NaOH straw	Chopped NaOH straw	NH_3 straw
A (Winter barley straw)				
Daily intake of straw	2.600	4.400	3.850	3.175
ME intake at: 5.8[1] MJ ME/kg DM	15.1 (49)	25.5 (82)	22.3 (72)	18.4 (59)
10% improvement in ME content	–	28.0 (90)	24.5 (79)	20.2 (65)
20% improvement in ME content	–	30.6 (99)	26.8 (86)	22.1 (71)
30% improvement in ME content	–	33.1 (107)	29.0 (94)	23.9 (77)
B (Spring barley straw)				
Daily intake of straw	2.925	4.600	3.450	–
ME intake at: 7.3[1] MJ ME/kg DM	21.4 (69)	33.6 (108)	25.2 (81)	–
10% improvement in ME content	–	37.0 (119)	27.7 (89)	–
20% improvement in ME content	–	40.3 (130)	30.2 (97)	–
30% improvement in ME content	–	43.7 (141)	32.8 (106)	–

Likely levels of improvement are shown outside the boxes

[1] Values for maintenance and composition of winter and spring barley straw.
From MAFF (1975)

and Table 19.3, and estimating the effect of improvements in ME content of 10-30%. The boxed areas are levels of improvement not obtained in the above trials, and, therefore, represent a challenge to the chemist and nutritionists. The increased contribution from straw to maintenance, could in practice have resulted in a saving of grain (up to 1 kg barley/day), rather than extra daily live-weight gain.

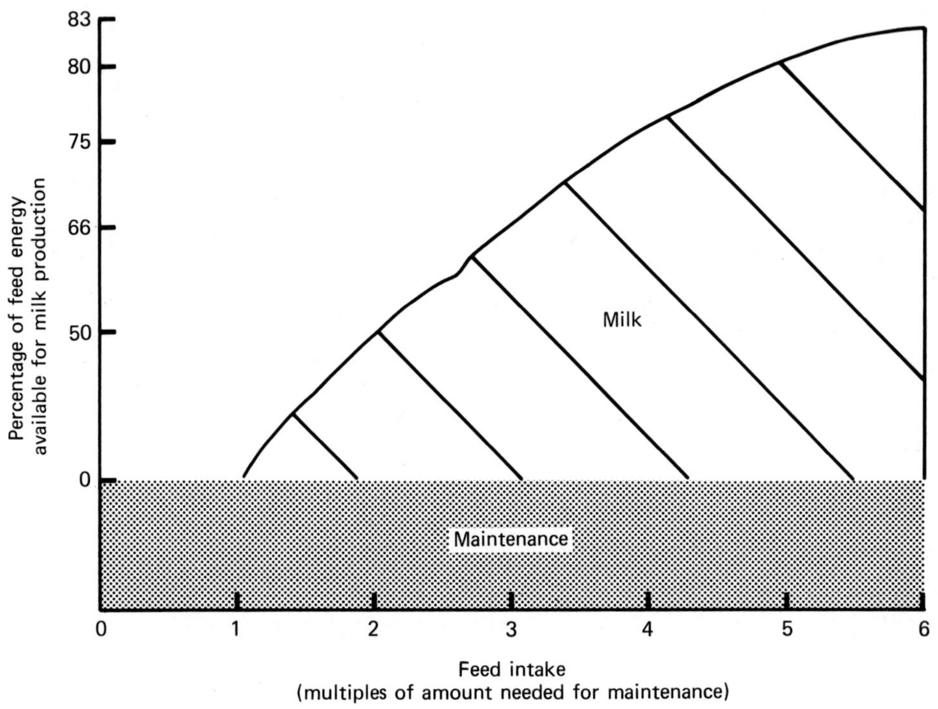

Figure 19.2. Effect of feed intake on percentage of energy intake available for milk production (Reid, 1970)

The benefits of a greater supply of feed energy to the animal are twofold. Firstly, there will be an increase in daily production, and secondly, with meat producing animals a reduction in the number of days to reach slaughter weight, which in turn will save a corresponding number of maintenance rations. Table 19.5 develops this argument using the experiment of Saadullah et al. (1981), and calculating the days and feed required for 10 kg of growth. In this instance the total feed used for an animal receiving untreated straw would have been sufficient for all the animals in the other three groups. Maintenance accounts for much of the energy consumed by the ruminant, the proportion being greatest with those of lower productivity. For the dairy cow excessive age at first calving and overlong dry periods both increase the proportion of nonproductive life and the amount of maintenance feed required by the animal. Figure 19.2 emphasises the large proportion of the diet that is used for maintenance in the dairy cow, when production of milk is low. These arguments apply equally to the species of domesticated ruminants (Chapters 14, 15 and 16). The effects of straw feeding on activity in the rumen are discussed in Chapter 13.

Table 19.5. Results and further calculations from an experiment with treated rice straw[1] fed to calves (Saadullah et al., 1981)

	Treatment			
	Untreated straw	5% urea	4% lime	3% NaOH 1% lime
Initial liveweight (kg)	54.0	57.8	56.8	56.0
DM intake of straw (kg/100 kg liveweight)	3.15	3.25	2.85	2.76
Daily liveweight gain (g)	35	110	99	120
Conversion (kg total DM/kg gain)	50.3	18.2	17.5	14.1
To gain 10 kg liveweight at the rates shown				
Days required	286	91	102	84
Total straw required (kg DM)	532	186	180	142

[1] 500 g green grass was supplied to all animals daily

Bone meal and common salt were supplemented at the rate of 1% of the straw supplied.

19.3.2. <u>Nitrogen supply</u>

Supplementation of a straw diet, especially with nitrogen, has been stressed both in this chapter and in preceeding chapters of this book. To maximise output from straw a combination of supplementation and processing is necessary. The form of the nitrogen supplementation is also important (see section 19.2). Table 19.6 shows the amounts of supplementary nitrogen required, to supply the rumen degradable protein necessary (ARC, 1980), either in soya bean meal or urea if the total amount of wheat straw fed is increased, or if the ME content of all the wheat straw used is increased. Figure 19.3 indicates the amounts of urea needed to meet the ruminants requirements for rumen degradable protein (ARC, 1980), in the hypothetical situation of the total straw crop (360.2 Mt, FAO, 1974) being used for ruminant feeding, both before and after processing. In Table 19.6 and Figure 19.3 the wheat straw has been assumed to have an ME value of 5.7 MJ/kg DM and a crude protein concentration of 40 g/kg DM (close to the value found by Bath, 1981 and Smith et al., 1981), with a degradability of 0.5. These amounts of NPN would result in an all straw diet with a crude protein concentration of 64 g/kg DM, close to the 60 g/kg DM found necessary with 85% roughage diets by Elliot and Topps (1963). As ME content of the diet rises so does the total rumen degradable protein requirement, hence the greater concentrations of crude protein found necessary with mixed diets and increasing levels of improvement. Table 19.7 gives the amounts of urea necessary to correct the diet when ME supply, from wheat straw, is increased, by either greater intake or digestibility of the straw.

Table 19.6. Nitrogen required to meet the rumen degradable protein requirement of cattle utilizing energy derived from wheat straw[1], both before and after processing (developed from Table 19.1; MAFF, 1975; ARC, 1980)

Nitrogen requirement as:	Rumen degradable protein, t	Soya bean meal, t	Urea t
Each 1% of total world crop fed without processing	88104	262214	38306
Each 1% improvement in ME content of crop for each 1% fed	1587	4724	690

[1]Wheat straw: 5.7 MJ ME/kg DM; CP = 40 g/kg DM, rumen degradability: 0.5

Soya bean meal: CP = 480 g/kg DM; rumen degrability 0.7.

Urea: 46.0% nitrogen, 80% efficiency in rumen.

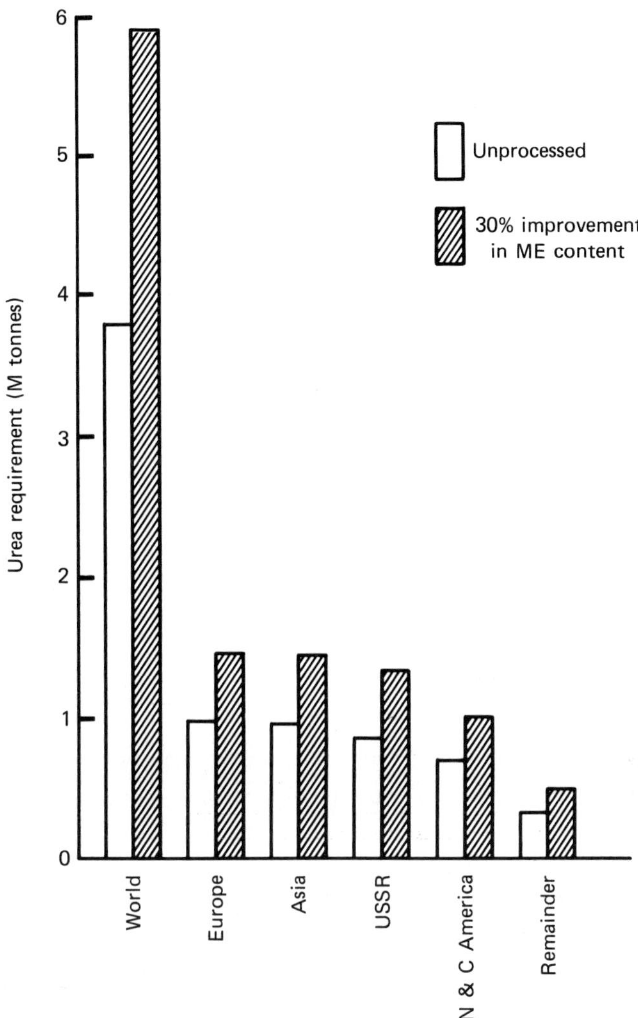

Figure 19.3. Urea needed to balance the rumen degradable nitrogen requirement of cattle, assuming the total world wheat straw crop to be used for feeding (Based on FAO, 1974 and ARC, 1980)

Table 19.7. Urea (g) needed to meet rumen degradable protein requirements (ARC, 1980) in a steer, as DM intake or digestibility of wheat straw[1] changes to increase ME intake (MJ)

Digestibility coefficient:		0.40		0.50	
		ME	Urea	ME	Urea
Intake kg DM/					
100 kg live weight	1.00	5.70	10.63	7.13	15.48
	1.25	7.13	13.31	8.91	19.35
	1.50	8.55	15.95	10.69	23.21
	1.75	9.98	18.63	12.48	27.11

[1] Wheat straw: ME MJ/kg DM = 5.7; CP g/kg DM = 40; degradability of straw 0.5.

The amounts shown for nitrogen requirement are theoretical, a large portion of the protein needed in mixed diets coming in practice from the other feeds, and where NH_3 or urea are used as improving agents they will also contribute. There has been no allowance in these calculations for any by-pass protein to further enhance the response from straw, the amounts necessary depending in part on the remainder of the diet and on the type and level of production required.

In the example presented above wheat straw has again been cited, but it accounts for about 20% of the total world supply of cereal straws (FAO, 1974; Chapter 2). Table 19.8 lists the major straw sources, their estimated digestibilities and crude protein concentrations (Göhl, 1975; MAFF, 1975). They are all currently regarded as feed energy sources for ruminants, and are all low in crude protein. The widespread use of these materials will affect the geographical distribution of supplementary nitrogen requirement. Overemphasizing the use of straw as a feed where other feeds are available could aggravate the world shortage of protein nitrogen beyond the limits of production. It should also be recognised that the utilization of straw and other by-products from tropical crops presents special problems (Preston, 1981; Leng et al., 1981).

The costs of supplementing an improved straw diet can be 40-60% of the total costs, especially if imported proteins are used, stressing the need to use locally produced protein and to assess potential for stepping up production where necessary (Pigden, 1981). Potential sources of extra nitrogen include animal excreta (Fontenot, 1981; Saadullah et al., 1981), and in grassland greater dependence on legumes (Rukanda and Lwoga, 1981). The use of extracted grass juice (Connell and Foot, 1972) as an ensiling additive to straw is another possibility.

The amount of nitrogen needed to supplement straw leads to one further consideration. What other uses are there for that nitrogen: direct feeding to human beings; protein feeds for species of farm livestock other than ruminants; nitrogen fertilizer for grazing and conservation areas?

Table 19.8. Crude protein and crude fibre content of the cereal straws listed by FAO (1974), compared with the same values for hay

	Crude protein	Crude fibre	Reference
	---- g/kg DM ----		
Straw source:			
Wheat	31	454	Göhl (1975)
Paddy rice	31	350	"
Barley	42	421	"
Maize	61	367	"
Millet	36	388	"
Sorghum	40	395	"
Rye	36	447	MAFF (1975)
Oats	28	398	"
Hay source:			
Grass (high digestibility)	101	320	"
Rhodes Grass	74	359	Göhl (1975)
Lucerne	187	344	"

19.4. ECONOMIC CONSIDERATIONS

Where straw, or other poor fibrous materials, are fed in large quantities because of the absence of other feedstuffs economic considerations of its use are probably not applicable. In many instances such materials will be plentiful, intake and productivity will be low, but such dietary regimes do ensure that domesticated ruminants and herbivores can be maintained, for example, from from one wet season's grazing to another. When adding straw to a balanced diet the desired level of productivity, the fibre requirement of the animal and the other feeds available must all be considered.

Most of the roughages utilized as fodder are either fed on the farms on which they are produced or traditionally purchased for use in established feeding systems. Where opportunities exist for upgrading, economics will probably dictate the best system to be adopted.

When capital investment in plant and machinery is planned several factors must be considered: throughput of straw; existing road network; supply of spare parts; technical expertise. If the proposed material is to be transported over long distances then densification, by grinding and pelleting may be desirable. With NH_3 the capital involvement for the farmer can be as low as the cost of a plastic sheet, or, for example, in Bangladesh the use of urine to provide NH_3 and family labour to seal the stack with banboo mats, dung and mud (Saadullah et al., 1981).

It is impossible here to calculate which method of straw improvement is economically right for each specific feeding system. In Chapter 18 the role of straw as a feed is related to stage of development and population density of the major geographical regions. It is possible, however, to use a simple equation to assist with the decision making, and in particular whether it is justified to substitute roughage (unprocessed or processed) into the diet in order to save expenditure on another component. The equation compares the energy value of the roughage with the cost per unit of energy of the feed to be replaced. By applying it to the same roughage, both before and after processing, then an allowable cost of processing can be calculated. If the goal is simply to increase intake then the cost of the extra input must not be greater than the value of the extra output.
The calculations are:

$$(1)\ \frac{x}{y} \times a = p \qquad (2)\ \frac{x}{y}_1 \times a = p_1 \qquad (3)\ p_1 - p = \text{allowable cost of processing}$$

where: x   = ME (MJ/kg DM) of roughage before processing

x_1 = ME (MJ/kg DM) of roughage after processing

y   = ME (MJ/kg DM) of feed to be replaced

a   = Cost per t of feed to be replaced

P   = Allowable cost of unprocessed roughage

P_1 = Allowable cost of processed roughage

For example, by substitution, where: $x = 7.3$, $x_1 = 9.0$, $y = 13.0$; $a = \$208$:

$$(1)\ \frac{7.3}{13.0} \times \$208 = \$116.8 \qquad (2)\ \frac{9.0}{13.0} \times \$208 = \$144.0$$

$$(3)\ \$144.0 - \$116.8 = \$27.2$$

Thus in this example $ 27.2 would be the allowable cost of processing one tonne of roughage. There are, however, a number of constraints to be applied to the values obtained from these equations:

1. The allowable costs of both unprocessed and processed roughage must include the extra charges involved in handling and storing bulky feeds.

2. The allowable costs must include the cost of any supplementary feed necessary. If changing from an energy-rich feed to a fibrous feed this can include the cost of protein and mineral supplementation.

3. The calculations are made on the assumption that the diet is within the intake capacity of the animal for the desired level of output.

4. Any diet containing processed roughage must remain economic within the input/output relationship dictated by feed costs and market returns.

The equations have the ability to increase the value to be placed on roughages when other energy feeds are scarce, and also take into account roughage quality. No adjustment is made for efficiency of use of ME from differing sources as in some feed tables (MAFF, 1975).

The values obtained for P and P_1 are the maximum allowable costs, roughage at less cost representing a relatively cheap sour-

594

ce of energy. When energy-rich feeds are expensive and fibrous foods cheap, then if the total cost of processed roughage is not above the P_1 value, then the value of P can probably be ignored.

Some flexibility can also be used on the allowable cost of processing where intake will be greatly increased, but again the total cost must not exceed the P_1 value. Table 19.9 demonstrates the allowable costs of roughages, before and after processing when replacing feed costing $ 192 or $ 240/t. The allowable cost of processing increases with the basic quality of the roughage, the amount of improvement expected and the cost of the feed to be replaced. This relationship offsets the greater improvements generally claimed for the poorer straws, in that to achieve equal actual improvement with a good or poor straw, the percentage improvement must be greater with the poor straw.

No attempt has been made to evaluate the input/output relationship of utilizing roughages in terms of energy exchange. It has been assumed that the price structure will broadly reflect energy costs. However, the actions involved in mechanical handling and processing require the use of finite fuels, whereas the material being processed is to a large extent infinite and of little value in its naturally occurring form.

19.5. SELECTION OF STOCK TO RECEIVE STRAW

In large areas of the world straws or poor quality herbage are practically the only feeds available for ruminant feeding. Low production associated with low energy intake is unavoidable. The benefits to be obtained from supplementation of the diet and processing of the roughage have already been discussed.

In mixed diets straw can either be a major energy source, a ration diluent or a conditioner, dependent on the type of production required. Where energy-rich feeds are plentiful slow-growing stock will most frequently be the ones to receive large amounts of straw, with beef cattle receiving amounts commensurate with the growth rate required. Straw for high yielding dairy cows, especially in early lactation, will be limited to that necessary to maintain long fibre intake and milk fat content. Kristensen (1981) concluded that dairy cows can consume alkali-treated straw up to 1-1.4% of liveweight/day if large amounts of concentrates are also given. Broster (1974) described the importance of allocating con-

Table 19.9.   Allowable cost of roughage ($), before (0% improvement), and after processing (10, 20 or 30% improvement), replacing energy-rich feeds costing $192 or $240/t

Cost of energy-rich feed (13.0 MJ ME/kg DM)	$192				$240			
% improvement of roughage	0	10	20	30	0	10	20	30
ME of roughage: 5.5								
Allowable cost of roughage	81.23	89.36	97.47	105.60	101.54	111.70	121.84	132.00
Allowable cost of processing		8.13	16.24	24.37		10.16	20.30	30.46
ME of roughage: 6.5								
Allowable cost of roughage	96.00	105.60	115.20	124.80	120.0	132.00	144.00	156.00
Allowable cost of processing		9.60	19.20	28.80		12.00	24.00	36.00
ME of roughage: 7.5								
Allowable cost of roughage	110.77	121.84	132.93	144.00	138.46	152.30	166.16	180.00
Allowable cost of processing		11.07	22.16	33.23		13.84	27.70	41.54

To spend the same money on processing, improvement must increase as native straw becomes poorer, eg to spend $22.16 to replace feed at $ 192/tonne:

ME of unprocessed straw	% improvement necessary
7.5	20.0
6.5	23.1
5.5	27.3

centrate feeding according to stage of lactation. Straw proportion should be increased or reduced in order to maximise straw use and minimise loss of milk yield. However, where alternative feeds are scarce and production is at a lower level the situation changes. In Tanzania, for instance, Edelsten and Lijongwa (1981) found roughage improvement was economic for intensive dairy production (milk yield 7-15 l/day over the first four months of lactation) and for goat fattening, but not for fattening beef cattle.

Alkali-treated straw has also been fed to buffalo calves Chaturvedi et al., 1973a and b). Owen (1981) concluded that for sheep and goats poor quality roughages, after alkali-treatment, can replace hay under maintenance and low production situations (see also Chapters 14 and 15). The needs of the draught animal (Chapter 13) must not be ignored.

The term "straw" is generally used to describe the residue of a cereal crop after the grain has been removed. However, if the description of a poor quality roughage (Balch, 1977) is applied to natural herbage then the diet of many wild ruminants and herbivores, including the browsers, fall within the scope of this book (see Chapter 16). Game cropping is a possbility in Africa (Reid, 1969; Pratt and Gwynne, 1977), where species of antelope are of particular interest, as, indeed, are deer in parts of Europe (Blaxter, 1975). Such animals presumably have the advantages of being resistant to indigenous disease and adapted to the local vegetation (Van Soest, 1982). The contribution of traditionally wild species should not be ignored, intensification of their management being applied to match the desired level of output (Payne, 1964).

Processed straw has also been included in the diets of commercially reared rabbits (Lindeman et al., 1982), although straw use by small species will only account for a limited portion of the crop.

19.6. LOGISTICS OF STRAW FEEDING

Fibrous feeds are bulky and the cost of transport is high (Edelstein and Lijongwa, 1981), and, therefore, they are probably of value only if used in the area of production. Lack of transport and inadequate road systems are features of many areas. Both the high density baler (Klinner and Chaplin, 1977) and densification during processing (Wilson and Brigstocke, 1977) are techniques developed to reduce bulk. The approach described by Squire and Creek (1973), in Kenya, of bringing store cattle into a centralized finishing complex, ideally with a slaughter house and meat processing facility, is a possibility. Such complexes in the grain growing areas (e.g. parts of the Sudan) would mean that the cattle, not the feed, would be moved, the grazier selling his animals before they reached slaughter weight. Such developments could lead to increased production from traditionally nomadic areas, for example in the northern half of Africa. However, one of the main objections to such policies is their potential disruptive effect on traditional ways of life. Great care would be necessary to ensure the advantages outweighed the disadvantages.

19.7. OTHER FIBROUS FEEDS AVAILABLE

In this chapter poor quality roughages such as cereal straws and coarse grasses have been considered. FAO (1974) (Table 19.1) estimated that there is a further 116.4 Mt of plant material from the sugar crop (equal to 24.4% of the total straw crop by weight). Dyer et al. (1975) concluded that 10^6 tonnes of fibrous materials and waste cellulosic materials were available annually.

Wood pulp, sawdust and paper pulp are all potential energy sources, but up to the present time successful use has been limited beyond the laboratory stage (Coombe and Briggs, 1974; Nicholson, 1981). A comprehensive list of the by-products available in California was given by Bath (1981), of the major tropical crop by-products by Preston (1981) and also in Chapter 2.

19.8. FURTHER RESEARCH NEEDS

1. The search must continue for more efficient treatment processes in order to improve the nutritive value of straws and increase the economic viability of the operation.

2. Alternative nitrogen sources are required. The use of animal wastes can only be developed with the necessary enforcable legislation to control their use. Further validation of protein requirements, based on the concepts of ARC (1980), are needed.

598

3.   The use of straw in mixed diets, the composition of these
     diets, the substitution effects on digestibility and intake,
     and the stock to receive them all need attention in order to
     assess the economic factors of processing.

4.   The risk to the animal of continued exposure to processed
     materials and the hazards of pollution must be monitored for
     both existing and new processes (Jackson, 1977, calculated
     that 100 animals fed NaOH straw for a year could excrete up to
     10 t of sodium).

5.   The use of cellulose-rich materials such as wood by-products
     and derivatives.

6.   The economic effects in the developed world of changing to
     fibre-rich diets, and, therefore, to lower levels of produc-
     tion, on animal numbers, fixed costs and farm income.

19.9.   CONCLUSIONS AND SUMMARY

1.   A number of treatment processes can increase the nutritive
     value of the straw crop. Protein supplementation is essential
     and source of protein used will affect the level of production
     achieved.

2.   The yield of feed energy from straw is potentially high, espe-
     cially when processing is efficient. This, together with in-
     creased dry matter intake by the animal, would result in eit-
     her a greater percentage of maintenance, or production, coming
     from straw. With meat producing animals saving in days to
     slaughter, and hence a saving of maintenance rations, could
     have a large impact on ruminant feed supply in the developing
     world.

3.   Supplementary protein needs will escalate if straw feeding in-
     creases, although processing is only fully effective in con-
     junction with correct supplementation. Other uses for the
     large quantities of nitrogen involved should be considered.
     Is the nutrient limitation to animal production either energy
     or protein?

4. The value of straw as a feed must be judged relative to its cost and energy content, compared to the feeds it could replace. The cost of extra supplementation must also be included. The amount of money expendable on processing will increase with the degree of improvement obtainable and the price of other energy feeds.

5. Other topics discussed have included: animals most suited to straw feeding; logistics of straw feeding; other fibrous foods available.

6. Further research is necessary to improve the effectiveness of treatments, to monitor new treatments, and to increase and ensure the efficient use of nitrogen supplies.

19.10. REFERENCES

Agricultural Research Council, 1980. The Nutrient Requirements of Ruminant Livestock. Commonwealth Agricultural Bureaux, Slough.

Andrews, R.P., Escuder-Volonte, J., Curran, M.K. and Holmes, W., 1972. The influence of supplements of energy and protein on the intake and performance of cattle fed on cereal straws. Anim. Prod. 15: 167-176.

Balch, C.C., 1977. The potential of poor quality agricultural roughages for animal feeding. FAO, Animal Production & Health Paper 4, New Feed Resources, 1-15.

Balch, C.C. and Cooke, G.W., 1982. The efficiency of nutrients and energy in plant and animal production systems. In: Proc. 12th Congr. International Potash Institute, Goslar, 47-69.

Bass, J.M., Fishwick, G. and Parkins, J.J., 1981. The effect of the method of presentation of a concentrated solution containing urea, minerals, trace elements and vitamins on the voluntary intake of oat straw by beef cattle. Anim. Prod. 33: 15-17.

Bath, D.I., 1981. Feed by-products and their utilization by ruminants. In: T.J. Huber (ed.): Upgrading residues and by-products for animals. CRC Press Inc., Florida, 1-16.

Blaxter, K.L., 1975. Conventional and uncoventional farmed animals. Proc. Nutr.Soc. 34: 51-56.

Blaxter, K.L., Wainman, F.W. and Wilson, R.S., 1961. The regulation of food intake by sheep. Anim. Prod. 3: 51-61.

Blaxter, K.L. and Wilson, R.S., 1963. The assessment of a crop husbandry technique in terms of animal production. Anim. Prod. 5: 27-42.

Broster, W.H., 1974. Response of the dairy cow to level of feeding. Bienn. Rev. 1974, Natn. Inst. Res. Dairy. 14-34.

Broster, W.H., Smith, T., Broster, V.J. and Siviter, J.W., 1978. Experiments on the nutrition of the dairy heifer. X. Effect on nitrogen utilization for growth of chemical and physical treatment of the principal source of protein in the diet. J. agric. Sci. (Camb.) 90: 299-310.

Campling, R.C., Freer, M. and Balch, C.C. 1961. Factors affecting the voluntary intake of food by cows. 2. The relationship between the voluntary intake of roughages, the amount of digesta in the reticulo-rumen and the rate of disappearance of digesta from the alimentary tract. Br. J. Nutr. 15: 531-540.

Campling, R.C., Freer, M. and Balch, C.C., 1962. Factors affecting the voluntary intake of food by cows. 3. The effect of urea on the voluntary intake of oat straw. Br. J. Nutr. 16: 115-124.

Campling, R.C. and Murdoch, J.C., 1966. The effect of concentrates on the voluntary intake of roughages by cows. J. Dairy Res. 33: 1-11.

Chaturvedi, M.L., Singh, U.B. and Ranjhan, S.K., 1973a. Effect of feeding water-soaked and dry wheat straw on food intake, digestibility of nutrients and VFA production in growing zebu and buffalo calves. J. agric. Sci. (Camb.) 80: 393-397.

Chaturvedi, M.L., Singh, U.B. and Ranjhan, S.K., 1973b. Effect of alkali treatment of wheat straw on feed consumption, digestibility and VFA production in cattle and buffalo calves. Indian J. Anim. Sci. 43: 677-683.

Connell, J. and Foot, A.S., 1972. Mechanical dewatering of forages prior to high temperature artificial drying. 1971-72 Ann. Rep., Natn. Inst. Res. Dairy, p. 51.

Coombe, J.B. and Briggs, A.L., 1974. Use of waste paper as a feedstuff for ruminants. Aust. J. exp. Agric. 14: 292-301.

Duckham, A.N., Jones, J.G.W. and Roberts, E.H., 1976. 1. The function and efficiency of human food chains. 2. An approach to the planning and administration of human food chains and nutrient cycles. In: A.N. Duckham, J.G.W. Jones and E.H. Roberts (eds.): Food Production and Consumption. North Holland Publishing Company, Amsterdam, 3-11 & 462-517.

Dyer, I.A., Riquelme, E., Baribo, L. and Couch, B.Y., 1975. Waste cellulose as an energy source for animal protein production. World Anim. Rev. 15: 39-43.

Edelsten, P. and Lijongwa, C., 1981. Utilization of low quality roughages - economical considerations in the Tanzanian context. In: J.A. Kategile , A.N. Said and F. Sundstøl (eds.): Utilization of Low Quality Roughages in Africa. AUN-Agricultural Development Report 1, Aas, Norway, 189-200.

Ekern, A., 1981. Results from feeding trials and practical experience concerning protein feeding of ruminants in Norway. In: E.L. Miller, I.H. Pike and A.J.H. van Es (eds.): Protein Contribution of Feedstuffs for Ruminants: application to feed formulation. Butterworth Scientific, London, 86-102.

Elliott, R.C. and Topps, J.C., 1963. Voluntary intake for low protein diets by sheep. Anim. Prod. 5: 269-276.

Ellis, F.B., 1979. Agronomic problems from straw residues with particular reference to reduced cultivation and direct drilling in Britain. In: E. Grossbard (ed.): Straw Decay and its Effect on Disposal and Utilization. John Wiley & Sons, Chichester, 11-20.

El-Shazly, K. and Naga, M.A., 1981. Supplementation of rations based on low quality roughages under tropical conditions. In: J.A. Kategile, A.N. Said and F. Sundstøl (eds.): Utilization of Low Quality Roughages in Africa. AUN-Agricultural Devlopment Report 1, Aas, Norway, 157-170.

Fontenot, J.P., 1981. The nutritive value of and methods of incorporating animal wastes into rations for ruminants. In: J.T. Huber (ed.): Upgrading residues and by-products for animals. CRC Press Inc., Florida, 17-37.

Food and Agriculture Organisation of the United Nations, 1974. Production Yearbook 1974. FAO, Rome.

Göhl, B., 1975. Tropical Feeds: Feeds Information Summaries and Nutritive Values. FAO, Rome.

Greenhalgh, J.F.D., 1980. Use of straw and cellulosic wastes and methods of improving their value. In: E.R. Ørskov (ed.): By-products and Wastes in Animal Feeding. Occasional Publ. No. 3, Br. Soc. Anim. Prod. 25-31.

Greenhalgh, J.F.D. and Wainman, F.W., 1972. The nutritive value of processed roughages for fattening cattle and sheep. Proc. Br. Soc. Anim. Prod. 61-72.

Grigera-Naón, J.J., Smith, T., Siviter, J.W. and Broster, W.H., 1982. Digestion of alkali-treated straws by yearling steers. Report 1982, Natn. Inst. Res. Dairy, p. 63.

Harper, S.H.T. and Lynch, J.M., 1979. The kinetics of the decomposition of straw in relation to the production of phytotoxins. In: E. Grossbard (ed.): Straw Decay and its Effect on Disposal and Utilization. John Wiley & Sons, Chichester, 289-291.

Holzer, Z., Levy, D., Tagari, H. and Volcani, R., 1975. Soaking of complete fattening rations high in poor-roughage. 1. The effect of moisture content and spontaneous fermentation on nutritional value. Anim. Prod. 21: 323-335.

Homb, T., Sundstøl, F. and Arnason, J., 1977. Chemical treatment of straw at commercial and farm levels. FAO, Animal Production & Health Paper 4, New Feed Resources, 25-37.

Horton, G.M.J., Nicholson, H.H. and Christensen, D.A., 1982. Ammonia and sodium hydroxide treatment of wheat straw in diets for fattening steers. Anim. Feed Sci. Technol. 7: 1-10.

Hughes, R.G., 1979. Arable farmers' problems with straw. In: E. Grossbard (ed.): Straw Decay and Its Effect on Disposal and Utilization. John Wiley & Sons, Chichester, 3-10.

Jackson, M.G., 1977. Review article: The alkali treatment of straws. Anim. Feed Sci. Technol. 2: 105-130.

Kay, M., 1972. Processed roughage in diets containing cereals for ruminants. In: Cereal Processing and Digestion. U.S. Feed Grains Council, London, 39-52.

Kay, M., MacDearmid, A. and MacLeod, N.A., 1970. Intensive beef production. 10. Replacement of cereals with chopped straw. Anim. Prod. 12: 261-266.

Kellner, O., 1915. The Scientific Feeding of Animals, translated by W. Goodwin. Duckworth & Co, London.

Klinner, W.E. and Chaplin, R.V., 1977. Developments in straw densification. In: A.R. Staniforth (ed.): Report on Straw Utilization Conference. Ministry of Agriculture, Fisheries and Food, Oxford, 14-20.

Kristensen, V.F., 1981. Use of alkali-treated straw in rations for dairy cows, beef cattle and buffaloes. In: J.A. Kategile, A.N. Said and F. Sundstøl (eds.): Utilization of Low Quality Roughages in Africa. AUN-Agricultural Development Report 1, Aas, Norway, 91-106.

Leng, R.A., Nolan, J.V. and Preston, T.R., 1981. Rumen bypass nutrients, manipulation and implications. Joint FAO/IAEA Meeting, Vienna.

Lindeman, M.A., Brigstocke, T.D.A. and Wilson, P.N., 1982. A note on the response of growing rabbits to varying levels of sodium hydroxide-treated straw. Anim. Prod. 34: 107-110.

McCullough, T.A., 1969. A study of factors affecting the voluntary intake of food by cattle. Anim. Prod. 11: 145-153.

McDowell, L.R., Conrad, J.H., Thomas, J.E. and Harris, L.E., 1974. Latin American Tables of Feed Composition. University of Florida, Gainesville, Florida.

Mellanby, K., 1975. Can Britain Feed Itself. Merlin Press, London.

Miles, T.R., 1977. Straw uses as feed, fibre and fuel in Oregon. In: A.R. Staniforth (ed.): Report on Straw Utilization Conference. Ministry of Agriculture, Fisheries and Food, Oxford, 75-81.

Ministry of Agriculture, Fisheries and Food, 1975. Technical Bulletin No 33, HMSO, London.

Montgomery, M.J. and Baumgardt, B.R., 1965. Regulation of food intake in ruminants. 2. Rations varying in energy concentration and physical form. J. Dairy Sci. 48: 1623-1628.

National Farmers Union, 1978. The straw and stubble burning code. (Revised 1978). NFU, London.

Nicholson, J.W.G., 1981. Nutrition and feeding aspects of the utilization of processed lignocellulosic waste materials by animals. Agric. Environm. 6: 205-228.

Osbourn, D.F., Beever, D.E. and Thomson, D.J., 1976. The influence of physical processing on the intake, digestion and utilization of dried herbage. Proc. Nutr. Soc. 35: 191-200.

Owen, E., 1976. Farm wastes: Straw and other fibrous materials. In: A.N. Duckham, J.G.W. Jones and E.H. Roberts (eds.): Food Production and Consumption. North Holland Publishing Co., Amsterdam, 299-318.

Owen, E., 1981. Use of alkali-treated low quality roughages to sheep and goats. In: J.A. Kategile, A.N. Said and F. Sundstøl (eds.): Utilzation of Low Quality Roughages in Africa. AUN-Agricultural Development Report 1, Aas, Norway, 131-150.

Palmer, F.G., 1976. The feeding value of straw to ruminants. ADAS Quart. Rev. 21: 220-234.

Payne, W.J.A., 1964. Specific problems of semi-arid environments. Proc. 6th int. Congr. Nutr., Edinburgh, 213-226.

Pigden, W.J., 1981. Use of low quality forages in the future-needs for research and implementation. In: J.A. Kategile, A.N. Said and F. Sundstøl (eds.): Utilization of Low Quality Roughages in Africa. AUN-Agricultural Development Report 1, Aas, Norway, 201-213.

Pratt, D.J. and Gwynne, M.D., 1977. Rangeland management and ecology in East Africa. Hodder and Stoughton, London.

Preston, T.R., 1981. Utilization of tropical crop residues and agroindustrial by-products in animal nutrition: constraints and perspective. Joint FAO/IAEA Meeting, Vienna.

Raven, A.M., Forbes, T.J. and Irwin, J.H.D., 1969. The utilization by beef cattle of concentrate diets containing different levels of milled barley straw and of protein. J. agric. Sci. (Camb.) 73: 355-363.

Reid, J.T., 1969. The future role of ruminants in animal production. In: A.T. Phillipson (ed.): Physiology of Digestion and Metabolism in the Ruminant. Oriel Press Ltd., Newcastle upon Tyne, 1-22.

Reid, J.T., 1970. Speculations on future production of meat, milk and eggs. In: Proc. 1970 Cornell Nutr. Conf. Feed Mfrs, Buffalo, New York, 50-63.

Rukanda, N.A.K. and Lwoga, A.B., 1981. The role of legumes in improving natural grasslands for livestock production. In: J.A. Kategile, A.N. Said and F. Sundstøl (eds.): Utilization of Low Quality Roughages in Africa. AUN-Agricultural Development Report 1, Aas, Norway, 185-188.

Saadullah, M., Haque, M. and Dolberg, F., 1981. Practical methods for chemical treatment of rice straw for ruminant feeding in Bangladesh. In: J.A. Kategile, A.N. Said and F. Sundstøl (eds.): Utilzation of Low Quality Roughages in Africa. AUN-Agricultural Development Report 1, Aas, Norway, 85-89.

Shepperson, G., Marchant, W.T.B., Wilkins, R.J. and Raymond, W.F., 1972. The tehcniques and technical problems associated with the processing of naturally and artificially dried forage. Proc. Br. Soc. Anim. Prod. 1: 41-51.

Smith, T., Broster, W.H., Broster, V.J. and Siviter, J.W., 1981. The effects of grinding, either with or without NaOH treatment, on the utilization of straw by yearling dairy cattle. J. agric. Sci. (Camb.) 96: 159-165.

Smith, T., Broster, V.J. and Hill, R.E., 1980. A comparison of sources of supplementary nitrogen for young cattle receiving fibre-rich diets. J. agric. Sci. (Camb.) 95: 687-695.

Smith, T., Broster, W.H. and Siviter, J.W., 1980. An assessment of barley straw and oat hulls as energy sources for yearling cattle. J. agric. Sci. (Camb.) 95: 677-686.

Smith, T., Broster, W.H., Siviter, J.W. and Grigera-Naón, J.J., 1982. NaOH v. NH_3 treatment of straw. Report 1982, Natn. Inst. Res. Dairy, p. 62.

Smith, T., Siviter, J.W. and Broster, W.H., 1983. A comparison of two methods of NaOH treatment of spring barley straw with untreated straw and hay. J. agric. Sci. (Camb.) 100: 343-350.

Squire, H.A. and Creek, M.J., 1973. Custom feeding of cattle - a proposal for a standardized package approach to project formulation. World Anim. Rev. 8: 8-16.

Staniforth, A.R., 1979. Industrial uses of straw and the relevance of decay to such uses. In: E. Grossbard (ed.): Straw Decay and its Effect on Disposal and Utilization. John Wiley and Sons, Chichester, pp. 209-216.

Sundstøl, F., 1981. Methods for treatment of low quality roughages. In: J.A. Kategile, A.N. Said and F. Sundstøl (eds.): Utilization of Low Quality Roughages in Africa. AUN-Agricultural Development Report 1, Aas, Norway, pp. 61-80.

Swan, H. and Clarke, V.J., 1974. The use of processed straw in rations for ruminants. In: H. Swan and D. Lewis (eds.): Proc. 8th Nutr. Conf. Feed Mfrs, Univ. Nottingham. Butterworth, London, pp. 205-233.

Van Soest, P.J., 1982. Nutritional ecology of the ruminant. O and B Books Inc., Oregon.

Waldo, D.R. and Tyrrell, H.F., 1980. The relation of insoluble nitrogen intake to gain, energy retention, and nitrogen retention in Holstein steers. In: H.J. Oslage and K. Rohr (eds.): Proc. 3rd EAAP Symp. on Protein Metabolism and Nutrition, European Association for Animal Production, Publ. No. 27, pp. 572-577.

Wilson, P.N. and Brigstocke, T.D.A., 1977. The commercial straw process. Process Biochem. 12: 17-21.